Cell-to-Cell Communication

Cell-to-Cell Communication

Edited by

Walmor C. De Mello
University of Puerto Rico
San Juan, Puerto Rico

Plenum Press • New York and London

Library of Congress Cataloging in Publication Data

Cell-to-cell communication.

Includes bibliographies and index.
1. Cell interaction. 2. Cell junctions. I. De Mello, Walmor C. [DNLM: 1. Cell Communication. 2. Intercellular Junctions—physiology. QH 603.C4 I61]
QH604.2.I554 1987 574.87′6 87-18573
ISBN 0-306-42623-4

A Division of Plenum Publishing Corporation
233 Spring Street, New York, N.Y. 10013

Printed in the United States of America

Contributors

Robert C. Berdan Department of Medical Physiology, The University of Calgary, Health Sciences Centre, Calgary, Alberta T2N 4N1, Canada

Peter R. Brink Department of Anatomical Sciences, Health Sciences Center, State University of New York, Stony Brook, New York 11794

E. E. Daniel Department of Neurosciences, McMaster University, Hamilton, Ontario L8N 3Z5, Canada

Walmor C. De Mello Department of Pharmacology, Medical Sciences Campus, University of Puerto Rico, San Juan, Puerto Rico 00936

Sarah C. Guthrie Department of Anatomy, University College London, London WC1E 6BT, United Kingdom. *Present address:* Department of Molecular Biology, Research Institute of Scripps Clinic, La Jolla, California 92037

Stewart W. Jaslove Department of Anatomical Sciences, Health Sciences Center, State University of New York, Stony Brook, New York 11794

Camillo Peracchia Department of Physiology, University of Rochester, Rochester, New York 14642

Judson D. Sheridan Department of Cell Biology and Neuroanatomy, University of Minnesota Medical School, Minneapolis, Minnesota 55455

Robert Weingart Department of Physiology, University of Berne, Berne, Switzerland

Guido Zampighi Department of Anatomy, UCLA School of Medicine, Los Angeles, California 90024

Preface

Grau, teurer Freund, ist alle Theorie
Und grün des Lebens goldner Baum

All theory, dear friend, is gray
But the golden tree of actual life springs ever green

—Goethe

Progress achieved in the last 25 years indicates that the establishment of intercellular junctions at the area of cell contact represents an important mechanism of intercellular communication.

Evidence is available that intercellular channels are essential for electrical synchronization in excitable tissues and for the diffusion of molecules from cell to cell.

This process of cell-to-cell communication is so reliable that it was preserved throughout the evolutionary scale. As evolution generates diversity it is not surprising that gap junctions are not the same in all systems. It is known, for instance, that junctional permeability is reduced by high free $[Ca^{2+}]_i$ in some cells or by a fall in pH_i in others, or enhanced by cAMP.

Our knowledge of the physiological modulators of junctional permeability is still scanty. Moreover, the compartmentalization of the intracellular milieu represents an additional complication in the interpretation of many studies of cell-to-cell communication.

The present volume represents an effort to provide the reader with an actualized view of the mechanisms of cell communication and of the physiological and pathological implications of junctional and nonjunctional communication.

Let us hope that the content of this book helps future studies in establishing a better picture of the cellular and molecular mechanisms involved in the process of intercellular communication.

I want to express my sincere thanks to all the authors for their interest and enthusiasm and to Plenum Press for their efficiency and professionalism.

Walmor C. De Mello

San Juan

Contents

Chapter 1
Gap Junction Structure
Guido Zampighi

Chapter 2
Modulation of Junctional Permeability
Walmor C. De Mello

Chapter 7
Intercellular Communication in Embryos
Sarah C. Guthrie

Chapter 8
Mechanisms of Cell-to-Cell Communication Not Involving Gap Junctions
Walmor C. De Mello

Chapter 9
Cell-to-Cell Communication in Salivary Glands
Robert Weingart

Chapter 10
Intercellular Communication in Arthropods: Biophysical, Ultrastructural, and Biochemical Approaches
Robert C. Berdan

Chapter 1

Gap Junction Structure

Guido Zampighi
Department of Anatomy
UCLA School of Medicine
Los Angeles, California 90024

1. INTRODUCTION

The significance of gap junctions in intercellular communication was first demonstrated in parallel structural and electrophysiological studies of synapses between neurons forming the giant escape mechanism of crayfish (Furshpan and Potter, 1959; Robertson, 1953, 1954, 1955, 1960; Hama, 1961; Watanabe and Grundfest, 1961). Electron microscopic studies of these synapses showed regions of close axolemma contact characterized by a narrow extracellular space (Robertson, 1953, 1954, 1955, 1960, 1963). Electrophysiological studies of these synapses demonstrated that action potentials propagated across the synapse with small delays and in the synapse between the lateral giant axons equally well in both directions. Cell communication through gap junctions was later observed in nerve tissue of several fishes (Bennett *et al.*, 1963, 1967; Furshpan, 1964), in smooth and cardiac muscles (Dewey and Barr, 1962, 1964), and between most epithelia and glands (Kanno and Loewenstein, 1966; Loewenstein and Kanno, 1964, 1967; Loewenstein *et al.*, 1965; Loewenstein, 1966). In addition to the observation that gap junctions allow the passage of ions carrying electrical impulses, physiological studies demonstrated that gap junctions also allow low-molecular-weight molecules to spread into adjacent cells without leaking into the extracellular space (Loewenstein, 1966; Loewenstein and Kanno, 1967). These observations had profound structural implications because they suggested that gap junctions must be constructed of large hydrophilic channels. Subsequently, it became possible to resolve these gateways between adjacent cytoplasms by structural methods.

The first complete structural study of a vertebrate gap junction was presented by Robertson (1963) in the Mauthner cell in goldfish medulla oblongata. Robertson demonstrated that gap junctions were disk-shaped contact regions of approximately 0.2–0.4 μm in diameter where the plasma membranes adopted a pentalamellar structure 14–16 nm in overall thickness. It was particularly significant that the plane of the junction was constructed of two-dimensional arrays of annuli, 8- to 9-nm center-to-center spacing. These annuli turn out to have been the first biological channel visualized by electron microscopy.

In the mid 1960s and early 1970s, there was considerable confusion regarding the types and basic morphology of these intercellular junctions. These problems were elegantly resolved by the work of Revel and Karnovsky (1967). They took advantage of the properties of lanthanum to fill the extracellular space of tissues with electron-dense precipitates. The contrast provided by lanthanum permitted the distinction between two different types of junctions. One was a punctate or tight junction (zonula adherens), which brings the apposed cells into a contact so close that it excludes lanthanum. In the other junction the plasma membranes came into close contact but the extracellular space contained lanthanum deposits. The extracellular space of this junction was resolved as a 2- to 4-nm-wide "gap" between the junctional membranes in tissues block stained with uranyl acetate. The presence of this "gap" was used as a diagnostic feature to distinguish communicating junctions ("gap junctions") from other cell contacts. However, this morphological feature is less meaningful than the presence of hexagonal arrays of channels demonstrated by Robertson (1963).

2. CLASSIFICATION OF GAP JUNCTIONS

Although the majority of gap junctions display similar structural characteristics when studied by electron microscopy, the gap junctions from mammalian lens and crayfish nerve tissues display substantially different structures. Therefore, in this chapter gap junctions will be classified into three types. The great majority of gap junctions belong to type I gap junctions and have been found in epithelia and glands (e.g., liver, pancreas), in smooth and cardiac muscle, and between the cells of connective tissues such as fibroblasts and osteocytes. Type I junctions are also present in nerve tissues of vertebrates and of some invertebrates. The most characteristic feature of type I junctions is that the channels are organized in hexagonal arrays, with 8- to 9-nm center-to-center spacing. The junctions designated type II have been seen thus far only in vertebrate mature lens fibers and are characterized by the presence of channels arranged in square arrays spaced about 6–7 nm apart. The junctions designated type III are found in crayfish and other anthropods and are characterized by the presence of channels of larger diameter (about 12 nm), also arranged in hexagonal arrays but spaced about 20 nm apart. In addition, these junctions have associated rows of synaptic vesicles, 40–60 nm in diameter, firmly attached to the cytoplasmic leaflets.

Although the structural organization of these three types of junctions is substantially different and there is mounting evidence that their chemical composition also differs, it is not understood how these differences translate into function, which, in every case, appears to be the passage of ions and small molecules between cells without involvement of the extracellular space.

3. MORPHOLOGICAL CHARACTERIZATION OF TYPE I GAP JUNCTIONS

Gap junctions designated type I in this review, are widely distributed among cells of epithelia and glands, muscle fibers (with the exception of mature skeletal muscle), connective tissue, and nerve tissue. Since the structure of the channels between hepatocytes has been solved to the quaternary structural level, gap junctions from this tissue will be used as the most representative example of this type.

3.1. Conventional Electron Microscopy

Gap junctions between hepatocytes are most frequently located on the lateral regions where adjacent cells face each other. Thin-section electron microscopy shows the junctions as straight, densely stained bands, formed by two closely apposed plasma membranes (Fig. 1A). At higher magnifications, the structure of the gap junctions is studied with respect to two planes separated by a 90° angle. One is the projection along the surface of the junction (the *en face* view) and the other is across the thickness of the junction (the transverse view). Figures 1G, 3, and 4 show *en face* views of rat liver gap junctions by thin sectioning, freeze-fracture, and negative staining techniques, respectively. In all three cases, the gap junction is constructed of a hexagonal array of units spaced 8–9 nm center-to-center apart. Each unit is annulus-shaped with a small (1- to 2-nm diameter) deposit of stain or metal at its geometrical center.

Transverse views (the view perpendicular to the hexagonal array) are presented in Fig. 1B–F. The most frequently observed transverse view is a pentalamellar structure (three dense bands separated by two light bands), 15–16 nm in overall thickness (Fig. 1B). In livers block-stained with uranyl acetate, after glutaraldehyde–osmium fixation, the transverse views of the junctions display a heptalamellar structure, because the central dense band is replaced by a narrow gap, 2–3 nm wide (arrows, Fig. 1C). Livers fixed in glutaraldehyde–tannic acid solutions (Zampighi *et al.*, 1980; Zampighi 1980) show the central band formed by rows of densities 4–5 nm in diameter, and having either spherical (curved arrows, Fig. 1D) or elongated shapes (straight arrows, Fig. 1D). These densities are spaced approximately 8 nm apart. The gap junction in Fig. 1E contains densities protruding partially into the hydrophobic cores (the light bands) of the membranes, spaced 4–5 nm apart. Figure 1F shows large densities spaced 8–9

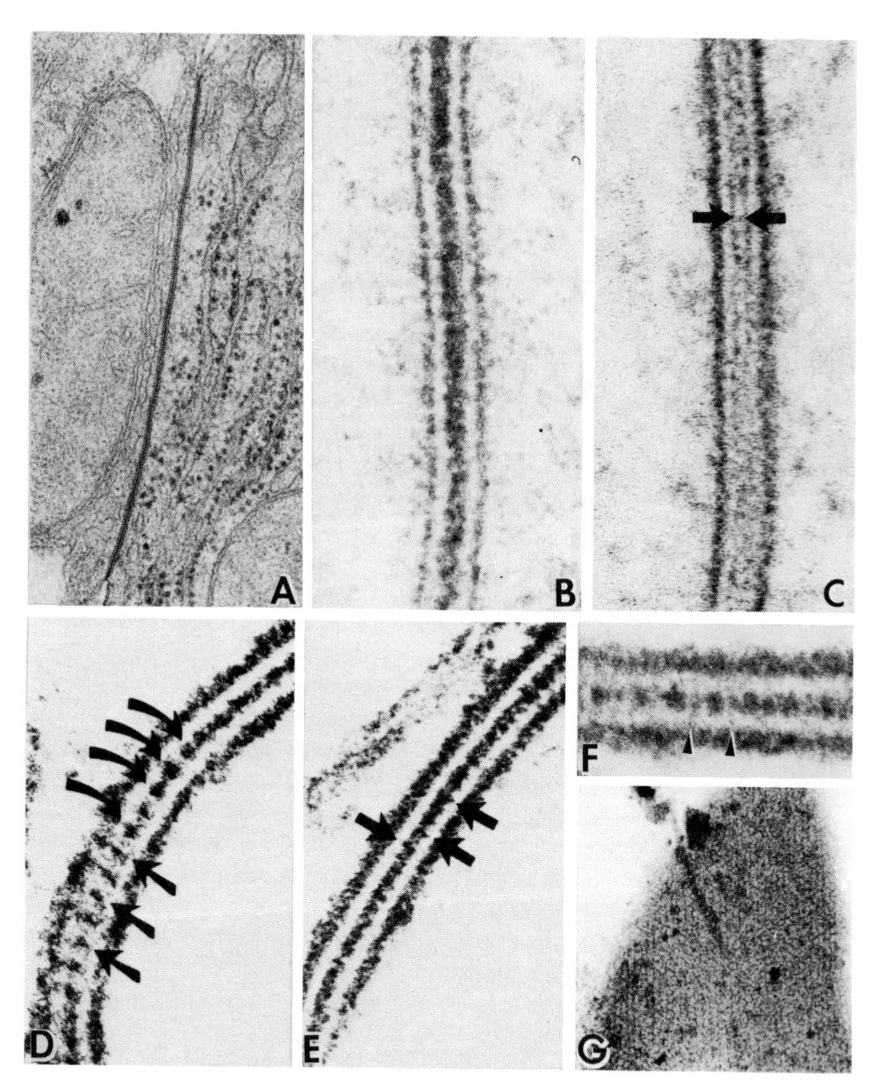

FIGURE 1. Thin-section electron micrograph showing type I gap junction in transverse and *en face* views. (A) A low-magnification view of an extensive region of close membrane apposition between two hepatocytes. ×60,000. (B) A higher magnification of a transverse view to demonstrate the pentalamellar structure of gap junctions. ×700,000. (C) A transverse view from liver stained in block with uranyl acetate to show the heptalamellar structure. In this method of tissue preparation, the central dense band in B is visualized as a 2- to 3-nm-wide gap (arrows). ×650,000. (D–F) Other transverse views of the region of membrane contact. Note the periodic arrangement of dense dots and the slender densities spanning the entire thickness of the junction, indicated by the arrowheads in F. D and E, ×550,000; F, ×900,000. G is an *en face* view of the junction to demonstrate the hexagonal lattice. ×175,000.

nm apart and midway between them smaller densities 1–2 nm in diameter (arrowheads) that extend across the entire thickness of the junction. These small transverse densities are consistent with the hypothesis that hydrophilic channels connect the cytoplasm of adjacent cells. Figure 1G shows the *en face* view of the junction displaying annular units in the membrane plane.

The *en face* and transverse views collected by thin sectioning electron microscopy can be combined to produce low-resolution models of the liver gap junction (Zampighi *et al.*, 1980). The model in Fig. 2 includes seven channels (cx) in the section; each channel is represented by a hollow cylinder about 6 nm in diameter with an inner pore of 1–2 nm. Stain molecules occupy the external domains of the channels (at the gap region) and a small cavity at the geometrical center of the channel (the hydrophilic pathway). The left side of the model emphasizes the coaxial alignment of the hemichannels from opposing membranes. The bottom part of the model indicates the expected projections of such a model in transverse sections (the *x*, *y*, *z* plane). The hexagonal array of channels was used to compute different transverse views at different lattice rotations. Therefore, computer analysis of views of the junction at different angles, obtained by conventional electron microscopy, allowed the interpretation of the slender densities detected in Fig. 1F (arrowheads) as part of the hydrophilic pore spanning the entire thickness of the gap junction.

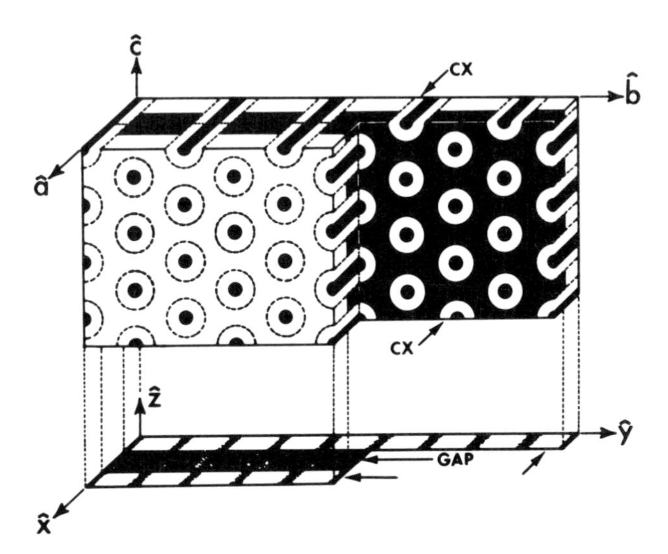

FIGURE 2. Stain distribution used to predict the effects of lattice rotation on transverse views of gap junctions. The model consists of seven channels (cx) arranged hexagonally. Two orthogonal coordinate systems are defined. The projection *x*, *y*, and *z* plane is perpendicular to the optical axis of the microscope. The *a*, *b*, and *c* plane defines the coordinate system of the junction (Zampighi *et al.*, 1980).

Additional information regarding the structure and distribution of gap junctions between cells has been derived from studies using freeze-fracture and freeze-fracture-etch techniques (Kreutziger, 1968; McNutt and Weinstein, 1970; Chalcroft and Bullivant, 1970; Goodenough and Revel, 1970; Goodenough and Gilula, 1974; Raviola *et al.*, 1980; Hirokawa and Heuser, 1982). Figure 3 shows the typical freeze-fracture pattern displayed by a type I gap junction. The protoplasmic (''P'') fracture face contains clusters of intramembrane particles 6–7 nm in diameter spaced 8–9 nm apart and the external (''E'') fracture face displays complementary pits. Frequently the particles (channels) form clusters without crystalline order but, on occasion, the E faces contain small domains of channels crystallized in hexagonal arrays, separated by smooth areas. The fracture through the extracellular ''gap'' is visualized as a small step separating the P from the E fracture faces. This small fracture step contrasts with the wider separation seen when the fracture plane jumps across nonjunctional regions (open arrows, Fig. 3). Another feature observed by freeze-fracture, in gap junctions from several tissues, is the presence of bands of plasma membrane, approximately 30 nm wide, forming the perimeter of the plaque (solid arrows, Fig. 3). This specialized band of plasma membrane, also called the ''halo,'' contains fewer intramembrane particles than the regular P fracture face (McNutt and Weinstein, 1973). Analysis of the height of the fracture step at the ''halo'' shows that the extracellular gap also measures 3–4 nm. Beyond this ''halo,'' the junctional membranes curve to adopt the normal separation between cells seen by electron microscopy (15–20 nm). The chemical composition and possible function of the membranes forming the ''halo'' are unknown.

Information about the organization of the true external and cytoplasmic surfaces of the junctions was obtained by freeze-fracture-etch methods (Goodenough and Gilula, 1974; Hirokawa and Heuser, 1982). In this method, pieces of tissue are slammed onto a copper block cooled to liquid helium temperatures. The surface of the tissue is fractured and controlled amounts of water are sublimated (''etched'') to expose the true surfaces of the junctions, thus permitting the direct study of the portions of the channel protruding into the cytoplasm and extracellular gap. Hirokawa and Heuser (1982) presented a complete study of the fracture and etched (''true'') faces of hepatocytes. They found that the channels protruded by 2–3 nm into the extracellular gap, but were not detected on the cytoplasmic surface of the cell. In addition, they found no evidence for the presence of cytoskeletal elements associated with the junctional surfaces. However, cytoskeletal elements in the form of a single-layered basket of actin filaments and coated vesicles and pits have been described in thin sections of several vertebrate gap junctions (Larsen *et al.*, 1979) and in crayfish junctions (Zampighi *et al.*, 1978).

The appearance of junctions by freeze-fracture methods have been correlated with the conductance state of the channel. Peracchia and Dulhunty (1976) and Peracchia and co-workers (Bernardini and Peracchia, 1981; Peracchia, 1977,

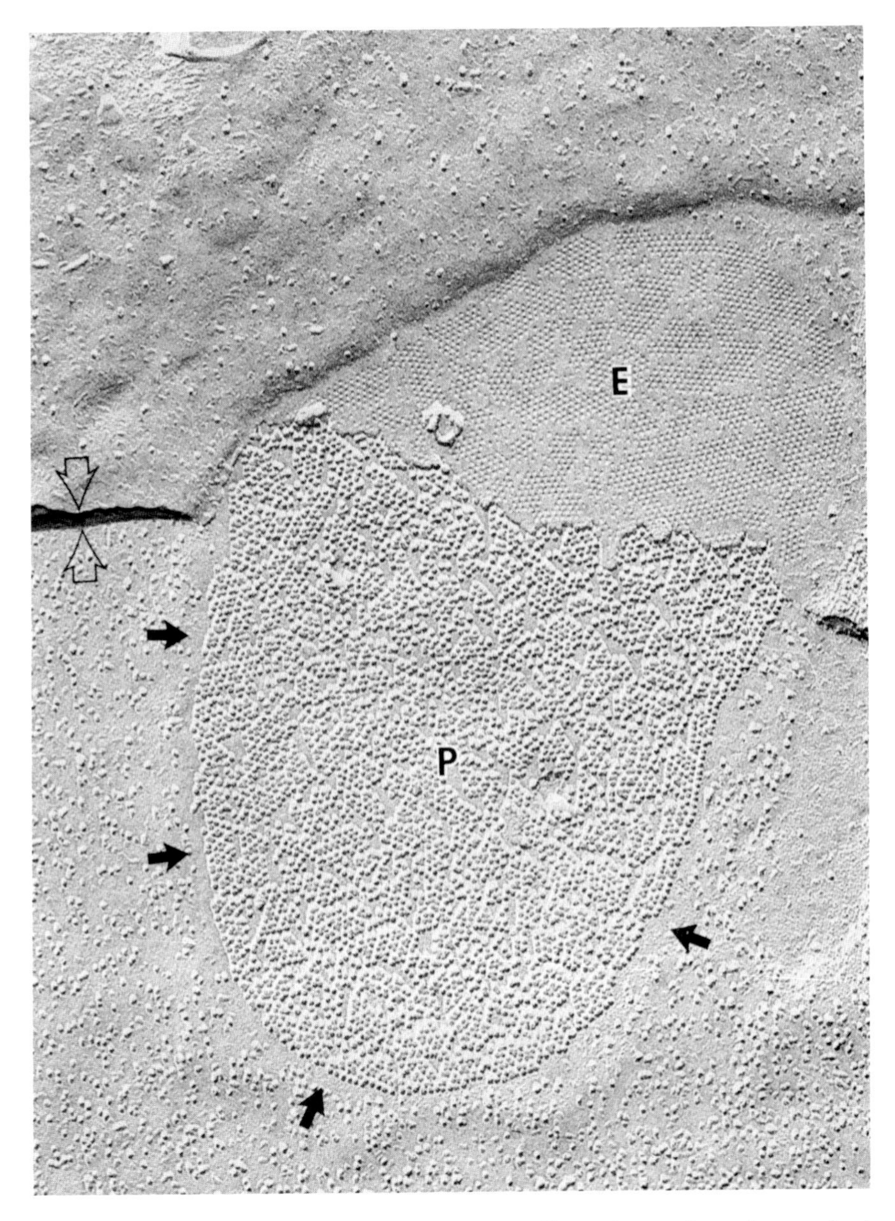

FIGURE 3. Freeze-fracture replica of type I gap junction. The cardiac muscle was frozen unfixed and uncryoprotected, by slamming onto a copper block at liquid helium temperature. P and E indicate the membrane fracture planes. The solid arrows point to the "halo" at the plaque perimeter. ×170,000.

1978, 1980; Peracchia and Peracchia, 1980a,b) reported that gap junction plaques (particularly those from tissues fixed with the junctions uncoupled) displayed particles arranged in crystalline hexagonal arrays. On the other hand, gap junction plaques from tissues fixed when coupled showed particles forming clusters, with no apparent crystalline order. Based on this correlation, Peracchia proposed a model wherein the state of aggregation of the particles in freeze-fracture replicas represented the conductance state of the channels. Similar correlations have been reported for a variety of tissues observed after uncoupling treatment (for a review see Peracchia and Bernardini, 1984). However, several recent investigations have found that the aggregation state of the particles forming the gap junctions seems unrelated to the conductance state of the channels (Raviola *et al.*, 1980; Green and Severs, 1984; Miller and Goodenough, 1985).

A striking specialization observed in gap junctions of some epithelia such as the enamel organ (Garant, 1972), granulosa cells (Merk *et al.*, 1973; Albertini and Anderson, 1975; Larsen *et al.*, 1979) as well as other tissues (McNutt and Weinstein, 1973) is the presence of circular profiles inside the cell cytoplasm which in some cases are disconnected from the plasma membrane. These profiles are called "annular" gap junctions. They appear as large vesicles (about 0.1–0.5 μm in diameter) with walls comprised of double membranes that have clearly defined gap junction channels. The spherical shape proposed for the annular gap junction is interesting because the two-dimensional hexagonal net must be periodically deformed to produce a sphere. Although the function of these annular gap junctions is still unknown, they have been proposed to be part of the mechanism of junctional degradation in which one of the partner cells internalizes the entire plaque (Larsen, 1985). However, these annular junctions are not seen in most tissues connected through gap junctions and their involvement in junctional channel turnover in several tissues has been questioned (Lee *et al.*, 1982; Yancey *et al.*, 1979).

3.2. Quaternary Organization of the Gap Junction Channel

The original work of DeRosier and Klug (1968) demonstrated that three-dimensional information obtained from electron micrographs of cylindrical viruses can be reconstructed by processing single micrographs by Fourier analysis (for a review of the methods see Amos *et al.*, 1982). However, planar objects such as membranes containing crystalline arrays cannot be reconstructed from single micrographs. To define their structure unequivocally, it is necessary to create different views by tilting the junctions at different angles with respect to the incident electron beam (Henderson and Unwin, 1975; Unwin and Henderson, 1975). These "tilted" views are then analyzed in a computer by Fourier methods and the Fourier terms (amplitudes and phase angles) are "merged" together along continuous lines. The resulting three-dimensional transforms are used to calculate maps of the structure by reverse Fourier transformation (Henderson and

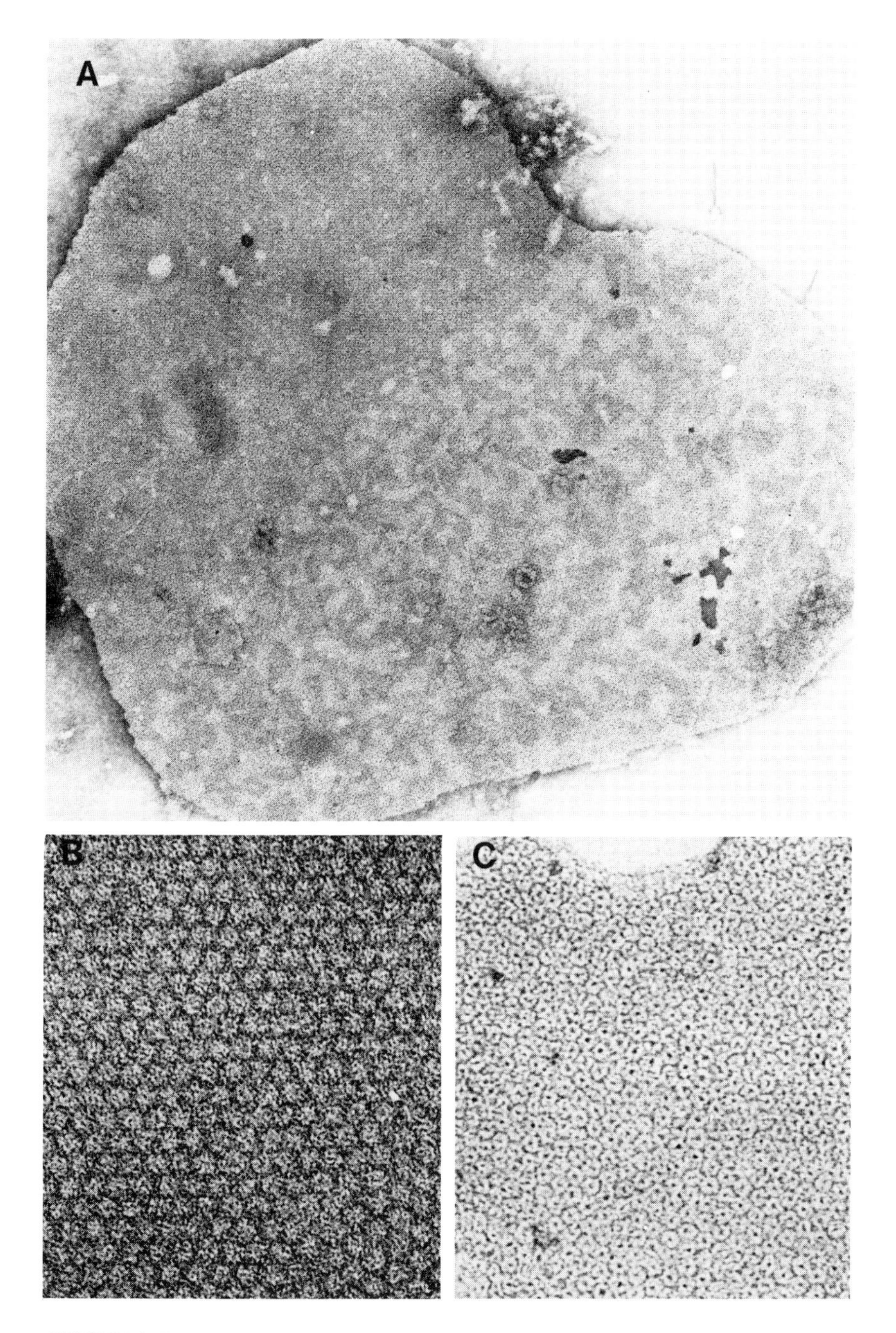

FIGURE 4. Type I gap junctions studied in isolated and negatively stained preparations. (A) A large plaque constructed of channels spaced approximately 8–9 nm apart. ×140,000. (B, C) Gap junction stained with phosphotungstic acid (B) and uranyl acetate (C). B, ×580,000; C, ×470,000.

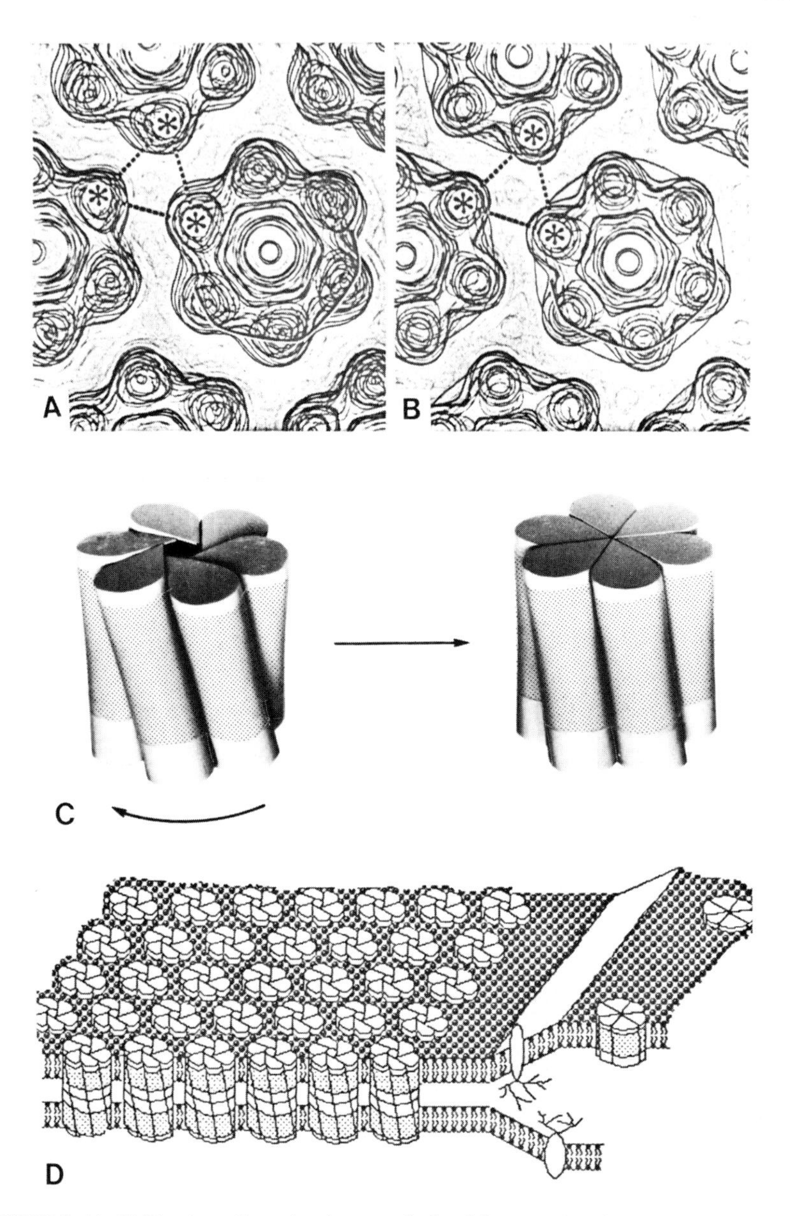

FIGURE 5. (A, B) The three-dimensional maps calculated from gap junctions embedded in frozen solutions (Unwin and Ennis, 1984). The map in A was obtained from junctions in low calcium and it may correspond to the channel in the opened configuration. The map in B was calculated from specimens to which calcium was added. It may correspond to the channel in the closed configuration. (C) Models derived from the three-dimensional maps. The hemichannel (hexamer) is oriented with the cytoplasmic domain to the top. The shadowed area corresponds to the region spanning the membrane. The hemichannel consists of six subunits that enclose a small hydrophilic cavity. The

Unwin, 1975; Unwin and Henderson, 1975). The resulting map depicts the distribution of matter, both perpendicularly and in the plane of the junction.

To calculate three-dimensional maps it is necessary to isolate the junctions and then to arrange the annular units (channels) in crystalline arrays (Benedetti and Emmelot, 1965, 1968). The crystalline junctional plaques are deposited on carbon-coated grids and washed with uranyl acetate solution (''negatively stained''; see Fig. 4) or frozen without staining by plunging the grid with the specimen into a liquid coolant (Unwin and Ennis, 1984). The frozen junctions are transferred to electron microscopes modified to maintain the specimen at liquid nitrogen temperatures during imaging. The images (from either method of preparation) that display the highest resolution are processed by Fourier analysis. Two-dimensional maps of the junctions are represented as series of lines (called contours) that connect points in the plane having the same density.

Figure 5A and B show two maps calculated from electron micrographs of junctions embedded in frozen solutions (Unwin and Ennis, 1984). In both maps the hexamer (hemichannel) is viewed from the cytoplasmic side of the junction. The maps depict matter as contoured regions, which are clustered around points having sixfold symmetry. The study of maps calculated from electron micrographs of isolated gap junctions embedded in negative stain (Zampighi and Robertson, 1973; Zampighi and Unwin, 1979; Baker *et al.*, 1983) or in frozen solutions (Unwin and Ennis, 1984) provided important details of the organization of gap junction channels. First, the complete channel is a 12-subunit arrangement (a dodecamer) formed by two hexamers aligned coaxially and connected through their external domains. Second, the hexamers are asymmetrically located in the bilayer since they protrude more on the external than on the cytoplasmic surfaces of the junctions. This characteristic was later independently observed by freeze-fracture-etch methods (Hirokawa and Heuser, 1982). Third, the hexamers contain a narrow cavity that spans the entire membrane (i.e., the pore of the channel) and widens to a diameter of 2 nm in the external domain. Fourth, each cylindrical subunit, approximately 2 nm in diameter and 7.5 nm long, is tilted relative to the plane of the membrane.

Isolated gap junction membranes stacked in partially oriented pellets have also been studied by low-angle x-ray diffraction methods (Makowski *et al.*, 1977, 1982, 1984; Makowski, 1985). These studies showed that the phospholipids of junctional membranes exhibit the typical bilayer configuration with the distance between polar groups of 4.0–4.2 nm. The junctional protein (the

---

change in conformation that closes and opens the channel, consists of a cooperative change in the angle of inclination of the subunits. These subunits become almost perpendicular to the plane of the membrane, after the addition of calcium (the closed state). (D) A model of the entire gap junction. The model includes the two hexamers aligned to form the complete channel, the halo surrounding the plaque, and the regions of high curvature connecting the junctional plaque with the nonjunctional plasma membrane.

channels) are depicted as dodecamers containing a hydrophilic cavity spanning the entire thickness of the junction. The main difference between the maps calculated from electron micrographs and those deduced from x-ray intensities of partially oriented pellets is centered around the size, shape, and location in the membrane of the protein subunit comprising the channels. Low-angle x-ray diffraction studies suggest that each subunit has a dumbbell shape with lobes protruding symmetrically into the cytoplasmic and external surfaces. In addition, the subunits are oriented perpendicularly across the membrane.

The differences in subunit size, shape, and orientation observed in maps calculated from electron microscopy and x-ray diffraction are the basis for larger disagreements in model gating of gap junction channels.

3.3. Possible Conformational Changes Involved in Gating

Cells can regulate communication via gap junctions by two possible mechanisms. They either change the number of channels comprising the gap junctions or they gate the channels from a high conductance (opened) to a low conductance (closed) state. Both mechanisms are present in tissues coupled via gap junctions. For example, hormones greatly increase the number of channels and level of coupling in tissues such as myometrium and granulosa cells. On the other hand, increase in the intracellular concentration of calcium and/or hydrogen ions can reversibly uncouple cells. Because the change in conductance of the junction upon increase of Ca^{2+} or acidification occurs with a time course of seconds or minutes, this type of uncoupling has been interpreted as *gating* of the channel. It is thought that these ions promote conformational changes of the channel from the opened to the closed state.

The mechanisms regulating channel number such as channel protein synthesis, assembly into hexamers and dodecamers, and internalization remain poorly understood. However, several models have attempted to explain possible conformational changes involved in channel gating based on structural, chemical, and physiological information.

The first model describing conformational changes responsible for channel gating was proposed by Makowski *et al.* (1977). The model was obtained from combined chemical data, electron microscopic information, and measurements of low-angle x-ray diffraction intensities of partially oriented liver gap junctions. The model proposed that the gate was located in the *external* lobe of the dumbbell-shaped subunit of the channel. The conformational change gating the channel between the two conductance states was visualized as movement of mass into the channel lumen. Such conformational change decreased the width of the extracellular gap from 3.5 to 2.7 nm (Makowski *et al.*, 1977). The experimental manipulation that triggered this conformational change was unclear, although it seems to be the result of treating the isolated junctional pellets with the exogenous proteolytic enzyme, trypsin.

The second model proposing a conformational change responsible for channel gating was advanced by Makowski *et al.* (1982, 1984). This model was also obtained by studying electron density profiles calculated from isolated liver gap junctions. The extent of the hydrophilic cavity (the pore of the channel) was mapped by comparing electron density profiles calculated from pellets equilibrated with solutions containing different concentrations of sucrose (Makowski *et al.*, 1984). These experiments permit, at least in principle, the differentiation between aqueous phase, lipid bilayers, and the walls of the channels. By comparing x-ray patterns of junctional pellets equilibrated in solutions of different sucrose concentrations, Makowski *et al.* (1984) proposed that a region of 10 nm in *the middle of the junction* (i.e., that portion of the channel that spans the extracellular gap and half of each junctional membrane) was inaccessible to sucrose, even after treatment with trypsin, that in the previous model appeared to gate the channel from the opened to the closed state. This second model of the gap junction channel proposed that isolated junctions had their channels in a closed configuration due to constrictions (gates) located near the *middle* of each bilayer, at the place where sucrose diffusion stopped. Therefore, the model proposed a funnel-shaped opening of the channel at the cytoplasmic domain and constrictions (i.e., gates) in the portion of the dumbbell-shaped subunit connecting the lobes (i.e., the middle of the membrane).

A third model of the gap junction channel proposing conformational changes responsible for gating has been advanced by Makowski (1985). Analysis of x-ray patterns of isolated liver gap junctions now suggests that the gate of the channel is located in the *cytoplasmic lobe* of the dumbbell-shaped subunit. This was the position where a funnel-shaped entrance was observed in the experiments replacing water by sucrose (Makowski *et al.*, 1984). The conformational change that closes the channel is a large synchronous movement (toward the center of the pore) of the small cytoplasmic lobes of the dumbbell-shaped subunit without notable rearrangement of the rest of the subunits. Thus, x-ray diffraction information has identified three possible locations of the gate in the hemichannel, at the external domain (Makowski *et al.*, 1977), at the membrane-spanning domain (Makowski *et al.*, 1982, 1984), and at the cytoplasmic domain (Makowski, 1985).

An alternative mechanism of channel gating was proposed by studying three-dimensional maps obtained by electron microscopy (Unwin and Zampighi, 1980; Unwin and Ennis, 1984), combined with low-angle x-ray diffraction studies of pellets suspended in solutions containing ions that date the channel in physiological conditions (Unwin and Ennis, 1983). The model resulting from these studies proposes that the change in the structure of the channel, upon going from the opened to the closed state, can be explained by assuming a coordinate rearrangement of the entire channel due to tilting and sliding of the individual subunits along their lines of contact. The subunits span the membrane inclined at angles in the opened conformation. The addition of calcium (which in cells closes the channel) reduces the average inclination of each subunit, making it more perpendicular to the plane of the membrane (from 14° to 9°). Because each subunit is long, the resulting displacement at the cytoplasmic end is large (about 0.9 nm). Thus, a pair of opposing subunits could move toward each other by up to 1.8 nm in the cytoplasmic region of each hemichannel.

Many questions and some controversies remain regarding the detailed structure of the gap junction channel, the exact location of the gate, and the nature of the conformational changes responsible for gating. Undoubtedly, these controversies will be resolved when higher-resolution models of the channel-forming protein become available. Nevertheless, electron microscopic studies at 2-nm resolution have supplied some sensible schemes for the gating of this important and ubiquitous channel.

4. MORPHOLOGICAL CHARACTERIZATION OF TYPE II GAP JUNCTIONS

Type II gap junctions have been observed only between mature lens fibers. They are classified as a separate group because there is compelling structural (Peracchia and Peracchia, 1980a,b; Simon *et al.*, 1982; Zampighi *et al.*, 1982), chemical (Gorin *et al.*, 1984), and immunological (Bok *et al.*, 1982; Fitzgerald *et al.*, 1983; Sas *et al.*, 1985) evidence that these junctions are different from those between hepatocytes, which have been described as the most representative example of type I gap junctions.

The role of the lens in vision is to focus the image on the retina. For this, the lens possesses two main properties: it is deformable and transparent to light. These properties are derived from one kind of cell of ectodermal origin, with an inward-oriented pattern of growth. Only the anterior portion of the lens maintains a simple cuboidal epithelium that is responsible for most of the movement of ions occurring in this tissue. These cuboidal cells are connected to each other through type I gap junctions. The epithelial cells proliferate and move toward the equator of the lens (the bow zone), where they elongate and differentiate into lens fibers, which have a flattened, hexagonal cross section. In addition to elongation, the lens fibers eliminate most of their organelles and synthesize large amounts of soluble "crystalline" proteins that provide the lens with its optical properties.

At this stage, the fibers become connected to each other by elaborate regions of interlocking contacts. The thin edges of the fibers are generally joined by ball-and-socket interdigitations, or overlapping flaps (Dickson and Crock, 1972; Kuwabara, 1975). The flat surfaces of the fibers, particularly in the deep cortex and nucleus, display undulating membrane pairs, the so-called tongue-and-groove interdigitations (Dickson and Crock, 1972; Kuwabara, 1975; Okinami, 1978; Lo and Harding, 1984). Thin-section electron microscopy also shows that the plasma membranes are in close contact over long distances, producing the pentalamellar appearance characteristic of gap junctions (Fig. 6). However, lens fiber junctions have a reduced overall thickness (13–14 nm) and they do not display the usual 2- to 4-nm-wide extracellular gap nor the characteristic hexagonal array of channels when studied by freeze-fracture or negative staining. Freeze-fracture replicas of lens fiber gap junctions show 9- to 11-nm-diameter particles loosely clustered on the P fracture face and complementary pits on the E face (Fig. 7). That these patches correspond to junctions and not to aggregates of other membrane proteins is demonstrated by the presence of "P to E" fracture steps of 2–3 nm in height which are characteristic of gap junction plaques (arrows, Fig. 7). In addition, freeze-fracture studies demonstrated the presence of square arrays of units spaced 6–7 nm apart (Okinami, 1978; Peracchia and Peracchia, 1980a,b; Zampighi *et al.*, 1982). The square arrays are small in the lens cortex but they become conspicuous in the region of the tongue-and-groove interdigitations (Fig. 8). The subunit organization (i.e., the hemichannels are constructed of four subunits), the exact location in the lens, and their possible relation to cell communication are topics of controversy.

The location of the junctional complexes in the lens has been addressed in recent electron microscopic studies (Lo and Harding, 1984; Costello *et al.*, 1985). These studies have shown that disordered gap junction plaques (as the one in Fig. 6) are common in the cortex. Also, these studies showed that small square

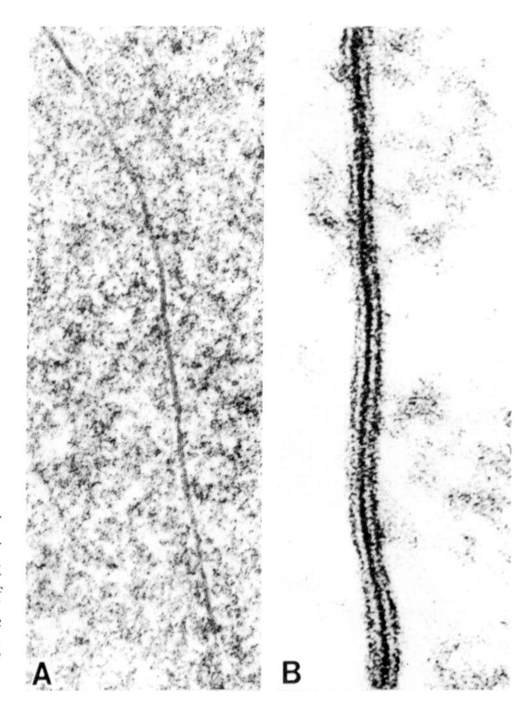

FIGURE 6. Thin-section electron micrographs of type II gap junctions between rabbit lens fibers at the cortex region. Note the extensive regions of close membrane apposition (A) and the pentalamellar structure (B). A, ×77,000; B, ×310,000.

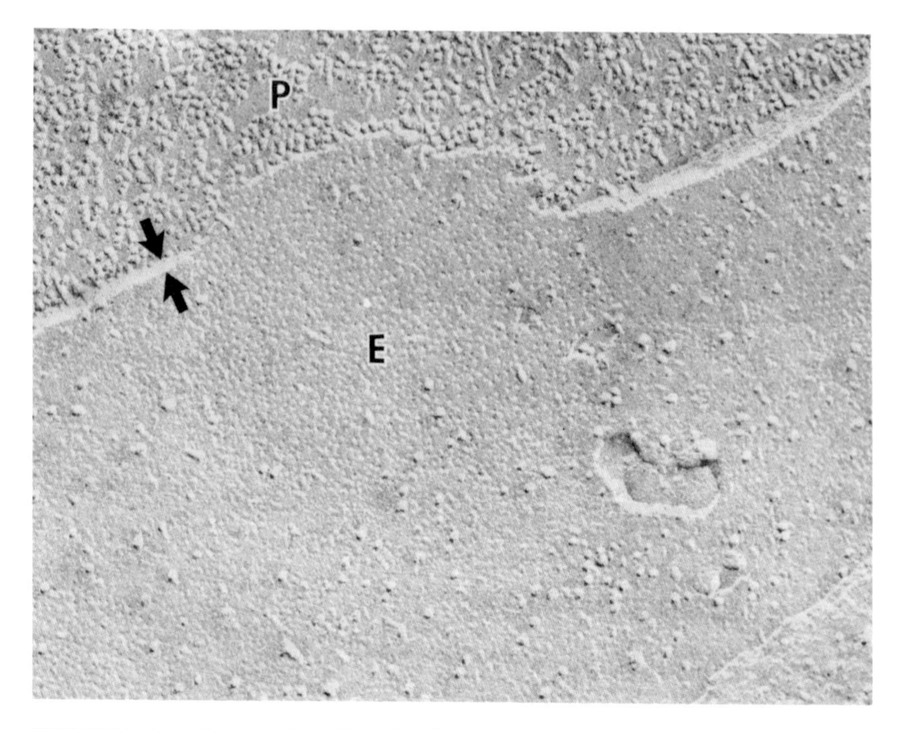

FIGURE 7. Freeze-fracture of type II gap junction of rabbit lens fibers (outer cortex). The junction forms a circular patch made of intramembrane particles (P fracture face) and complementary pits (E fracture face). The close contact between both membranes is represented in this figure by the small size of the fracture step separating the E and P fracture faces. Note that the perimeter of type II gap junctions does not seem to contain a halo. ×210,000.

arrays start appearing in the deep cortex and that their size and frequency greatly increase toward the lens nucleus (Costello *et al.*, 1985). These studies also suggest that the undulating nature of the tongue-and-groove interdigitations (Fig. 8) is the result of regions of square crystalline arrays, in one fiber cell, matched with regions of naked bilayers on the opposing cell (Costello *et al.*, 1985; Lo and Harding, 1984). This structural organization of undulating junctions (Fig. 9) is inconsistent with a function in cell communication.

The composition and structural organization of square arrays in the undulating, thinner junctions of the tongue-and-groove interdigitations were analyzed in isolated fractions of lens plasma membranes. Lens fiber junctions can be isolated in large quantities and high purity without using detergents or proteases (Benedetti *et al.*, 1976; Simon *et al.*, 1982; Zampighi *et al.*, 1982). SDS–gel electrophoresis shows that these isolated lens membrane fractions are comprised of a major intrinsic polypeptide of apparent molecular weight of 26,000–27,000 (MIP-26) and two other minor proteins of 21,000 and 16,000, respectively.

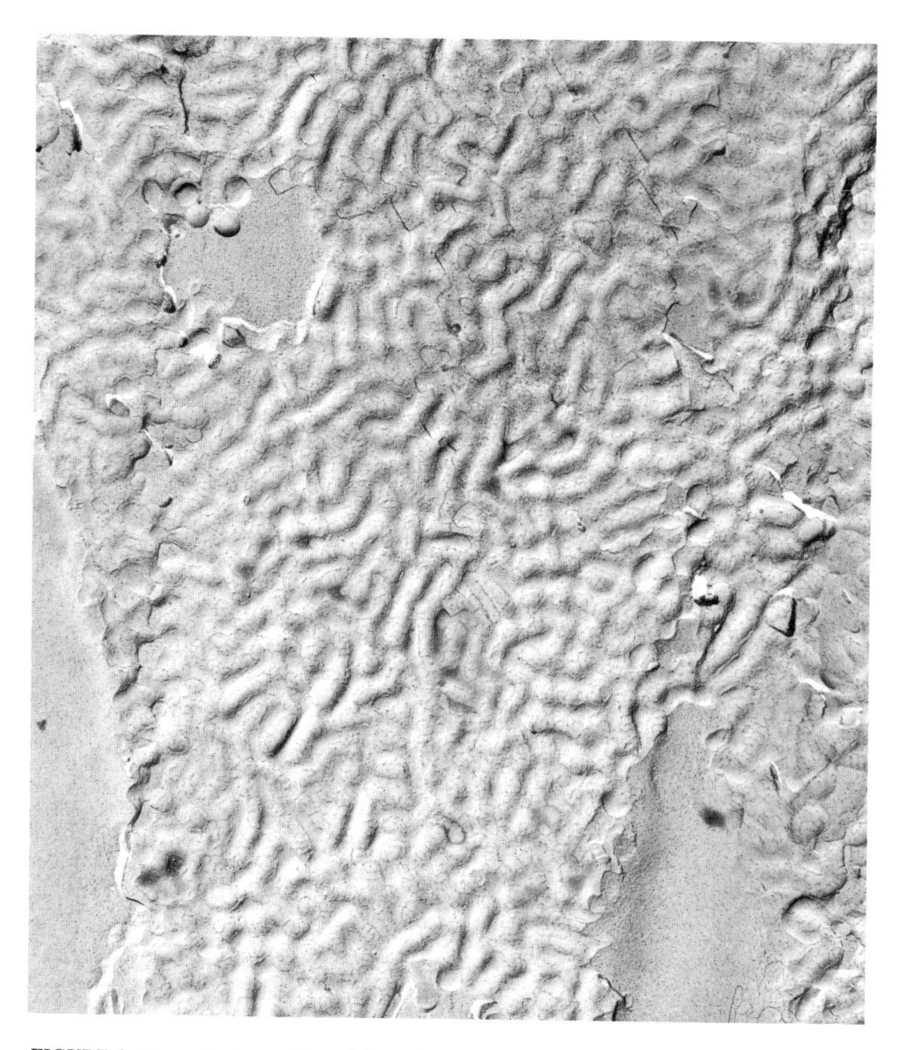

FIGURE 8. Freeze-fracture replica of the region of contact between lens fibers in the inner cortex and nucleus to demonstrate the undulating organization of the tongue-and-groove interdigitations. ×25,000.

Thin-section electron microscopy of these isolated fractions shows that they contain large undulating junctions of approximately 13 nm in overall thickness and single membranes. Occasionally the presence of typical pentalamellar structures, 16–18 nm in overall thickness, is observed intermingled with the most abundant undulating junctions (Fig. 10). These undulating junctions most likely correspond to the tongue-and-groove interdigitations described between the flat surfaces of the lens fibers. Freeze-fracture electron microscopy (Fig. 9) and low-

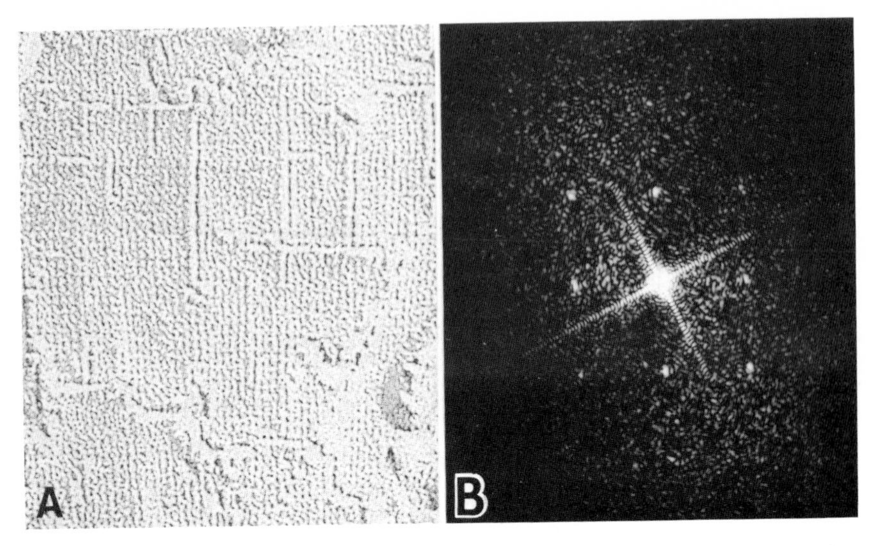

FIGURE 9. Higher magnification of the organization of particles characteristic of the undulating, thinner gap junctions in the lens fibers. The units are arranged in square arrays spaced approximately 6–7 nm apart. ×230,000. From Zampighi *et al.* (1982).

angle x-ray diffraction studies of these undulating junctions show the presence of extensive regions of square arrays (Zampighi *et al.*, 1982).

At present there is good agreement that patches of square arrays, spaced 6–7 nm apart, do exist from the outer cortex to the nucleus of the lens. It is also well documented that these square arrays are the dominant organization of the undulating junctions (Costello *et al.*, 1985; Lo and Harding, 1984; Zampighi *et al.*, 1982). There is considerable disagreement, however, about the functional significance of these square arrays in the lens fiber. For example, Benedetti *et al.*, (1981) suggested that the square arrays were involved in transport through a single membrane. Goodenough (1979) dismissed them as insignificant and artifactual, and designated lens junctions to be regular type I (liver) gap junctions. Peracchia and co-workers (Bernardini and Peracchia, 1981; Peracchia and Peracchia, 1980a,b) observed extensive square arrays, but also interpreted these as type I. They proposed that the channels could be aggregated or disaggregated, depending on the ionic (i.e., Ca^{2+} and H^{+}) composition of the medium in which they are suspended. Kistler and Bullivant (1980a,b) interpreted these arrays as artifacts of isolation, chemical fixation, or proteolysis. On the other hand, Zampighi *et al.* (1982) interpreted the square arrays to be the main organizational component of isolated lens membranes both junctional and single, and in addition pointed out that these membranes were unrelated to the junctions isolated from liver hepatocytes.

The different chemical composition and structural organization adopted by

MIP-26 in the lens raise the question of whether the polypeptide is involved in cell communication in the lens. Several conflicting hypotheses have been proposed in order to answer this question. Goodenough (1979) suggested that all the junctions in the lens have structures similar to those described between hepatocytes (type I in this review). The structural differences between lens and liver junctions resulted from the anaerobic metabolic conditions in the lens (Good-

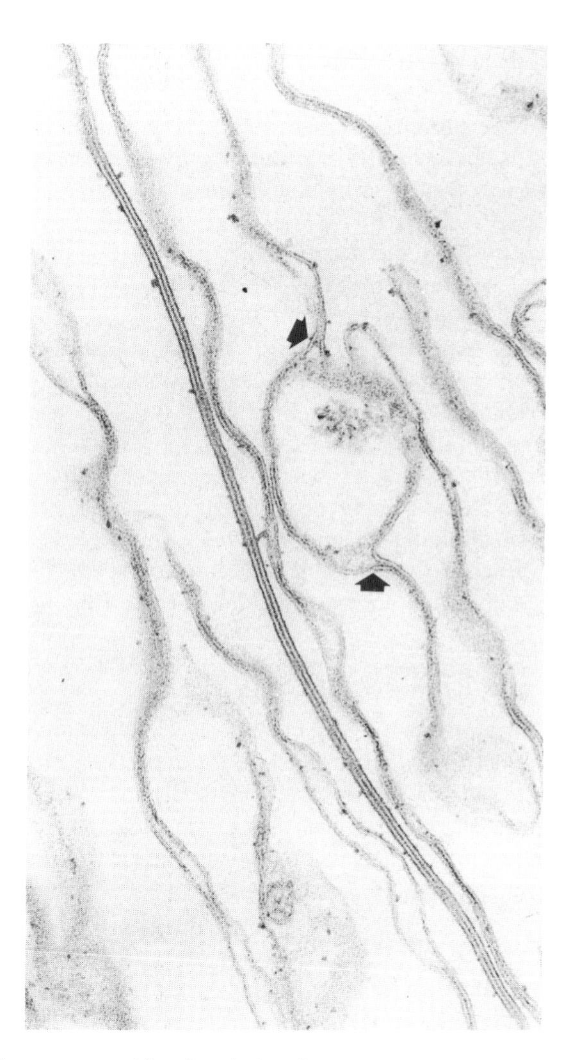

FIGURE 10. The two types of junctions in lens fibers are seen here side by side in an isolated preparation. The large flat junction at the center has a greater overall thickness than the undulating junctions also present in this figure. Note that the thinner, undulating junctions frequently separate their membranes to include large spaces. ×110,000. From Zampighi *et al.* (1982).

enough, 1979). In this hypothesis, it was proposed that lens gap junction channels were nonregulated, low-resistance pathways, permanently locked in the opened state and thus lacking the characteristic hexagonal arrays seen in liver gap junctions. This hypothesis is in conflict with most of the experimental evidence available.

The involvement of MIP-26 in the organization of lens gap junctions has also been explored using antibodies as specific markers. The result of this line of research is highly controversial. For example, anti-MIP-26 antibodies have been reported to bind to junctions alone (Sas *et al.*, 1985), to junctions and single membranes (Bok *et al.*, 1982; Fitzgerald *et al.*, 1983), and to nonjunctional membranes only (Paul and Goodenough, 1983). In addition, Kistler *et al.* (1985) reported that antibodies against MIP-26 bind to single membranes only but that antibodies raised against another membrane protein of lens surface of apparent molecular weight of 70,000 (MP70) bound specifically to macular domains on the lens fiber surfaces that they interpreted as gap junction plaques.

There is evidence suggesting that MIP-26, the main protein of lens fiber plasma membranes, forms gap junctions that are involved in cell communication. The protein has been localized at junctional plaques by immunocytochemistry at least by most investigators (Bok *et al.*, 1982; Fitzgerald *et al.*, 1983; Sas *et al.*, 1985). The amino acid composition of MIP-26 is consistent with that of a channel-forming protein (Gorin *et al.*, 1984) and reconstitution of this protein in planar lipid films produces single conductance units consistent with gap junction channels (Hall and Zampighi, 1985; Zampighi *et al.*, 1986). However, this hypothesis has difficulties in explaining the binding of anti-MIP-26 antibodies in single membranes and the functional significance of the structural organization proposed for the undulating junctions (Costello *et al.*, 1985; Lo and Harding, 1984).

Although the present view of the structural organization of lens fiber junctions remains controversial, the studies performed on this type of gap junction strongly suggest that a protein, different in its amino acid composition from that forming the type I gap junction (either MIP-26 or MP70), can construct large channels connecting the cytoplasm of adjacent cells and associate to form anatomically recognizable gap junctions. Thus, the protein forming type I gap junctions does not seem to be a unique solution used by cells to solve the problem of cell communication.

5. MORPHOLOGICAL CHARACTERIZATION OF TYPE III GAP JUNCTIONS

Crayfish commonly escape from threats by tail-flips rapidly initiated by giant command neurons, or slowly initiated by nongiant circuitry. Gap junctions form part of the giant escape circuitry of the animal. The giant system includes two pairs of interneurons with giant axons, the lateral (LG) and medial (MG) giant axons. These act as command neurons for two slightly different types of tail-flips: the upward-moving and backward-moving, respectively (Krasne *et al.*, 1977). Stimulation of the giant axons generates motor responses directly by virtue of their patterns of monosynaptic connections with two major classes of motoneurons, the motor giants (MoGs) and the nongiant fast flexor motoneurons.

It has been known, from the original work of Johnson (1924), that the LG axons are constructed of independent axonal segments extending from ganglion to ganglion and connected to each other through prominent septa. Electron microscopic studies of the septa have determined the presence of gap junctions that are designated type III in this chapter (Hama, 1961; Pappas *et al.*, 1971; Peracchia, 1973a,b; Zampighi *et al.*, 1978). The medial axons, on the other hand, extend continuously from head to tail making gap junctions with the MoG axon just caudally of each ganglion. It is of interest that gap junctions at these two synapses differ functionally; the one forming the lateral axons is bidirectional (impulses pass equally well rostrally or caudally), whereas the gap junction between the interneuron (MG or LG) and the giant motoneuron (MoG) is unidirectional or rectifying (impulses pass from the medial or lateral giants to the giant motor, but not in the reverse direction). The structural organization of the bidirectional synapse (between the segments forming the LG) will be described first and then compared with what is known of the rectifying synapse (Hanna *et al.*, 1978).

The bidirectional gap junction connecting the axonal segments, forming the LG, is located at the contact region, the *septum*. Most of the septa (about 95%) are constructed of extracellular fibers and ground substance, covered with slabs of glial cells. The synapses consist of interruptions of the septa where the axolemmae come into direct contact. These synaptic regions contain thousands of gap junction plaques with hundreds of thousands of channels.

The structure of the bidirectional gap junction has been described previously (Pappas *et al.*, 1971; Peracchia, 1973a,b; Zampighi *et al.*, 1978). Their main structural characteristics are shown in Fig. 11. It consists of two closely apposed axolemmae, approximately 18–20 nm in overall thickness, separated by a narrow and unstained extracellular gap, 4–5 nm wide (Fig. 11B,C). In transverse sections, the gap junctions display discrete transverse densities spaced about 20 nm apart. These densities (or channels) traverse the plasma membranes, bridge the extracellular gap, and protrude on the cytoplasmic surface by 3–4 nm. *En face* views of the gap junctions show the densities arranged in two-dimensional lattices 20–22 nm center-to-center apart (Fig. 11D). This appearance contrasts sharply with the 8- to 9-nm separation of the channels in type I gap junctions. In addition, gap junctions in crayfish synapses have rows of *synaptic vesicles* approximately 40–60 nm in diameter, symmetrically placed in close proximity to the cytoplasmic surfaces. The function of these vesicles is still unknown.

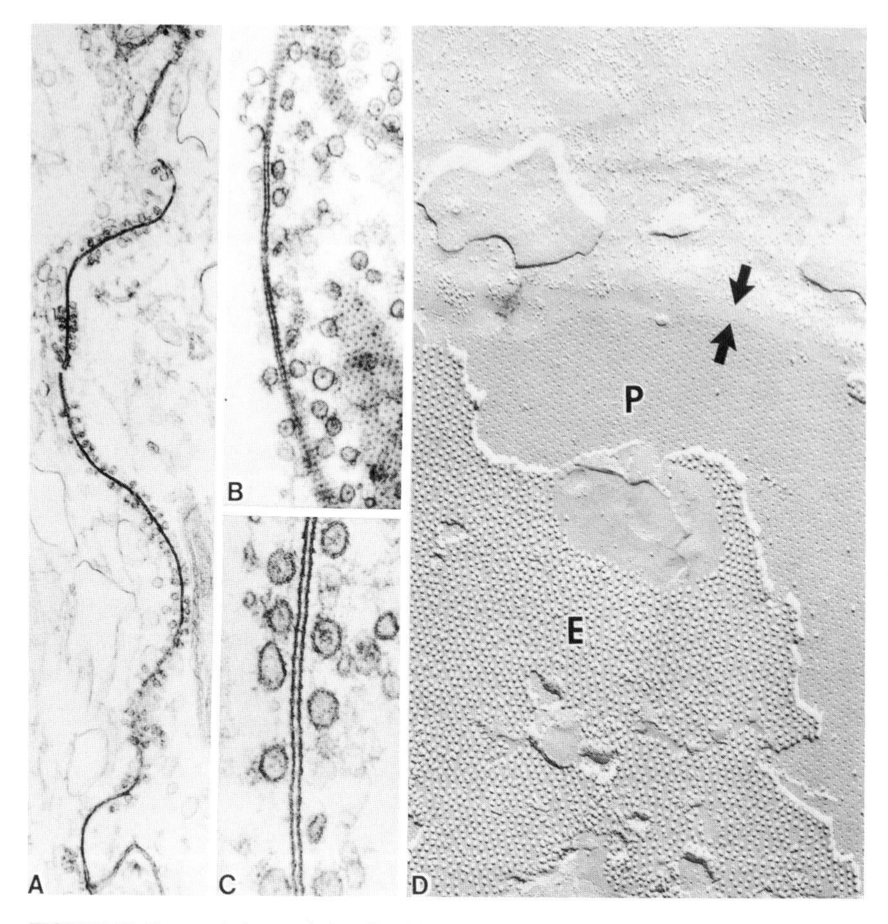

FIGURE 11. Structural characteristics of type III gap juncions. (A) A large portion of the synapse between both axonal segments comprising the lateral giant axon. Note the arrangement of vesicles at each side of the synapse. (B, C) Transverse and *en face* views, respectively. Note that the extracellular gap is always unstained and that the channels (circular densities in B and transverse densities in C) are separated by a 20-nm spacing (contrast this view with Figs. 1 and 3 of type I and Figs. 8 and 10 of type II gap junctions). (D) A replica of type III gap junction. Note that the particles are on the E fracture face and that pits are on the P fracture faces. A, ×40,000; B, ×100,000; C, ×200,000; D, ×120,000.

Freeze-fracture replicas of the synapses (i.e., regions of contact between axolemmae) between the giant axons show fracture faces that contain patches of intramembrane particles. The particles (channels) are located on the E-fracture face (instead of on the P-fracture face that characterizes type I junctions). The channels are about 12 nm in diameter and are spaced 20–22 nm apart. Each patch may contain as few as five to seven channels or as many as several thousand. These type III gap junction patches also have a particle-free "halo" at their perimeter similar to that described for type I gap junctions (see Figs. 3 and 11). The channels in each patch can be packed in hexagonal arrays or clustered in irregular-shaped aggregates. In the regions between the patches, there are also single particles that have an appearance similar to that of the channels. However, it is impossible to say whether they are hemichannels or single gap junction channels that are in the process of being added to the existing plaques.

Gap junctions that rectify are rare in nature. Curiously, the first gap junction described by Furshpan and Potter (1959) between the motor giant fiber (MoG) and the lateral axon (LG), has the property of rectification. Although the mechanisms of transmission at the bidirectional gap junctions are now well understood, the mechanism by which rectification is accomplished has remained a puzzle. In this section, I will compare briefly the structure of rectifying gap junctions to those between the LG axons that transmit bidirectionally.

The rectifying and bidirectional gap junctions will be compared, first with respect to the geometry of their region of contact and second, with respect to the structure of the gap junction itself. The bidirectional gap junction in the lateral giants is contained in a flat septum at the proximal and distal ends of the axonal segment. The region of contact containing rectifying gap junctions (MoG with either LG or MG) is not flat but consists of slender dendrites from the MoG that branches extensively before establishing contact. This geometry complicates attempts to voltage-clamp this synapse (Giaume and Korn, 1985; Jaslove and Brink, 1986). On the other hand, the gap junctions of rectifying and bidirectional synapses display similar structural organization. The channels of rectifying and bidirectional gap junctions have identical overall thickness, dimensions and location across the membranes. In addition, the extracellular gap separating the junctional membranes is 4–5 nm wide and unstained. Differences between rectifying and bidirectional gap junctions exist when the organelles located in the axoplasms are compared. The presynaptic element of the rectifying gap junction (LG or MG) contains clusters of 40- to 60-nm-diameter vesicles similar to those described in presynaptic terminals of chemical synapses. On the other hand, the postsynaptic element (the MoG) contains few large vesicles associated with the junctional membranes and numerous large mitochondria specifically located at the terminal.

The anatomical differences observed between both types of gap junctions do not explain by themselves the process of rectification. This puzzling property of some gap junctions could be explained by assuming a voltage-dependent gate, located only in the hemichannel of the postsynaptic element, as suggested by voltage-clamp studies (Jaslove and Brink, 1986). Unfortunately, the remarkably complex anatomy of the contact region between MoG and LG or MG contributes to the difficulties in collecting electrophysiological data from the rectifying electrical synapse.

6. CONCLUSIONS

Gap junctions are specializations of the cell surface, made up of clusters of channels that provide a direct pathway connecting the cytoplasm of adjacent cells. Structural studies of many tissues, both vertebrate and invertebrate, suggest that different types of gap junctions can be defined. In this review, gap junctions are classified into three types, although the list is by no means complete.

The structure of gap junctions has been solved at low resolution and the primary structure of one type has been reported. Low-resolution structural studies performed on gap junctions isolated from hepatocytes have resulted in the construction of reasonable models for the gating of this important and ubiquitous channel.

ACKNOWLEDGMENTS. I wish to thank Dr. J. Frank for providing the image used as Fig. 3, and Dr. E. Dirksen for useful comments on the manuscript. Supported by Grants EY-04110 from the National Institutes of Health and by an MDA Jerry Lewis Neuromuscular Research Center grant.

7. REFERENCES

Albertini, D. F., and Anderson, E., 1975, Structural modifications of lutein cell gap junctions during pregnancy in the rat and the mouse, *Anat. Rec.* **181:**171–194.

Amos, L. A., Henderson, R., and Unwin, P. N. T., 1982, Three-dimensional structure determination by electron microscopy of two-dimensional crystals, *Prog. Biophys. Mol. Biol.* **39:**183–231.

Baker, T. S., Caspar, D. L. D., Hollingshead, C. J., and Goodenough, D. A., 1983, Gap junction structures. IV. Asymmetric features revealed by low irradiation microscopy, *J. Cell Biol.* **96:**204–216.

Benedetti, E. L., and Emmelot, P., 1965, Electron microscopic observations of negatively stained plasma membranes isolated from rat liver, *J. Cell Biol.* **26:**299–305.

Benedetti, E. L., and Emmelot, P., 1968, Hexagonal array of subunits in tight junctions separated from isolated rat liver plasma membranes, *J. Cell Biol.* **38:**15–24.

Benedetti, E. L., Dunia, I., Bentzel, C. J., Vermorken, A. J. M., Kibbelaar, M., and Bloemendal, H., 1976, A portrait of plasma membrane specializations in eye lens epithelium fibers, *Biochim. Biophys. Acta* **457:**353–384.

Benedetti, E. L., Dunia, I., Ramaekers, F. C. S., and Kibbelaar, M. A., 1981, Lenticular plasma membrane and cytoskeleton, in: *Molecular and Cellular Biology of the Eye Lens* (H. Bloemendal, ed.), pp. 137–188, John Wiley, N.Y.

Bennett. M. V. L., Aljure, E., Nakajima, Y., and Pappas, G. D., 1963, Electrotonic junctions between teleost spinal neurons, *Science* **141:**262–264.

Bennett, M. V. L., Nakajima, Y., and Pappas, G. D., 1967, Physiology and ultrastructure of electrotonic junctions. I. Supramedullary neurons, *J. Neurophysiol.* **30:**148–161.

Bernardini, G., and Peracchia, C., 1981, Gap junction crystallization in lens fiber after an increase in cell calcium, *Invest. Ophthalmol.* **21:**291–299.

Bok, D., Dockstader, J., and Horwitz, J., 1982, Immunocytochemical localization of the lens main intrinsic polypeptide (MIP) in communicating junctions, *J. Cell Biol.* **92:**213–220.

Chalcroft, J. P., and Bullivant, S., 1970, An interpretation of liver cell membrane and junction structure based on observations of freeze-fracture replicas of both sides of the fracture, *J. Cell Biol.* **47:**49–60.

Costello, M. J., McIntosh, T. J., and Robertson, J. D., 1985, Membrane specializations in mammalian lens fiber cells: Distribution of square arrays, *Curr. Eye Res.* **4:**1183–1201.

DeRosier, D. J., and Klug, A., 1968, Reconstruction of three-dimensional structures from electron micrographs, *Nature* **217:**1–5.

Dewey, M. M., and Barr, L., 1962, Intercellular connection between smooth muscle cells: The nexus, *Science* **137:**670–672.

Dewey, M. M., and Barr, L., 1964, A study of the structure and distribution of the nexus, *J. Cell Biol.* **23:**553–585.

Dickson, D. H., and Crock, G. W., 1972, Interlocking patterns on primate lens fibers, *Invest. Ophthalmol.* **11:**809–815.

Fitzgerald, P. G., Bok, D., and Horwitz, J., 1983, Immunocytochemical localization of the main intrinsic polypeptide (MIP) in ultrathin frozen sections of rat lens, *J. Cell Biol.* **97:**1491–1499.

Furshpan, E. J., 1964, Electrical transmission in an excitatory synapse in a vertebrate brain, *Science* **144:**878–880.

Furshpan, E. J., and Potter, D. D., 1959, Transmission at the giant motor synapses of the crayfish, *J. Physiol. (London)* **145:**289–325.

Garant, P. R., 1972, The demonstration of complex gap junctions between the cells of the enamel organ with lanthanum nitrate, *J. Ultrastruct. Res.* **40:**333–340.

Giaume, C., and Korn, H., 1985, Junctional voltage-dependence at the crayfish rectifying synapse, in: *Gap Junctions* (M. V. L. Bennett and D. C. Spray, eds.), pp. 367–379, Cold Spring Harbor Laboratory, Cold Spring Harbor, N.Y.

Goodenough, D. A., 1979, Lens gap junctions: A structural hypothesis for nonregulated low-resistance intercellular pathways, *Invest. Ophthalmol. Vis. Sci.* **18:**1104–1122.

Goodenough, D. A., and Gilula, N. B., 1974, The splitting of hepatocyte gap junctions and zonulae occludentes with hypertonic solutions, *J. Cell Biol.* **61:**575–590.

Goodenough, D. A., and Revel, J.-P., 1970, A fine structural analysis of intercellular junctions in the mouse liver, *J. Cell Biol.* **45:**272–290.

Gorin, M. B., Yancey, S. B., Cline, J., Revel, J. P., and Horwitz, J., 1984, The major intrinsic protein (MIP) of the bovine fiber membrane: Characterization and structure based on cDNA cloning, *Cell* **39:**49–59.

Green, C. R., and Severs, N. J., 1984, Gap junction connexon configuration in rapidly frozen myocardium and isolated intercalated disks, *J. Cell Biol.* **99:**453–470.

Hall, J. E., and Zampighi, G., 1985, Protein from purified lens junctions induces channels in planar lipid bilayers, in: *Gap Junctions* (M. V. L. Bennett and D. C. Spray, eds.), pp. 117–190, Cold Spring Harbor Laboratory, Cold Spring Harbor, N.Y.

Hama, K., 1961, Some observations on the fine structure of the giant fibers of the crayfishes (*Cambarus virilus* and *Cambarus clarkii*) with special reference to the submicroscopic organization of the synapse, *Anat. Rec.* **141:**275–280.

Hanna, R. B., Keeter, J. S., and Pappas, G. D., 1978, Fine structure of a rectifying electrotonic synapse, *J. Cell Biol.* **79:**764–773.

Henderson, R., and Unwin, P. N. T., 1975, Three-dimensional model of purple membrane obtained by electron microscopy, *Nature* **257:**28–32.

Hirokawa, N., and Heuser, J., 1982, The inside and outside of gap junction membranes visualized by deep etching, *Cell* **30:**395–406.

Jaslove, S. W., and Brink, P. R., 1986, The mechanism of rectification at the electronic motor giant synapse of the crayfish *Nature* **323:**63–65.

Johnson, G. E., 1924, Giant nerve fibers in crustacean with special reference to *Cambarus* and *Palaemonetes, J. Comp. Neurol.* **36:**323–333.
Kanno, Y., and Loewenstein, W. R., 1966, Cell-to-cell passage of large molecules, *Nature* **212:**629–630.
Kistler, J., and Bullivant, S., 1980a, The connexon order in isolated lens gap junctions, *J. Ultrastruct. Res.* **72:**27–38.
Kistler, J., and Bullivant, S., 1980b, Lens gap junctions and orthogonal arrays are unrelated, *FEBS Lett.* **111:**73–78.
Kistler, J., Kirkland, B., and Bullivant, S., 1985, Identification of a 70,000-D protein in lens membrane junctional domains, *J. Cell Biol.* **101:**28–35.
Krasne, F. B., Wine, J. J., and Kramer, A., 1977, The control of the crayfish escape behaviour, in: *Identified Neurons and Behaviour of Arthropods* (G. Hoyle, ed.), pp. 275–292, Plenum Press, New York.
Kreutziger, G. O., 1968, Freeze-fracture of intercellular junctions of mouse liver, in: *Proc. 26th Annual Meeting EMSA*, pp. 138–234, Claitor's Publishing Division, Baton Rouge, La.
Kuwabara, T., 1975, The maturation of the lens cell: A morphologic study, *Exp. Eye Res.* **20:**427–443.
Larsen, W. J., 1985, Relating the population dynamics of gap junctions to cellular functions, in: *Gap Junctions* (M. V. L. Bennett and D. C. Spray, eds.), pp. 289–306, Cold Spring Harbor Laboratory, Cold Spring Harbor, N.Y.
Larsen, W. J., Tung, H., Murray, S. A., and Swenson, C. A., 1979, Evidence for the participation of actin microfilaments and bristle coats in the internalization of gap junction membranes, *J. Cell Biol.* **83:**576–587.
Lee, M. L., Cran, D. G., and Lane, N. J., 1982, Carbon dioxide induced disassembly of gap junctional plaques, *J. Cell Sci.* **57:**215–228.
Lo, W.-K., and Harding, C. V., 1984, Square arrays and their role in ridge formation in human lens fibers, *J. Ultrastruct. Res.* **86:**228–245.
Loewenstein, W. R., 1966, Permeability of membrane junctions, *Ann. N.Y. Acad. Sci.* **137:**441–472.
Loewenstein, W. R., and Kanno, Y., 1964, Studies on an epithelial (gland) cell junction. I. Modification of surface membrane permeability, *J. Cell Biol.* **22:**565–586.
Loewenstein, W. R., and Kanno, Y., 1967, Intercellular communication and tissue growth. I. Cancerous growth, *J. Cell Biol.* **33:**225–234.
Loewenstein, W. R., Socolar, S. J., Higashino, S., Kanno, Y., and Davidson, N., 1964, Intercellular communication: Renal, urinary bladder, sensory, and salivary gland cells, *Science* **149:**295–297.
McNutt, N. S., and Weinstein, R. S., 1970, The structure of the nexus: A correlated thin section and freeze-cleave study, *J. Cell Biol.* **47:**666–688.
McNutt, N. S., and Weinstein, R. S., 1973, Membrane structures at mammalian cell junctions, *Prog. Mol. Biol.* **26:**46–101.
Makowski, L., 1985, Structural domains in gap junctions: Implications for the control of intercellular coupling, in: *Gap Junctions* (M. V. L. Bennett and D. C. Spray, eds.), pp. 5–12, Cold Spring Harbor Laboratory, Cold Spring Harbor, N.Y.
Makowski, L., Caspar, D. L. D., Phillips, W. C., and Goodenough, D. A., 1977, Gap junction structures. II. Analysis of the x-ray diffraction data, *J. Cell Biol.* **74:**629–645.
Makowski, L., Caspar, D. L. D., Goodenough, D. A., and Phillips, W. C., 1982, Gap junction structures. III. The effect of variations in the isolation procedure, *Biophys. J.* **37:**188–191.
Makowski, L., Caspar, D. L. D., Phillips, W. C., and Goodenough, D. A., 1984, Gap junction structures. V. Structural chemistry inferred from x-ray diffraction measurements on sucrose accessibility and trypsin susceptibility, *J. Mol. Biol.* **174:**449–481.

Merk, F. B., Albright, J. T., and Botticelli, C. R., 1973, The fine structure of granulosa cell nexuses in rat ovarian follicles, *Anat. Rec.* **175:**107–118.
Miller, T. M., and Goodenough, D. A., 1985, Gap junction structures after experimental alteration of junctional channel conductance, *J. Cell Biol.* **101:**1741–1748.
Okinami, S., 1978, Freeze-fracture replica of the primate lens fibers, *Albrecht von Graefes Arch. Klin. Exp. Ophthalmol.* **209:**51–58.
Pappas, G. D., Asada, Y., and Bennett, M. V. L., 1971, Morphological correlates of increased coupling resistance at an electrotonic synapse, *J. Cell Biol.* **49:**173–188.
Paul, D. L., and Goodenough, D. A., 1983, Preparation, characterization, and localization of antisera against bovine MP26, an integral protein from lens fiber plasma membrane, *J. Cell Biol.* **96:**625–632.
Peracchia, C., 1973a, Low resistance junctions in crayfish. I. Two arrays of globules in junctional membranes, *J. Cell Biol.* **57:**54–65.
Peracchia, C., 1973b, Low resistance junctions in crayfish. II. Structural details and further evidence for intercellular channels by freeze-fracture and negative staining, *J. Cell Biol.* **57:**66–76.
Peracchia, C., 1977, Gap junctions—Structural changes after uncoupling procedures, *J. Cell Biol.* **72:**628–641.
Peracchia, C., 1978, Calcium effects on gap junction structure, *Nature* **271:**669–671.
Peracchia, C., 1980, Junctional correlates of gap junction permeation, *Int. Rev. Cytol.* **66:**81–146.
Peracchia, C., and Bernardini, G., 1984, Gap junction structure and cell-to-cell coupling regulation: Is there a calmodulin involvement? *Fed. Proc.* **43:**2681–2691.
Peracchia, C., and Dulhunty, A. F., 1976, Low resistance junctions in crayfish: Structure changes with functional uncoupling, *J. Cell Biol.* **70:**419–439.
Peracchia, C., and Peracchia, L. L., 1980a, Gap junction dynamics: Reversible effects of hydrogen ions, *J. Cell Biol.* **87:**719–727.
Peracchia, C., and Peracchia, L. L., 1980b, Gap junction dynamics: Reversible effects of divalent cations, *J. Cell Biol.* **87:**708–718.
Raviola, E., Goodenough, D. A., and Raviola, G., 1980, Structures of rapidly frozen junctions, *J. Cell Biol.* **87:**273–279.
Revel, J.-P., and Karnovsky, M. J., 1967, Hexagonal arrays of subunits in intercellular junctions of the mouse heart and liver, *J. Cell Biol.* **33:**C7–C12.
Robertson, J. D., 1953, Ultrastructure of two invertebrate synapses, *Proc. Soc. Exp. Biol. Med.* **82:**219–223.
Robertson, J. D., 1954, Electron microscope study of an invertebrate synapse, *Fed. Proc.* **13:**119–131.
Robertson, J. D., 1955, Recent electron microscope observations on the ultrastructure of the crayfish median-to-motor giant synapse, *Exp. Cell Res.* **8:**226–229.
Robertson, J. D., 1960, The molecular structure and contact relationship of cell membranes, in: *Progress in Biophysics* (B. Katz and J. A. V. Butler, eds.), pp. 343–418, Pergamon Press, Elmsford, N.Y.
Robertson, J. D., 1963, The occurrence of a subunit pattern in the unit membrane of club endings in Mauthner cell synapses in goldfish brains, *J. Cell Biol.* **19:**201–221.
Sas, D. F., Sas, M. J., Johnson, K. R., Menko, A. S., and Johnson, R. G., 1985, Junctions between lens fiber cells are labelled with a monoclonal antibody shown to be specific for MP26, *J. Cell Biol.* **100:**216–225.
Simon, S. A., Zampighi, G., McIntosh, T. J., Costello, M. J., Ting-Beall, H. P., and Robertson, J. D., 1982, *Biosci. Rep.* **2:**333–341.
Unwin, P. N. T., and Ennis, P. D., 1983, Calcium-mediated changes in gap junction structure: Evidence from low-angle x-ray pattern, *J. Cell Biol.* **97:**1459–1466.
Unwin, P. N. T., and Ennis, P. D., 1984, Two configurations of a channel forming membrane protein, *Nature* **307:**609–613.

Unwin, P. N. T., and Henderson, R., 1975, Molecular structural determination by electron microscopy of unstained crystalline specimens, *J. Mol. Biol.* **94:**425–440.
Unwin, P. N. T., and Zampighi, G., 1980, Structure of junctions between communicating cells, *Nature* **283:**545–549.
Watanabe, A., and Grundfest, H., 1961, Impulse propagation at the septal and commissural junctions of crayfish lateral giant axons, *J. Gen. Physiol.* **45:**267–308.
Yancey, S. B., Easter, D., and Revel, J.-P., 1979, Cytological changes in gap junctions during liver regeneration, *J. Ultrastruct. Res.* **67:**229–242.
Zampighi, G., 1980, On the structure of isolated junctions between communicating cells, *In Vitro* **16:**1018–1028.
Zampighi, G., and Robertson, J. D., 1973, Fine structure of the synaptic discs separated from goldfish medulla oblongata, *J. Cell Biol.* **56:**92–105.
Zampighi, G., and Unwin, P. N. T., 1979, Two forms of isolated gap junctions. *J. Mol. Biol.* **135:**451–464.
Zampighi, G., Ramon, F., and Duran, W., 1978, Fine structure of the electrotonic synapse of the lateral giant axons in a crayfish (*Procambarus clarkii*), *Tissue Cell* **10:**413–426.
Zampighi, G., Corless, J. M., and Robertson, J. D., 1980, On gap junction structure, *J. Cell Biol.* **86:**190–198.
Zampighi, G., Simon, S. A., Robertson, J. D., McIntosh, T. J., and Costello, M. J., 1982, On the structural organization of isolated bovine lens fiber junctions, *J. Cell Biol.* **93:**175–189.
Zampighi, G., Hall, J. E., and Kreman, M., 1986, Purified lens junctional protein forms channels in planar lipid films, *Proc. Natl. Acad. Sci. USA* **82:**8468–8472.

Chapter 2

Modulation of Junctional Permeability

Walmor C. De Mello
Department of Pharmacology
Medical Sciences Campus
University of Puerto Rico
San Juan, Puerto Rico 00936

1. INTRODUCTION: PHYSIOLOGICAL CONSIDERATIONS

Cell-to-cell coupling through low-resistance junctions represents a very old mechanism of intercellular communication. In sponges and medusae, for instance, without nervous tissue, the epithelia receive external stimuli and convert them into electrical pulses that are conducted in all directions through low-resistance junctions (Mackie, 1964).

The evidence available that intercellular junctions are essential for the electrical synchronization in excitable tissues of invertebrates and vertebrates, indicates that the establishment of intercellular channels was a reliable mechanism of cell-to-cell coupling finally incorporated into the evolutionary scheme.

The inability of medusae to canalize all the information in one direction was probably the main reason for the appearance of neurons with elongated axons in higher levels of the evolution. It is interesting to add that in tissues like heart muscle in which the cells are connected through gap junctions, there are mechanisms for the control of the internal resistivity.

As there is no monotone value in biology, an optimum size for the gap junctions is expected for each tissue. Indeed, a variation in size and number of gap junctions seems to represent a dynamic process of regulation of intercellular communication. During early stages of embryonic development, intercellular junctions are widespread. Potter *et al.* (1966) showed that in squid embryo the tissues are coupled to each other, but by 4 days before hatching, cell-to-cell communication disappears. In chick embryo studies of the proliferation of neural retinal cells, the number of gap junctions reaches a maximum just before cessation of cell proliferation (Fujisawa *et al.*, 1976).

With respect to the role of gap junctions in electrical synchronization, different aspects can be distinguished. In cardiac pacemaker cells, for instance, the electrical coupling is gradually augmented during diastole, reaching a maximum just before the firing of the action potential (De Mello, 1982). This increase in cell-to-cell coupling, which is in part due to a decrease in internal resistance and part due to a rise in the resistance of the non-junctional membrane (De Mello, 1986a), leads to a gradual enhancement of intercellular coupling transforming the pacemaker area into a single and large functional unit whose firing power guarantees the stimulation of neighboring nonpacemaker cells.

In the command nucleus of the electric fish, gap junctions also play an important role in the electrical synchronization. In this nucleus two giant neurons located in the first spinal segment control the whole electric organ (Bennett *et al.*, 1967). Each neuron innervates the entire organ on one side of the body in such a way that the firing of each neuron leads to a discharge of the organ on that side. As the giant neurons are electrically connected, the excitation of one cell initiates an action potential in the other cell and consequently the whole electric organ is discharged (Bennett *et al.*, 1967). The presence of electrical synapse in this case makes possible a synchronized firing of the organ not achievable if the cells were connected through chemical synapses.

In the central nervous system, electrical synapses are essential for the spread of electrotonic activity in the cerebellar cortex and in the inferior olive where dendrodendritic electrical synapses make possible the synchronization of the electrical activity (Llinas, 1975; Sotelo *et al.*, 1974).

In cultured heart cells the coupling resistance is quite high ($> 100\ \mathrm{M\Omega}$) at the moment of cell contact but begins to fall immediately after, reaching a value of about 20 MΩ at the moment of electrical synchronization (Clapham *et al.*, 1980). Assuming a resistance of $1 \times 10^{10}\ \Omega$ for a single hydrophilic channel, the rate of channel synthesis was found to be one channel per cell per minute (Clapham *et al.*, 1980).

In syncytial tissues the most probable mechanism of impulse propagation is local circuit current flow. During the passage of an electrical impulse action current flows in local circuit from the active areas along the myoplasm and gap junctions, outward through the resting adjacent cell membrane, back along the extracellular fluid and inward through the active region, so completing the local circuit (see Fig. 1).

The potential change across the nonjunctional cell membrane (V_m) is given by the difference between the potential on the inside (V_1) and the potential on the outside of the cell (V_o): $V_m = (V_1 - V_o)$. The current flowing along the outside of the cell membrane is $i_o = (1/r_o)\ (dV_o/dx)$, where dV_o/dx is the extracellular potential gradient and r_o the resistance per unit length of the extracellular fluid.

The current flowing along the core of the cardiac fiber is given by $i_i = (-1/r_i)\ (dV_i/dx)$, where dV_i/dx is the intracellular gradient and r_i the intracellular resistance per unit length, which represents the myoplasmic and junctional resistance.

Contrary to nerve cells, cardiac myocytes have a highly structured intra-

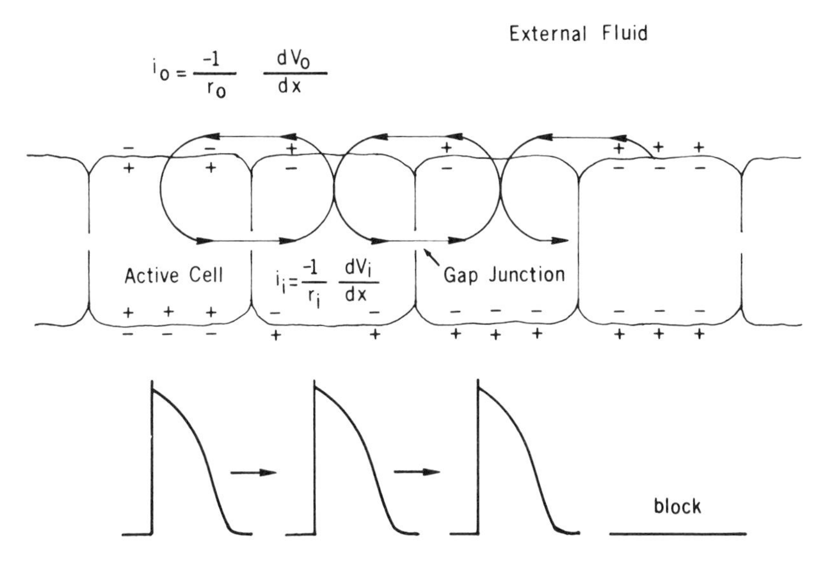

FIGURE 1. Diagram illustrating the flow of local circuit current in a cardiac fiber and the role of gap junctions in the spread of propagated activity. From De Mello (1983c) with permission.

cellular medium in which the sarcoplasmic reticulum is probably the most finely organized system. Although it is reasonable to think that these structures may enhance the myoplasmic resistance (Kushmerick and Podolsky, 1969), there is experimental evidence that the cytoplasm of muscle fibers has a specific resistance of the same order of magnitude as the resistance of the extracellular fluid (Fatt and Katz, 1951; Falk and Fatt, 1964).

The hypothesis that activity spreads in cardiac muscle by local circuit flow has been intensively investigated (see Weidman, 1970). A conclusive test of the hypothesis is the injection of a current pulse into the cell and the study of its effect on the membrane potential of neighboring cells. When subthreshold current pulses are used, electrotonic potentials can be recorded from cells located nearby. In cardiac Purkinje fibers, evidence has been provided that electrical current flows freely between the cells (Weidmann, 1952; De Mello, 1975). The electrotonus in these fibers can be precisely described by the cable equations and the steady-state distribution of the electrotonic potentials closely follows the predicted results for a uniform cable. The core resistivity in Purkinje fibers is quite low (105 Ω-cm; Weidmann, 1952) and the space constant is large (1.9 nm; Weidmann, 1952) in comparison to the length of a single cell (125 μm), which means that the electrical resistance of the intercellular junctions is very low (Kamiyama and Matsuda, 1966; Weidmann, 1970).

In rat atrium, the depolarization of one cell causes appreciable changes in the membrane potential of adjacent cells (Woodbury and Crill, 1961), but a steep decrement of the electrotonic potentials is found (λ = 130 μm) when the cell is polarized through an intracellular microelectrode. A high intracellular or extracellular resistance, or a low nonjunctional resistance might explain these results. The enormous decrement of the electrotonic potential, however, is related to the three-dimensional characteristics of the rat trabecula. The decay is abrupter than exponential and is better fitted by a Bessel function (Woodbury and Crill, 1961; Noble, 1962).

Widespread electrotonic potentials, however, can be observed in myocardial fibers when the current is applied to the tissue through large extracellular electrodes (Trautwein *et al.*, 1956; Sakamoto, 1969; Weidmann, 1970). When the polarizing current influences many cells synchronously, the space constant ranges from 880 μm (Weidmann, 1970) to 1300 μm (Kamiyama and Matsuda, 1966; Sakamoto, 1969).

Barr *et al.* (1965) demonstrated that impulse conduction along a thin bundle of atrial muscle is blocked by immersing the central portion in isotonic sucrose solution while keeping the two ends of the muscle strip on Ringer's solution. The suppression of impulse conduction is due to the increase of the extracellular resistance produced by the sucrose solution. Consequently, the longitudinal flow of current through the myoplasm and extracellular fluid is abolished. The conduction is reestablished, however, if an electrical shunt is produced between the two pools of Ringer's solution. These findings indicate that low internal and external resistances are necessary for the propagation of the action potential.

The velocity of impulse conduction (θ) depends on the intracellular longitudinal resistance (r_i) and is closely related to fiber diameter (a):

$$\theta \ \alpha \ 1/\sqrt{r_i a}$$

Therefore, one way of achieving a higher speed of impulse conductance is by reducing r_i and this can be obtained by increasing the gap junction size or the conductance of the intercellular channels. The role of r_i in the control of conduction velocity in heart has been neglected in the past because it was assumed that the intracellular resistance was low and fixed. We know today that the junctional resistance can be changed by variations of free $[Ca^{2+}]_i$, pH_i (De Mello, 1975, 1980a, 1982; Rose and Loewenstein, 1975; Spray *et al.*, 1982), or cAMP (De Mello, 1983a, 1984a).

Evidence has been provided, for instance, that ouabain decreases the junctional conductance in Purkinje fibers (De Mello, 1976) or myocardial fibers (Weingart, 1977) and concurrently reduces the conduction velocity. Complete suppression of impulse conduction can be accomplished by markedly enhancing the junctional resistance (De Mello, 1975; see Fig. 1).

The slow conduction in the atrioventricular node has been associated with a high intracellular resistance (De Mello, 1977; Ikeda *et al.*, 1980). In this tissue, the cells are electrically coupled but the space constant is quite small (430 μm; De Mello, 1977). The high value of r_i is probably related to the small diameter of the node cells and also to the small number of gap junctions (Maekawa *et al.*, 1967; James and Scherf, 1968; De Felice and Challice, 1969).

2. ARE GAP JUNCTIONS REALLY INVOLVED IN CELL-TO-CELL COMMUNICATION?

The electrical coupling is usually considered as circumstantial evidence for the presence of gap junctions between the apposing cells.

Some observations are considered relevant with regard to the role of these junctions in cell-to-cell coupling: (1) the treatment of cardiac muscle with hyperosmotic sucrose solutions disrupts the gap junctions and causes cell decoupling (Barr *et al.*, 1965; Dreifuss *et al.*, 1966; De Mello *et al.*, 1969); (2) in giant axons of the crayfish exposed to low chloride solutions the gap junctions disappear transiently and the coupling resistance increases simultaneously; on returning to normal solution, gap junctions reappear and the electrical coupling is reestablished (Asada and Bennett, 1971; Pappas *et al.*, 1971); (3) the participation of gap junctions in the process of cell-to-cell communication gained strong support when biochemical approaches were used to study the diffusion of molecules between cells. It is known, for instance, that mutant cell lines deficient in enzymes such as hypoxanthine-guanine-phosphoribosyltransferase (HGPRTase) are unable to incorporate hypoxanthine. When these cells are cultivated together with wild-type cells containing [^{3}H]hypoxanthine, they become labeled but only if gap junctions are established between the cells (Pitts and Finbow, 1977). Further studies indicate that the mutant cells stimulate the wild-type cells to produce more nucleotides in such a way that the number of resistant cells determines the incorporation of label into the coculture ("true metabolic cooperation"—see Sheridan *et al.*, 1979).

The question remains whether intercellular movement of molecules and the electrical coupling follow the same pathway. In *Chironomus* salivary gland (Oliveira-Castro and Loewenstein, 1971; Simpson *et al.*, 1977), in septate axon (Pappas and Bennett, 1966), in cancer cells (Johnson and Sheridan, 1971), and in heart muscle (De Mello, 1979a), the intercellular movement of molecules is interrupted when the electrical coupling is abolished. The estimation of the coupling coefficient by electrophysiological methods, however, does not provide information on the existence of a chemical pathway between cells (see Bennett and Goodenough, 1978). In the natural pacemaker of the mammalian heart, as well as in the atrioventricular node, the cells are electrically coupled (De Mello, 1977, 1980b), but cell-to-cell diffusion of fluorescein is not seen (Pollack, 1976; De Mello, 1980b).

3. JUNCTIONAL PERMEABILITY

Radioactive sucrose (mol. wt. 342), fluorescein (mol. wt. 330), neutral red (mol. wt. 252), and Procion Yellow (mol. wt. 697) were found to cross the intercellular junctions in dipteran salivary gland (Kanno and Loewenstein, 1966; Rose and Loewenstein, 1971). In several tissues, SO_4^{2-}, I^-, Cl^-, K^+, and Na^+ can diffuse through gap junctions (Bennett, 1973).

In cardiac muscle the gap junctions are permeable to K^+ (Weidmann, 1966), fluorescein (mol. wt. 330) (Pollack and Huntmann, 1973; De Mello, 1979a), TEA (mol. wt. 130) (Weingart, 1974), ^{14}C-labeled cAMP (mol. wt. 328) (Tsien and Weingart, 1976), and Lucifer Yellow CH (mol. wt. 473) (De Mello *et al.*, 1983, 1985). Lucifer Yellow CH also diffuses through gap junctions in embryonic cells (Bennett *et al.*, 1978b).

There is reasonable agreement that gap junctions are permeable to molecules as large as 1000 daltons with the exception of junctions of arthropods in which larger molecules can flow through the intercellular channels (Rose *et al.*, 1977; Simpson *et al.*, 1977; Schwarzmann *et al.*, 1981).

Lucifer Yellow CH is a suitable compound for studies of cell-to-cell communication because, as a substituted four-amino naphthalimide with two sulfonate groups (Stewart, 1978), it does not diffuse through the surface cell membrane of nerve cells (Stewart, 1978), embryonic cells (Bennett *et al.*, 1978a), and cardiac cells (De Mello *et al.*, 1983, 1985) but crosses the intercellular channels.

The junctional permeability to Lucifer Yellow CH in mammalian cardiac fiber (De Mello *et al.*, 1985) was studied with the cut-end method (Imanaga, 1974). In such studies, trabeculae from dog ventricle are gently mounted into a chamber divided by a rubber membrane. After exposing half of the preparation to Ca-free solution plus EDTA for 20 min to avoid healing-over (De Mello *et al.*, 1969), the preparation is cut about 1 mm from the rubber partition and immediately exposed to Lucifer Yellow CH (0.1%) for 10 min. After this time, Ca is readmitted to the bath, sealing is accomplished, and the fluorescent probe is trapped inside. Contrary to fluorescein, Lucifer Yellow CH does not cross the nonjunctional membrane (see above) and its internal diffusion through the myoplasm and gap junctions can be followed by measuring the fluorescence in the muscle segment not exposed to the dye (Fig. 2). From measurements of the diffusion coefficient (D) for Lucifer Yellow CH, P_{nexus} can be estimated (see Weingart, 1974). Values of 4×10^{-7} cm^2/sec for D and 3×10^{-4} cm/sec for P_{nexus} were found (De Mello *et al.*, 1985). Fluorescence microscopy performed on the segment not exposed to the dye showed Lucifer Yellow CH located exclusively inside the heart cells (De Mello, unpublished observations).

Similar studies of junctional permeability in cardiac muscle performed with other compounds indicate that the molecular diffusion along a cardiac fiber is restricted at the gap junctions. For Lucifer Yellow CH, for instance, the value

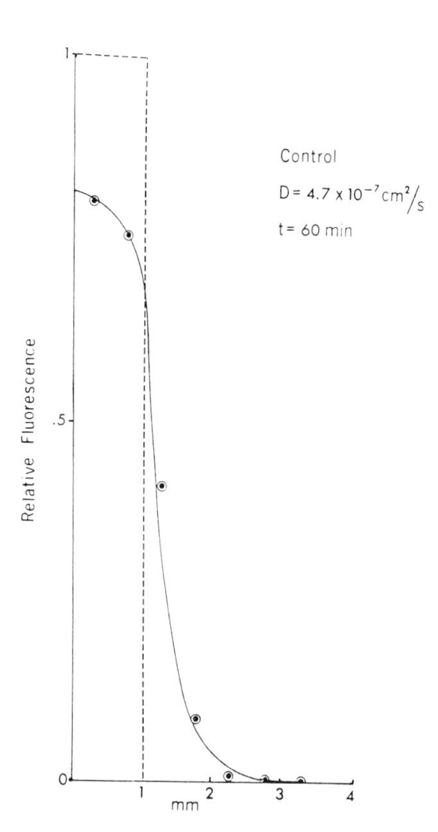

FIGURE 2. Longitudinal redistribution of Lucifer Yellow CH along a dog trabecula immersed in normal Tyrode solution. The fluorescence from 0.5-mm slices was normalized by total fluorescence and plotted against distance from the cut end. The points represent an experimental diffusion curve with $D = 4.7 \times 10^{-2}$ cm²/sec. The dashed line separated the loaded (left) from the unloaded segment. Diffusion period, 60 min. Length of the preparation, 4 mm. Temperature, 37°C.

found for D (4×10^{-7} cm²/sec) is smaller than that for the dye in the sarcoplasm (D_s) (2×10^{-6} cm²/sec).

The concept of a restrictive diffusion is supported by the finding that P_{nexus} is inversely proportional to the molecular weight of the molecule (Fig. 3). For example, in cardiac muscle P_{nexus} for K^+ is 7.68×10^{-3} cm/sec (Weidmann, 1966), for TEA (mol. wt. 130) 1.27×10^{-3} cm/sec (Weingart, 1974), for cAMP (mol. wt. 328) 1.33×10^{-6} cm/sec (Tsien and Weingart, 1976), and for Lucifer Yellow CH (mol. wt. 473) 3×10^{-4} cm/sec (De Mello *et al.*, 1985).

The permeability of gap junctions probably varies within a physiological range because of turnover gap junctions (19 hr) (Yancey *et al.*, 1981). The presence of gap junctional precursors in the cell membrane (Epstein and Gilula, 1977; Johnson *et al.*, 1974) might indicate that they can be quickly assembled to establish new gap junctions. The finding that inhibitors of protein synthesis (e.g., cycloheximide) abolish the electrical coupling in cardiac myocytes (Griepp and Bernfield, 1975) suggests that the synthesis of junctional or extrajunctional proteins is necessary for the formation of gap junctions.

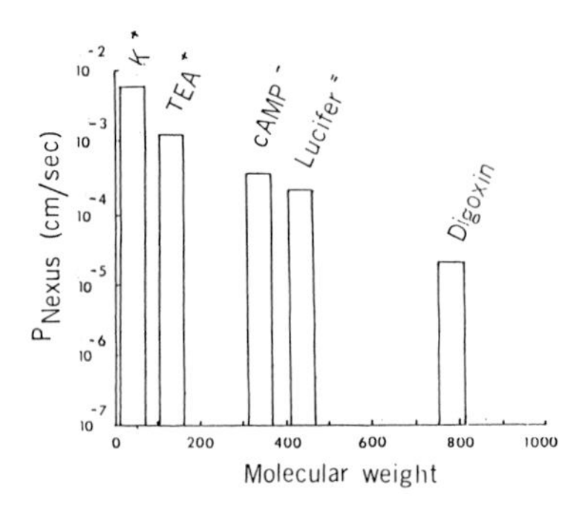

FIGURE 3. Relationship between the molecular weight of different substances and the permeability of the nexal membrane of mammalian heart muscle. Modified from Weingart (1981).

Although small molecules and ions can flow through gap junctions, macromolecules such as RNA, DNA, proteins, glycoproteins, and large polypeptides do not cross the intercellular channels (see Pitts and Finbow, 1977; Table I).

3.1. On the Regulation of Junctional Permeability

From the very first observations of Engelmann (1877) on the spontaneous reversibility of injury potentials in cardiac muscle, it became clear that ionic barriers are quickly established after cellular damage. As these potentials can be reestablished by a new lesion near the previous one, Engelmann's reasoning was that the rapid disappearance of the injury potential (healing-over) cannot be ascribed to a depolarization of the undamaged cells. These findings led to an important conclusion: "the death of a heart cell does not result in the death of adjacent cells" ("der Tod schreited nicht von Zelle auf Zelle fort") (Engelmann, 1877).

Rothschuh (1951) proposed that the quick disappearance of the injury potentials was due to the presence of preestablished transversal barriers between the heart cells, i.e., lesioning a single cell would not elicit depolarization in nearby cells. Rothschuh's hypothesis was probably influenced by the notion that cardiac muscle is organized by the fusion of "muscle territories" (Muskelterritorium) (Werner, 1910).

The presence of preestablished ionic barriers was discarded when Weidmann (1952) found that the core resistivity in cardiac fibers is quite low (105 Ω-cm). A plausible explanation for the healing-over process is that an ionic barrier is formed near the site of injury immediately after lesioning. Heilbrunn (1956) found that in damaged egg cells a new membrane is established (surface precipitation reaction) insulating the cytoplasm from the extracellular fluid.

The so-called surface precipitation reaction occurs in skeletal muscle fibers (Heilbrunn, 1956) but cannot prevent whole-fiber depolarization when the cell membrane is damaged (De Mello, 1973).

It is known, indeed, that the healing-over process does not occur in skeletal

Table I
Permeability of Gap Junctions

Ions and molecules that cross gap junctions	
Na^+	Bennett (1973)
Cl^-	Bennett (1973)
SO_4^{2-}	Bennett (1973)
I^-	Bennett (1973)
K^+	Weidmann (1966)
Co^{2+}	Politoff *et al.* (1969)
TEA	Weingart (1974)
Procion Yellow	Bennett (1973), Imanaga (1974)
Lucifer Yellow CH	Stewart (1978), Bennett *et al.* (1978b), De Mello *et al.* (1983)
^{14}C-labeled cAMP	Tsien and Weingart (1976)
Fluorescein	Loewenstein (1966), Bennett (1973), Pollack (1976), De Mello (1977)
Nucleotides from	
Hypoxanthine	Cox *et al.* (1970)
Uridine	Pitts (1976)
Thymidine	Pitts (1971)
Thioguanine	Fujisomoto *et al.* (1971)
Sucrose	Bennett (1973)
Derivatives of glucose	Pitts and Finbow (1977)
Derivatives of 2-deoxyglucose	Pitts and Finbow (1977)
Proline	Pitts and Finbow (1977)
Molecules that are unable to cross gap junctions	
RNA	Pitts (1976)
DNA	Pitts (1976)
Glycoproteins	Simms (1973)
Phospholipids	Pitts and Finbow (1977)

muscle (Rothschuh, 1951; De Mello, 1973). Small lesions of the surface cell membrane of frog sartorius muscle are followed by quick sealing only if the fibers are immersed in isotonic Ca solution (De Mello, 1973; see Fig. 4). Sealing has also been described in *Chironomus* salivary gland cells when the surface cell membrane is damaged (Oliveira-Castro and Loewenstein, 1971) and in *Ascaris* somatic muscle cells which are coupled through low-resistance junctions (De Mello, 1971).

It is then conceivable that a change in the ionic concentration at the gap junctions located near the lesion would increase the junctional resistance and promote healing-over. It is known, for instance, that Ca^{2+} ions are essential for the healing-over process in cardiac muscle (Délèze, 1965; De Mello *et al.*, 1969; De Mello, 1972; Fig. 5). The hypothesis that a change of junctional resistance produced by Ca^{2+} is involved in the healing-over process was investigated by injecting Ca^{2+} into the cytosol and searching for changes in the electrical coupling. The results showed that a rise in free $[Ca^{2+}]_i$ leads to cell decoupling (De Mello, 1975; Fig. 6). In *Chironomus* salivary gland the use of aequorin made it possible for Rose and Loewenstein (1975) to demonstrate that calcium injection caused a decrease in the electrical coupling when the light emission was seen to spread all the way to the intercellular junction. In heart muscle the injection of calcium is seen concurrently with an appreciable increase in the input resistance (V_o/I_o) of the injected cell—a finding that rules out the possibility that the fall in intercellular communication is due to a decrease in resistance of the nonjunctional membrane (see Fig. 6).

The suppression of cell-to-cell coupling produced by calcium injection is totally reversible, and reestablishment of the electrical coupling depends on homeostatic mechanisms involved in the maintenance of a low free $[Ca^{2+}]_i$ such as the uptake of calcium by the sarcoplasmic reticulum and mitochondria or the extrusion of the ion through the nonjunctional cell membrane (De Mello, 1975). Drugs (e.g., caffeine, 6–10 mM) that inhibit the uptake of calcium by the

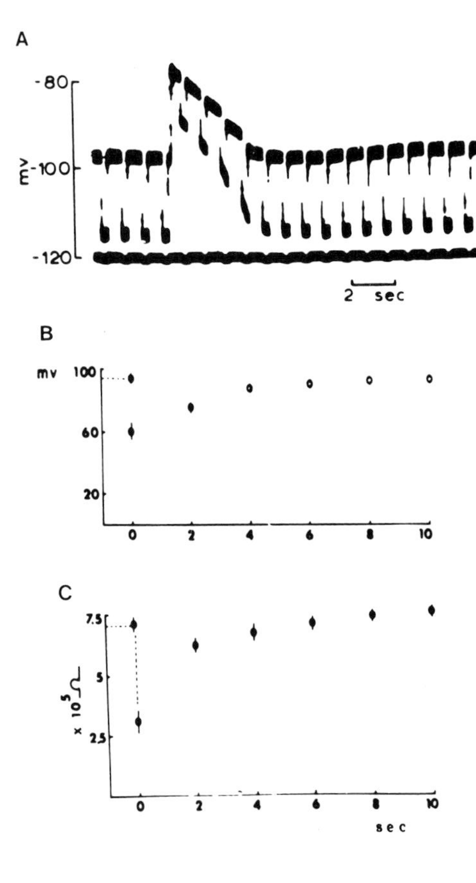

FIGURE 4. (A) Quick recovery of resting potential and resistance after puncture of a single muscle fiber immersed in isotonic calcium solution. Downward deflections of resting potential (top) are electrotonic potentials elicited by injecting inward current pulses (0.02 μA)—lower record—into the fiber. Panels B (resting potential) and C (input resistance) show the average rate of sealing recorded from 25 muscle fibers immersed in isotonic calcium solution. Dotted line indicates moment of lesion. Vertical line at each point is S.E.M. Temperature, 24°C. From De Mello (1973) with permission.

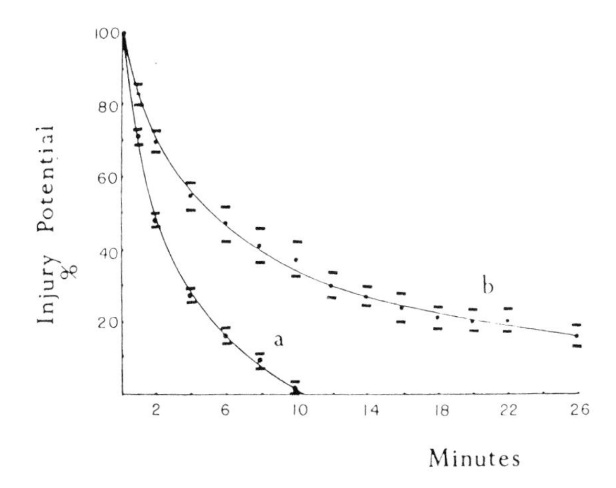

FIGURE 5. Effect of EDTA on the rate of healing of toad myocardium. *a*, average decay of the injury potentials [nine muscles obtained from preparations immersed in EDTA (2.5 mM) solution for 20 min]. *b*, healing curve obtained from the same preparations, immersed previously in normal Ringer's solution. The two horizontal lines of each point indicate the S.E.M. From De Mello *et al.* (1969) with permission.

sarcoplasmic reticulum reduce the rate of cell recoupling seen after the interruption of calcium injection (De Mello, 1975). The effect of calcium on the intercellular coupling, also described in other systems (see Baux *et al.*, 1978; Spray and Bennett, 1985), indicates that the conductance of the intercellular channels can be varied by calcium—a finding that, by itself, influences our view of the process of impulse propagation in excitable tissues and the diffusion of molecules between nonexcitable cells.

The evolutionary changes that culminated in the choice of cell-to-cell communication through intercellular junctions are probably stochastic in nature because in the selection process it would be simpler to have cells communicate by fusing their cytoplasms in a single junctional unit, i.e., a heart as a single large myocyte, a salivary gland as a single gland cell, the epithelia as a single and large epithelial cell, than to generate intercellular channels.

Such a design, however, is extremely inefficient and highly vulnerable because a lesion applied to the surface cell membrane would represent the death of the whole tissue. Survival seems to depend on compartmentalization, i.e., the damage inflicted on one of the compartments leads to quick isolation of this compartment from the others. This is precisely the meaning of the healing-over process in heart and in other tissues in which the cells communicate through gap junctions. But even under physiological conditions the cell identity must be preserved and the degree of cell-to-cell communication must be varied according to the needs of the whole cell population.

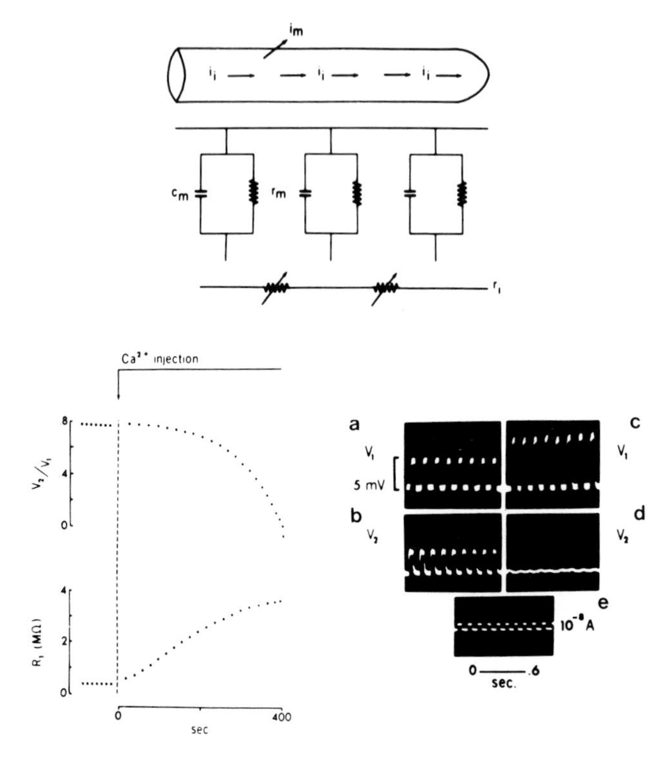

FIGURE 6. Effect of intracellular Ca injection on the electrical coupling of cardiac Purkinje cells. (Top) Cable model of a cardiac Purkinje fibre showing its electrical equivalent circuit with the variation in intracellular longitudinal resistance (r_i) produced by the effect of high free $[Ca^{2+}]_i$ on junctional resistance. (Bottom) right: typical effect of intracellular Ca injection on the electrical coupling. (a) and (b) show V_1 and V_2; controls (c) and (d) recorded after 410 sec of Ca injection showing cell decoupling; (e) outward current pulses (60-msec duration, 5 Hz, 10^{-8} A). Left: influence of intracellular Ca injection on the coupling coefficient (V_1/V_2) and input resistance (R_i) of Purkinje cells (average from six experiments). Temperature, 37°C. From De Mello (unpublished findings).

3.2. Na/Ca Exchange and Metabolic Inhibitors Alter the Electrical Coupling

In excitable tissue the extrusion of Ca from the cytosol depends in part on the energy provided by the Na concentration gradient (Reuter and Seitz, 1968; Baker *et al.*, 1969). Indeed, in these tissues the inward movement of Ca is extremely sensitive to increments in the intracellular Na concentration. In squid axon, for instance, the increase in Ca influx is proportional to the square of the internal free $[Na^+]$ (Baker *et al.*, 1969).

When Na ions are injected iontophoretically into a heart cell, the input resistance of the injected cell is greatly enhanced and the electrical coupling is abolished (De Mello, 1974, 1976; see Fig. 7). Inasmuch as Na injection does not produce cell uncoupling in fibers immersed in a Ca-free solution (De Mello, 1976), it is assumed that the rise in $[Na^+]_i$ activates a Na/Ca exchange with consequent increase in free $[Ca^{2+}]_i$ and cell decoupling. The major implication of these observations is that the inhibition of the Na pump leads to accumulation of intracellular Na and cell decoupling through the rise in free $[Ca^{2+}]_i$ (discussed in De Mello, 1976).

Ouabain, which inhibits the Na/K pump, also increases the intracellular resistance (Weingart, 1977). The role of Ca in the decoupling action of ouabain is supported by the finding that the suppression of the cell-to-cell coupling produced by the drug is accelerated by enhancing the extracellular $[Ca^{2+}]$ (Weingart, 1977) or by stimulation of the muscles at a high rate (De Mello, 1976). It is then possible to conclude that the Na pump is indirectly involved in the maintenance of high junctional conductance in cardiac muscle.

Metabolic inhibition can also impair the cell-to-cell coupling through a rise in the intracellular Na and Ca concentration or through a fall in pH_i. 2,4-Dinitrophenol (DNP), an uncoupler of oxidative phosphorylation, causes electrical uncoupling in *Chironomus* salivary glands (Politoff *et al.*, 1969; Rose and

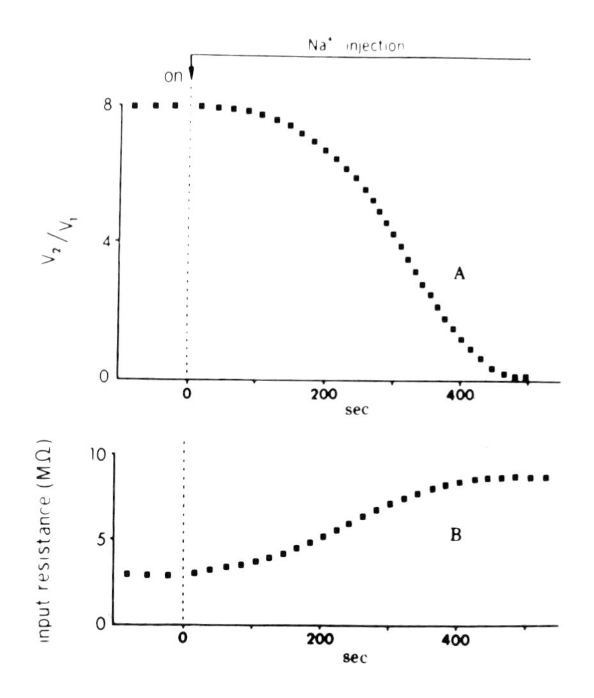

FIGURE 7. Influence of intracellular Na injection on the coupling coefficient (V_2/V_1) and input resistance of a canine Purkinje fiber. From De Mello (1976) with permission.

Table II
Influence of Intracellular Injection of cAMP on the Electrical Coupling (V_2/V_1) of Cardiac Cells and on the Time Constant of the Cell Membrane (τ_m)

	V_2/V_1		τ_m (msec)	
	Control	cAMP	Control	cAMP
M-3-3	0.60	1	24	16
A-5-2	0.68	0.98	—	—
A-10-3	0.70	0.96	26	18
M-1-3	0.66	0.99	—	—
J-2-6	0.65	0.92	—	—
J-10-1	0.72	0.94	25	20
Mean	0.66	0.96	25	18
± S.E.M.	0.024	0.012		

Rick, 1978), in crayfish septate axons (Peracchia and Dulhunty, 1976), and in cardiac Purkinje fibers (De Mello, 1979a; see Table II and Figs. 8 and 9). This effect of DNP seems related to the increase of free $[Ca^{2+}]_i$ because the resting tension is increased (see Fig. 8). The possibility exists, however, that a fall in pH_i, under these conditions, is also implicated in the decoupling action of the compound.

The mechanism by which Ca increases the junctional resistance is not known. As discussed elsewhere (De Mello, 1982), Ca can trigger enzyme reactions (through a Ca-ATPase?) closing the channels through a conformational change of gap junction proteins or the ion can bind to negative polar groups of gap junction phospholipids and suppress the permeability of the hydrophilic channels. This hypothesis is supported by the fact that a strongly charged cation such as La^{3+} when injected intracellularly is more effective than Ca^{2+} in

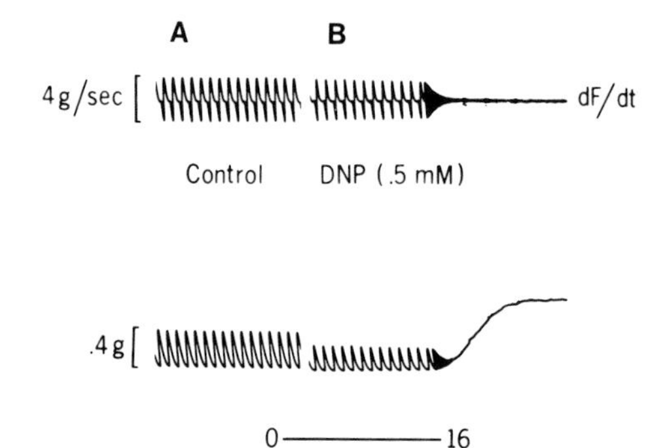

FIGURE 8. Effect of DNP on peak tension, rate of tension development, and resting tension of a strip dissected from guinea pig right ventricle. A, control; B, effect of DNP (0.5 mM). Bottom traces: twitch; top traces: *df/dt*. Calibration of time just for fast speed. Temperature, 37°C.

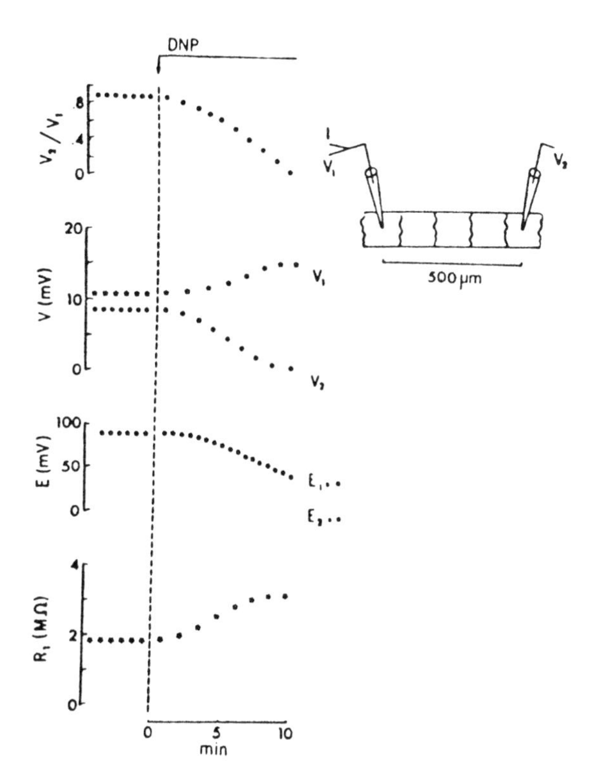

FIGURE 9. Effect of DNP on intercellular communication in dog Purkinje fiber. Time course of R_i (input resistance of cell 1), V_1 and V_2 (changes in membrane potential in cells 1 and 2 at zero current taking as outside minus inside potential), and coupling coefficient (V_2/V_1). Arrow indicates start of treatment with DNP (0.5 mM). Right: drawing (not to exact scale) showing experimental procedure used during this experiment. Cells impaled were not located at the end of a strand. Temperature, 37°C. From De Mello (1979a) with permission.

abolishing the electrical coupling of cardiac cells (De Mello, 1979b). Similar results were obtained with Sr^{2+} and Mn^{2+}, in cardiac cells (De Mello, 1975, 1979b) and in *Chironomus* salivary glands (Oliveira-Castro and Loewenstein, 1971). Recently, healing-over was accomplished in damaged cardiac muscle immersed in a Ca-free solution containing Co^{2+} (2 mM), Mn^{2+} (2 mM), or verapamil (10^{-5} M) (De Mello, 1985), supporting the view that the site(s) controlling the junctional conductance is not specific for Ca.

3.3. Is Calcium a Physiological Modulator of Junctional Permeability?

Although under pathological conditions Ca is a good cell "decoupler," the question whether the ion is a modulator of junctional conductance under physiological conditions is difficult to answer. Part of the difficulty is to precisely identify the minimum Ca concentration able to modify the junctional permeability. Indeed, the free $[Ca^{2+}]_i$ needed to cause cell decoupling is difficult to assess because the ion is transported actively into the sarcoplasmic reticulum and mitochondria, is bound to contractile proteins, or is extruded from the cell. The estimation of free $[Ca^{2+}]_i$ with the use of Ca-sensitive microelectrodes might help but the interpretation of these results is difficult because the precise location of the microelectrode tip is not known.

In *Chironomus* salivary gland cells a concentration of $5\text{–}8 \times 10^{-5}$ M seems to be required to suppress the electrical coupling; in cardiac muscle the threshold concentration of Ca necessary to abolish the electrical coupling seems to be greater than that required for the activation of the contractile process (Weingart, 1977). If these are the Ca concentrations that interrupt the flow of ions and molecules through gap junctions, they are certainly not within the physiological range. This is indeed expected because part of the cell's energy is used precisely to keep the free $[Ca^{2+}]_i$ very low.

The high buffer capacity of the cytosol for Ca, as well as the low diffusion of the ion in the cytoplasm, constantly protect the gap junctions from changes in free $[Ca^2]_i$. Certainly, we cannot conclude that Ca is not a modulator of junctional permeability on the basis of the Ca concentration necessary to uncouple cells. Further studies must provide information whether the Ca concentration in the cytosol near the gap junction can be changed under physiological conditions and whether slight variations of free $[Ca^{2+}]_i$ can modify the channel conductance.

3.4. Influence of Protons on Cell-to-Cell Communication

In 1977 Turin and Warner demonstrated that when embryonic cells of *Xenopus* were exposed to 100% CO_2 the intracellular pH was reduced from 7.7 to 6.4, the membrane potential was decreased, and the electrical coupling was abolished. Similar results were reported by Bennett *et al.* (1978a) in embryonic cells of *Fundulus*. In cardiac muscle the intracellular injection of H^+ also causes cell decoupling (De Mello, 1980a), but no simultaneous measurements of pH_i were made.

Further studies of Spray *et al.* (1981) demonstrated that the pH_i/g_j relation in fish embryonic cells is steep and occurs over a physiological range of pH_i. In earthworm axons (pH_i 7.1) a fall in pH_i can produce cell decoupling but the pK is 6.7 (Brink *et al.*, 1984) while in embryonic cells the pK is 7.3 (Spray *et al.*, 1981, 1984).

In rat hepatocytes (pH_i 7.4) a decrease in pH_i of about 1 unit is required to produce a decrease in junctional conductance (Meyer and Revel, 1981) and in other tissues such as differentiated lens fibers of the chicken the sensitivity of g_j to pH_i is extremely small (Scheutze and Goodenough, 1982).

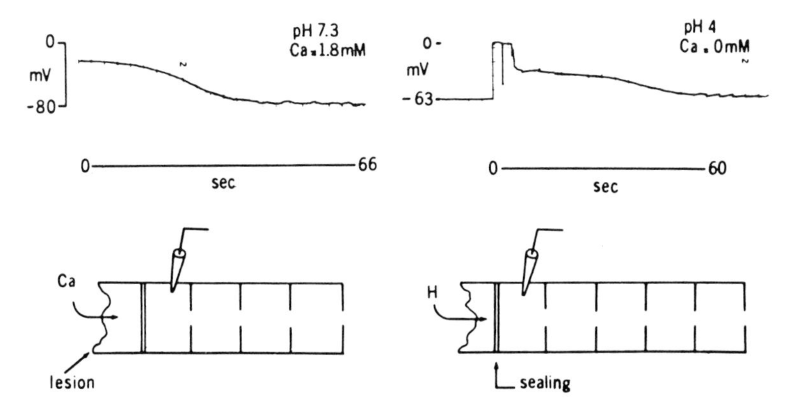

FIGURE 10. Healing-over of cardiac muscle promoted by Ca^{2+} (left) and H^{+} ions (right). From De Mello (1984b) with permission.

Protons and Ca probably act upon the same gating structure in *Chironomus* (Obaid *et al.*, 1983) and in cardiac muscle (De Mello, 1986b). It is interesting to add that protons interact with Ca for the healing-over process, reducing the effect of Ca on the rate of healing (De Mello, 1985, 1986b).

The healing-over process, which is considered to be related to an appreciable increase in junctional resistance near the lesion (see De Mello, 1972), can be promoted by protons in the absence of Ca^{2+} ions but only if the extracellular pH is reduced to 5.5 or lower (De Mello, 1983b; see Fig. 10). On the assumption that cell uncoupling and healing-over are related phenomena these findings are in a quantitative contrast to those obtained with *Xenopus* and *Fundulus* embryonic cells (Turin and Warner, 1977, 1980; Spray *et al.*, 1982) where acidification to pH 6.5 completely uncouples.

Although the effectiveness of protons and Ca compared on a molar basis is in a range quite similar [3.2–10 $\times$ 10^{-6} M for protons (De Mello, 1983b) and 4.3–8.2 $\times$ 10^{-5} M for Ca (Nishiye, 1977)], it is important to emphasize that "the range at which H ions exert a major effect is well outside the pH values which are found in a living heart cell."

The noneffectiveness of pH shifts to the region of 6.5 for the healing-over process is in agreement with data from cable analysis made on sheep Purkinje fibers by Reber and Weingart (1982). These authors found a 30% increase in the intracellular longitudinal resistance when the pH_i was reduced from 7.3 (control) to 6.8.

On analyzing the discrepant results, care must be exercised in generalizing the process of regulation of junctional conductance. Although it is true that basic mechanisms are preserved throughout evolution and development, one must never forget that evolution generates diversity.

3.5. Cyclic AMP—A Physiological Regulator of Junctional Permeability?

Ca^{2+} ions and cAMP are intimately related in the modulation of cell function induced by hormones (see Rasmussen, 1975) and it seems reasonable that cAMP might be involved in the control of junctional conductance.

In salivary glands of *Drosophila* cAMP enhances the cell-to-cell coupling (Hax *et al.*, 1974) while in mammalian liver cells in culture dB-cAMP increases the transfer of molecules from cell to cell—a phenomenon related to an increase in the number of gap junction particles (Flagg-Newton *et al.*, 1981).

In cardiac muscle, evidence exists that cAMP enhances the intercellular communication (Estapé and De Mello, 1982, 1983). Thus, when the nucleotide is injected intracellularly the electrical coupling is appreciably increased while the time constant of the cell membrane is reduced (see Fig. 11 and Table II; De Mello, 1984a, 1984b), indicating that the increment in V_2/V_1 cannot be ascribed to a rise in resistance of the nonjunctional membrane. Epinephrine, which increases the intracellular concentration of cAMP in heart muscle, increases the cell-to-cell coupling (De Mello, 1986a).

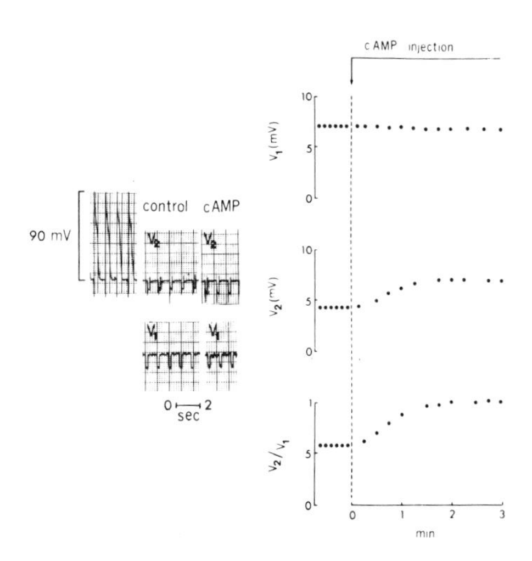

FIGURE 11. Influence of intracellular injection of cAMP on the electrical coupling of canine Purkinje cells. Left: typical experiment showing normal transmembrane action potentials and the amplitudes of the electrotonic potentials (V_1 and V_2) recorded before and 90 sec after cAMP injection. The efflux of cAMP from the microelectrodes was produced by hyperpolarizing current pulses (200-msec duration, 1 Hz, 10^{-8} A). Right: also from a typical experiment, time course of the changes in membrane potential recorded at the site of cAMP injection (V_1) and far (350 μm) from that site (V_2). As can be seen, the coupling coefficient (V_2/V_1) was enhanced (45%) by cAMP injection. Temperature, 37°C. From De Mello (1984a) with permission.

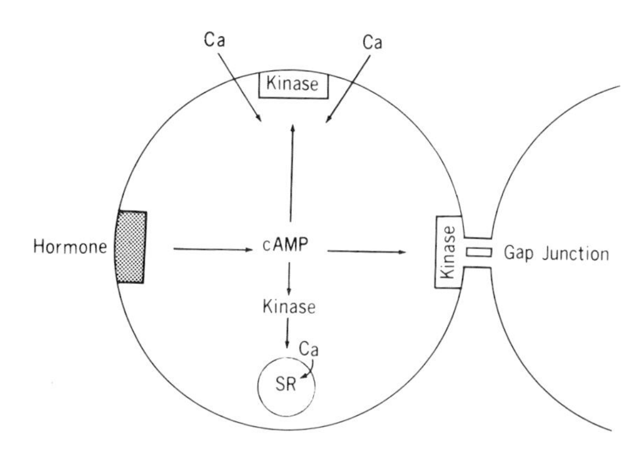

FIGURE 12. General diagram showing the different effects of cAMP including possible activation of a kinase located at gap junctions and consequent phosphorylation of gap junction proteins. This effect of the nucleotide would produce an increase in junctional conductance as is shown below. From De Mello (1983a) with permission.

Considering the short time required to increase the electrical coupling (42% in 50 sec) (see De Mello, 1984a), it is possible to conclude that cAMP is increasing the electrical coupling of heart cells through a rise in junctional conductance. The cAMP hypothesis (De Mello, 1983a) proposes that the effect of the nucleotide on cell-to-cell coupling is explained by the activation of kinases specifically related to the phosphorylation of gap junction proteins and consequent increase in junctional conductance (see Fig. 12).

It is known that an increased concentration of cAMP enhances the permeability of the surface cell membrane to calcium (see Rasmussen, 1975; Reuter and Scholz, 1977) and epinephrine is a strong positive inotropic agent precisely because it increases the inward movement of Ca secondary to the rise in cAMP concentration (Reuter and Scholz, 1977). Despite the increase in free $[Ca^{2+}]_i$, epinephrine or cAMP injection does not impair the electrical coupling—on the contrary, the intercellular communication is increased and consequently the conduction velocity is augmented.

The question of how a hormone that increases free $[Ca^{2+}]_i$ can simultaneously reduce the junctional resistance is probably explained by the way the free $[Ca^{2+}]_i$ is controlled, i.e., the increased free $[Ca^{2+}]_i$ produced by physiological concentrations of epinephrine is not able to reach the gap junctions because of the low diffusion coefficient of calcium in the cytosol (Kushmerick and Podolsky, 1969) and an effective buffering system for calcium.

It thus seems possible to visualize two different effects of epinephrine in cardiac muscle: one increasing the strength of the heart beat by incrementing the free $[Ca^{2+}]_i$ and the other facilitating the spread of the electrical current through an increase in junctional conductance mediated by cAMP.

Additional support to the cAMP hypothesis came from studies of the influence of dB-cAMP on the cell-to-cell diffusion of Lucifer Yellow CH in cardiac muscle. When muscle trabeculae from dog atrium are exposed to dB-cAMP (5×10^{-4} M) plus theophylline (0.4 mM), a phosphodiesterase inhibitor, the D value for Lucifer Yellow CH is increased from 4×10^{-7} cm²/sec (control) to 1.6×10^{-6} cm²/sec (De Mello *et al.*, 1985; see Fig. 13) and the estimated permeability of the gap junctions (P_{nexus}) is enhanced from 3×10^{-4} cm/sec (control) to 9.1×10^{-4} cm/sec after 80 min of exposure to dB-cAMP plus theophylline (De Mello *et al.*, 1985). Although no information exists as to whether the number or size of gap junctional particles was increased in this particular experiment, the relatively short time required to increase the junctional permeability suggests that the major event was an increase in unit channel permeability produced by cAMP.

The interaction between cAMP and Ca is involved in many aspects of the hormonal action and might also be important for the control of cell-to-cell coupling. One example of this interaction is the finding that the increment in the electrical coupling produced by intracellular injection of cAMP in cardiac cells is

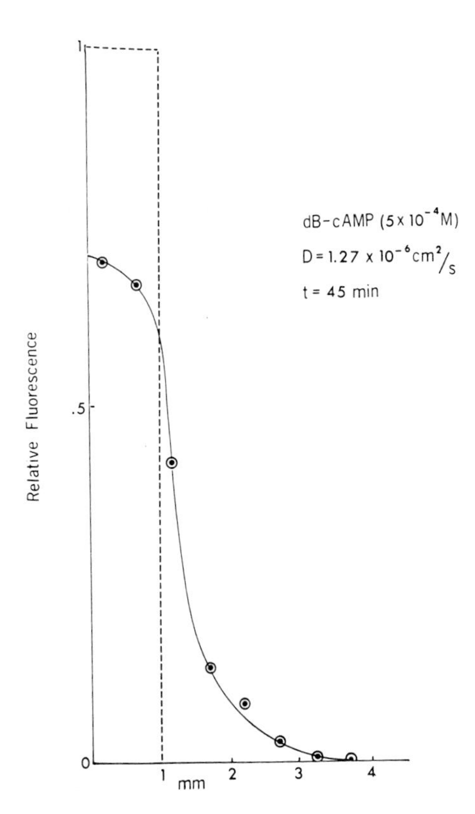

FIGURE 13. Longitudinal redistribution of Lucifer Yellow CH along a dog trabecula exposed to dB-cAMP (5×10^{-4} M). The points represent experimental values fitted by a theoretical diffusion curve with $D = 1.27 \times 10^{-6}$ cm²/sec and diffusion period of 45 min. Length of the preparation, 4.5 mm. Temperature, 37°C.

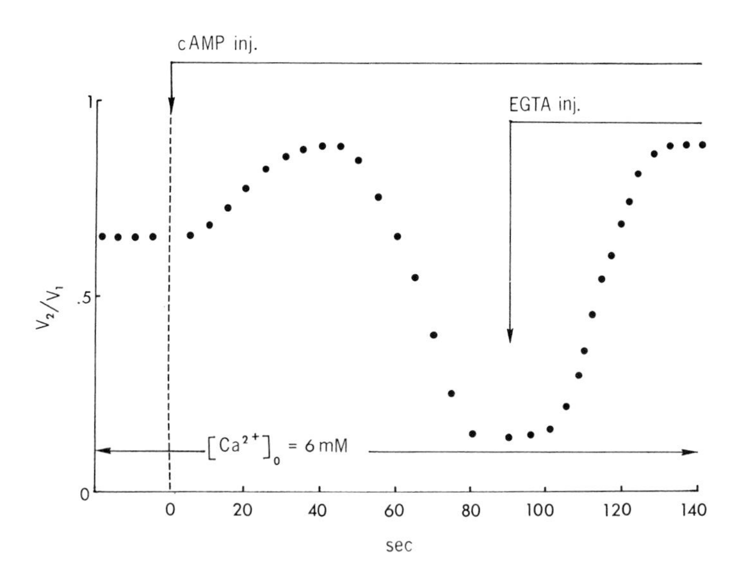

FIGURE 14. Effect of intracellular injection of cAMP on the electrical coupling of canine Purkinje cells exposed to high (6 mM) Ca solution. An initial increase in V_2/V_1 is followed by a drastic decline in coupling which is reversed by EGTA injection into the same cell. From De Mello (1986c).

followed by a marked decline in cell-to-cell coupling if the extracellular $[Ca^{2+}]$ is increased (see Fig. 14; DeMello, 1986c). This decrease in electrical coupling seems related to a rise in free $[Ca^{2+}]_i$ because it can be reversed by injection of EGTA (buffered to pH 7.3) into the same cell (De Mello, 1986c). This observation indicates that if too much Ca is moved into the cell by cAMP injection (through phosphorylation of nonjunctional membrane proteins), an impairment of cell-to-cell coupling can be produced.

A feedback exists between Ca and cAMP (Rasmussen, 1975) in such a way that a rise in free $[Ca^{2+}]_i$ reduces the [cAMP] through the activation of phosphodiesterase or the inhibition of adenylate cyclase while cAMP decreases free $[Ca^{2+}]_i$ by increasing the uptake of Ca by the sarcoplasmic reticulum. It is not known, however, if the interaction between Ca and cAMP plays a role in the modulation of junctional permeability under physiological conditions, and further studies will be required to clarify this point.

4. SYNAPTIC TRANSMISSION AND CELL-TO-CELL COUPLING

In cardiac pacemaker cells the electrical stimulation of the vagus nerve causes the release of acetylcholine (ACh) at the nerve endings. The interaction of

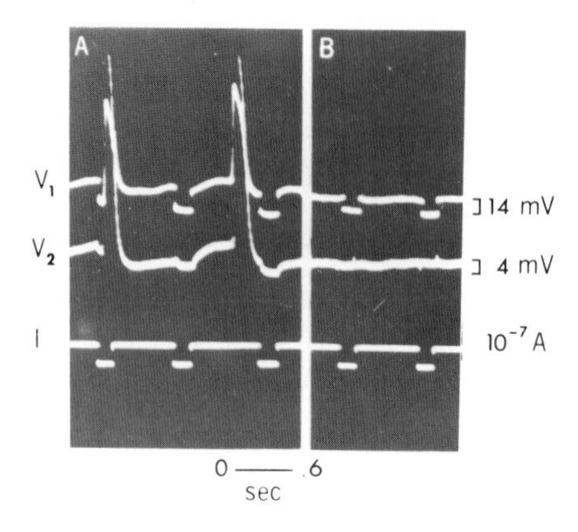

FIGURE 15. Effect of acetylcholine on the electrical coupling and input resistance of cardiac pacemaker cells of the SA node located near crista terminals. (A) Control; (B) immediately after addition of ACh (10^{-5} g/ml) to the bath. Distance between site of current injection and V_2 = 100 μm. From De Mello (1980b) with permission.

the neurotransmitter with muscarinic receptors increases the K permeability of the nonjunctional membrane, with consequent hyperpolarization and suppression of spontaneous activity (Hutter, 1957). When ACh (10^{-5} M) is added to the bath the input resistance of the pacemaker cells falls and the electrical coupling between neighboring cells is abolished (see Fig. 15). The time course of the electrotonic potential was greatly reduced by ACh, indicating that the resistance of the nonjunctional membrane is decreased by ACh.

In the atrioventricular node of the rabbit the space constant is quite small (430 μm) due to a high intracellular resistance (40.9 MΩ/cm) (De Mello, 1977). In this tissue ACh decreases the space constant by 38%. As the time constant of the cell membrane (3.4 msec) is reduced (33%) by ACh, it is reasonable to assume that the fall in space constant is due to a decrease in the resistance of the nonjunctional membrane (see De Mello, 1977).

In the neuromuscular system of *Ascaris lumbricoides* the giant muscle cells are electrically coupled through low-resistance junctions, while excitatory and inhibitory nerves modulate the frequency of the spontaneous activity (see Chapter 8). γ-Aminobutyric acid (GABA) (10^{-5} M) hyperpolarizes the muscle cells through an increase in chloride conductance of the nonjunctional membrane (see Chapter 8) (del Castillo *et al.*, 1964).

Recently, it was found that the electrical coupling of *Ascaris* muscle cells is greatly impaired by GABA (De Mello and Maldonado, 1985) (see Fig. 16 and Table III; De Mello and Maldonado, 1985). Although one cannot rule out the possibility that the transmitter increases the junctional resistance, the fall in the time constant (see Table III) suggests that a fall in the resistance of the surface cell membrane is the main explanation for the decreases in electrical coupling (see Fig. 17).

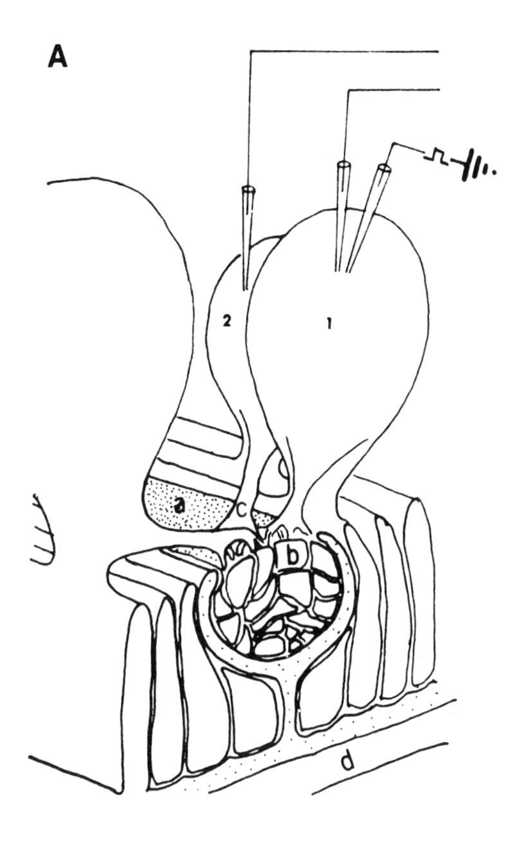

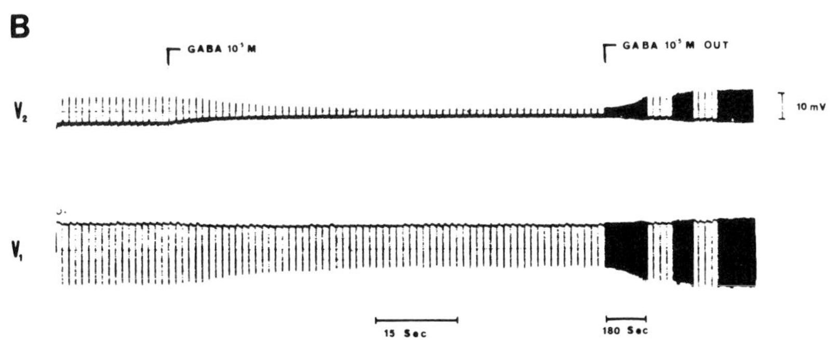

FIGURE 16. (A) Drawing illustrating the experimental arrangement used in our experiments with giant somatic muscle cells of *Ascaris lumbricoides* showing the bellies (1 and 2), the arms (c) with fingers which are electrically coupled establishing a syncytial layer (a). Conventional neuromuscular junctions are established between the syncytium and nerve fibers; one of them is represented in (b). The cuticle is represented in (d). Modified from del Castillo *et al.* (1967); from De Mello and Maldonado (1985) with permission. (B) Reversible effect of GABA (10^{-5} M) on the electrical coupling of giant somatic muscle cells of *A. lumbricoides*. Rectangular pulses of inward current (5×10^{-7} A, 300-msec duration, 1 Hz) were injected into cell 1 throughout the experiment.

Table III

Effect of GABA on the Coupling Coefficient (V_2/V_1), Input Resistance (R_{in}), Membrane Potential (E), and Time Constant of Cell Membrane (τ_m) or Giant Muscle Cells of *Ascaris lumbricoides*[a]

	Control				GABA (10^{-5} M)			
Expt	E (mV)	R_{in} (kΩ)	V_2/V_1	τ_m (msec)	E (mV)	R_{in} (kΩ)	V_2/V_1	τ_m (msec)
1	31	43	0.71	4.34	35	22	0.31	3.13
2	34	40	0.68	4.20	38	25	0.32	3.03
3	31	38	0.65	3.58	38	24	0.26	2.58
4	32	44	0.70	4.40	37	27	0.34	3.20
5	33	42	0.72	4.0	38	25	0.30	2.88
6	30	39	0.64	3.30	34	23	0.29	2.31
7	32	43	0.69	4.01	37	26	0.32	2.81
Mean	31.8	41.2	0.68	3.97	36.7	25.2	0.30	2.84
± S.E.M	0.5	0.86	0.011	0.15	0.6	0.58	0.011	0.11

[a]Differences between all parameters in control and experimental groups were statistically significant ($p < 0.05$).

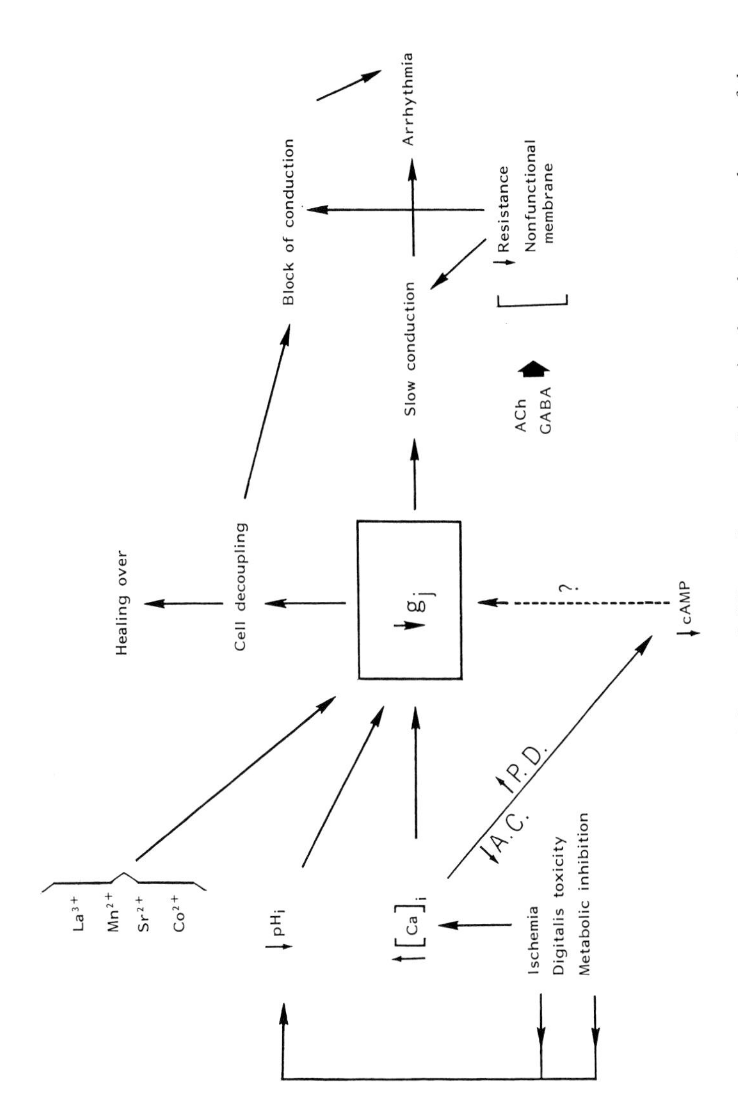

FIGURE 17. Diagram illustrating the influence of different factors on the junctional conductance and some of the possible pathological implications of cell decoupling in heart muscle. From De Mello (1982) with permission.

In *Navanax* the motoneurons that control the expansion of muscles of the pharynx are electrically coupled (Spira and Bennett, 1972) and the mechanism of cell decoupling seems to be the decline in nonjunctional membrane resistance.

5. VOLTAGE DEPENDENCE

Furshpan and Potter (1959) described in the crayfish an electrical synapse with rectifying properties and high dependence on voltage.

More recently, voltage-dependent gap junctions were reported in embryonic cells of amphibia (Spray *et al.*, 1979; Spray and Bennett, 1985). In this system, pulses of enough duration and intensity injected into the cell lead to a fall in the transfer resistance in an adjacent cell while the input resistance of the injected cell is enhanced. The junctional conductance in this system is reduced symmetrically by the changes in transjunctional potential of either sign.

In leech neurons, rectifying synapses have also been reported (Nichols and Purves, 1972), while in *Chironomus* salivary gland the junctional conductance is greatly influenced by the voltage between the extracellular fluid and the cytoplasm in such a way that a hyperpolarization of a pair of cells increases the junctional conductance and a depolarization decreases it (Obaid *et al.*, 1983).

In cardiac muscle, hyperpolarization of a cell previously decoupled by intracellular Na injection reestablishes the electrical coupling (De Mello, 1976), but a decrease in free $[Ca^{2+}]_i$ produced by the increment in resting potential might in part explain these results.

Transjunctional potential does not seem to be involved in the regulation of junctional permeability in cardiac muscle (Spray and Bennett, 1985). Indeed, in dog trabeculae mounted into a chamber separated by a rubber membrane, the superfusion of the right segment of the muscle with 60 mM KCl solution (resting potential -20 mV) does not alter the diffusion of Lucifer Yellow CH added to the left segment (cut-end method) exposed to normal Tyrode solution (resting potential -80 mV) (De Mello, unpublished findings).

6. ANTIBODIES

Antibody prepared against liver gap junctions causes a rapid and irreversible block of dye transfer and electrical uncoupling when injected intracellularly (Hertzberg, 1985; Hertzberg *et al.*, 1984). These observations open new avenues for the study of the physiological role of gap junctions in different tissues (see Chapter 7).

7. INFLUENCE OF TEMPERATURE ON g_j

The healing-over process, which is related to the diffusion of Ca through the cut end and the increase in junctional resistance, is highly dependent on temperature (De Mello and Motta, 1969; De Mello, 1972; Ochi and Nishiye, 1973). This finding suggests that the establishment of high-resistance barriers promoted by Ca^{2+} or H^+ (De Mello, 1983b) is dependent on metabolic processes.

In *Chironomus* salivary gland, cooling causes cell decoupling (Politoff *et al.*, 1969). Similar results were described in crayfish septate axon (Payton *et al.*, 1969; Ramon and Zampighi, 1980). More recently, it was shown that the junctional permeability in earthworm septate axon is suppressed at 4°C (Brink *et al.*, 1984).

Cooling of solvent with consequent increase in hydration shell size or a decrease in diameter of the intercellular channels have been considered as possible explanations for the decline in g_j (Brink *et al.*, 1984).

An alternative explanation for these results might be an increase in free $[Ca^{2+}]_i$ or a decrease of pH_i caused by low temperatures.

8. PATHOLOGICAL IMPLICATIONS OF JUNCTIONAL CONDUCTANCE

8.1. Increased Junctional Resistance—A Cause for Slow Conductance, Reentry, and Cardiac Arrhythmias

Abnormalities in impulse conduction and automaticity are usually associated with the generation of cardiac arrhythmias (Hoffman and Cranefield, 1960). Since the beginning of the century, atrial fibrillation and flutter have been attributed to a circular movement through a rather long pathway (Mines, 1913; Lewis, 1925). The requirements for reentry are: (1) unidirectional conduction block; (2) slow conduction; (3) a short refractory period, or the time taken to conduct the impulse around the circuit must be greater than the refractory period; (4) the circuit path must be long enough for an action potential to return to the starting point when the refractory period has already disappeared.

Membrane depolarization, especially in Purkinje fibers, can severely impair impulse conduction. It is known, indeed, that anoxia, low temperature, or low extracellular $[K^+]$ depolarizes the heart cells (see Trautwein, 1964).

The role of changes in the intracellular resistance in the generation of slow conduction or even block of impulse conduction has not been seriously considered in the past (De Mello, 1982) probably because the intercalated disks containing gap junctions were considered as fixed low-resistance pathways between heart cells. We know today that the junctional conductance in heart can be changed by intracellular factors such as H^+ or Ca^{2+} (De Mello, 1975, 1982).

As discussed by Jennings and Steenbergen (1985), severe myocardial ischemia is characterized by depletion of ATP, acidosis, and increase in AMP. An increase in free $[Ca^{2+}]_i$ to 100 μM can induce cell death by activation of endogenous proteases and phosphilipases (Jennings and Reimer, 1981).

The usual gain of Ca seen in myocardial ischemia (2–10 μmole/g dry wt) (see Shen and Jennings, 1972) is probably enough to increase the junctional resistance, to cause cell decoupling and unidirectional conduction block in some areas (see Fig. 17) and slow conduction in others. These effects of the disease on junctional resistance might produce reentry and cardiac arrhythmias.

The total closure of intercellular channels between ischemic and nonischemic cells (see Fig. 18) (healing-over) represents an important mechanism of survival because it suppresses the spread of injury current, thus avoiding the depolarization of large masses of normal cardiac cells (De Mello, 1972). Intracellular and extracellular acidity also occurs during myocardial ischemia. In totally ischemic rat hearts the pH_o and pH_i are reduced from 7.25 to 6.8 (Neely *et al.*, 1975). As judged from other findings, this value of pH_i (6.8) is not low enough to abolish the electrical coupling in intact cardiac fibers, but an increase

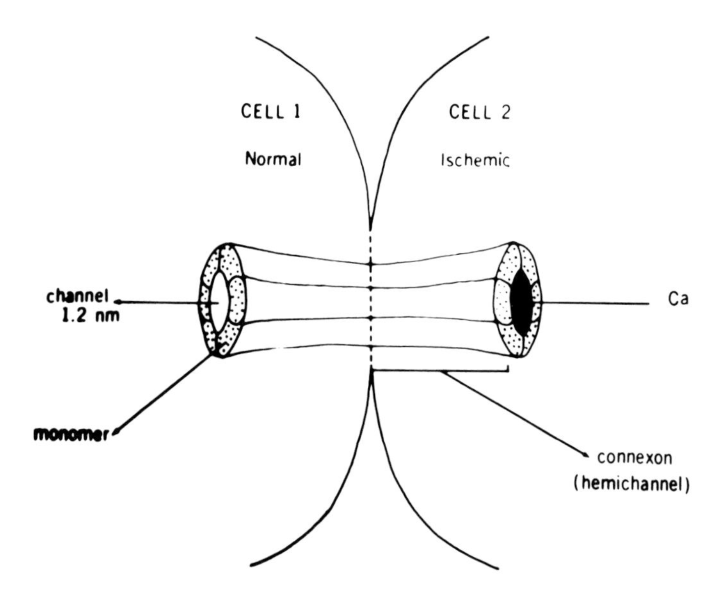

FIGURE 18. Connexons from two apposing cells showing the intercellular channel of a hypothetical ischemic cell closed, probably by the high free $[Ca^{2+}]_i$; the hemichannel from a normal cell is open. Each connexon is composed of six monomers. From De Mello (1985) with permission.

in r_i of 30% has been reported (Reber and Weingart, 1982). Therefore, the possibility exists that intracellular acidity during myocardial ischemia contributes to the generation of slow conduction and reentry.

The decline in nonjunctional resistance is certainly an important mechanism of cell decoupling. As stressed above in areas such as the atrioventricular node, the fall in membrane resistance caused by ACh is probably the main explanation for the decrease in space constant and the block of impulse conduction seen with the drug (De Mello, 1977, 1982). The presence of a high intracellular resistance makes the impulse conduction in the atrioventricular node extremely vulnerable to a fall in resistance of the nonjunctional membrane.

Digitalis toxicity reduces the Na gradient across the surface cell membrane and increases the free $[Ca^{2+}]_i$ and the junctional resistance in cardiac fibers (De Mello, 1976; Weingart, 1977). The block of atrioventricular conduction seen with high doses of digitalis, as well as the ventricular arrhythmias produced by the drug are certainly explained by both depolarization of nonjunctional membranes and a rise in junctional resistance (De Mello, 1982).

8.2. Uncoupling Leads to Decreased Strength of Heartbeat

The strength of heart contractions is dependent not only on the process of activation of contractile proteins, but also on the electrical synchronization which enables large populations of cells to contract simultaneously. Electrical uncoupling produced by anoxia, ischemia, or even by drugs can lead to lack of mechanical synchronization and a decline in the strength of muscle contraction.

The possibility that cardiac failure can be produced by impairment of electrical synchronization has not been emphasized in clinical cardiology and is a subject for further investigation.

9. CONCLUSIONS

There has been increasingly conclusive evidence that a substantial rise in free $[Ca^{2+}]_i$ mainly achieved under pathological or toxicological conditions, can produce cell decoupling in some systems but not in others. The same seems to be true with respect to protons.

Contrary to previous conclusions, Ca^{2+} or H^+ ions can alter g_j by themselves. In embryonic cells g_j is appreciably reduced by H^+ (Hill coefficient $n = 4.5$) but seems much less sensitive to Ca^{2+} (Hill coefficient $n = 2$) (Spray *et al.*, 1981).

In other systems such as cardiac muscle, rat hepatocytes, and differentiated lens fibers of the chicken (see above), an appreciable and probably unphysiological decline in pH_i is required to decouple cells. The healing-over process in cardiac muscle, which is related to a rise in junctional resistance, is much more sensitive to Ca^{2+} than to H^+ (De Mello, 1986b).

Although Ca^{2+} is a good cell decoupler in many systems, it is not clear if the ion is a modulator of g_j under physiological conditions. It is important to emphasize, however, that it is not feasible to argue if Ca^{2+} or H^+ ions are physiological modulators of g_j on the basis of the concentrations required to produce cell decoupling. In cardiac muscle and other excitable tissues, cell decoupling is not a physiological occurrence. In other systems and in the differentiation process, however, complete suppression of cell-to-cell communication seems to be an essential control mechanism.

Evidence is available that cAMP increases g_j in cardiac muscle when the nucleotide is injected inside the cell (De Mello, 1984a). The quick action of the compound might indicate that the increase in g_j is due to activation of kinases and phosphorylation of junctional proteins with consequent increase in unit channel conductance (De Mello, 1983a; see Fig. 19).

Chronic exposure of cultured liver cells to dB-cAMP enhances the junctional permeability (Flagg-Newton *et al.*, 1981), a phenomenon that requires

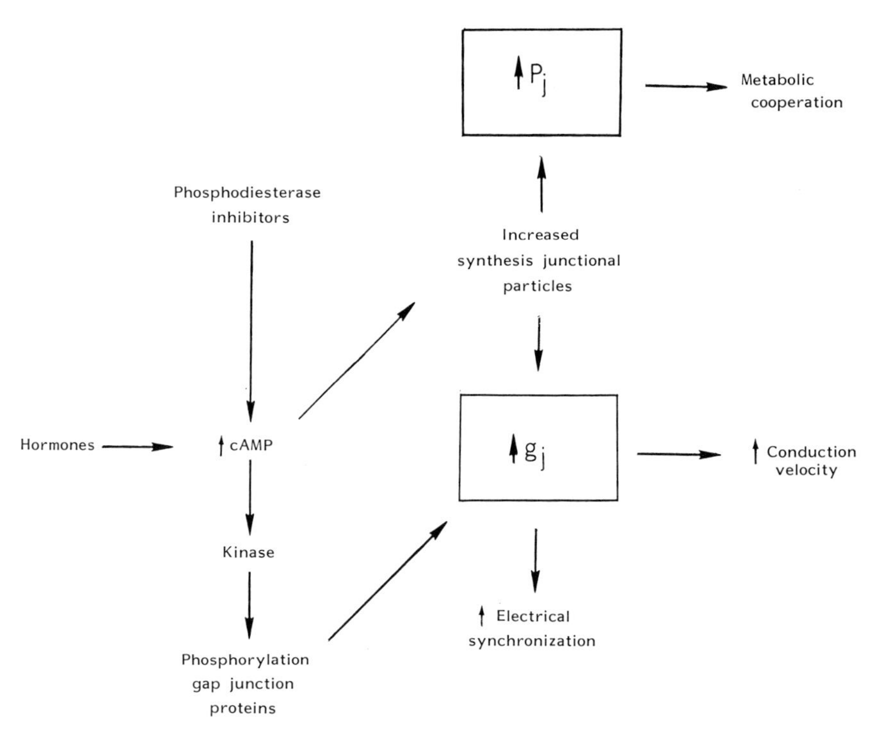

FIGURE 19. Diagram showing the influence of cAMP and hormones on junctional conductance and some of the physiological consequences.

several hours and is probably related to an increase in synthesis of junctional particles. Although these results might indicate that cAMP is involved in the long-term regulation of junctional communication, considering the inherent limitations of the tissue culture method, it is advisable to test this hypothesis in other tissues and under different experimental situations.

Despite the fact that gap junctions seem related to the establishment of metabolic cooperation in some systems, we are still unsure whether a similar situation occurs under physiological or pathological conditions. Considering the enormous progress achieved in recent years in the field of intercellular communication, many of the questions raised in this volume will likely be answered within the next 10 years.

ACKNOWLEDGMENT. Supported by Grants HL-34353, HL-34148, and RR-08102.

10. REFERENCES

Asada, Y., and Bennett, M. V. L., 1971, Experimental alteration of coupling resistance at an electrotonic synapse, *J. Cell Biol.* **49:**159–172.

Baker, P. F., Blaustein, M. P., Hodgkin, A. L., and Steinhardt, R. A., 1969, The influence of calcium on sodium efflux in squid axons, *J. Physiol. (London)* **200:**431–458.

Barr, L., Dewey, M. M., and Berger, W., 1965, Propagation of action potentials and the nexus in cardiac muscle, *J. Gen. Physiol.* **48:**797–823.

Baux, G., Simmoneau, M., Tauc, L., and Segundo, J. P., 1978, Uncoupling of electrotonic synapses by calcium, *Proc. Natl. Acad. Sci. USA* **75:**4577–4581.

Bennett, M. V. L., 1973, Function of electrotonic junctions in embryonic and adult tissues, *Fed. Proc.* **32:**65–75.

Bennett, M. V. L., and Goodenough, D. A., 1978, Gap junctions, electrotonic coupling and intercellular communication, *Neurosci. Res. Program Bull.* **16:**373–486.

Bennett, M. V. L., Pappas, G. D., Aljure, E., and Nakajima, Y., 1967, Physiology and ultrastructure of electrotonic junctions. Spinal and medullary electromotor nuclei in mormyrid fish, *J. Neurophysiol.* **30:**180–208.

Bennett, M. V. L., Brown, J. E., Harris, A. L., and Spray, D. C., 1978a, Electrotonic junctions between *Fundulus* blastomeres: Reversible block by low intracellular pH, *Biol. Bull.* **155:**442.

Bennett, M. V. L., Spira, M. E., and Spray, D. C., 1978b, Permeability of gap junctions between embryonic cells of *Fundulus:* A reevaluation, *Dev. Biol.* **65:**114–125.

Brink, P. R., Verselis, V., and Barr, L., 1984, Solvent solute interactions within the nexal membrane, *Biophys. J.* **45:**121–124.

Clapham, D. E., Schrier, A., and DeHaan, R. L., 1980, Junctional resistance and action potential delay between embryonic cell aggregates, *J. Gen. Physiol.* **75:**633–654.

Cox, R. P., Krauss, M. J., Balis, M. E., and Daucis, J., 1970, Evidence for transfer of enzyme product as a basis of metabolic cooperation between tissue culture fibroblasts of Lesch–Nyhan disease and normal cells, *Proc. Natl. Acad. Sci. USA* **67:**1573.

De Felice, L. J., and Challice, C. E., 1969, Anatomical and ultrastructural study of the electrophysiological atrioventricular node of the rabbit, *Circ. Res.* **24:**457–474.

del Castillo, J., De Mello, W. C., and Morales, T., 1964, Inhibitory action of γ-aminobutyric acid (GABA) on *Ascaris* muscle, *Experientia* **20:**1–15.

Délèze, J., 1965, Calcium ions and the healing-over of heart fibers, in: *Electrophysiology of the Heart* (B. Taccardi and G. Marchetti, eds.), pp. 147–148, Pergamon Press, Elmsford, N.Y.
De Mello, W. C., 1971, The sealing process in heart and other muscle fibers, in: *Research in Physiology* (F. F. Kao, K. Koisumi, and M. Vassalle, eds.), pp. 275–288, Aulo Gaggi Pub, Bologna.
De Mello, W. C., 1972, The healing-over process in cardiac and other muscle fibers, in: *Electrical Phenomena in the Heart* (W. C. De Mello, ed.), pp. 323–351, Academic Press, New York.
De Mello, W. C., 1973, Membrane sealing in frog skeletal-muscle fibres, *Proc. Natl. Acad. Sci. USA* **70:**982–984.
De Mello, W. C., 1974, Electrical uncoupling in heart fibers produced by intracellular injection of Na or Ca, *Fed. Proc.* **17:**3.
De Mello, W. C., 1975, Effect of intracellular injection of calcium and strontium on cell communication in heart, *J. Physiol. (London)* **250:**231–245.
De Mello, W. C., 1976, Influence of the sodium pump on intercellular communication in heart fibres: Effect of intracellular injection of sodium ion on electrical coupling, *J. Physiol. (London)* **263:**171–197.
De Mello, W. C., 1977, Passive electrical properties of the atrioventricular node, *Pfluegers Arch.* **371:**135–139.
De Mello, W. C., 1979a, Effect of 2-4 dinitrophenol on intercellular communication in mammalian cardiac fibres, *Pfluegers Arch.* **380:**267–276.
De Mello, W. C., 1979b, Effect of intracellular injection of La^{3+} and Mn^{2+} on electrical coupling of heart cells, *Cell Biol. Int. Rep.* **3:**113–119.
De Mello, W. C., 1980a, Influence of intracellular injection of H^+ on the electrical coupling in cardiac Purkinje fibres, *Cell Biol. Int. Rep.* **4:**51–57.
De Mello, W. C., 1980b, Intercellular communication and junctional permeability, in: *Membrane Structure and Function,* Vol. 3 (E. E. Bittar, ed.), pp. 128–164, Wiley, New York.
De Mello, W. C., 1982, Cell-to-cell communication in heart and other tissues, *Prog. Biophys. Mol. Biol.* **39:**147–182.
De Mello, W. C., 1983a, The role of cAMP and Ca on the modulation of junctional conductance: An integrated hypothesis, *Cell Biol. Int. Rep.* **7:**1033–1040.
De Mello, W. C., 1983b, The influence of pH on the healing-over of mammalian cardiac muscle, *J. Physiol. (London)* **339:**299–307.
De Mello, W. C., 1983c, Modulation of functional permeability in cardiac fibers, in: *Myocardial Injury* (J. J. Spitzer, ed.), pp. 37–59, Plenum Publishing Co., N.Y.
De Mello, W. C., 1984a, Effect of intracellular injection of cAMP on the electrical coupling of mammalian cardiac cells, *Biochem. Biophys. Res. Commun.* **119:**1001–1007.
De Mello, W. C., 1984b, Modulation of junctional permeability, *Fed. Proc.* **43:**2692–2696.
De Mello, W. C., 1985, Ca-blocking agents promote healing-over in cardiac muscle, *Physiologist* **28:**4.
De Mello, W. C., 1986a, Increased spread of electrotonic potentials during diastolic depolarization in cardiac muscle, *J. Mol. Cell. Cardiol.* **18:**23–29.
De Mello, W. C., 1986b, Healing-over process in heart; interaction between Ca and protons, *Biophys. J.* **49:**339a.
De Mello, W. C., 1986c, Interaction of cyclic AMP and Ca in the control of electrical coupling in heart fibers, *Biochim. Biophys. Acta.* **888:**91–99.
De Mello, W. C., and Maldonado, H., 1985, Synaptic inhibition and cell communication; impairment of cell-to-cell coupling produced by γ-aminobutyric acid (GABA) in the somatic musculature of *Ascaris lumbricoides, Cell Biol. Int. Rep.* **9:**803–813.
De Mello, W. C., and Motta, G., 1969, Temperature and myocardial healing-over, *Fed. Proc.* **28:**2.
De Mello, W. C., Motta, G., and Chapeau, M., 1969, A study on the healing-over of myocardial cells of toads, *Circ. Res.* **24:**475–487.

De Mello, W. C., Gonzalez Castillo, M., and van Loon, P., 1983, Intercellular diffusion of Lucifer Yellow CH in mammalian cardiac fibers, *J. Mol. Cell. Cardiol.* **15:**637–643.

De Mello, W. C., van Loon, P., and Vizcarra, N., 1985, Increased cell-to-cell diffusion of Lucifer Yellow CH produced by dB-cAMP in heart fibers, *Biophys. J.* **47:**505a.

Dreifuss, J. J., Girardier, L., and Forsmann, W. G., 1966, Etude de la propagation de l'excitation dans le ventricule de rat au moyen de solutions hypertoniques, *Pfluegers Arch.* **292:**13–33.

Engelmann, T. W., 1877, Vergleichende Untersuchungen zur Lehre von der Muskel-und Nervenelektricitat, *Pfluegers Arch.* **15:**116–148.

Epstein, M. L., and Gilula, N. B., 1977, A study of communication specificity between cells in culture, *J. Cell Biol.* **75:**769–787.

Estapé, E., and De Mello, W. C., 1982, Effect of theophylline on the spread of electrotonic activity in heart, *Fed. Proc.* **41:**1505.

Estapé, E., and De Mello, W. C., 1983, Cyclic nucleotides and calcium: Their role in the control of cell communication in the heart, *Cell Biol. Int. Rep.* **7:**91–97.

Falk, G., and Fatt, P., 1964, Linear electrical properties of striated muscle fibres observed with intracellular electrodes, *Proc. R. Soc. (London) Ser. B* **160:**69–123.

Fatt, P., and Katz, B., 1951, An analysis of the end-plate potential recorded with an intracellular microelectrode, *J. Physiol. (London)* **115:**320–370.

Flagg-Newton, J. L., Dahl, G., and Loewenstein, W. R., 1981, Cell junction and cyclic AMP. I. Upregulation of junctional membrane permeability and junctional membrane particles by administration of cyclic nucleotide or phosphodiesterase inhibitor, *J. Membr. Biol.* **63:**105–121.

Fujisawa, H., Morioka, H., Watanabe, H., and Nakamura, H., 1976, A decay of gap junction in association with cell differentiation of neural retina in chick embryonic development, *J. Cell Sci.* **22:**585–596.

Fujimoto, W. Y., Subak-Sharpe, J. H., and Seegmiller, J. E., 1971, Hypoxanthine–guanine phosphoribosyltransferase deficiency: Chemical agents selective for mutant or normal cultured fibroblasts in mixed and heterozygote cultures, *Proc. Natl. Acad. Sci. USA* **68:**1516–1518.

Furshpan, E. J., and Potter, D. D., 1959, Transmission at the giant motor synapses of the crayfish, *J. Physiol. (London)* **245:**289–325.

Griepp, E. B., and Bernfield, M. R., 1975, Acquisition of ionic coupling in beating embryonic myocardial cells, *Circulation* **52**(Suppl. 2)**:**54.

Hax, W. M. A., van Venrooij, G. E. P. M., and Vossenberg, J. B. J., 1974, Cell communication: A cyclic-AMP mediated phenomenon, *J. Membr. Biol.* **19:**253–266.

Heilbrunn, L. V. (ed.), 1956, *Dynamics of Living Protoplasm,* Academic Press, New York.

Hertzberg, E., 1985, Antibody probes in the study of gap junctional communication, *Annu. Rev. Physiol.* **47:**305–318.

Hertzberg, E., Spray, D. C., and Bennett, M. V. L., 1984, An antibody to gap junctions blocks gap junctional conductance, *J. Cell Biol.* **99:**343a.

Hoffman, B. F., and Cranefield, P., 1960, *Electrophysiology of the Heart,* McGraw–Hill, New York.

Hutter, O. F., 1957, Mode of action of autonomic transmitters on the heart, *Br. Med. Bull.* **13:**176–180.

Ikeda, N., Toyama, J., Shimizu, T., Kodama, I., and Yamada, K., 1980, The role of electrical uncoupling in the genesis of atrioventricular conduction disturbance, *J. Mol. Cell. Cardiol.* **12:**809–816.

Imanaga, I., 1974, Cell-to-cell diffusion of Procion Yellow in sheep and calf Purkinje fibres, *J. Membr. Biol.* **16:**381–388.

James, T. N., and Scherf, L., 1968, Ultrastructure of the atrioventricular node, *Circulation* **37:**1049–1070.

Jennings, R. B., and Reimer, K. A., 1981, Lethal myocardial ischemic injury, *Am. J. Pathol.* **102:**241–255.

Jennings, R. B., and Steenbergen, C., 1985, Nucleotide metabolism and cellular damage in myocardial ischemia, *Annu. Rev. Physiol.* **47:**727–749.

Johnson, R. G., and Sheridan, J. D., 1971, Junctions between cancer cells in culture: Ultrastructure and permeability, *Science* **174:**717–719.

Johnson, R. G., Hammer, J. D., Sheridan, J. J. D., and Revel, J.-P., 1974, Gap junction formation between reaggregated Novikoff hepatoma cells, *Proc. Natl. Acad. Sci. USA* **71:**4536–4540.

Kamiyama, A., and Matsuda, K., 1966, Electrophysiological properties of the canine ventricular fiber, *Jpn. J. Physiol.* **16:**407–420.

Kanno, Y., and Loewenstein, W. R., 1966, Cell-to-cell passage of large molecules, *Nature* **212:**629–630.

Kushmerick, M. J., and Podolsky, R. J., 1969, Ionic mobility in muscle cells, *Science* **166:**1297–1298.

Lewis, T., 1925, *Mechanism and Graphic Registration of the Heart Beat,* Shaw, London.

Llinas, R., 1975, Electrical synaptic transmission in the mammalian central nervous system, in: *Perspectives in Neurobiology* (M. Santini, ed.), pp. 379–386, Raven Press, New York.

Loewenstein, W. R., 1966, Permeability of membrane functions, *Ann. N.Y. Acad. Sci.* **137:**441–472.

Mackie, G. O., 1964, Analysis of locomotion in a siphonophore colony, *Proc. R. Soc. London Ser. B* **159:**366–391.

Maekawa, M., Nohara, Y., Kawamura, K., and Hayashi, K., 1967, Electron-microscope study of the conduction system in mammalian hearts, in: *Electrophysiology and Ultrastructure of the Heart* (T. Sano, V. Mizuhira, and K. Matsuda, eds.), pp. 41–54, Grune & Stratton, New York.

Meyer, D. J., and Revel, J.-P., 1981, CO_2 does not uncouple hepatocytes in rat liver, *Biophys. J.* **30:**105A.

Mines, G. R., 1913, On dynamic equilibrium in the heart, *J. Physiol. (London)* **45:**350–383.

Neely, J. R., Whitmer, J. T., and Robetto, M. J., 1975, Effect of coronary flow on glycolytic efflux and intracellular pH in isolated rat hearts, *Circ. Res.* **37:**733–741.

Nichols, J. G., and Purves, D., 1972, A comparison of chemical and electrical synaptic transmission between single sensory cells and a motoneurone in the central nervous system of the leech, *J. Physiol. (London)* **225:**637–656.

Nishiye, H., 1977, The mechanism of Ca^{2+} action on the healing-over process in mammalian cardiac muscles: A kinetic analysis, *Jpn. J. Physiol.* **27:**451–466.

Noble, D., 1962, The voltage dependence of the cardiac membrane conductance, *Biophys. J.* **2:**381–393.

Obaid, A. L., Socolar, S. J., and Rose, B., 1983, Cell-to-cell channels with two independent regulated gates in series: Analysis of junctional channel modulation by membrane potential, calcium and pH, *J. Membr. Biol.* **73:**69–89.

Ochi, R., and Nishiye, H., 1973, Temperature dependence of the healing-over in mammalian cardiac muscle, *Proc. Jpn. Acad.* **49:**372–375.

Oliveira-Castro, G. M., and Loewenstein, W. R., 1971, Junctional membrane permeability: Effects of divalent cations, *J. Membr. Biol.* **5:**51–77.

Pappas, G. D., and Bennett, M. V. L., 1966, Specialized junctions involved in electrical transmission between neurons, *Ann. N.Y. Acad. Sci.* **137:**495–508.

Pappas, G. D., Asada, Y., and Bennett, M. V. L., 1971, Morphological correlates of increased coupling resistance at an electrotonic synapse, *J. Cell Biol.* **49:**173–182.

Payton, B. W., Bennett, M. V. L., and Pappas, G. D., 1969, Temperature dependence of resistance at an electrotonic junction, *Science* **165:**594–597.

Peracchia, C., and Dulhunty, A. F., 1976, Low resistance junctions in crayfish: Structural changes with functional uncoupling, *J. Cell Biol.* **70:**419–439.

Pitts, J. D., 1971, Molecular exchange and growth control in tissue culture, in: *Ciba Foundation*

Symposium on Growth Control in Cell Cultures (G. E. Wolstenholme and J. Knight, eds.), pp. 89–105, Livingstone, London.
Pitts, J. D., 1976, Junctions as channels of direct communications between cells, in: *The Development of Biology of Plants and Animals* (C. F. Graham and E. F. Wareing, eds.), pp. 96–110, Blackwell, Oxford.
Pitts, J. D., and Finbow, M. E., 1977, Junctional permeability and its consequence, in: *Intercellular Communication* (W. C. De Mello, ed.), pp. 61–86, Plenum Press, New York.
Politoff, A. L., Socolar, S. J., and Loewenstein, W. R., 1969, Permeability of a cell membrane junction: Dependence on energy metabolism, *J. Gen. Physiol.* **53:**498–515.
Pollack, G. H., 1976, Intercellular coupling in the atrioventricular node and other tissues of the heart, *J. Physiol. (London)* **255:**275–298.
Pollack, G. H., and Huntmann, L. L., 1973, Intercellular pathways in the heart: Direct evidence for low resistance channels, *Experientia* **29:**1501.
Potter, D. D., Furshpan, E. J., and Lennox, E. S., 1966, Connections between cells of the developing squid as revealed by electrophysiological methods, *Proc. Natl. Acad. Sci. USA* **55:**328.
Ramon, F., and Zampighi, G., 1980, On the electrotonic coupling mechanism of crayfish segmented axons: Temperature, dependence of junctional conductance, *J. Membr. Biol.* **54:**165–171.
Rasmussen, H., 1975, Ions as "second messengers," in: *Cell Membranes, Biochemistry, Cell Biology and Pathology* (G. Weismann and R. Claiborne, eds.), pp. 203–212, HP Publishing, New York.
Reber, W., and Weingart, R., 1982, Ungulate cardiac Purkinje fibres: The influence of intracellular pH on the electrical cell-to-cell coupling, *J. Physiol. (London)* **328:**87–104.
Reuter, H., and Scholz, H., 1977, The regulation of the calcium conductance of cardiac muscle by adrenaline, *J. Physiol. (London)* **264:**49–62.
Reuter, H., and Seitz, N., 1968, The dependence of calcium efflux from cardiac muscle on temperature and external ion composition, *J. Physiol. (London)* **195:**451–470.
Rose, B., and Loewenstein, W. R., 1971, Junctional membrane permeability, *J. Membr. Biol.* **5:**20–50.
Rose, B., and Loewenstein, W. R., 1975, Calcium ion distribution in cytoplasm visualized by aequorin: Diffusion in cytosol restricted by energized sequestering, *Science* **190:**1204–1206.
Rose, B., and Rick, R., 1978, Intracellular pH, intracellular free Ca, and junctional cell-to-cell coupling, *J. Membr. Biol.* **44:**377–415.
Rose, B., Simpson, I., and Loewenstein, W. R., 1977, Calcium ion produces graded changes in permeability of membrane channels in cell junction, *Nature* **267:**625–627.
Rothschuh, K. E., 1951, Ueber den funkionellen Aufbau des Herzens aus elektrophysiologischen Elementen and ueber den Mechanisms der Erregungsleitung in Herzen, *Pfluegers Arch.* **253:**238–251.
Sakamoto, Y., 1969, Membrane characteristic of the canine papillary muscle fiber, *J. Gen. Physiol.* **54:**765–781.
Scheutze, S. M., and Goodenough, D. A., 1982, Dye transfer between cells of embryonic chick lens becomes less sensitive to CO_2 treatment with development, *J. Cell Biol.* **92:**694–705.
Schwarzmann, G., Wiegandt, H., Rose, B., Zimmerman, A., Ben-Haim, D., and Loewenstein, W. R., 1981, Diameter of the cell-to-cell junctional membrane channels as probed with neutral molecules, *Science* **213:**551–553.
Shen, A. C., and Jennings, R. B., 1972, Kinetics of calcium accumulation in acute myocardial ischemic injury, *Am. J. Pathol.* **67:**441–452.
Sheridan, J. D., Finbow, M. E., and Pitts, J. D., 1979, Metabolic interactions between animal cells through permeable intercellular junctions, *Exp. Cell Res.* **123:**111–117.
Simpson, I., Rose, B., and Loewenstein, W. R., 1977, Size limit of molecules permeating the junctional membrane channels, *Science* **195:**294–296.

Sotelo, C., Llinas, R., and Baker, R., 1974, Structural study of inferior olivary nucleus of the cat, morphological correlates of electrotonic coupling, *J. Neurophysiol.* **37:**541–559.

Spira, M. E., and Bennett, M. V. L., 1972, Synaptic control of electrotonic coupling between neurons, *Brain Res.* **37:**294–300.

Spray, D. C., and Bennett, M. V. L., 1985, Physiology and pharmacology of gap junctions, *Annu. Rev. Physiol.* **47:**281–303.

Spray, D. C., Harris, A. L., and Bennett, M. V. L., 1979, Voltage dependence on junctional conductance in early amphibian embryos, *Science* **204:**432–434.

Spray, D. C., Harris, A. L., and Bennett, M. V. L., 1981, Gap junctional conductance is a simple and sensitive function of intracellular pH, *Science* **211:**712–715.

Spray, D. C., Stern, J. H., Harris, A. L., and Bennett, M. V. L., 1982, Gap junctional conductance: Comparison of sensitivities to H and Ca ions, *Proc. Natl. Acad. Sci. USA* **79:**441–445.

Spray, D. C., White, R. L., Campos de Carvalho, A., Harris, A. L., and Bennett, M. V. L., 1984, Gating of gap junction channels, *Biophys. J.* **45:**219–230.

Stewart, W. C., 1978, Functional connections between cells as revealed by dye-coupling with a high fluorescent naphthalimide tracer, *Cell* **14:**741–759.

Trautwein, W., 1964, Pathophysiologie des Herzflimmerns, *Verh. Dtsch. Ges. Kreislaufforsch.* **30:**40–56.

Trautwein, W., Kuffler, S. W., and Edwards, C., 1956, Changes in membrane characteristics of heart muscle during inhibition, *J. Gen. Physiol.* **40:**135–145.

Tsien, R., and Weingart, R., 1976, Inotropic effect of cyclic AMP in calf ventricular muscle studied by a cut-end method, *J. Physiol. (London)* **260:**117–141.

Turin, L., and Warner, A. E., 1977, Carbon dioxide reversibly abolishes ionic communication between cells of early amphibian embryos, *Nature* **270:**56–57.

Turin, L., and Warner, A. E., 1980, Intracellular pH in early *Xenopus* embryos: Its effect on current flow between blastomeres, *J. Physiol. (London)* **300:**489–504.

Weidmann, S., 1952, The electrical constants of Purkinje fibres, *J. Physiol. (London)* **118:**348–360.

Weidmann, S., 1966, The diffusion of radiopotassium across intercalated discs of mammalian cardiac muscle, *J. Physiol. (London)* **187:**323–342.

Weidmann, S., 1970, Electrical constants of trabecular muscle from mammalian heart, *J. Physiol. (London)* **210:**1041–1054.

Weingart, R., 1974, The permeability to tetraethylammonium ions of the surface membrane and the intercalated disks of the sheep and calf myocardium, *J. Physiol. (London)* **240:**741–762.

Weingart, R., 1977, The action of ouabain on intercellular coupling and conduction velocity in mammalian ventricular muscle, *J. Physiol. (London)* **264:**341–365.

Weingart, R., 1981, Cell-to-cell coupling in cardiac tissue, *Proc. 28th Int. Cong. Physiol. Sci.* **8:**59–68, Budapest, Akademiai Kiado.

Werner, M., 1910, Besteht die Herzmuskulatur der Säugetiere aus allseitz scharf begrentzten Zeller o der nicht?, *Arch. Mikrosk. Anat.* **71:**101–129.

Woodbury, J. W., and Crill, W. E., 1961, On the problem of impulse conduction in the atrium, in: *Nervous Inhibition* (E. Florey, ed.), pp. 124–125, Pergamon Press, Elmsford, N.Y.

Yancey, S. B., Nicholson, B. J., and Revel, J.-P., 1981, The dynamic state of liver gap junctions, *J. Supramol. Struct. Cell Biochem.* **16:**221–232.

Chapter 3

Permeability and Regulation of Gap Junction Channels in Cells and in Artificial Lipid Bilayers

Camillo Peracchia
Department of Physiology
University of Rochester
Rochester, New York 14642

1. INTRODUCTION

Many cell communities, in spite of the apparent structural individuality of their members, behave in some respects like syncytia, due to the existence of well-defined cell-to-cell channels of communication. Cooperative functions such as the synchronous spread of electrical impulse in heart, smooth muscle, and some areas of the nervous system, equilibration of ionic and metabolic pools, and coordinated responses of cell communities to hormonal or transmitter-mediated stimuli are among some of the many functions of direct cell-to-cell communication (cell coupling), a mechanism that enables tissues to respond to external and internal signals as integrated systems. Cell coupling represents the ability of cells to freely exchange with neighboring cells in direct contact with them, ions, metabolites, and messengers, while maintaining their individuality regarding macromolecules; the syncytiumlike feature being restricted to molecules smaller than M_r 1000 (M_r 2000 in certain invertebrates).

There is general agreement that cell coupling is due to the presence of transcellular membranous channels located at gap (communicating) junctions; however, in the past, evidence for this was mostly circumstantial (Peracchia, 1985). Recently, data showing the capacity of isolated gap junction proteins to form channels in liposomes (Girsch and Peracchia, 1985a,b), in lipid bilayers formed at the tip of pipettes (Girsch *et al.*, 1986), and in planar bilayers (Hall and Zampighi, 1985), and evidence for the capacity of intracellularly injected antibodies to liver gap junction protein to uncouple *Xenopus* (Warner *et al.*, 1984) and liver (Hertzberg *et al.*, 1985) cells have given a convincing and more direct support to this notion.

Like other membranous channels, the cell-to-cell channels are gated pathways, but the gating mechanism is still poorly understood. While the principal factors involved in the control of channel permeability have been recognized, many physical and chemical changes leading to channel occlusion have been determined, and some agents transmitting the effects of uncoupling treatments to the channel gates have been identified, other aspects of cell coupling like the site of action of uncouplers, the possible interaction among uncouplers, the possible involvement of soluble intermediates, and the molecular mechanisms of channel gating are still poorly understood.

In an effort to simplify the research approach for studying coupling regulation, our interest has gradually shifted from whole tissue to two-cell systems and eventually to internally perfused systems, the result having been a continuous improvement in our ability to bring into focus individual elements of cell-to-cell transfer regulation. Along this line, we have recently concerted our efforts in developing and adapting *in vitro* reconstituted systems to cell coupling studies. These systems provide several advantages as they allow one to limit the number of variables and to control more effectively the channel environment; nevertheless, they also carry the obvious limitations of artificial systems, requiring continuous comparison of the results with data derived from intact cells.

2. EVIDENCE FOR CELL-TO-CELL CHANNELS AT GAP JUNCTIONS

2.1. Structural Data

Circumstantial evidence for the location of coupling channels at gap junctions has come from a variety of structural and functional data. First suggested by the observation of structures resembling channel openings at the surface of junctional particles in freeze-fracture replicas and in negatively stained preparations (reviewed in Peracchia, 1980), the presence of channels at gap junctions was confirmed by x-ray diffraction and low-dose-microscopy studies (reviewed in Peracchia, 1985). Recently, a remarkable study of the structure of isolated liver gap junctions has provided a detailed portrait of the channel structure (Fig. 1), through images of electron density distribution derived from x-ray diffraction (Makowski *et al.*, 1982, 1984; Makowski, 1985). The reconstructed image of the junction profile shows three regions occupied by solvent: one, interpreted as the intercellular channel, extends vertically along the sixfold rotation axis of the junctional particle, another corresponds to the extracellular gap, and a third to the cytoplasmic side of the junction. The channel has a diameter of 2–3 nm along

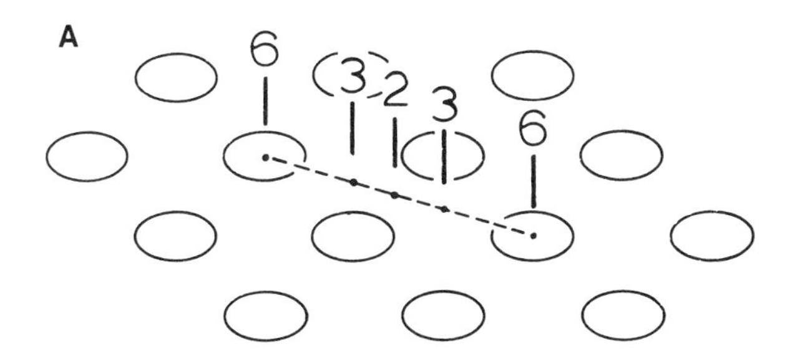

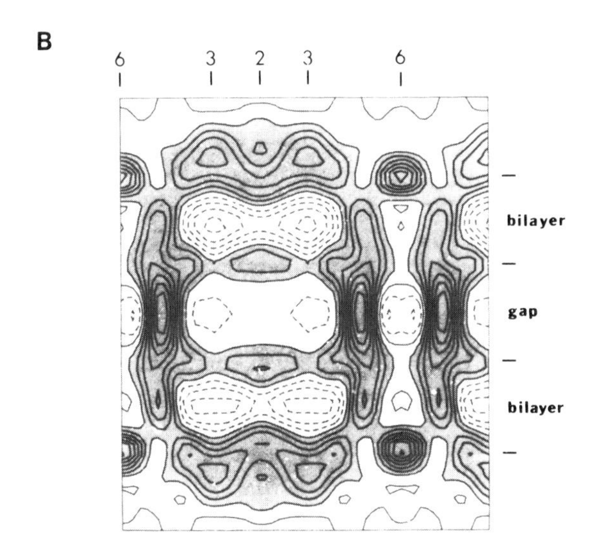

FIGURE 1. Electron density map of rat liver gap junctions calculated from x-ray diffraction. (A) Diagram of the hexagonal lattice of the junction indicating the position of the vertical section drawn in (B). (B) Cross section through the gap junction structure with positions of the six-, three-, and twofold axes marked. Electron density greater than that of solvent is indicated by shading and solid contours. Electron density less than the solvent level is indicated by broken contours. Protein and lipid polar head groups have high electron density and will be confined largely to the regions indicated by shading in the map. Lipid hydrocarbon chains have electron density lower than solvent and occupy regions indicated by broken contours within the bilayers. The remainder of the volume is occupied by solvent. There are three regions occupied by solvent: the cell-to-cell channel (6); the gap; and the solvent external to the junction (cytoplasm in intact cells). The channel (6) is 20–30 Å in diameter for most of its length, but narrows to 15 Å in the extracellular half of the bilayer. The channel appears to be closed near the cytoplasmic surface by a blocking structure, possibly the gate (arrow). From Makowski (1985).

most of its length, narrowing to 1.5 nm at the extracellular half of the bilayer, and appears isolated from the solvent lining of the cytoplasmic side of the junction by a blocking structure (Makowski, 1985). The blocking structure is believed to provide the functional gate of the cell-to-cell channels, consistent with data suggesting that the C-terminal arm of junctional proteins participates in channel gating at the cytoplasmic end of the channel (Peracchia and Girsch, 1985b).

2.2. Structure–Function Studies

The presence of coupling channels at gap junctions has been indirectly suggested by data on the correlation between the occurrence of gap junctions and cell-to-cell coupling. Already in the 1950s and early 1960s, regions of close plasma membrane apposition between electrically coupled cells were suspected of being sites of cell-to-cell communication, but data on the relationship between gap junctions and cell coupling were first obtained in experiments showing that treatments that alter gap junction integrity also result in functional uncoupling (reviewed in Peracchia, 1980).

Finer modifications in gap junction structure with functional uncoupling have also been described (Peracchia and Dulhunty, 1976; Peracchia, 1977). The structural changes, characterized by an increase in packing order (crystallization) and density of junctional particles following uncoupling treatments, suggested that channel occlusion results from conformational changes in the channel protein. However, questions have been raised about the existence of a relationship between particle packing arrangement and channel function.

Although a number of studies have supported the correlation between particle packing arrangement and functional coupling, recently conflicting reports have caused controversy and disagreement. While most data support an increase in tightness and order of particle array with cell uncoupling (reviewed in Peracchia and Bernardini, 1984), some studies suggest that particle dispersion rather than particle crystallization is the structural correlate of functional uncoupling (Lee *et al.*, 1982; Green and Severs, 1984; Mazet *et al.*, 1985; Atkinson and Sheridan, 1985). However, it seems clear that liver gap junctions, which contain disordered particle arrays when viewed after rapid freezing in a coupled state (Peracchia and Girsch, 1985a), uncouple and assume crystalline particle arrays when the cells are disrupted for junction isolation. In fact, Makowski *et al.* (1984) have clearly shown by x-ray diffraction that the channels of crystalline liver gap junctions are inaccessible to sucrose, a channel permeant, demonstrating that indeed they are in a high-resistance configuration and, more recently, have produced the first convincing images of the actual gating structures closing the channels of these crystalline junctions at their cytoplasmic ends (Makowski, 1985) (Fig. 1). On the other hand, a recent study on chick embryonal lens has reported no changes in particle arrangement with dye uncoupling induced by a variety of treatments in rapidly frozen gap junctions, warning that particle arrangements in gap junctions may not bear any relationship to the functional state of the channels (Miller and Goodenough, 1985).

A line of study on cultured myocardial cells has given an elegant demonstration of the necessity of gap junction formation for the development of synchronous contractions, and thus cell-to-cell electrical coupling (Goshima, 1969, 1970; Hyde *et al.*, 1969; DeHaan and Hirakow, 1972; Lawrence *et al.*, 1978), and other studies on cell cultures have reached the same conclusion by showing that both metabolic and electrical coupling are strictly dependent on the capacity of cells to form gap junctions (Gilula *et al.*, 1972; Azarnia *et al.*, 1974).

2.3. Data from Intracellular Injection of Antibodies to Gap Junctions

While methods for gap junction isolation have been available for almost two decades (reviewed in Peracchia, 1980, 1985) only recently have efforts been made to prepare antibodies to junctional protein (Traub and Willecke, 1982; Hertzberg, 1984; Paul, 1985). With polyclonal antibodies to rat liver gap junctions (Hertzberg, 1984), Warner *et al.* (1984) and Hertzberg *et al.* (1985) have produced new evidence for the participation of gap junctions in cell coupling.

One study, involving intracellular injection of gap junction antibody into a *Xenopus* embryo blastomere, showed loss of electrical coupling and dye transfer between the injected cell and its neighbors (Warner *et al.*, 1984). In addition, embryos injected with gap junction antibodies generated a high proportion of tadpoles with developmental defects such as absence of eye and trigeminal ganglia formation at the injected side. While some caution should be exercised in interpreting these results, as the antibodies were not produced from the purified junctional protein but rather from membrane fractions enriched in gap junctions, this study is indeed the first to give some substance to the hypothesis that gap junction communication is important for cell differentiation.

The other study also showed an inhibition of electrical coupling and cell-to-cell dye diffusion, following antibody injection into various cultured cells including hepatocytes, myocardial cells, and neurons. In addition, it suggested that a homologous segment of the junctional protein may exist in cells of different origin (Hertzberg *et al.*, 1985). Indeed, 27,000-dalton (27k) polypeptides cross-reacting with antibodies to rat liver gap junctions were detected in a variety of other rat tissues, as well as in liver from other vertebrates (Hertzberg and Skibbens, 1984), and homologies in tryptic peptide map were detected among 28k components isolated from liver, heart, and myometrium gap junctions (Zervos *et al.*, 1985). However, some differences in amino acid sequence between the amino-terminal residues of liver and heart proteins have been found (Nicholson *et al.*, 1985), indicating that junctional proteins may in part be tissue specific.

2.4. Junctional Proteins Make Channels in Artificial Lipid Systems

Strong evidence for the presence of channels in gap junctions comes from data on the capacity of junctional proteins, purified from mammalian lens and liver, to form channels in liposomes (Girsch and Peracchia, 1983, 1985a,b; Nikaido and Rosenberg, 1985; Gooden *et al.*, 1985) in planar bilayers (Hall and Zampighi, 1985; Zampighi *et al.*, 1985) and in bilayers formed at the tip of patch clamp pipettes (Girsch *et al.*, 1986; Spray *et al.*, 1986; Wojtczak *et al.*, 1987).

Our studies employed first the protein of eye lens gap junctions because this protein can easily be purified in sizable quantities and is stable after detergent extraction. In the lens, while the surface epithelial cells are coupled by typical gap junctions, the fiber cells are extensively joined (~ 60% of their surface) by junctions that differ from typical (liver type) gap junctions in many respects. Morphologically, they are thinner and crystallize into pseudohexagonal, rhombic, and orthogonal arrays with a 6.5-nm periodicity (Peracchia and Peracchia, 1980a,b; Bernardini and Peracchia, 1981). Biochemically, they are composed of a 28.2k protein (MIP26) (Fig. 2) different in amino acid composition from the 28k protein of liver gap junctions (Nicholson *et al.*, 1983). Immunologically, they have only a weak, if any, antigenic cross-reactivity with the liver protein (Traub and Willecke, 1982). Nevertheless, their physiological properties are thought to be basically the same as those of other communicating junctions because they are the only junctions between cells that communicate electrically with each other, pass dyes and metabolites, and can be uncoupled with treatments effective in a variety of cell systems (Bernardini *et al.*, 1981; Rae *et al.*, 1982; Jacob, 1983).

For liposome incorporation, MIP26 proved ideal because unlike other junctional proteins it can be purified in large quantities by a simple procedure (Russell *et al.*, 1981; Girsch and Peracchia, 1985b). This involves removal of soluble and extrinsically bound proteins from lens fiber junctions by successive washes in Tris buffer, urea, and NaOH and either extraction with octylpolyoxyethylene or solubilization of the junctions with sodium dodecyl sulfate (SDS) and isolation of MIP26 with SDS–polyacrylamide thick-gel electrophoresis (SDS–PAGE) and electroelution. Liposome incorporation was obtained by sonicating a mixture of MIP26 and brain phospholipids. Incorporation efficiency was monitored with freeze-fracture, by determining the appearance of intramembrane particles (Fig. 3B) similar to those of intact junctions (Fig. 3A). For studying channel permeability, both incorporated and protein-free liposomes were loaded with a solution of a channel impermeant, dextran T-10 (10k), and suspended into solutions of channel permeants, such as KCl, sucrose (SUC), or polyethyleneglycol (PEG; M_r 1450 ± 300), 50% hypertonic to the T-10 solution (Girsch and Peracchia, 1985b).

The presence of open channels was determined spectrophotometrically by measuring the decrease in optical density (scattering, $OD_{500\ nm}$) caused by

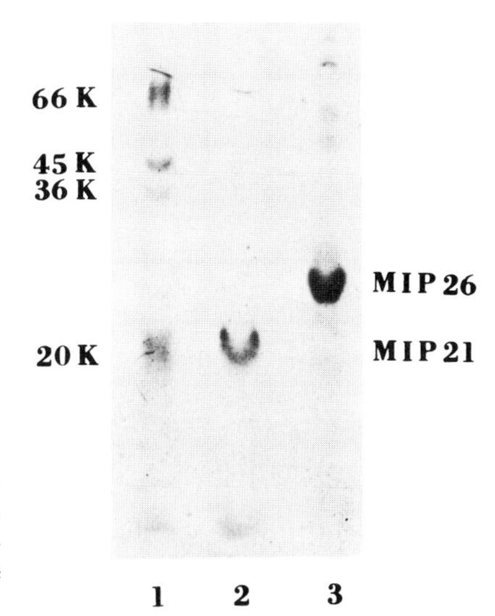

FIGURE 2. Electrophoretic profiles of octylpolyoxyethylene-extracted lens junctional proteins: MIP26 (lane 3) and its trypsin-cleaved product, MIP21 (lane 2). Standards are shown in lane 1. From Peracchia and Girsch (1985c).

liposome swelling, as permeants and water diffuse through the channels while T-10 does not (Luckey and Nikaido, 1980). In hypertonic solutions of permeants there is a biphasic change in optical density, with open channels: a rapid increase that peaks briefly, followed by a slow decrease (Fig. 4). The rapid increase in optical density is due to the initial osmotic gradient which causes water efflux and liposome shrinkage; the slow decrease results from a channel-mediated influx of permeants, causing a progressive increase in the internal tonicity and consequently water influx and liposome swelling. In contrast, in the absence of channels, as with unincorporated liposomes (Fig. 4, PLAIN), or with closed channels (Fig. 11), the initial rapid increase in optical density is not followed by a slow decrease, because only water diffuses through the liposome membrane, causing liposome shrinkage up to osmotic equilibrium.

Liposomes in which MIP26 is incorporated swell, following the brief initial shrinkage, in hypertonic KCl, sucrose, or PEG (Fig. 4), suggesting the presence of channels permeable to molecules as heavy as approximately 1.5k (Girsch and Peracchia, 1983, 1985b). By using the initial rate of change in OD as a measure of probe permeance, the log–log plot of OD rate versus molecular weight of probe size is linear (Fig. 10) (Peracchia and Grisch, 1985c). Independently, Nikaido and Rosenberg (1985) have also incorporated MIP26 into liposomes and from the permeability rates of the resulting channels have estimated a channel diameter of ~ 1.4 nm.

Preliminary experiments with liposomes incorporated with the junctional protein isolated from rat liver (28k component) show that this protein forms large channels as well. The liver channels may be smaller than those of the lens, because the incorporated liposomes swell in KCl and sucrose but not in PEG (Girsch and Peracchia, 1985a).

Recently, Gooden *et al.* (1985) have further confirmed the capacity of MIP26 to form large channels in reconstituted proteoliposomes by using an elegant assay system in which reconstituted liposomes, loaded with cytochrome *c*, are exposed to ascorbate (M_r 176). The presence of channels permeable to ascrobate is monitored spectrophotometrically by measuring color change as the intraliposomal cytochrome *c* is progressively reduced by incoming ascorbate.

High-conductance channels have also been reconstituted in planar lipid bilayers using lens junctional protein (Hall and Zampighi, 1985; Zampighi *et al.*, 1985). In this study, channels were formed either by adding the octylglucoside-extracted protein to the aqueous solutions bathing the planar bilayer or by first incorporating the lens protein into liposomes and then fusing the liposomes with

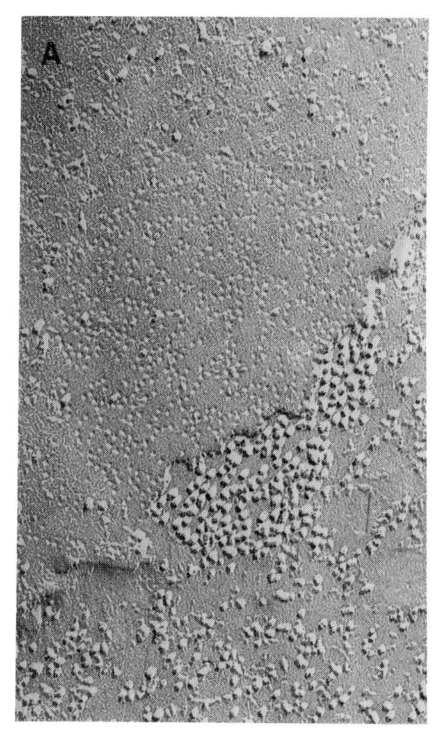

FIGURE 3. Freeze-fracture replicas of a lens fiber gap junction (A) and two liposomes (B) in which the lens junctional protein (MIP26) has been incorporated. Lens fiber junctions show a disordered array of intramembrane particles (~ 8 nm in size) and complementary pits. Liposomes in which MIP26 is incorporated show similar size particles on both the convex and the concave (not shown) fracture face, suggesting a bilateral protein orientation. ×123,600. From Peracchia and Girsch (1985c).

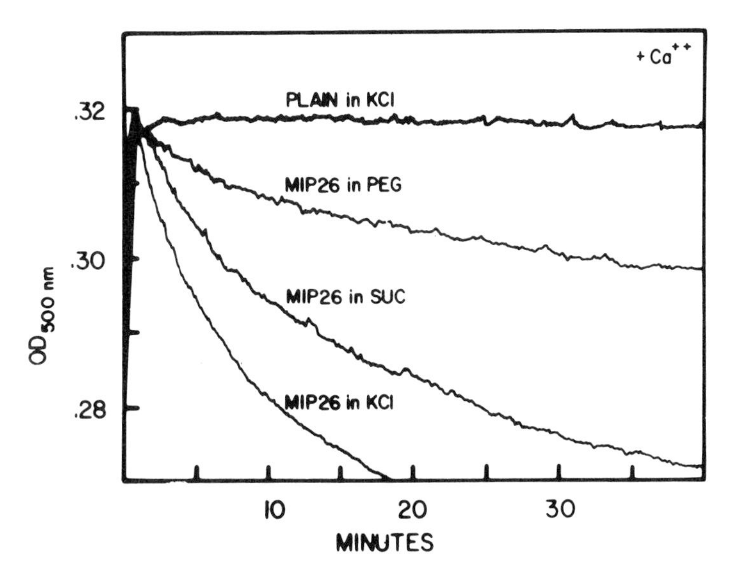

FIGURE 4. Permeability of liposomes in which MIP26 has been incorporated. Channel permeability is determined spectrophotometrically by measuring the decrease in optical density ($OD_{500\ nm}$) as the liposomes loaded with an impermeant (dextran T-10) swell, after a brief initial shrinkage, when suspended in solutions of permeants such as KCl, sucrose (SUC), or polyethylene glycol (PEG) 50% hyperosmotic to the T-10 solution. The liposomes containing MIP26 swell (decrease in OD) in KCl, sucrose, and PEG with progressively lower rates, while protein-free (PLAIN) liposomes do not, indicating the presence of channels permeable to probes as heavy as 1500 daltons (PEG). From Peracchia and Girsch (1985c).

the planar bilayer. The conductance of single channels (~ 200 psec in 100 mM KCl) is consistent with that expected for junctional channels, but the channels appear to be voltage-dependent and insensitive to Ca^{2+} and H^+. In contrast, in intact cells voltage dependence is seen only in embryonic gap junctions and the channels close with Ca^{2+} and H^+, although different sensitivity to these ions has been reported in different cells (Spray and Bennett, 1985).

In our reconstituted system the channels acquire sensitivity to Ca^{2+} when reconstituted in the presence of calmodulin (CaM) (Figs. 10 and 11) (Girsch and Peracchia, 1985b); thus, it is possible that the channels formed in planar bilayers lack gating properties because of the absence of a CaM-like protein in the medium. Indeed, the channels reconstituted by Hall and Zampighi (1985) appear to be sensitive to octanol, a long-chain alcohol found to cause uncoupling in various cell systems (Johnston *et al.*, 1980). However, the reported disappearance of channel events occurred only after unusually long latencies and at high octanol concentrations (Hall and Zampighi, 1985).

Another *in vitro* approach for studying permeability and gating of single gap junction channels has recently been developed in our laboratory (Girsch *et al.*,

1986; Wojtczak *et al.*, 1987) as a modification of a technique first described by Coronado and Latorre (1983). This approach consists of incorporating into bilayers formed at the tip of patch-clamp pipettes, junctional protein purified from isolated gap junctions with chloroform–methanol or detergent extraction. Cholesterol is added to lecithin in chloroform and the lipid mixture is dried under nitrogen. The lipids are resuspended in pentane and mixed (10 : 1) with the junctional protein, also in pentane. An aliquot of lipid–protein mixture sufficient to produce a monolayer is layered at the air–water interface of a small glass petri dish filled with saline solution in which a patch-clamp pipette was previously immersed. Following pentane evaporation, the pipette is withdrawn and reimmersed through the monolayer, a process that causes the formation of a bilayer at the tip of the pipette. The bilayer adheres tightly to the glass pipette, forming a seal of 2–5 gΩ. Preliminary experiments show that large channels often form in the clamped bilayer both with lens and liver junctional protein. The channel conductance is greater than 100 psec in 150 mM saline, in agreement with the conductance values reported for junction channels in intact cells (Neyton and Trautmann, 1985).

With a different method that employs suspensions of intact liver gap junction fragments, Spray *et al.* (1986) have succeeded in studying reconstituted gap junction channels as well. This study has reported the formation of channels with a conductance of 150 psec in 150 mM electrolyte solutions that close at low pH and with octanol and appear to be blocked by polyclonal antibodies to the liver junctional protein.

3. CELL-TO-CELL CHANNEL GATING AND PERMEABILITY MODULATION

For many years since direct cell communication was discovered, the cell-to-cell channel was considered by most as a passive and rather uninteresting structure. Being viewed as a large and nonselective conduit locked in a virtually permanent state of patency, this channel seemed quite isolated from the many dynamic changes in cellular function and rather insensitive to the effects of regulatory mechanisms. In recent years, this view has been rapidly changing as the cell-to-cell channel has shown up more and more as an active participant of cell function closely linked to a variety of control processes.

For over 20 years we have known that cell-to-cell channels can close, causing contiguous cells to uncouple electrically and metabolically from each other, but only recently have we learned that channel permeability can also be finely modulated. Data showing that channel proteins can be phosphorylated both *in vitro* and *in vivo* by cyclic nucleotide- and Ca^{2+}-dependent kinases, evidence for the effects of cAMP and protein kinase C activators on cell-to-cell communication, and knowledge of the involvement of CaM-like proteins in channel permeability regulation have opened a new exciting chapter in cell communication research.

3.1. Uncouplers

Direct cell-to-cell communication ceases after a variety of treatments (reviewed in Peracchia, 1980, 1985) such as exposure of cells to Ca^{2+} chelators, cell damage in the presence of Ca^{2+}, intracellular injection of H^+, Na^+, Ca^{2+}, and a variety of other divalent cations, inhibition of the metabolism, removal of external Na^+, exposure to agents that lower pH_i, hypertonic solutions, proteolytic enzymes, aldehyde fixatives, secretagogues, lectins, alkanols, hypoxia, changes in either membrane potential or transjunctional potential, and intracellular injection of cAMP (horizontal cells). In spite of the diversity of uncoupling treatments, it is now generally agreed that only a few factors mediate the effect of these treatments on the cell-to-cell channel gates. Of the uncouplers presently known, Ca^{2+}, H^+, and alkanols appear to be effective in all cells tested, while voltage, as a change in either membrane or transjunctional potential, seems to be effective mainly in embryonal cells, an exception being the crayfish rectifying junctions.

3.1.1. Calcium Ions

The relationship between Ca^{2+} and uncoupling was first described in "healing-over" experiments. Healing-over is the capacity of cells to isolate themselves from injured neighboring cells. Délèze (1964) noticed that healing-over occurs only if the extracellular medium contains Ca^{2+}, suggesting for the first time the possible connection between increased $[Ca^{2+}]_i$ in the injured cells and cell-to-cell channel occlusion at the junctions coupling these cells to their undamaged neighbors. The relationship between Ca^{2+} and uncoupling was reported in several experiments of intracellular Ca^{2+} injection in insect gland cells (Loewenstein *et al.*, 1967; Rose and Loewenstein, 1975a), cardiac myocytes (De Mello, 1975), lymphocytes (Oliveira-Castro and Barcinski, 1974), and other cells, but questions about the directness of the Ca^{2+} effects were raised, in view of the fact that an increase in $[Ca^{2+}]$ triggers a variety of changes in the intracellular medium. In some cells, for example, an increase in $[Ca^{2+}]_i$ causes a decrease in pH_i (Meech and Thomas, 1977) which by itself is believed to cause uncoupling (Turin and Warner, 1977, 1980; Spray *et al.*, 1981, 1982). However, experiments in which one of two coupled dimer cells was perfused with Ca^{2+}-containing solutions buffered to pH 7.6 showed that Ca^{2+} indeed induces channel occlusion independently from H^+ (Spray *et al.*, 1982).

The $[Ca^{2+}]_i$ necessary to cause uncoupling is still uncertain. While $[Ca^{2+}]_i$ ranging from 40 to 400 μM were reported to be effective in ruptured cells, in which Ca^{2+} was allowed to diffuse into the cytoplasm from the extracellular medium (Oliveira-Castro and Loewenstein, 1971; Nishiye, 1977), as well as in cells intracellularly perfused with buffered Ca^{2+} solutions (Spray *et al.*, 1982), lower concentrations appear to initiate channel gating in intact cells.

Délèze and Loewenstein (1976) and Rose *et al.* (1977) noticed that small intracellular injections of Ca^{2+}, insufficient to cause membrane depolarization, block dye coupling. They concluded that channel permeability is affected by Ca^{2+} elevation in the junctional area in the range of 0.1–100 μM.

Weingart (1977) noticed that the increase in longitudinal resistance of cardiac muscle fibers exposed to ouabain and the parallel increase in tension have a similar time course, implying that a $[Ca^{2+}]_i$ in the low micromolar range is sufficient for initiating the uncoupling process. After ouabain, the increase in $[Ca^{2+}]_i$ is believed to result mainly from competition between $Ca^{2+}{}_i$ and $Na^+{}_i$ (increased by the inhibition of the Na^+ pump) for sites at the Na^+/Ca^{2+} exchanger.

Iwatsuki and Petersen (1978) reported that cells of exocrine pancreas and lacrimal glands uncouple momentarily with application of secretagogues at concentrations lower than those causing maximal secretion, depolarization, and cyclic nucleotide accumulation. Since in pancreas $[Ca^{2+}]_i$ increases only from 0.2 to 0.9 μM, with the secretagogue at concentrations capable of eliciting maximal secretory activity (Ochs *et al.*, 1983), the study of Iwatsuki and Petersen (1978) indicates that these cell-to-cell channels are sensitive to $[Ca^{2+}]_i$ in the low micromolar range or even less. However, one could argue that secretagogues cause a change in pH_i sufficient to cause uncoupling independently. Indeed, the ACh-evoked uncoupling of pancreatic cells was reduced by cytoplasmic alkalinization with 10 mM NH_4Cl (Iwatsuki and Petersen, 1979), but the mere fact that uncoupling was not completely abolished by this treatment supports the involvement of Ca^{2+}. In fact, exposure to NH_4Cl is known to increase pH_i far above the values reported to affect coupling; in squid axon, for example, exposure to 10 mM NH_4Cl raises the pH_i from 7.3 to 7.8 (DeWeer, 1978). The possibility that secretagogues uncouple via a change in membrane potential is unlikely first of all because some secretagogues cause slight depolarization while others induce hyperpolarization, and secondly because voltage-dependent gating of mammalian cell-to-cell channels occurs only in embryonal cells (Spray and Bennett, 1985). Supporting the involvement of Ca^{2+} in the ACh-induced uncoupling of pancreatic acinar cells is also the finding that uncoupling is strongly inhibited by Ca^{2+} channel blockers such as Mn^{2+}, Ni^+, and Co^{2+} (Iwatsuki and Petersen, 1978).

A high Ca^{2+}-sensitivity of the cell-to-cell channel gate is also supported by the observation that in voltage-clamped cell pairs isolated from *Chironomus* salivary glands the curve relating junctional conductance to membrane potential is shifted toward an increase in junctional conductance by intracellular injection of EGTA (pH 7.2) (Obaid *et al.*, 1983). This observation suggests that even the normal $[Ca^{2+}]_i$ (in the 10^{-7} M range) has some effects on channel conductance that are abolished by a further reduction of $[Ca^{2+}]_i$ in the absence of changes in voltage and pH_i.

Recently, a remarkable study using whole cell clamp of two isolated heart muscle cells, one of which was pierced to allow intracellular diffusion of external media, has given the most convincing data on the high sensitivity of gap junction channels to Ca^{2+} (Noma and Tsuboi, 1987). In this study, the channels were found to start closing at $[Ca^{2+}]_i$ as low as 0.3 μM, at $[Mg^{2+}]_i > 1$ mM, and at $pH < 6.5$. These data are consistent with our study showing that $[Ca^{2+}]_i = 0.5$ μM, $[Mg^{2+}]_i > 1$ mM, and $pH < 6.5$ cause crystallization of isolated lens gap junctions (Peracchia and Peracchia, 1980a,b).

The idea that cell-to-cell channels are sensitive to micromolar $[Ca^{2+}]_i$ seems at first in contradiction with the lack of any detectable change in coupling in heart muscle during action potential (Weidmann, 1970) or twitch (Weingart, 1977). A possible interpretation is that the calcium transient during twitch is too short for the uncoupling to develop (Weingart, 1977) and the large buffering capacity of sarcoplasmic reticulum membranes in regions neighboring gap junctions may protect cell-to-cell channels from momentary, even though sizable, changes in $[Ca^{2+}]_i$. Indeed, Keith *et al.* (1985) have shown that in endosperm cells injected with Quin-2 the transient increase in $[Ca^{2+}]_i$ during mitosis can be just local, demonstrating that long-lasting Ca^{2+} gradients exist in dividing cells; there is no reason to believe that sizable Ca^{2+} gradients do not occur also in nondividing cells. This study confirms previous evidence for the limited diffusivity of Ca^{2+} in the cytoplasm (Rose and Loewenstein, 1975b).

From this variety of data one may conclude that cell-to-cell channels are sensitive to Ca^{2+} in the low micromolar range or less, but higher concentrations (10–100 μM) are necessary to close all the channels and cause complete cell-to-cell uncoupling. A possible reason for the apparent "Ca^{2+}-insensitivity" (Spray *et al.*, 1985) of the channels in internally perfused (Spray *et al.*, 1982) or ruptured (Oliveira-Castro and Loewenstein, 1971) cells is the loss of a soluble Ca^{2+}-intermediate from the cytoplasm. Indeed, in recent years the existence of a soluble intermediate (Johnston and Ramón, 1981) likely to be a CaM-like protein (Peracchia *et al.*, 1981, 1983; Peracchia and Girsch, 1985a) has been proposed and will be discussed below (Section 3.2.1).

3.1.2. Hydrogen Ions

Evidence for the rapid development of cell-to-cell electrical uncoupling in parallel with a decrease in pH_i was first reported by Turin and Warner (1977, 1980) in amphibian embryos. In these studies, H^+ ions were believed to affect channel gating independently from Ca^{2+} because no signs of contraction of the cortical cytoplasm, a Ca^{2+}-mediated phenomenon, were detected. The capacity of H^+ to uncouple independently from Ca^{2+} was demonstrated more directly by data from dimer cells, one of which was internally perfused with EGTA-Ca^{2+}

solutions buffered to different pH values (Spray *et al.*, 1982). A similar conclusion was reached by Reber and Weingart (1982) in mammalian heart cells in which, in contrast to other cell systems (Lea and Ashley, 1976; Rose and Rick, 1978; Rink *et al.*, 1980; Mullins *et al.*, 1983; Requena *et al.*, 1986), a decrease in pH_i causes a decrease in $[Ca^{2+}]_i$ (Hess and Weingart, 1980).

Spray *et al.* (1981) have proposed that cell-to-cell channel conductance is "a simple and sensitive function of intracellular pH" and that its changes are a direct effect of protons on channel macromolecules. This conclusion was based on the observation that the relation between pH_i and junctional conductance is well fitted by a Hill curve. Furthermore, there appeared to be no hysteresis and the effects of fast or slow pH_i changes were identical.

However, pH sensitivity varies considerably among cell systems. In amphibian embryos, the relationship between pH_i and junctional conductance (G_j), using the Hill equation $G_j = K^n/(K^n + [H^+]^n)$ where K is the apparent dissociation constant and n the Hill coefficient, gave a pK = 7.3 and $n = 4.5$, suggesting the presence of 4–5 cooperative proton sites (Spray *et al.*, 1981). In squid blastomere (Spray *et al.*, 1984b) and in *Chironomus* salivary gland cells (Obaid *et al.*, 1983) the pK was lower than 7.3, and much lower values were reported for molluscan neurons (pK $\approx$ 6.8; Bodmer and Spray, 1985), cultured cardiac myocytes (pK $\approx$ 6.8; White *et al.*, 1985), crayfish septate axons (pK = 6.75; Campos de Carvalho *et al.*, 1984), earthworm septate axons (pK = 6.5; Brink *et al.*, 1984; Verselis and Brink, 1984), and liver (pK = 6.3; Spray *et al.*, 1984a). The Hill curve was not as steep in mammalian embryos (n = 2–3) as in amphibian embryos and it was much less steep ($n \sim 1$) in mammalian heart fibers (Spray and Bennett, 1985). Moreover, in tunicate embryos both pK and n values differed between uncoupling and recoupling (Knier, personal communication; Knier *et al.*, 1986). Lowering pH_i from 7 to 6.6 caused only a 30% reduction in coupling in mammalian heart fibers (Reber and Weingart, 1982) and CO_2 exposure or NH_4Cl washout had no effect on longitudinal resistance in cat papillary muscle (Wojtczak, 1985). In addition, H^+ has little effect on heart healing-over (De Mello, 1983), does not change the junctional conductance of internally perfused crayfish axons (Johnston and Ramón, 1981), and, in the absence of CaM, does not alter the permeability of lens junction channels incorporated into liposomes (Girsch and Peracchia, 1985b). Thus, in spite of solid evidence for a tight link between $[H^+]_i$ and junctional conductance, it is still premature to conclude that conductance changes are a direct effect of H^+ on cell-to-cell gating structures.

3.1.3. Voltage

Furshpan and Potter (1959) produced the first clear example of direct electrical coupling between cells, by demonstrating a negligible delay in the transmission of electrotonic potentials at the giant-to-motor junctions of crayfish.

Curiously, this system was soon recognized in many respects as an exception rather than the rule. First of all, differently from virtually all other low-resistance junctions, the median-to-motor junctions are rectifying junctions, as they allow current to diffuse in one direction only; secondly, the rectifying property of these junctions results from their voltage-sensitive gates, a characteristic absent in other mature cells of vertebrates and invertebrates but present only in some embryonic cells (Spray *et al.*, 1985).

In amphibian embryonic cells (Spray *et al.*, 1979, 1981; Harris *et al.*, 1981) and *Fundulus* blastomeres (Spray and Bennett, 1985), channel gating is sensitive to transjunctional voltage (V_j) applied in either direction. In *Chironomus* salivary gland cells, gating is sensitive only to membrane potential (V_m), the junctional conductance increasing with polarization and decreasing with depolarization (Obaid *et al.*, 1983), and in squid blastomeres it is sensitive to both V_j and V_m (Spray *et al.*, 1984b).

Whether or not pH, Ca^{2+}, and voltage act on the same gate is still unclear. In amphibian embryos, Spray *et al.* (1984b) have shown that cytoplasmic acidification does not alter the voltage dependence of uncoupling, although the two uncouplers enhance the effects of one another. This would suggest two independent gating systems. However, in *Chironomus* salivary gland cells, Obaid *et al.* (1983) have reported that voltage (V_m) dependence is shifted in the depolarizing direction without change in slope by an increase in either $[H^+]_i$ or $[Ca^{2+}]_i$, indicating that V_m, H^+, and Ca^{2+} act on the same gating mechanism. A similar conclusion was reached by Giaume and Korn (1985) for the rectifying giant-to-motor junctions of crayfish; in these junctions, loss of voltage sensitivity (rectification) was observed in some experiments after brief periods of acidification, suggesting that also in this system voltage and protons may share a common site of action. Recently, a loss of voltage sensitivity has also been observed in these junctions after prolonged intracellular perfusion of the two axons (Jaslove and Brink, 1986). In this case the loss of rectification occurred in all experiments after internal perfusion and never recovered even after several hours, suggesting the possible washout of axoplasmic gating intermediates (Jaslove and Brink, 1986). These results are quite intriguing because they raise questions, at least in crayfish rectifying junctions, about the directness of voltage gating as well.

3.1.4. Other Uncouplers

Aside from Ca^{2+}, H^+, and voltage, several divalent and trivalent cations such as Mg^{2+}, Mn^{2+}, Sr^{2+}, Ba^{2+} (Oliveira-Castro and Loewenstein, 1971; Peracchia and Peracchia, 1980a; Noma and Tsuboi, 1987), Co^{2+} (Politoff *et al.*, 1974), and La^{3+} (De Mello, 1979), as well as glutaraldehyde (Bennett *et al.*, 1972), anesthetics such as alkanols [heptanol and octanol, (Johnston *et al.*, 1980; Bernardini *et al.*, 1984)] and halothane (Hauswirth, 1968; Wojtczak, 1985), and cAMP (Teranishi *et al.*, 1983; Piccolino *et al.*, 1984; Lasater and Dowling,

1985) have been suggested to cause uncoupling by affecting directly the junctional gating mechanism. In most cases, however, evidence for a direct uncoupling effect is still inconclusive.

Johnston and Ramón (1981) proposed an extracellular site of action for alkanols in view of the fact that uncoupling only followed their extracellular application. However, in view of the capacity of these anesthetics to diffuse rapidly across lipid bilayers, one may wonder why the presumed extracellular site should be out of reach after intracellular injection. Recently Requena *et al.* (1985) have shown an increase in Ca^{2+} following octanol application to squid axons. In cardiac muscle, however, uncoupling by halothane is not inhibited by the CaM inhibitor trifluoperazine (TFP) and both halothane and octanol have a negative inotropic effect, indicative of a decrease in $Ca^{2+}{}_i$ (Wojtczak, 1985).

The effects of cAMP are poorly understood. An increase in cAMP causes uncoupling in horizontal cells of the retina (Teranishi *et al.*, 1983; Piccolino *et al.*, 1984; Lasater and Dowling, 1985) but in many other cells it enhances coupling either rapidly (De Mello, 1984; Saez *et al.*, 1986) or after a long delay (Flagg-Newton *et al.*, 1981), the latter effect being the result of an increase in the number of junctional particles (cell-to-cell channels).

In pairs of horizontal cells studied with whole cell clamp, the anion composition of the patch pipette was found to have an effect on both coupling and cAMP-induced uncoupling. In particular, the presence of F^- resulted in good electrical coupling and weak uncoupling effects of dopamine (DA), while with aspartate, gluconate, or Cl^- coupling was weak. A possible interpretation of these results is that high concentrations of F^- maintain a low $[Ca^{2+}]_i$, and thus good electrical coupling. Consequently, cAMP may cause only a moderate increase in $[Ca^{2+}]_i$ in the presence of F^-, resulting in weak uncoupling effects.

3.2. Uncoupling Intermediates

The possible participation of a soluble intermediate in the mechanism of cell-to-cell channel gating was proposed by Johnston and Ramón (1981) to explain the lack of H^+- or Ca^{2+}-induced uncoupling in internally perfused crayfish septate axons. Independently, Peracchia *et al.* (1981, 1983) have proposed that a CaM-like protein could be an uncoupling intermediate. This hypothesis was based on the observed inhibitory effect of TFP, a CaM inhibitor, on electrical uncoupling of amphibian embryonic cells and data for the binding of CaM to lens and liver junctional proteins (Welsh *et al.*, 1981, 1982; Hertzberg and Gilula, 1981).

3.2.1. Inhibition of Electrical Uncoupling by CaM Inhibitors

The hypothesis suggesting the involvement of CaM-like proteins as mediators of the action of some uncouplers on the channel gates was tested in amphibian embryonic cells and crayfish axons using various CaM inhibitors. In *Xenopus* and *Rana* embryos at the 16- to 64-cell stage, both TFP (Peracchia *et al.*, 1981, 1983) and calmidazolium (R24571; CDZ) (Peracchia, 1984) (Figs. 5 and 6) strongly inhibited CO_2-induced uncoupling.

With 50 μM TFP, 15- to 20-min treatments with 100% CO_2 had virtually no effect on the coupling ratio (V_2/V_1), while in controls even CO_2 treatments as short as 5 min lowered the coupling ratio from 0.9 to less than 0.01. The TFP-induced inhibition of uncoupling was partially reversible. A lower TFP concentration (20 μM) also caused a significant and partially reversible uncoupling inhibition. At this concentration exposure to 100% CO_2 only lowered the coupling ratio by ~ 50%.

Similar results were obtained using CDZ (Fig. 5), a more specific CaM inhibitor. An inhibition of CO_2-induced uncoupling was seen after 30, 50, and 60 min of treatment with 100, 70, and 50 nM CDZ, respectively (Fig. 6). After 72 min of treatment with 100 nM CDZ, a 6-min exposure to 100% CO_2 only lowered the coupling ratio by 40%, while during the control period a 5-min exposure to 100% CO_2 lowered the coupling ratio by 96%. The CDZ effect was also only partially reversible and neither CDZ nor TFP had a sizable effect of nonjunctional membrane resistance. This was shown indirectly by the absence of any obvious increase in the amplitude of the electrotonic potentials (V_1 and V_2), and thus of input and transfer resistances, and more directly, by the absence of significant changes in the membrane resistance of individual embryonic cells, isolated from 16- to 64-cell embryos, even after long exposures to 100 nM CDZ.

Recently, the effects of another CaM inhibitor,[*N*-(6-aminohexyl)-5-chloro-1-naphthalene-sulfonamide] (W7) and, as control, of its nonchlorinated form (W5) which is only 20% effective in inhibiting CaM (Hidaka *et al.*, 1981), were studied in septate axons of crayfish (Peracchia, 1986, 1987) (Fig. 9). These neurons provide a two-cell system where, with a four-microelectrode setup, both the junctional and the surface membrane resistances of both cells can be accurately and independently measured, and thus provide a way to quantitate the effectiveness of uncoupling treatments on junctional communication.

In this study, four microelectrodes with a final resistance of 8–12 MΩ were inserted into a lateral giant axon, two on each side of the septum (Fig. 7A). Hyperpolarizing current pulses were injected through current microelectrodes (I_1, I_2) at 10-sec intervals alternatively into the caudal (C_1) and the rostral (C_2) axonal segment (Fig. 7B,C). The resulting electrotonic potentials V_1 and V_2 (from current injection in C_1), V_{1*} and V_{2*} (from current injection in C_2), and the membrane potentials (E_1 and E_2) were recorded with two voltage microelectrodes. The surface membrane (R_{m_1}, R_{m_2}) and junctional (R_{j_1}, R_{j_2}) resistances were calculated from current (I_1, I_2) and voltage (V_1, V_2, V_{1*}, V_{2*}) records (Fig. 8) using the relationships described by Watanabe and Grundfest (1961) and Bennett (1966).

Briefly superfusing crayfish axons with acetate saline (Ac) caused an in-

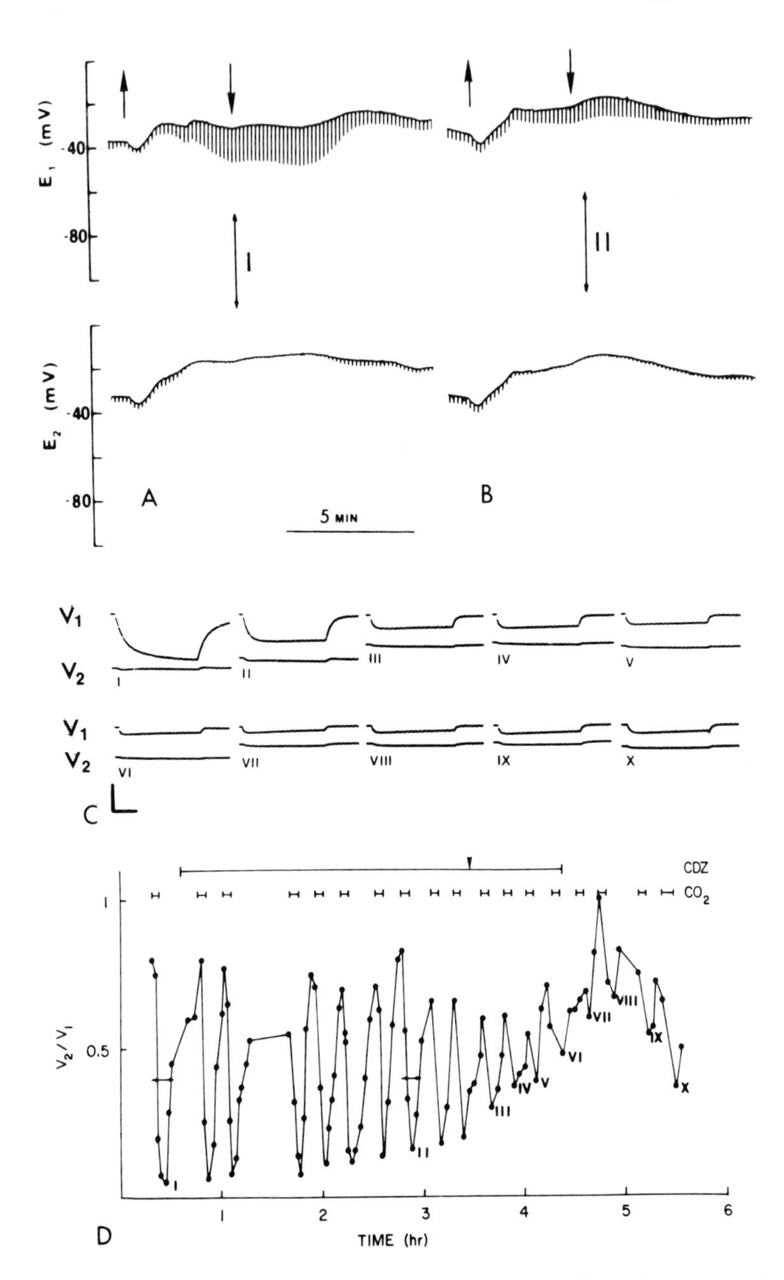

FIGURE 5. Effects of 50 and 100 nM calmidazolium (R24571; CDZ) on CO_2-induced electrical uncoupling in two adjacent cells of a *Xenopus* embryo (morula stage). A and B show low-speed chart recordings of membrane (E_1, E_2) and electrotonic (V_1, V_2) potentials in cell 1 and 2, respectively, during the periods indicated in D by the double-headed arrows (left without CDZ and right with CDZ). The large arrows mark beginning (↑) and end (↓) of exposures to 100% CO_2, and the double-headed

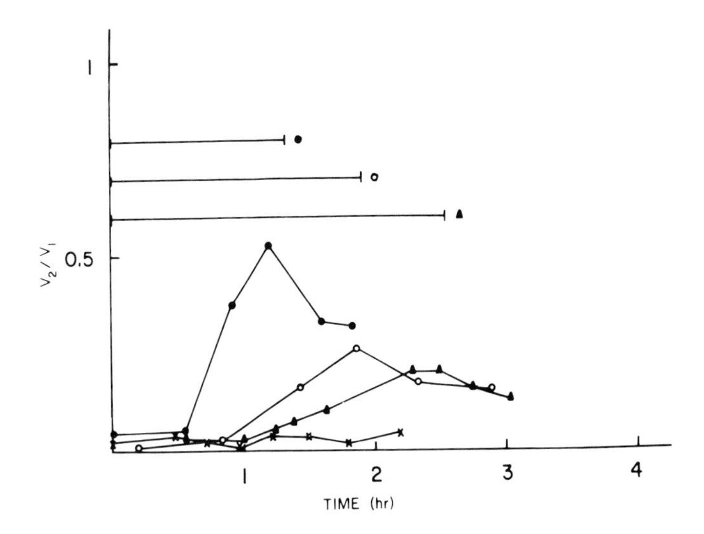

FIGURE 6. Time course of minimum coupling ratios (V_2/V_1) measured between neighboring *Xenopus* embryo cells upon exposure to 100% CO_2 in the presence of 100 nM (●), 70 nM (○), 50 nM (▲) CDZ and in the absence of CDZ (x). Duration of CDZ treatments is indicated by corresponding horizontal bars. From Peracchia (1984).

crease in V_1 and V_{2*} and a decrease in V_2 and V_{1*}, reflecting a rapid increase in R_j. W7 (100 μM) strongly inhibited the increase in R_j caused by Ac (Fig. 9). Inhibition developed rapidly, but recovery was slow and usually only partial. In contrast to the strong inhibitory effect of 100 μM W7, similar concentrations of W5 were completely ineffective (Fig. 9). Both W7 and W5 caused an increase in R_m (Fig. 9), but this effect was usually transient; it affected mostly or only one axonal segment (Fig. 9B) and had no relationship to the development of uncoupling inhibition.

CaM inhibitors have been shown to affect coupling regulation also in other systems. TFP has been shown to prevent the increase in junctional resistance of

---

arrows labeled with roman numerals refer to the corresponding oscilloscope recordings (I and II) shown in C. In D is shown the time course of the coupling ratio (V_2/V_1) sampled from the continuous chart recording shown in part in A and B. The bars represent the periods of exposure to 100% CO_2 and CDZ. The location of the oscilloscope potentials seen in C is indicated in D (with roman numerals). With 50 nM CDZ (D, left of arrowhead), inhibition of CO_2 uncoupling starts 60 min after the beginning of the treatment and proceeds fairly linearly at a slow rate (D). Upon increase to 100 nM CDZ (D, arrowhead), the rate of inhibition increases sharply such that in 55 min, V_2/V_1 decreases with CO_2 only to 0.48 [C(VI) and D(VI)]. After CDZ removal, the inhibition continues for approximately 30 min, then decreases sharply. In C, vertical bar = 10 mV, horizontal bar = 100 msec. From Peracchia (1984).

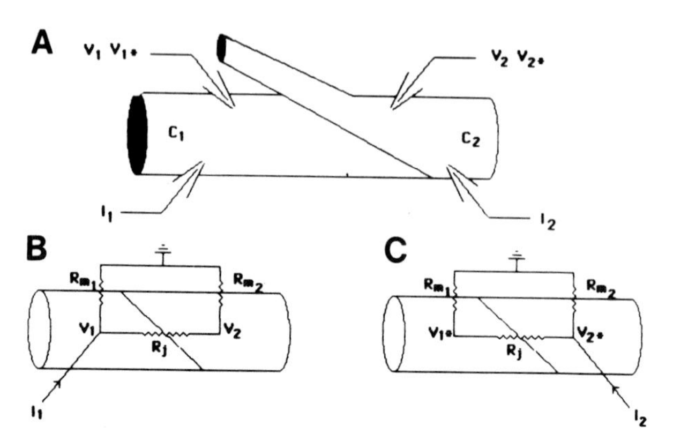

FIGURE 7. Diagram of electrode position (A) and equivalent circuit (B, C) at a septum of a crayfish lateral giant axon. Hyperpolarizing current pulses injected through current microelectrodes (I_1, I_2) alternatively in C_1 and C_2 (A) give rise to electrotonic potentials V_1, V_2 (B) and V_2*, V_1* (C), respectively. Membrane (R_{m_1}, R_{m_2}) and junctional (R_j) resistances (B and C) are calculated from the current and voltage records.

cardiac cells by the Ca^{2+}-loading solutions and CDZ inhibits the Ca^{2+}-induced healing-over in the same cells (Wojtczak, 1985). CDZ and chlorpromazine prevent the uncoupling of myometrial cells induced by the Ca^{2+} ionophore A23187 (Cole and Garfield, 1985). On the other hand, CDZ and chlorpromazine induce metabolic uncoupling in myometrium and tissue culture cells (Cole and Garfield, 1985) and high concentrations of both TFP and chlorpromazine prevent cell-to-cell dye spreading in insect cells (Lees-Miller and Caveney, 1982).

These studies, showing that CaM blockers affect electrical uncoupling in both vertebrates and invertebrates, indicate a CaM involvement in coupling regulation. However, one may question whether its involvement is direct or indirectly mediated by a cascade of events. A possibility is that CaM participates in a mechanism involving cyclic nucleotides and junctional protein phosphorylation. Following this hypothesis, one could suggest that the inhibitory effects of CaM blockers on electrical uncoupling are the result of an inhibition of phosphodiesterases and consequently of an increase in cyclic nucleotide concentration. This hypothesis is attractive, in view of recent data for cAMP involvement in cell-to-cell coupling. An increase in cAMP concentration has been found to improve coupling in some systems (Flagg-Newton *et al.*, 1981; De Mello, 1984) and to cause (Teranishi *et al.*, 1983; Piccolino *et al.*, 1984; Lasater and Dowling, 1985) or accelerate (Wojtczak, 1982) uncoupling in others.

To test the cyclic nucleotide hypothesis, the effects of an increase in the intracellular concentration of cyclic nucleotides were studied by superfusing septate axons with salines containing 50 μM–1 mM db-cAMP or db-cGMP (Peracchia, 1986, 1987). However, neither db-cAMP nor db-cGMP affected R_m,

R_j, or E in the presence or absence of Ac. db-cAMP also did not affect the inhibitory action of W7, when superfused prior to it. Thus, these negative results do not support an indirect CaM action via changes in cyclic nucleotides, nor a short term effect of cyclic nucleotides on crayfish cell coupling in general.

Alternatively, CaM-like proteins could be involved indirectly by participating in the phosphorylation of channel proteins, as several kinases are known to be CaM-dependent and there is evidence that junctional proteins can be phosphorylated (Garland and Russell, 1985; Louis *et al.*, 1985a; Johnson *et al.*, 1986; Saez *et al.*, 1986). On the other hand, there are data for a direct action of CaM on the channel proteins. Aside from evidence for binding of CaM to vertebrate lens and liver junctional protein and to crayfish hepatopancreas junctional protein (Hertzberg and Gilula, 1981; Welsh *et al.*, 1981, 1982; van Eldik *et al.*, 1985; van den Eijnden-van Raaij *et al.*, 1985), there is evidence for CaM participation in the gating mechanism of lens junctional channels incorporated in liposomes (Girsch and Peracchia, 1985b).

3.2.2. CaM Participation in Gating of Junctional Channels Reconstituted into Liposomes

To test the possible CaM involvement in gating reconstituted junctional channels, Girsch and Peracchia (1983, 1985b) have incorporated into liposomes the lens junctional protein (MIP26) in the presence of CaM at equimolar concentration and have studied the permeability and gating of the resulting channels spectrophotometrically, as previously described (Section 2.4).

Without added Ca^{2+}, MIP26–CaM liposomes swelled, following rapid initial shrinkage, when suspended in hypertonic KCl, sucrose (Fig. 11), or PEG, with an initial swelling rate only slightly lower than that of MIP26 liposomes (Fig. 10). In the presence of Ca^{2+} (100 μM), MIP26–CaM liposomes did not swell in any of the probes, indicating complete channel closure (Figs. 10 and 11). Addition of 500 μM EGTA to Ca^{2+}-treated MIP26–CaM liposomes reinitiated swelling (Fig. 11), demonstrating the reversibility of the channel gating mechanism. Preliminary data showed that lowering the pH from 7.4 to 6.5 sizably reduces the permeability of MIP26–CaM channels and a further reduction to pH 5.5 completely closes the channels. Addition of CaM to the external medium of MIP26 liposomes had only a small effect on channel gating, while exposing both inner and outer surfaces of MIP26 liposomes to CaM, by loading them with T-10 and CaM, restored full gating competency. This indicated that CaM must be on both liposome membrane surfaces to provide effective gating, consistent with the idea that channel proteins are incorporated into liposomes from either surface. The bilateral protein incorporation was indeed suggested by freeze-fracture images of intramembrane particles on both liposome fracture surfaces.

These data indicate that the addition of CaM is sufficient to render func-

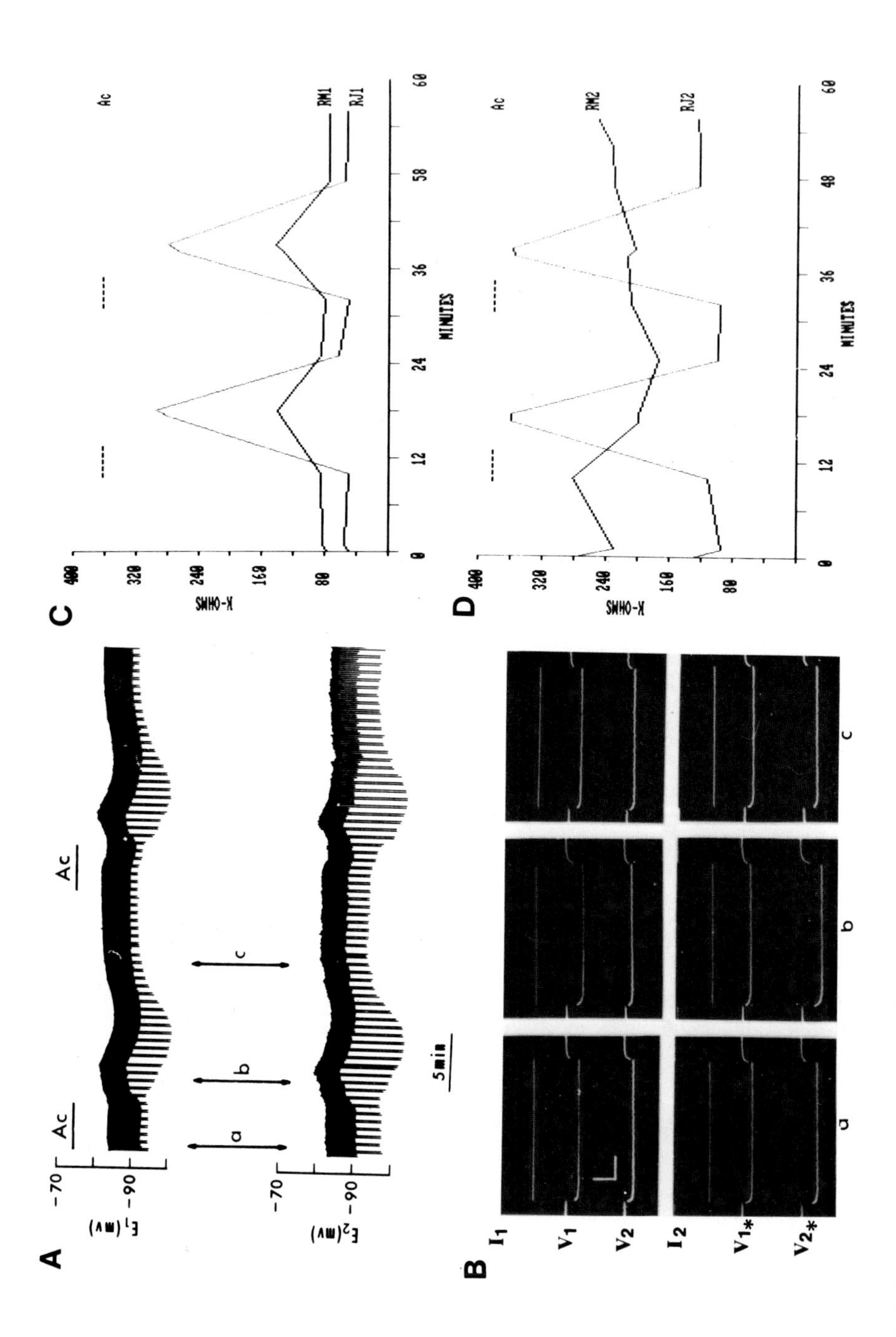
A
Ac
Ac
E_1(mV)
-70
-90
E_2(mV)
-70
-90
a
b
c
5 min
B
I_1
V_1
V_2
I_2
V_{1*}
V_{2*}
a
b
c
C
K-OHMS
400
320
240
160
80
0
12
24
36
48
60
MINUTES
Ac
RM1
RJ1
D
K-OHMS
400
320
240
160
80
0
12
24
36
48
60
MINUTES
Ac
RM2
RJ2

FIGURE 8. Effects of acetate saline solution (Ac) on electrical coupling at a septum between two segments of a crayfish lateral giant axons. (A) Low-speed chart recording of membrane (E_1, E_2) and electrotonic potentials in the caudal and rostral axon segment, respectively. Current is injected every 10 sec alternating caudal and rostral axon segments every 20 sec. The double-headed arrows labeled a, b, and c refer to the corresponding current (I_1, I_2) and voltage (V_1, V_2, $V_1{}^*$, $V_2{}^*$) oscilloscope records (B). (C, D) Time course of changes in membrane (R_{m_1}, R_{m_2}) and junctional (R_{j_1}, R_{j_2}) resistances. Following Ac superfusion, there is a reversible increase in V_1 and $V_2{}^*$ and a decrease in V_2 and $V_1{}^*$ (A and B) which reflect an increase in R_j (C and D). Note that the changes in R_j following similar Ac superfusions are consistent in magnitude (C and D). In B, vertical bar = 10 mV (= 100 nA for current records); horizontal bar = 20 msec.

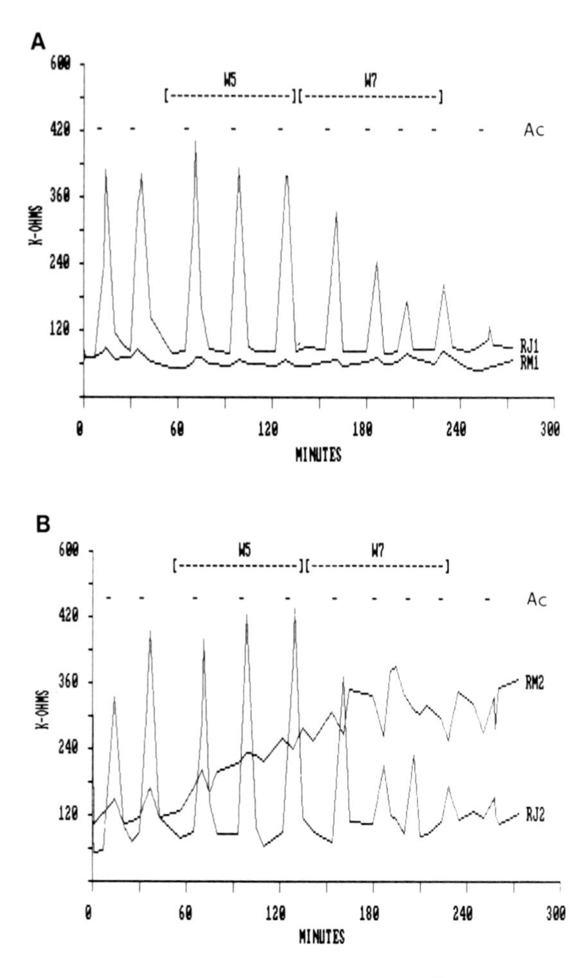

FIGURE 9. Effects of 100 μM W5 and W7 on Ac-induced uncoupling in septate axons. (A, B) Time course of changes in membrane (R_{m_1}, R_{m_2}) and junctional (R_{j_1}, R_{j_2}) resistances with Ac in the presence and absence of W5 or W7. Note that the Ac-induced increase in R_j is similar to controls, in the presence of W5 (Ac #3–5), but is gradually reduced more and more by W7 (Ac #6–9). In this experiment R_{m_2} increased progressively with W5 and W7 (B), while R_{m_1} did not change (A). In other experiments the changes in R_m were transient and less pronounced.

tional the gating mechanism of reconstituted lens channels. Thus, in this system, gating seems to require only MIP26, CaM, and phospholipids, in addition to uncoupling agents (Ca^{2+} or H^+), suggesting that all the gating information arises from tertiary and quaternary interactions of the two proteins and the lipid matrix.

Channel gating in lens junctions is likely to result from a conformational

FIGURE 10. Dependence of liposome swelling rate (channel permeation rate) on the size of solute in four liposome types. For MIP26 liposomes (1) the rate of swelling decreases linearly with the log of probe molecular weight with or without Ca^{2+} (100 μM). MIP21 liposomes (2) behave in a similar manner but with slower kinetics. MIP26–CaM (3) and MIP21–CaM (4) liposomes swell, but less rapidly than either type without CaM. Addition of Ca^{2+} closes the channels of MIP26–CaM liposomes (6) completely but only partially those of MIP21–CaM liposomes (5). From Peracchia and Girsch (1985c).

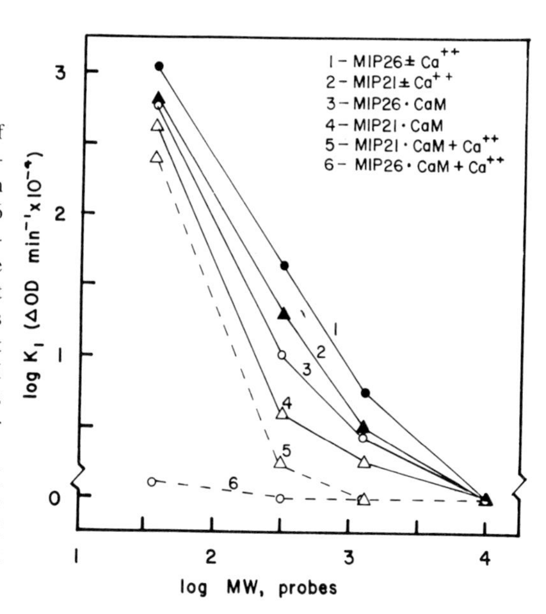

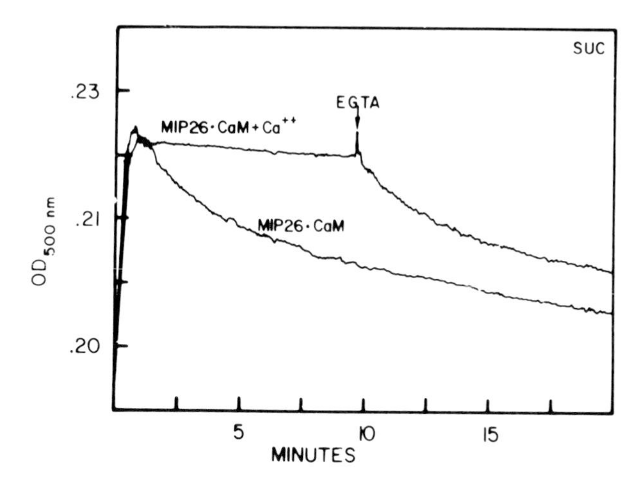

FIGURE 11. Ca^{2+}-induced gating of channels formed in liposomes by the incorporation of MIP26 in the presence of CaM. MIP26–CaM liposomes swell in 50% hyperosmotic sucrose (SUC) only in the absence of Ca^{2+}. Addition of 100 μM Ca^{2+} closes the channels. The Ca^{2+}-CaM gated channels reopen with addition of EGTA.

change in channel protein because changes in intrinsic fluorescence emission (ϕ_{em}) and circular dichroism (CD) were detected in MIP26–CaM complexes exposed to Ca^{2+} (Girsch and Peracchia, 1985c). MIP26 showed identical ϕ_{em} (342 nm, primarily tryptophan) and CD spectra with or without added Ca^{2+} (up to 0.1 mM). With 1 μM Ca^{2+} CaM showed a 32% increase in ϕ_{em} (322 nm, solely tyrosine) and a 44% increase in α-helicity. In the absence of Ca^{2+}, the MIP26–CaM complex displayed ϕ_{em} and CD spectra slightly different from the sum of the spectra of the two components, indicating an MIP26–CaM interaction, while in the presence of Ca^{2+} (> 1 μM), markedly different ϕ_{em} and CD spectra were obtained. The fluorescence emission changes included a 5-nm blueshift in λ_{max} and a 50% increase in tyrosine ϕ_{em}, which occurred about an isosbestic point ($\lambda_{max} = 370$ nm). The observed CD changes of the complex were characterized by a 5% increase in α-helicity over and above that calculated for the sum of the two components, if no interaction is assumed. All the changes were fully reversible.

The data from fluorescence and CD spectra indicate that CaM is able both to interact with MIP26 in the absence of Ca^{2+}, confirming previous evidence of its ability to bind to MIP26 in gel overlay experiments (Welsh *et al.*, 1981, 1982; Hertzberg and Gilula, 1981; van Eldik *et al.*, 1985; van den Eijnden-van Raaij *et al.*, 1985), and to change the MIP26 conformation in the presence of Ca^{2+}.

In contrast to these results are data showing the apparent inability of [^{125}I]-CaM to bind to MIP26 in isolated lens membrane fragments (Louis *et al.*, 1985b). In this study [^{125}I]-CaM was cross-linked to lens membranes and the lens membrane proteins were isolated by gel electrophoresis. Electrophoretic analysis showed two major ^{125}I-containing products of M_r 49,000 and 36,000, believed to represent a combination of CaM (M_r 17,000) with the M_r 32,000 and 19,000 components, respectively. However, the possibility that [^{125}I]-CaM did not bind to MIP26 because the binding site was already occupied by intrinsic CaM was not taken into consideration. Nevertheless, evidence for CaM binding to gap junctions in intact cells is still lacking.

3.2.3. What Makes the Channel Gate?

There are many ways in which a membrane channel may be gated. Hille (1984) has proposed 11 possible mechanisms for channel gating, but certainly even more models could be drawn. Unfortunately, we still have only a vague idea on how any biological channel may open and close. While it seems likely that tetrodotoxin closes the Na^+ channels by plugging their extracellular end, other mechanisms may involve protein conformational changes, translated into "swinging doors," "slider movements," or "subunit twisting," as well as assembly–disassembly of subunits or swinging of channels all the way in and out of the membrane (Hille, 1984).

In gap junctions, the molecular events that result in channel closure are still unclear. Recent data from low-dose microscopy and x-ray diffraction (Unwin and Ennis, 1983, 1984), showing Ca^{2+}-induced changes of subunit orientation in isolated liver gap junctions, may support the model of major protein conformational change. However, in these studies the relationship between subunit movement and channel patency was not established and no structural change was detected with lowered pH.

Recent studies on the molecular structure of MIP26 and the M_r 28,000 liver junctional protein indicate that these proteins have both N- and C-terminal arms on the cytoplasmic side of the membrane, the C-terminal arm being the major side chain (Nicholson *et al.*, 1981, 1983; Gorin *et al.*, 1984). Peracchia and Girsch (1985b) have tested the possiblity that the C-terminal arm participates in channel gating by studying the gating properties of channels reconstituted into liposomes from either the intact lens junctional protein (MIP26) or its trypsin-cleaved product (MIP21), an M_r 21,000 component that has lost most of its C-terminal arm (Fig. 2). Channel permeability to sucrose and PEG and channel gating by Ca^{2+} in the presence or absence of CaM were studied spectrophotometrically as previously described (Section 2.4).

Liposomes in which MIP21 was incorporated, swelled when exposed to hypertonic sucrose or PEG, both in the presence and absence of Ca^{2+} (up to 100 μM), indicating the presence of large water-filled channels permeable to molecules as heavy as ~ 1500 daltons (PEG). However, the swelling rates of MIP21-containing liposomes were lower than those containing MIP26 (Fig. 10), either because of less efficient incorporation or because MIP21 channels are less permeable than those of MIP26.

Without Ca^{2+}, the swelling rates of MIP26–CaM and MIP21–CaM liposomes were only slightly lower than those of liposomes incorporated with MIP26 or MIP21 alone (Fig. 10). Upon addition of Ca^{2+} (100 μM), MIP26–CaM liposomes did not swell in either sucrose (Fig. 11 and 13) or PEG (Fig. 12), indicating that the channels became completely impermeable to both molecules. In contrast, similar Ca^{2+} additions to MIP21–CaM liposomes prevented swelling in PEG (Fig. 12) but not in sucrose (Fig. 13), indicating that the channels close only partially, becoming impermeable to PEG while remaining permeable to a smaller molecule (sucrose). A partial occlusion of the channels is also suggested by the fact that the swelling rate of MIP21–CaM liposomes in sucrose decreases upon Ca^{2+} addition (Fig. 10).

These data indicate that the removal of the C-terminal arm from the protein subunits of lens channels impairs the channel gating mechanism, such that the channels close only partially with Ca^{2+}–CaM. One may suggest that the 7k C-terminal arm represents the major portion of the channel gate and that channel occlusion may result from a Ca^{2+}–CaM-induced conformational change in MIP26 causing the movement of the C-terminal arm toward the channel opening (Fig. 14).

Data showing an inhibition of the H^+-induced electrical uncoupling by a

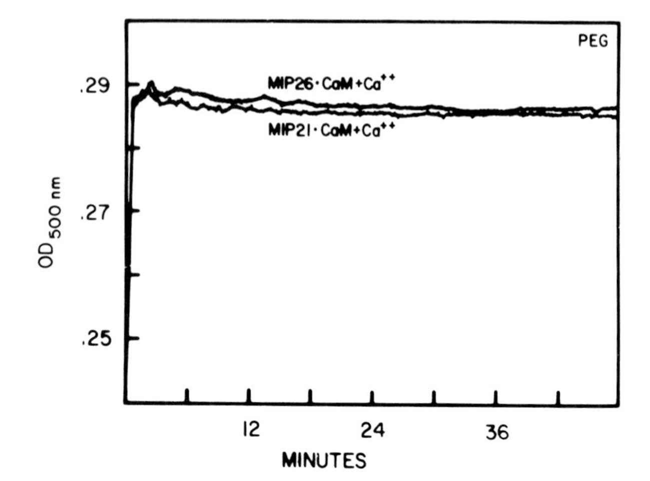

FIGURE 12. Ca^{2+}-induced gating of channels formed in liposomes by the incorporation of either MIP26 or MIP21 in the presence of CaM. Neither MIP26–CaM nor MIP21–CaM liposomes swell in 50% hyperosmotic polyethylene glycol (PEG) in the presence of 100 μM Ca^{2+}, indicating complete channel closure. From Peracchia and Girsch (1985c).

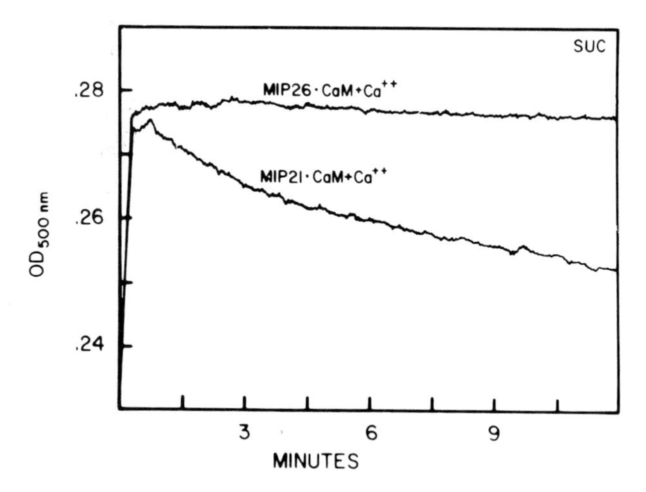

FIGURE 13. Ca^{2+}-induced gating of channels formed in liposomes by the incorporation of either MIP26 or MIP21 in the presence of CaM. In the presence of Ca^{2+} (100 μM) only MIP26–CaM liposomes do not swell in 50% hyperosmotic sucrose (SUC), indicating that the MIP21–CaM channels close only partially, becoming impermeable to the larger probe (PEG, Fig. 12) while remaining permeable to SUC. From Peracchia and Girsch (1985c).

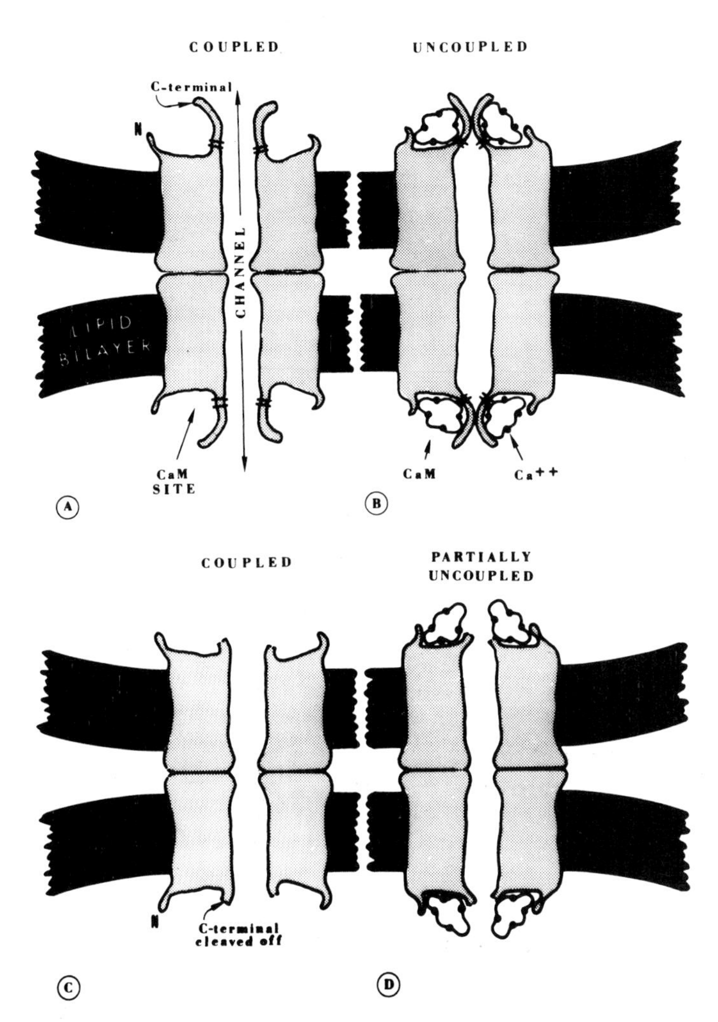

FIGURE 14. Hypothetical model of channel gating via CaM. The channel protein is believed to have CaM sites (A). CaM would bind to the sites and cause protein conformational changes resulting in channel occlusion (uncoupled) near the cytoplasmic end (B). Channel occlusion would result from a movement of the trypsin-cleavable C-terminal component (the gate) into the channel lumen (B). This is suggested by the observation that the channels of liposomes incorporated with MIP21 (C), the trypsin-cleaved lens channel protein which has lost the 5–7k C-terminal component, lose part of their gating properties, closing only partially with Ca^{2+}–CaM (D). With Ca^{2+}–CaM the MIP21 channels become impermeable to polyethylene glycol (M_r 1500), while remaining permeable to sucrose (M_r 342). From Peracchia and Girsch (1985a).

carboxyl reagent (Spray *et al.*, 1984b) are consistent with the hypothesis of an involvement of the C-terminal arm in channel gating. Indeed, the location of the C-terminal arm in MIP26 is also consistent with its involvement in gating. A recent representation of MIP26 derived from cDNA sequence data (Gorin *et al.*, 1984) shows that the molecule traverses the lipid bilayer six times and has both N- and C-terminal arms on the cytoplasmic side. The C-terminal arm appears to be very close to the channel opening because it is an extension of the only amphiphilic transmembrane segment, believed to constitute the aqueous surface structure of the channel. Moreover, the C-terminal arm contains serine and threonine residues in a configuration favorable to cAMP-dependent phosphorylation (Gorin *et al.*, 1984), and there is evidence both for a cAMP effect on channel gating (see Section 3.2.1) and for MIP26 phosphorylation (Garland and Russell, 1985; Louis *et al.*, 1985a). However, at present alternative interpretations of the phenomenon cannot be discarded. The fact that part of the channel gating property is still present, after C-terminal removal, only when CaM is present, indicates that CaM interaction with channel protein still takes place. However, CaM could be bound less tightly and/or inappropriately to MIP21, resulting in less efficient gating. Indeed, CaM has been shown to bind to MIP21 in gel overlay experiments (van Eldik *et al.*, 1985; van den Eijnden-van Raaij *et al.*, 1985).

It is interesting that also in the case of Na^+ channels a proteolytic enzyme with a specificity close to that of trypsin has been found to prevent gating (eliminate inactivation) (Armstrong *et al.*, 1973; Rojas and Rudy, 1976). In this system, however, the N-terminal peptide appears to be the cleavable gating structure (Armstrong and Bezanilla, 1977). In MIP26, only five amino acid residues are removed by trypsin from the N-terminus (Nicholson *et al.*, 1983), virtually all the cleavage occurring at the C-terminus. Thus, the possibility that the N-terminal arm participates in cell-to-cell channel gating cannot be completely ruled out yet.

3.2.4. Is CaM a Satisfactory Candidate for Uncoupling Intermediate?

While there are several reasons to believe that the gating action of uncouplers may be mediated by soluble intermediates, evidence that CaM is indeed the one and only intermediate is still inconclusive. As previously mentioned, evidence for the existence of uncoupling intermediates comes from a number of observations including: the apparent difference in sensitivity to Ca^{2+}, H^+, and, in one case, voltage between intact and ruptured or internally perfused cells; the variability in H^+ sensitivity between different cell systems, shown by differences in pK and n values of the Hill relationship between pH and junctional conductances; the difference (in one system) in n and pK values between uncoupling and recoupling process; the effects of CaM inhibitors on the uncoupling mechanism; the lack of Ca^{2+} and H^+ sensitivity of junctional channels incorporated into artificial membranes in the absence of CaM; the inability of Ca^{2+} and H^+ to induce conformational changes in isolated MIP26 proteins in the absence of CaM.

Experiments with isolated junctional components have shown that the presence of CaM is necessary for channel gating and have demonstrated that CaM binds to the proteins of mammalian lens and liver gap junctions as well as to those of crayfish hepatopancreas gap junctions. However, there is evidence that other CaM-like proteins can substitute for CaM in gating MIP26 channels in liposomes (Peracchia and Girsch, 1985a) and are able to bind to junctional proteins from different species (van Eldik *et al.*, 1985). Moreover, electrophysiological experiments testing the effects of CaM inhibitors on uncoupling in intact cells could not entirely rule out the possibility that intermediates other than CaM are involved, as none of the inhibitors is absolutely specific for CaM.

The ability of TFP, CDZ, and W7 to inhibit uncoupling by acidifying agents such as CO_2 and acetate is indeed puzzling as there is no direct evidence that CaM can be activated by H^+. On the other hand, there are data indicating that CaM-like proteins are affected by lowered pH. In troponin C, conformational and tyrosyl fluorescence emission changes similar to those caused by Ca^{2+} result from lowering the pH below 6.5 (Lehrer and Leavis, 1974) and similar changes have been reported for CaM (Steiner *et al.*, 1983). Recently, Pundak and Roche (1984) have shown that Tyr-99 of CaM exhibits a pH-driven tyrosine to tyrosinate transition which alters the binding of Ca^{2+} to CaM. This indicates the presence of a pH-dependent fine tuning of Ca^{2+}–CaM-modulated processes in cell regulation. Similarly, Blumenthal and Stull (1982) have shown that the activation of myosin light chain kinase by Ca^{2+}–CaM has a pH optimum of 6.5–6.7, activation being inhibited below pH 6 and above pH 7.5. Furthermore, Calhoon and Gillette (1983) have reported that the Ca^{2+}–CaM-dependent activity of a cAMP phosphodiesterase of molluscan neurons is substantially inhibited above pH 7.5. In this system pH also affects the ratio of soluble to total CaM, indicating that $[H^+]$ affects CaM binding (Wilson and Gillette, 1985).

In view of these data, while CaM should be kept in consideration as a possible intermediate of channel gating, search for other possible candidates, with a sensitivity to Ca^{2+} and H^+ in the range believed to affect junctional conductance, should continue. An intriguing possibility is that a variety of CaM-like proteins substitute for each other in modulating junctional conductance. Indeed, troponin C substitutes well for CaM in gating lens junctional channels reconstituted into liposomes (Peracchia and Girsch, 1985a) and various CaM-like proteins such as troponin C, bovine brain $s100_\alpha$ and $s100_\beta$ components, parvalbumin, and the vitamin D-dependent calcium-binding protein bind as well as CaM to both liver and lens junctional proteins in gel overlay experiments (van Eldik *et al.*, 1985). Interestingly, the binding of troponin C, $S100_\alpha$, and $S100_\beta$

shows no calcium dependence of binding to either the liver or lens protein, while the binding of parvalbumin and the vitamin D-dependent calcium-binding protein is enhanced in the presence of EDTA.

Following this hypothesis, one could conceive of a mechanism for modulating junctional communication based on competition among various CaM-like proteins for binding sites at the channel proteins. Differences in pH and Ca^{2+} sensitivity among CaM-like proteins could explain the variable H^{+} and Ca^{2+} sensitivity of uncoupling reported in different cell systems.

ACKNOWLEDGMENTS. The author wishes to thank Lillian L. Peracchia for her help in preparing the manuscript. The work of the author was supported by NIH Grant GM 20113.

4. REFERENCES

Armstrong, C. M., and Bezanilla, F., 1977, Inactivation of the sodium channel. II. Gating current experiments, *J. Gen. Physiol.* **70:**567–590.

Armstrong, C. M., Bezanilla, F., and Rojas, E., 1973, Destruction of sodium conductance activation in squid axons perfused with pronase, *J. Gen. Physiol.* **62:**375–391.

Atkinson, M. M., and Sheridan, J. D., 1985, Reduced junctional permeability in cells transformed by different viral oncogenes, in: *Gap Junctions* (M. V. L. Bennett and D. C. Spray, eds.), pp. 205–213, Cold Spring Harbor Laboratory, Cold Spring Harbor, N.Y.

Azarnia, R., Larsen, W., and Loewenstein, W. R., 1974, The membrane junctions in communicating and non-communicating cells, their hybrids and segregants, *Proc. Natl. Acad. Sci. USA* **71:**880–884.

Bennett, M. V. L., 1966, Physiology of electrotonic junctions, *Ann. N.Y. Acad. Sci.* **137:**509–539.

Bennett, M. V. L., Spira, M. E., and Pappas, G. D., 1972, Properties of electrotonic junctions between embryonic cells of Fundulus, *Dev. Biol.* **29:**419–435.

Bernardini, G., and Peracchia, C., 1981, Gap junction crystallization in lens fibers after an increase in cell calcium, *Invest. Ophthalmol. Vis. Sci.* **21:**291–299.

Bernardini, G., Peracchia, C., and Venosa, R. A., 1981, Healing over in rat crystalline lens, *J. Physiol. (London)* **320:**187–192.

Bernardini, G., Peracchia, C., and Peracchia, L., 1984, Reversible effects of heptanol on gap junction structure and cell-to-cell electrical uncoupling, *Eur. J. Cell Biol.* **34:**307–312.

Blumenthal, D. K., and Stull, J. T., 1982, Effects of pH, ionic strength, and temperature on activation by calmodulin and catalytic activity of myosin light chain kinase, *Biochemistry* **22:**2386–2391.

Bodmer, R., and Spray, D. C., 1985, Permeability and electrophvsiological properties of Aplysia neurons *in situ* and in culture, *Biophys. J.* **47:**504a.

Brink, P. R., Verselis, V., and Barr, L., 1984, Solvent–solute interactions within the nexal membrane, *Biophys. J.* **45:**121–124.

Calhoon, R. D., and Gillette, R., 1983, Ca^{++} activated and pH sensitive cyclic AMP phosphodiesterase in the nervous system of the mollusc pleurobranchaea, *Brain Res.* **271:**371–374.

Campos de Carvalho, A. C., Spray, D. C., and Bennett, M. V. L., 1984, pH dependence of transmission of electrotonic synapses of the crayfish septate axon, *Brain Res.* **321:**279–286.

Cole, W. C., and Garfield, R. E., 1985, Alterations in coupling in uterine smooth muscle, in: *Gap*

Junctions (M. V. L. Bennett and D. C. Spray, eds.), pp. 215–230, Cold Spring Harbor Laboratory, Cold Spring Harbor, N.Y.

Coronado, R., and Latorre, R., 1983, Phospholipid bilayers made from monolayers on patch-clamp pipettes, *Biophys. J.* **43:**231–236.

DeHaan, R. L., and Hirakow, R., 1972, Synchronization of pulsation rates in isolated cardiac myocytes, *Exp. Cell Res.* **70:**214–220.

Délèze, J., 1964, Calcium ions and the healing-over of heart fibers, in: *Electrophysiology of the Heart* (B. Taccardi and C. Marchetti, eds.), p. 147, Pergamon Press, Elmsford, N.Y.

Délèze, J., and Loewenstein, W. R., 1976, Permeability of a cell junction during intracellular injection of divalent cations, *J. Membr. Biol.* **28:**71–86.

De Mello, W. C., 1975, Effect of intracellular injection of calcium and strontium on cell communication in heart, *J. Physiol. (London)* **250:**231–245.

De Mello, W. C., 1979, Effect of intracellular injection of La^{3+} and Mn^{2+} on electrical coupling of heart cells, *Cell Biol. Int. Rep.* **3:**113–119.

De Mello, W. C., 1983, The influence of pH on the healing-over of mammalian cardiac muscle, *J. Physiol. (London)* **339:**299–307.

De Mello, W. C., 1984, Effect of intracellular injection of cAMP on the electrical coupling of mammalian cardiac cells, *Biochem. Biophys. Res. Commun.* **119:**1001–1007.

DeWeer, P., 1978, Intracellular pH transients induced by CO_2 on NH_3, *Respir. Physiol.* **33:**41–50.

Flagg-Newton, J. L., Dahl, G., and Loewenstein, W. R., 1981, Cell junction and cyclic AMP. 1. Upregulation of junctional membrane permeability and junctional membrane particles by administration of cyclic nucleotide or phosphodiesterase inhibitor, *J. Membr. Biol.* **63:**105–121.

Furshpan, E. J., and Potter, D. D., 1959, Transmission of the giant synapse of the crayfish, *J. Physiol. (London)* **145:**289–325.

Garland, D., and Russell, P., 1985, Phosphorylation of lens fiber cell membrane proteins, *Proc. Natl. Acad. Sci. USA* **82:**653–657.

Giaume, C., and Korn, H., 1985, Junctional voltage-dependence at the crayfish rectifying synapse, in: *Gap Junctions* (M. V. L. Bennett and D. C. Spray, eds.), pp. 367–379, Cold Spring Harbor Laboratory, Cold Spring Harbor, N.Y.

Gilula, N. B., Reeves, O. R., and Steinbach, A., 1972, Metabolic coupling, ionic coupling and cell contacts, *Nature* **235:**262–265.

Girsch, S. J., and Peracchia, C., 1983, Lens junction protein (MIP26) self-assembles into liposomes forming large channels regulated by calmodulin (CaM), *J. Cell Biol.* **97:**83a.

Girsch, S. J., and Peracchia, C., 1985a, Liposome-incorporated liver gap junction channels are less permeable than lens channels, *Biophys. J.* **47:**507a.

Girsch, S. J., and Peracchia, C., 1985b, Lens cell-to-cell channel proteins. I. Self-assembly into liposomes and permeability regulation by calmodulin, *J. Membr. Biol.* **83:**217–225.

Girsch, S. J., and Peracchia, C., 1985c, Lens cell-to-cell channel protein. II. Conformational changes in the presence of calmodulin, *J. Membr. Biol.* **83:**227–233.

Girsch, S. J., Shrager, P., and Peracchia, C., 1986, Channel formation following incorporation of lens junction protein (MIP26) in lipid bilayers on patch pipettes, *Proc. Int. Soc. Eye Res.* **4:**128.

Gooden, M., Rintoul, D., Takehana, M., and Takemoto, L., 1985, Major intrinsic polypeptide (MIP26K) from lens membrane: Reconstitution into vesicles and inhibition of channel-forming activity by peptide antiserum, *Biochem. Biophys. Res. Commun.* **128:**993–999.

Gorin, M. B., Yancey, S. B., Cline, J., Revel, J.-P., and Horwitz, J., 1984, The major intrinsic protein (MIP) of the bovine lens fiber membrane: Characterization and structure based upon cDNA cloning, *Cell* **39:**49–59.

Goshima, K., 1969, Synchronized beating of and electrotonic transmission between myocardial cells, mediated by heterotypic strain cells in monolayer culture, *Exp. Cell Res.* **58:**420–426.

Goshima, K., 1970, Formation of nexuses and electrotonic transmission between myocardial and FL cells in monolayer culture, *Exp. Cell Res.* **63:**124–130.

Green, C. R., and Severs, N. J., 1984, Gap junction connexon configuration in rapidly frozen myocardium and isolated intercalated disks, *J. Cell Biol.* **99:**453–463.

Hall, J. E., and Zampighi, G. A., 1985, Protein from purified lens junctions induces channels in planar bilayers, in: *Gap Junctions* (M. V. L. Bennett and D. C. Spray, eds.), pp. 177–189, Cold Spring Harbor Laboratory, Cold Spring Harbor, N.Y.

Harris, A. L., Spray, D. C., and Bennett, M. V. L., 1981, Kinetic properties of a voltage-dependent junctional conductance, *J. Gen. Physiol.* **77:**95–117.

Hauswirth, O., 1968, Influence of halothane on electrical properties of cardiac Purkinje fibers, *J. Physiol. (London)* **201:**42P–43P.

Hertzberg, E. L., 1984, A detergent-independent procedure for the isolation of gap junctions from rat liver, *J. Biol. Chem.* **259:**9936–9943.

Hertzberg, E. L., and Gilula, N. B., 1981, Liver gap junctions and lens fiber junctions: Comparative analysis and calmodulin interaction, *Cold Spring Harbor Symp. Quant. Biol.* **46:**639–645.

Hertzberg, E. L., and Skibbens, R. V., 1984, A protein homologous to the 27,000 dalton liver gap junction protein is present in a wide variety of species and tissues, *Cell* **39:**61–69.

Hertzberg, E. L., Spray, D. C., and Bennett, M. V. L., 1985, Reduction of gap junction conductance by microinjection of antibodies against the 27-kDa liver gap junction polypeptide, *Proc. Natl. Acad. Sci. USA* **82:**2412–2416.

Hess, P., and Weingart, R., 1980, Intracellular free calcium modified by pH in sheep cardiac Purkinje fibers, *J. Physiol. (London)* **307:**60P–61P.

Hidaka, H., Sasaki, Y., Tamaka, T., Endo, T., Ohno, S., Fujii, Y., and Nagata, T., 1981, *N*-(6-Aminohexyl)-5-chloro-1-naphthalene-sulfonamide, a calmodulin antagonist, inhibits cell proliferation, *Proc. Natl. Acad. Sci. USA* **78:**4354–4357.

Hille, B., 1984, *Ionic Channels of Excitable Membranes,* Sinauer, Sunderland, Mass.

Hyde, A., Blondel, B., Matter, A., Cheneval, J. P., Filloux, B., and Girardier, L., 1969, Homo- and heterocellular junctions in cell cultures: An electrophysiological and morphological study, *Prog. Brain Res.* **31:**283–311.

Iwatsuki, N., and Petersen, O. H., 1978, Electrical coupling and uncoupling of exocrine acinar cells, *J. Cell Biol.* **79:**533–545.

Iwatsuki, N., and Petersen, O. H., 1979, Pancreatic acinar cells: The effect of carbon dioxide, ammonium chloride and acetylcholine on intercellular communication, *J. Physiol. (London)* **291:**317–326.

Jacob, T. J., 1983, Raised intracellular free calcium within the lens causes opacification and cellular uncoupling in the frog, *J. Physiol. (London)* **341:**595–601.

Jaslove, S. W., and Brink, P. R., 1986, The mechanism of rectification at the electronic motor giant synapse of the crayfish, *Nature* **323:**63–65.

Johnson, K. R., Lampe, P. D., Hurk, C., Louis, C. F., and Johnson, R. G., 1986, A lens intercellular junction protein, MP26, is a phosphoprotein, *J. Cell Biol.* **102:**1334–1343.

Johnston, M. F., and Ramón, F., 1981, Electrotonic coupling in internally perfused crayfish segmented axons, *J. Physiol. (London)* **317:**509–518.

Johnston, M. F., Simon, S. A., and Ramón, F., 1980, Interaction of anesthetics with electrical synapses, *Nature* **286:**498–500.

Keith, C. H., Ratan, R., Maxfield, F. R., Bajer, A., and Shelanski, M. L., 1985, Local cytoplasmic calcium gradients in living mitotic cells, *Nature* **316:**848–850.

Knier, J., Verselis, V., Spray, D. C., and Bennett, M. V. L., 1986, Gap junctions in tunicate embryos: pH and voltage gating mechanisms, *Biophys. J.* **49:**203a.

Lasater, E. M., and Dowling, J. E., 1985, Dopamine decreases conductance of the electrical junctions between cultured retinal horizontal cells, *Proc. Natl. Acad. Sci. USA* **82:**3025–3029.

Lawrence, T. S., Beers, W. H., and Gilula, N. B., 1978, Transmissions of the hormonal stimulation by cell-to-cell communication, *Nature* **272:**501–506.

Lea, T. J., and Ashley, C. C., 1978, Increase in free Ca^{++} in muscle after exposure to CO_2, *Nature* **275:**236–238.

Lee, W. M., Cran, D. C., and Lane, N. J., 1982, Carbon dioxide induces disassembly of gap junctional plaque, *J. Cell Sci.* **57:**215–228.

Lees-Miller, J. P., and Caveney, S., 1982, Drugs that block calmodulin activity inhibit cell-to-cell coupling in the epidermis of *Tenebrio molitor, J. Membr. Biol.* **69:**233–245.

Lehrer, S. S., and Leavis, P. C., 1974, Fluorescence and conformational changes caused by proton binding to troponin C, *Biochem. Biophys. Res. Commun.* **58:**159–165.

Loewenstein, W. R., Nakas, M., and Socolar, S. J., 1967, Junctional membrane uncoupling: Permeability transformation at a cell membrane junction, *J. Gen. Physiol.* **50:**1865–1891.

Louis, C. F., Johnson, R., Johnson, K., and Turnquist, J., 1985a, Characterization of the bovine lens plasma membrane substrates for cAMP-dependent protein kinase, *Eur. J. Biochem.* **150:**279–286.

Louis, C. F., Johnson, R., and Turnquist, J., 1985b, Identification of the calmodulin-binding components in bovine lens plasma membranes, *Eur. J. Biochem.* **150:**271–278.

Luckey, J., and Nikaido, H., 1980, Specificity of diffusion channels produced by phage receptor protein of *Escherichia coli, Proc. Natl. Acad. Sci. USA* **77:**167–171.

Makowski, L., 1985, Structural domains in gap junctions: Implications for the control of intercellular communication, in: *Gap Junctions* (M. V. L. Bennett and D. C. Spray, eds.), pp. 5–12, Cold Spring Harbor Laboratory, Cold Spring Harbor, N.Y.

Makowski, L., Caspar, D. L. D., Goodenough, D. A., and Phillips, W. C., 1982, Gap junction structures. III. The effects of variations in the isolation procedure, *Biophys. J.* **37:**189–191.

Makowski, L., Caspar, D. L. D., Phillips, W. C., and Goodenough, D. A., 1984, Gap junction structures. V. Structural chemistry inferred from X-ray diffraction measurements on sucrose accessibility and trypsin susceptibility, *J. Mol. Biol.* **174:**449–481.

Mazet, F., Dunia, I., Vassort, G., and Mazet, J. L., 1985, Ultrastructural changes in gap junctions associated with CO_2 uncoupling in frog atrial fibers, *J. Cell Sci.* **74:**51–63.

Meech, R., and Thomas, R. C., 1977, The effect of calcium injection on the intracellular sodium and pH of snail neurones, *J. Physiol. (London)* **265:**867–879.

Miller, T. M., and Goodenough, D. A., 1985, Gap junction structures after experimental alteration of junctional channel conductance, *J. Cell Biol.* **101:**1741–1748.

Mullins, L. J., Tiffert, T., Vassort, G., and Whittenbury, J., 1983, Effects of internal sodium and hydrogen ions and of external calcium ions and membrane potential on calcium entry in squid axons, *J. Physiol. (London)* **378:**295–319.

Neyton, J., and Trautmann, A., 1985, Single-channel currents of an intercellular lens junction, *Nature* **317:**331–335.

Nicholson, B. J., Hunkapiller, M. W., Grim, L. B., Hood, L. E., and Revel, J.-P., 1981, Rat liver gap junction protein: Properties and partial sequence, *Proc. Natl. Acad. Sci. USA* **78:**7594–7598.

Nicholson, B. J., Takemoto, L. J., Hunkapiller, M. W., Hood, L. E., and Revel, J.-P., 1983, Differences between liver gap junction protein and lens MIP26 from rat: Implications for tissue specificity of gap junctions, *Cell* **32:**967–978.

Nicholson, B. J., Gros, D. B., Kent, S. B. H., Hood, L. E., and Revel, J.-P., 1985, The Mr 28,000 gap junction proteins from rat heart and liver are different but related, *J. Biol. Chem.* **260:**6514–6517.

Nikaido, H., and Rosenberg, E. Y., 1985, Functional reconstitution of lens gap junction proteins into proteoliposomes, *J. Membr. Biol.* **85:**87–92.

Nishiye, H., 1977, The mechanism of Ca^{++} action on the healing-over process in mammalian cardiac muscles: A kinetic analysis, *Jpn. J. Physiol.* **27:**451–466.

Noma, A., and Tusboi, N., 1987, Dependence of junctional conductance on proton, calcium and magnesium ions in cardiac paired cells of guinea pig, *J. Physiol.* **382:**193–211.

Obaid, A. L., Socolar, S. J., and Rose, B., 1983, Cell-to-cell channels with two independently regulated gates in series: Analysis of junctional conductance modulation by membrane potential, calcium and pH, *J. Membr. Biol.* **73:**69–89.

Ochs, D. S., Korenbrot, J. I., and Williams, J. A., 1983, Intracellular free calcium concentrations in isolated pancreatic acini: Effects of secretagogues, *Biochem. Biophys. Res. Commun.* **117:**122–128.

Oliveira-Castro, G. M., and Barcinski, M. A., 1974, Calcium-induced uncoupling in communicating human lymphocytes, *Biochim. Biophys. Acta* **352:**338–343.

Oliveira-Castro, G. M., and Loewenstein, W. R., 1971, Junctional membrane permeability: Effects of divalent cations, *J. Membr. Biol.* **5:**51–77.

Paul, D. L., 1985, Antibody against liver gap junction 27-kD protein is tissue specific and cross-reacts with a 54-kD protein, in: *Gap Junctions* (M. V. L. Bennett and D. C. Spray, eds.), pp. 107–122, Cold Spring Harbor Laboratory, Cold Spring Harbor, N.Y.

Peracchia, C., 1977, Gap junctions—Structural changes after uncoupling procedures, *J. Cell Biol.* **72:**628–641.

Peracchia, C., 1980, Structural correlates of gap junction permeation, *Int. Rev. Cytol.* **66:**81–146.

Peracchia, C., 1984, Communicating junctions and calmodulin: Inhibition of electrical uncoupling in Xenopus embryo by calmidazolium, *J. Membr. Biol.* **91:**49–58.

Peracchia, C., 1985, Cell coupling, in: *The Enzymes of Biological Membranes* (A. Martonosi, ed.), pp. 81–130, Plenum Press, New York.

Peracchia, C., 1986, Effects of a calmodulin inhibitor (W-7) and cyclic nucleotides on electrical uncoupling of crayfish septate axons, *Biophys. J.* **49:**338a.

Peracchia, C., 1987, Calmodulin-like proteins and communicating junctions--Electrical uncoupling of crayfish septate axons is inhibited by the calmodulin inhibitor W7 and is not affected by cyclic nucleotides, *Pflug. Arch. Eur. J. Physiol.* **408:**379–385.

Peracchia, C., and Bernardini, G., 1984, Gap junction structure and cell-to-cell coupling regulation: Is there a calmodulin involvement? *Fed. Proc.* **43:**2681–2691.

Peracchia, C., and Dulhunty, A. F., 1976, Low resistance junctions in crayfish: Structural changes with functional uncoupling, *J. Cell Biol.* **70:**419–439.

Peracchia, C., and Girsch, S. J., 1985a, Functional modulation of cell coupling: Evidence for a calmodulin-driven channel gate, *Am. J. Physiol.* **248:**H765–H782.

Peracchia, C., and Girsch, S. J., 1985b, Is the C-terminal arm of lens gap junction channel protein the channel gate? *Biochem. Biophys. Res. Commun.* **133:**688–695.

Peracchia, C., and Girsch, S. J., 1985c, An in vitro approach to cell coupling: Permeability and gating of gap junction channels incorporated into liposomes, in: *Gap Junctions* (M. V. L. Bennett and D. C. Spray, eds.), pp. 191–203, Cold Spring Harbor Laboratory, Cold Spring Harbor, N.Y.

Peracchia, C., and Peracchia, L. L., 1980a, Gap junction dynamics: Reversible effects of divalent cations, *J. Cell Biol.* **87:**708–718.

Peracchia, C., and Peracchia, L. L., 1980b, Gap junction dynamics: Reversible effects of hydrogen ions, *J. Cell Biol.* **87:**719–727.

Peracchia, C., Bernardini, G., and Peracchia, L. L., 1981, A calmodulin inhibitor prevents gap junction crystallization and electrical uncoupling, *J. Cell Biol.* **91:**124a.

Peracchia, C., Bernardini, G., and Peracchia, L. L., 1983, Is calmodulin involved in the regulation of gap junction permeability? *Pfluegers Arch.* **399:**152–154.

Piccolino, M., Neyton, J., and Gerschenfeld, H. M., 1984, Decrease of gap junction permeability induced by dopamine and cyclic adenosine monophosphate in horizontal cells of turtle retina, *J. Neurosci.* **4:**2477–2488.

Politoff, A., Pappas, G. D., and Bennett, M. V. L., 1974, Cobalt ions cross an electrotonic synapse if cytoplasmic concentration is low, *Brain Res.* **76:**343–346.

Pundak, S., and Roche, R. S., 1984, Tyrosine and tyrosinate fluorescence of bovine testes calmodulin: Calcium and pH dependence, *Biochemistry* **23:**1549–1555.

Rae, J. L., Thompson, R. D., and Eisenberg, R. S., 1982, The effect of 2-4 dinitrophenol on cell-to-cell communication in the frog lens, *Exp. Eye Res.* **35:**598–609.

Reber, W. R., and Weingart, R., 1982, Ungulate cardiac Purkinje fibers: The influence of intracellular pH on the electrical cell-to-cell coupling, *J. Physiol. (London)* **328:**87–104.

Requena, J., Whittenbury, J., Tippert, T., Eisner, D. A., and Mullins, L. J., 1985, The influence of chemical agents on the level of ionized [Ca^{++}] in squid axon, *J. Gen. Physiol.* **85:**789–804.

Requena, J., Mullins, L. J., Whittenbury, J., and Brinley, F. J., Jr., 1986, Dependence on ionized and total Ca in squid axons on Na_o-free or high K_o conditions, *J. Gen. Physiol.* **87:**143–159.

Rink, T. J., Tsien, R.-Y., Warner, A., 1980, Free calcium in Xenopus embryos measured with ion-selective microelectrodes, *Nature* **283:**658–660.

Rojas, E., and Rudy, B., 1976, Destruction of the sodium conductance inactivation by a specific protease in perfused nerve fibers from Loligo, *J. Physiol. (London)* **262:**501–531.

Rose, B., and Loewenstein, W. R., 1975a, Permeability of cell junction depends on local cytoplasmic calcium activity, *Nature* **254:**250–252.

Rose, B., and Loewenstein, W. R., 1975b, Calcium ion distribution in cytoplasm visualized by aequorin: Diffusion in the cytosol is restricted due to energized sequestering, *Science* **190:**1204–1206.

Rose, B., and Rick, R., 1978, Intracellular pH, intracellular free Ca, and junctional cell–cell coupling, *J. Membr. Biol.* **44:**377–415.

Rose, B., Simpson, I., and Loewenstein, W. R., 1977, Calcium ion produces graded changes in permeability of membrane channels in cell junction, *Nature* **267:**625–627.

Russell, P., Robison, G., and Kinoshita, J., 1981, A new method for rapid isolation of the intrinsic membrane proteins of the lens, *Exp. Eye Res.* **32:**511–516.

Saez, J. C., Spray, D. C., Nairn, A. C., Hertzberg, E. L., Greengard, P., and Bennett, M. V. L., 1986, cAMP increases junctional conductance and stimulates phosphorylation of the 27 kD principal gap junction polypeptide, *Proc. Natl. Acad. Sci. USA* **83:**2473–2477.

Spray, D. C., and Bennett, M. V. L., 1985, Physiology and pharmacology of gap junctions, *Annu. Rev. Physiol.* **47:**281–303.

Spray, D. C., Harris, A. L., and Bennett, M. V. L., 1979, Voltage dependence of junctional conductance in early amphibian embryos, *Science* **204:**432–434.

Spray, D. C., Harris, A. L., and Bennett, M. V. L., 1981, Gap junctional conductance is a simple and sensitive function of intracellular pH, *Science* **211:**712–715.

Spray, D. C., Stern, J. H., Harris, A. L., and Bennett, M. V. L., 1982, Gap junctional conductance: Comparison of sensitivities to H and Ca ions, *Proc. Natl. Acad. Sci. USA* **79:**441–445.

Spray, D. C., Ginzberg, R. D., Morales, E. A., Bennett, M. V. L., and Babayatsky, M., 1984a, Physiological and pharmacological properties of gap junctions between dissociated pairs of rat hepatocytes, *J. Cell Biol.* **99:**344a.

Spray, D. C., White, R. L., Campos de Carvalho, A., Harris, A. L., and Bennett, M. V. L., 1984b, Gating of gap junction channels, *Biophys. J.* **45:**219–230.

Spray, D. C., White, R. L., Verselis, V., and Bennett, M. V. L., 1985, General and comparative physiology of gap junction channels, in: *Gap Junctions* (M. V. L. Bennett and D. C. Spray, eds.), pp. 139–153, Cold Spring Harbor Laboratory, Cold Spring Harbor, N.Y.

Spray, D. C., Saez, J. C., Brosius, D., Bennett, M. V. L., and Hertzberg, E. L., 1986, Isolated liver gap junctions: Gating of transjunctional currents is similar to that in intact pairs of rat hepatocytes, *Proc. Natl. Acad. Sci. USA* **83:**5494–5497.

Steiner, R. F., Lamboy, P. K., and Sternberg, H., 1983, The dependence of molecular dynamics of calmodulin upon pH and ionic strength, *Arch. Biochem. Biophys.* **222:**158–169.

Teranishi, T., Negishi, K., and Kato, S., 1983, Dopamine modulates S-potential amplitude and dye-coupling between external horizontal cells in carp retina, *Nature* **301:**243–246.

Traub, O., and Willecke, K., 1982, Cross-reaction of antibodies against liver gap junction protein (26 K) with lens fiber junction protein (MIP) suggests structural homology between these tissue specific gene products, *Biochem. Biophys. Res. Commun.* **109:**895–901.

Turin, L., and Warner, A. E., 1977, Carbon dioxide reversibility abolishes ionic communication between cells of early amphibian embryo, *Nature* **270:**56–57.

Turin, L., and Warner, A. E., 1980, Intracellular pH in early Xenopus embryos: Its effect on current flow between blastomeres, *J. Physiol. (London)* **300:**489–504.

Unwin, P. N. T., and Ennis, P. D., 1983, Calcium-mediated changes in gap junction structure: Evidence from the low angle X-ray patterns, *J. Cell Biol.* **97:**1459–1466.

Unwin, P. N. T., and Ennis, P. D., 1984, Two configurations of a channel-forming membrane protein, *Nature* **307:**609–613.

van den Eijnden-van Raaij, A. J. M., de Leeuw, A. L. M., and Broekhuise, R. M., 1985, Bovine lens calmodulin—Isolation, partial characterization and calcium-independent binding to lens membrane proteins, *Curr. Eye Res.* **4:**905–912.

van Eldik, L. J., Hertzberg, E. L., Berdan, R. C., and Gilula, N. B., 1985, Interaction of calmodulin and other calcium-modulated proteins with mammalian and arthropod junctional membrane proteins, *Biochem. Biophys. Res. Commun.* **126:**825–832.

Verselis, V., and Brink, P. R., 1984, Voltage clamp of earthworm septum, *Biophys. J.* **45:**147–150.

Warner, A. E., Guthrie, S. C., and Gilula, N. B., 1984, Antibodies to gap-junctional protein selectively disrupt junctional communication in the early amphibian embryo, *Nature* **311:**127–131.

Watanabe, A., and Grundfest, H., 1961, Impulse propagation at the septal and commissural junctions of crayfish lateral giant axons, *J. Gen. Physiol.* **45:**267–308.

Weidmann, S., 1970, Electrical constants of trabecular muscle from mammalian heart, *J. Physiol. (London)* **210:**1041–1054.

Weingart, R., 1977, Action of ouabain on intercellular coupling and conduction-velocity in mammalian ventricular muscle, *J. Physiol. (London)* **264:**341–365.

Welsh, M. J., Aster, J., Ireland, M., Alcala, J., and Maisel, H., 1981, Calmodulin and gap junctions: Localization of calmodulin and calmodulin binding sites in chick lens cells, *J. Cell Biol.* **91:**123a.

Welsh, M. J., Aster, J. C., Ireland, M., Alcala, J., and Maisel, H., 1982, Calmodulin binds to chick lens gap junction protein in a calcium-independent manner, *Science* **216:**642–644.

White, R. L., Spray, D. C., Carvalho, A. C., Wittenberg, B. A., and Bennett, M. V. L., 1985, Some physiological and pharmacological properties of cardiac myocytes dissociated from adult rat, *Am. J. Physiol.* **249:**C447–C455.

Wilson, M. A., and Gillette, R., 1985, pH sensitivity of calmodulin distribution in nervous tissue fractions, *Brain Res.* **331:**190–193.

Wojtczak, J., 1982, Influence of cyclic nucleotides on the internal longitudinal resistance and contractures in the normal and hypoxic mammalian cardiac muscle, *J. Mol. Cell. Cardiol.* **14:**259–265.

Wojtczak, J. A., 1985, Electrical uncoupling induced by general anesthetics: A calcium-independent process? in: *Gap Junctions* (M. V. L. Bennett and D. C. Spray, eds.), pp. 167–175, Cold Spring Harbor Laboratory, Cold Spring Harbor, N.Y.

Wojtczak, J., Girsch, S., Shrager, P., and Peracchia, C., 1987, Large conductivity channels in patch pipette lipid bilayers formed by lens junction protein (MIP26), *Biophy. J.* **51:**141a.

Zampighi, G. A., Hall, J. E., and Kreman, M., 1985, Purified lens junctional protein forms channels in planar lipid films, *Proc. Natl. Acad. Sci. USA* **82:**8468–8472.

Zervos, A. S., Hope, J., and Evans, H., 1985, Preparation of a gap junction fraction from uterus of pregnant rats: The 28-KD polypeptides of uterus, liver, and heart gap junctions are homologous, *J. Cell Biol.* **101:**1363–1370.

Chapter 4

Electrotonic Coupling in the Nervous System

Stewart W. Jaslove and Peter R. Brink
Department of Anatomical Sciences
Health Sciences Center
State University of New York
Stony Brook, New York 11794

1. INTRODUCTION

The topic of cell-to-cell communication in the nervous system primarily concerns the electrotonic synapse. This is a low-resistance intercellular pathway that permits the direct flow of electrical current between coupled cells, thus obviating the need for a neurochemically mediated step. The first description of a definitive electrotonic synapse, the crayfish motor giant synapse, appeared in 1957 (Furshpan and Potter, 1957) and numerous other examples from nervous systems of many phyla have since been demonstrated. By *definitive* we mean a specific synaptic structure mediating neurotransmission, as opposed to a fusion or close apposition of neighboring cells. Several lines of anatomical and pharmacological evidence have shown that this structure is the gap junction intercellular channel (see Section 4).

Perhaps the simplest nervous system in which electrotonic coupling has been identified is that of the jellyfish (a coelenterate) in which the cells of the nerve-net appear to be coupled as a syncytium (Spencer and Satterlie, 1980). At the other end of the phylogenetic scale is the mammalian central nervous system in which electrotonic coupling has been identified in neocortex (Gutnick and Prince, 1981; Connors *et al.*, 1983), hippocampus (MacVicar and Dudek, 1980, 1981; Taylor and Dudek, 1982a), inferior olivary nucleus (Llinas *et al.*, 1974), mesencephalic nucleus (Baker and Llinas, 1971), vestibular nucleus (Korn *et al.*, 1973; Wylie, 1973), abducens nucleus (Gogan *et al.*, 1974), hypothalamus (Andrew *et al.*, 1981; Cobbett and Hatton, 1984), newborn rat spinal cord (Fulton *et al.*, 1980), and among neuroglia (Gutnick *et al.*, 1981). Additional areas of coupling found in other vertebrates are the avian ciliary ganglion (Martin and Pilar, 1963), turtle retina (Baylor *et al.*, 1971; Detwiler and Hodgkin, 1979), amphibian adult spinal cord (Grinnell, 1966; Sotelo and Taxi, 1970; Shapovalov and Shiriaev, 1980; Westerfield and Frank, 1982), and teleost oculomotor nucleus (Kriebel *et al.*, 1969; Korn and Bennett, 1975), medulla, and spinal cord (Bennett *et al.*, 1963, 1967a–d; Furshpan, 1964) especially pertaining to electromotor activities.

Coupling is generally difficult to study in vertebrate preparations because neurons tend to be small, irregularly shaped, deeply embedded, and complexly interconnected. Much of the quantitative understanding of electrotonic synapses, therefore, comes from invertebrate preparations, in which coupling between giant neurons is widespread and more accessible to study. Important model synapses are found among the identified ganglionic cells of the gastropod molluscs including the pond snails *Planorbis* (Berry, 1972), *Lymnaea* (Audesirk *et al.*, 1982), and *Helisoma* (Kater, 1974; Bulloch and Kater, 1982), and the opisthobranchs *Aplysia* (Arvanitaki and Chalazonitis, 1959; Tauc, 1959, 1969; Kaczmarek *et al.*, 1979; Rayport and Kandel, 1980), *Navanax* (Levitan *et al.*, 1970; Spira *et al.*, 1980), and *Tritonia* (Getting and Willows, 1974). Several well-studied electrotonic synapses also occur among the cells and axons of the ventral nerve cord of the annelids, earthworm (Brink and Barr, 1977) and leech (Baylor and Nicholls, 1969; Muller and Scott, 1981), and the crustacean, crayfish (Furshpan and Potter, 1959; Watanabe and Grundfest, 1961).

Electrotonic synapses, like chemical ones, appear in a variety of cytologic configurations including axodendritic, axosomatic, somatosomatic, and dendrodendritic (Pappas *et al.*, 1975; Sotelo and Korn, 1978). Some other interesting arrangements are the segmented axons, interposed neurons, and mixed synapses. The segmented axons of crayfish (Watanabe and Grundfest, 1961) and earthworm (Brink and Barr, 1977) are chains of cylindrically shaped cells joined end-to-end by gap junction-containing septa to create structures that function as single giant axons (Figs. 1A, 8A,B). These are important preparations for biophysical experimentation because of their simple neuronal geometry. Interposed neurons are electrotonically coupled to two or more other cells, thus coupling those cells to each other as well. If the interposed cell is presynaptic (Fig. 1B), then this arrangement can serve to fire the other cells simultaneously (Meszler *et al.*, 1972; Korn *et al.*, 1973; Pappas *et al.*, 1975). If one of the other cells is presynaptic (Fig. 1C), as with the leech coupling interneuron (Muller and Scott, 1981), then the interposed cell may act to regulate the strength of the coupling pathway (see Section 3.2). Mixed synapses are common structures in vertebrates in which electrotonic and chemical synapses lie side-by-side between the same cells (Fig. 1D) (Martin and Pilar, 1963; Sotelo and Korn, 1978; Shapovalov, 1980). The role of this dual synapse is not clear. The electrotonic component may serve to speed the onset of the chemical EPSP (see Fig. 18). In teleost

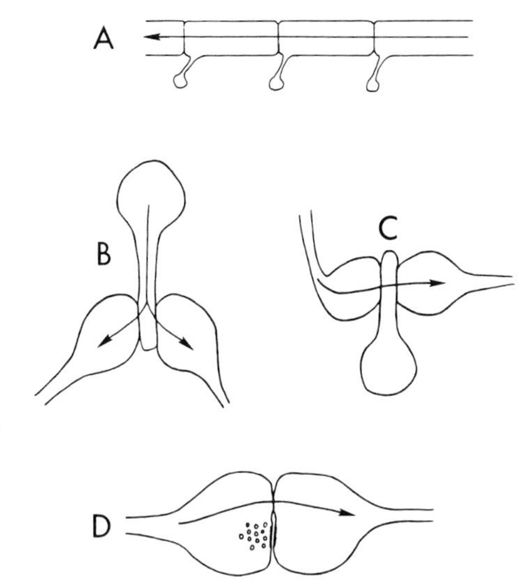

FIGURE 1. Diagrammatic representations of four novel configurations of electrotonically coupled cells. Arrows mark normal direction of current flow. (A) A segmented axon; (B) an interposed presynaptic neuron; (C) an interposed interneuron; (D) a mixed (chemical and electronic) synapse.

electromotor tissues the chemical component has been identified ultrastructurally, but apparently does not function physiologically (Meszler *et al.*, 1972).

Review of the comparative physiology of electrotonic and chemical synapses (Bennett, 1968, 1972, 1977) suggests that the main advantage of the chemical synapses is their ability to provide mechanisms for a wide range of plasticity, inhibition, and amplification (the last because chemical synaptic potentials largely depend upon postsynaptic membrane mechanisms). The distinctive physiological characteristics of electrotonic synapses are their potential for high speed and mutual excitation (Fig. 2). The former arises from the lack of a neurochemical delay and the latter from the inherent bidirectionality of the gap junction channel. But a millisecond or so of synaptic delay might not be significant for a behavior that lasts seconds or longer, and bidirectionality would certainly be disadvantageous in a circuit that requires precise specificity of connections for processing. So chemical synapses remain the primary mechanism of neurotransmission in the nervous system.

What then is the role of the electrotonic synapse? Based upon a higher incidence of coupling among spinal neurons of lower than higher vertebrates, it has been suggested that electrotonic synapses are simply relics of a more primitive evolutionary state (Shapovalov, 1980, 1982). We would take the position that even "lower" vertebrates have had ample evolutionary time to optimize their synapses so that when junctions do appear, there must be some particular function that they serve, even if we are slow to recognize what it is. Observations of a similar unequal distribution of electrotonic coupling in the vestibular nucleus

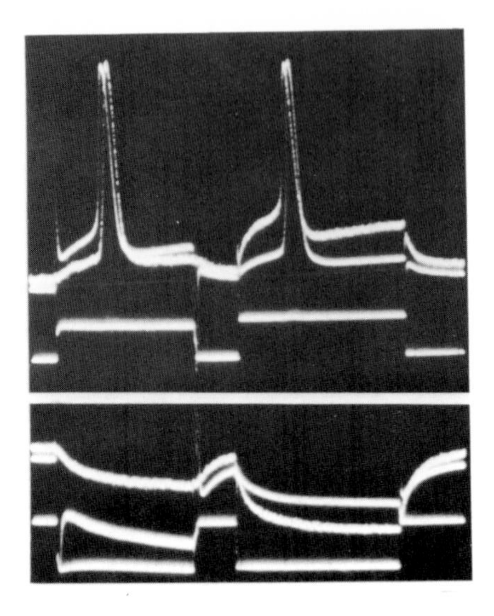

FIGURE 2. Current clamp records from a crayfish septal junction, a linear electrotonic synapse linking lateral giant axons of adjoining ventral nerve cord segments. The chief characteristics demonstrated are short synaptic delay and bidirectional transmission of both depolarizations and hyperpolarizations. Two microelectrodes were placed in each cell, as in Fig. 9, top. Top panel: depolarization. Bottom panel: hyperpolarization. First one cell was pulsed, then the other. In each panel, total injected current is shown in the bottom trace and voltage records from each cell in the top two traces. Calibrations: 20 mV; 500 nA; 0.5 msec. 15°C.

(responsible for controlling posture) of various phyla led to one suggestion that non-temperature-regulating animals rely more on electrotonic synapses because they might be less sensitive to changes in ambient temperature than are chemical synapses (Korn *et al.*, 1977).

Electrotonic synapses seem primarily designated for use in neural circuits designed for escape or synchronization. For example, they are used in mediating the fast tail flip of the crayfish (Furshpan and Potter, 1959) and the coordinated pharyngeal contractions of *Navanax* (Spira *et al.*, 1980). Neurons of the cat inferior olive are extensively coupled and the pattern of that coupling has been suggested to provide a basis for bands of synchronously active Purkinje cells in the cerebellar cortex (Llinas, 1985). Intuitively, it makes sense that nuclei controlling the finely coordinated muscles of eye motion (including oculomotor and abducens) should utilize electrotonic synapses. In many cases, however, the role of coupling is not clear. Among hippocampal pyramidal cells it could provide a basis for slow-wave potentials (and also for epileptic activity), but the incidence of coupling identified in that tissue is only 5–10% so synchronization might better be effected by extracellular electric fields (Traub *et al.*, 1985). Coupling among neurons of the rat mesencephalic nucleus is also about 10% (Baker and Llinas, 1971). It should not be overlooked that gap junctions in the nervous system may play a role in metabolic cooperation and molecular signaling (see Section 4.5), as they do in nonexcitable tissues. This may be particularly important for neuroglia and in development (see Section 3.5).

Electrotonic synapses are capable of exhibiting many of the same kinds of properties as chemical synapses (Bennett, 1968, 1972). Although examples of this tend to be special cases, electrotonic synapses are certainly less rigid in their behavior than simple intercellular bridges. In the next two sections we will review current understanding of two areas relevant to this point: mechanisms of electrotonic inhibition, and plasticity or modulation of electrotonic coupling. Various experimental approaches have been developed in an attempt to overcome the difficulties of studying coupling in the central nervous system and these have sometimes led to ambiguous and conflicting results. In Section 4 we will discuss these techniques and their limitations.

2. ELECTROTONIC INHIBITION

Inhibition has two related but distinct connotations. One involves a sign inversion of a presynaptic depolarization into a postsynaptic hyperpolarization; the other simply implies a reduction in postsynaptic excitability. Inhibition is an essential aspect of neural processing and is most easily and commonly performed by chemical synapses, for electrotonic synapses cannot on their own generate membrane polarizations. However, there do exist some examples of electrotonic inhibition in which coupling acts in conjunction with other membrane processes.

The trivial case would be the simple electrical loading down or shunting of one neuron by the current sink created by another, electrotonically coupled neuron. Since simultaneously activated neurons would not load each other down, this arrangement is an important mechanism of synchronization (Bennett, 1972; Getting and Willows, 1974). A more specific inhibitory design would be the selective transmission of hyperpolarizing afterpotentials following spikes (Fig. 3) (Tauc, 1969; Getting, 1974). This could result from spike attenuation by the low-pass filtering property of the electrotonic junction (see Section 3.3) or simply from the longer total duration of many afterpotentials.

An example of inhibition by loading comes from the control of swimming in *Tritonia* (Getting and Willows, 1974) in which both the initiation and termination of bursting of groups of command neurons appear to be mediated by cou-

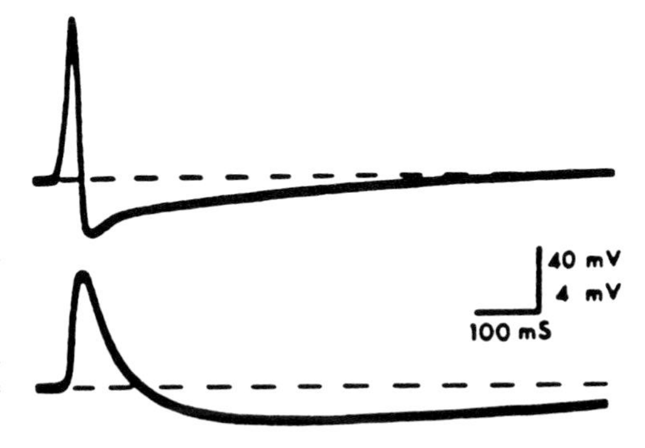

FIGURE 3. The low-pass filtering property of an electrotonic synapse coupling pleural ganglionic neurons of *Tritonia*. A fast presynaptic spike with a prolonged posthyperpolarization (top) is transformed into a small EPSP followed by a relatively large hyperpolarization (bottom) due to the charging time of the postsynaptic capacitance. From Getting (1974) with copyright permission of the American Physiological Society.

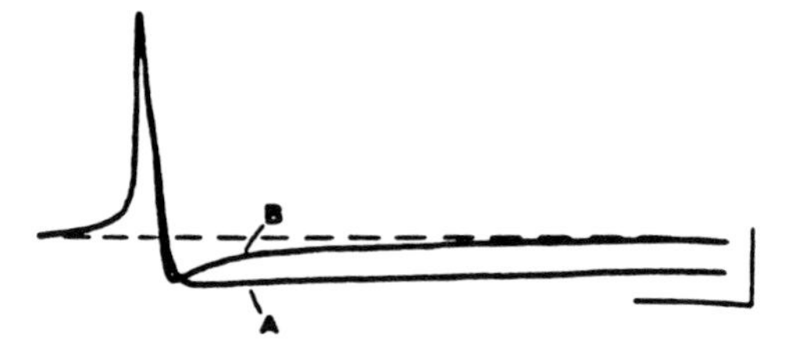

FIGURE 4. The effect of loading by electrotonic coupling on spikes in *Tritonia* pleural ganglion. Trace A shows the prolonged posthyperpolarization created by the synchronized firing of bursting coupled neurons. Trace B shows an unsynchronized spike with its posthyperpolarization shortened because of current shunting into nonfiring coupled cells. The depolarizing phase of the spike is little affected by loading because of the low plasma membrane resistance during its time course. The large posthyperpolarization of trace A tends to inhibit further bursting, and so is self-limiting. Calibrations: 20 mV; 100 msec. From Getting and Willows (1974) with copyright permission of the American Physiological Society.

pling (Fig. 4). Initiation occurs as excitation regeneratively spreads through the coupled network. As the rate of firing increases, spikes tend to synchronize, thus unloading the junctions, increasing the amplitude of the spike afterhyperpolarizations, and terminating the bursts.

Sign inversion in a system of electrotonic synapses has been reported to occur in the buccal ganglion of *Navanax* (Fig. 5) (Spira *et al.*, 1976; Bennett *et*

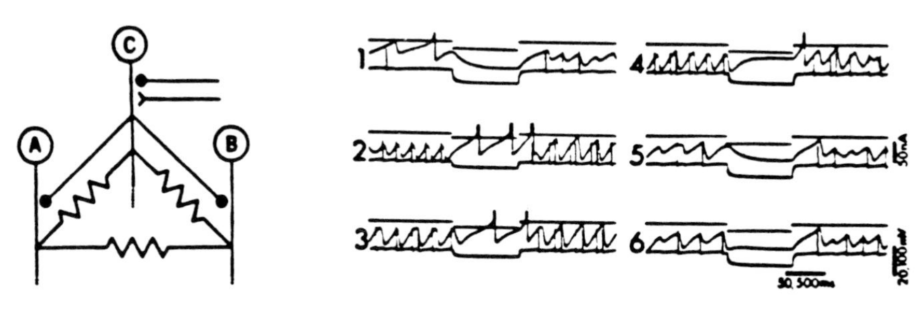

FIGURE 5. Sign inversion in a system of electrotonic synapses from the buccal ganglion of *Navanax*. On the left is a model of two circumferential motoneurons (A, B) and an interneuron (C) that are coupled by electrotonic (resistors) and inhibitory (filled endings) and excitatory (open endings) chemical synapses. Groups of traces on the right show voltage recorded in the two motoneurons (bottom two traces) and current injected into one of them (top trace). A spontaneous spike in the bottom trace neuron usually is followed by an IPSP in the other neuron. When the interneuron is quiescent (group 1), a hyperpolarization of one neuron results in a typical electrotonic hyperpolarization of the other neuron. When the interneuron is active (2–6), the motoneurons are tonically inhibited by it. When one cell is then weakly hyperpolarized (2), the interneuron is inhibited and the other motoneuron is depolarized by release from chemical inhibition. As the amplitude of the hyperpolarizing pulse is increased (3–6), the depolarizing disinhibition summates with the electrotonic hyperpolarization. From Spira *et al.* (1976) with copyright permission of the American Association for the Advancement of Science. Recent work on this preparation has shown that only one of the motoneurons is electrotonically coupled to the interneuron so that sign inversion occurs in one direction only between motoneurons (Bennett *et al.*, 1985).

al., 1985). Ingestion of food requires rapid pharyngeal contraction followed by peristaltic swallowing. The rapid contraction appears to be controlled by the synchronized firing of coupled circumferential motoneurons. These are also coupled to chemically inhibitory interneurons. Once contraction is complete, the interneurons become activated and tonically inhibit the motoneurons. Thereafter, motoneuron polarizations spreading to interneurons result in polarizations of opposite sign in other motoneurons (due to relief from or increase of chemical inhibition), thus providing a basis for peristalsis. The novel design of this system, normal coupling when interneurons are not activated and inverted coupling when they are, allows a single circuit to mediate multiple functions.

Another kind of electrotonic inhibition, more like that of the chemical synapse, is found on the goldfish Mauthner cell (Furukawa and Furshpan, 1963; Korn and Faber, 1975). In this case, extracellular currents from a closely apposed axonal ending are generated through a restricted extracellular space to create a small, but effective, hyperpolarization of the spike-initiating region of the Mauthner cell. This is the only example of this type of inhibition thus far described. It is an example of ephaptic coupling, not involving gap junctions.

3. MODULATION OF ELECTROTONIC COUPLING

Modifiability of synaptic coupling is an essential aspect of plasticicty in the nervous system. Many examples of modulation of electrotonic synapses have now been described, ranging in speed from the rapid gating of voltage-dependent channels to long-term changes in development.

3.1. Voltage-Dependent Gating (Rectification)

Where bidirectionality of electrotonic coupling would be disadvantageous, it can be overcome by adding voltage-dependent gating to junctional channels. Unidirectional electrotonic synapses created in this way are known as rectifiers and only a few examples of them have been described. Usually, the direction of rectification is arranged so that electrical current, and therefore action potentials, can be transmitted peripherally, but not centrally. This produces a high conduction velocity system without the hazard of action potential rebound.

The most well-studied rectifying electrotonic synapse is the crayfish motor giant synapse which mediates the fast tail-flip escape response. This junction is interposed between the giant axons of the ventral nerve cord and the motor giant axon innervating the fast flexor muscles of the tail. Depolarizing current can only flow from the giant axons of the cord into the motor axon, not back (Fig. 6A,B) (Furshpan and Potter, 1957, 1959; Margiotta and Walcott, 1983; Giaume and Korn, 1983). Lucifer Yellow, a fluorescent dye with a size-limiting dimension of 1.26 nm, can permeate the junction in either direction when the synapse is in a

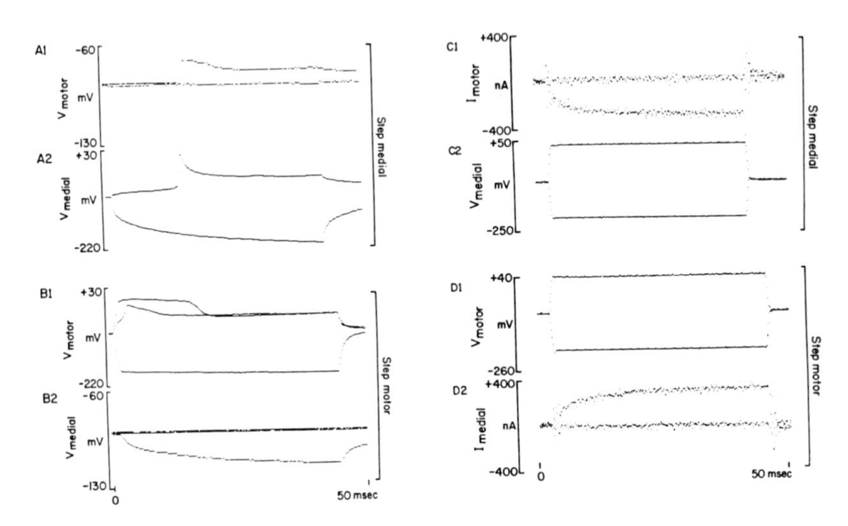

FIGURE 6. Data recorded from the rectifying electrotonic motor synapse of the crayfish. (A, B) Current clamp records. Only depolarizations can spread from medial to motor axon (A); only hyperpolarizations can spread from motor to medial axon (B). (C, D) Voltage clamp records. Current can only flow in the direction from medial to motor axon, as indicated by an inward current of the motor axon (C1) and an outward current of the medial axon (D2). The rising phase of the current records represents channel gating kinetics at 12°C. From Jaslove and Brink (1986) with copyright permission of Macmillan Journals Limited.

conducting state (Giaume and Korn, 1984). Voltage clamp studies have shown that these junctions contain gates that open or close as a function of transjunctional potential (Fig. 6C,D) so that the synapse conducts only when the giant axons of the cord are depolarized with respect to the motor axon (Jaslove and Brink, 1986). Thus, a presynaptic spike would activate the junction whereas a motor axon spike would deactivate it. A precedent for transjunctional voltage-dependent gating of gap junction channels exists in certain early embryo preparations (Spray *et al.*, 1979), but it is substantially different. The embryonic gates are symmetrical (closing with applied transjunctional potentials of either polarity) and one or two orders of magnitude slower than the crayfish junction.

The ventral nerve cord of the leech contains three examples of rectifying electrotonic junctions between neurons with multiply impalable somata. The synapses themselves, however, are located at the ends of long, thin processes. The three junctions are between a touch sensory neuron and a motoneuron (Nicholls and Purves, 1970), between a touch neuron and an s-cell interneuron (Muller and Scott, 1981), and between a pair of touch cells (Baylor and Nicholls, 1969). The latter two connections are actually formed by way of an interposed coupling interneuron (Muller and Scott, 1981), so all of these rectifying junctions are at a touch cell interface and all appear to be quite similar in function to the crayfish motor synapse. The inter-touch cell connection has the interesting property of double rectification: action potentials can pass in either direction between them, but hyperpolarizations cannot. This ability seems to arise from the incomplete block of depolarization spreading from the coupling interneuron to the touch cell (Muller and Scott, 1981).

The remaining examples of rectifying electrotonic synapses are between retinular and eccentric cells of the horseshoe crab lateral eye (Smith *et al.*, 1965; Smith and Baumann, 1969), between giant fibers and motoneurons of the hatchetfish (Auerbach and Bennett, 1969), and between Muller fibers and lateral interneurons of the lamprey spinal cord (Ringham, 1975). Each of these also appears to be similar to the crayfish motor synapse, but is difficult to study because access is limited to only a single microelectrode.

3.2. Extrajunctional Membrane Effects (Functional Decoupling)

Functional decoupling is a process whereby the postsynaptic input resistance is set too low to be depolarized by synaptic activity even though the junction itself is in a high conductance state. This process may utilize either permanent extrajunctional membrane properties or else transient chemical synaptic resistance changes.

The electrotonic synapse between the mucus secretion control motoneurons, R_2 and LPl_2, in the cerebral ganglion of *Aplysia* (Rayport and Kandel, 1980), is conductive throughout life, yet in the adult it does not appear capable of transmitting action potentials. The disproportionately large growth of axonal membrane compared to synaptic membrane during maturation apparently lowers the postsynaptic input resistance below the current-carrying capability of the synapse. Similar results have been produced experimentally in crayfish lateral giant axons by using alkalinization to lower the extrajunctional membrane resistance and thus produce conduction failure at the septal junctions (Giaume and Korn, 1982).

Nonlinear nonjunctional membrane resistance–voltage relations can result in effectively rectifying electrotonic synapses. For example, a cell with a typical delayed rectifier potassium channel is more easily hyperpolarized than depolarized (unless the synaptic transmission is significantly shorter than the onset of the potassium conductance). The opposite effect is also possible, as described between the lateral and rostral penile evertor motoneurons of the leech (Fig. 7) (Zipser, 1979). Here, the rostral cell input resistance varies anomalously: it increases when depolarized and decreases when hyperpolarized. Thus depolarization is transmitted more effectively than hyperpolarization from lateral to rostral cells.

An example of modulation of electrotonic coupling by chemical synapses is

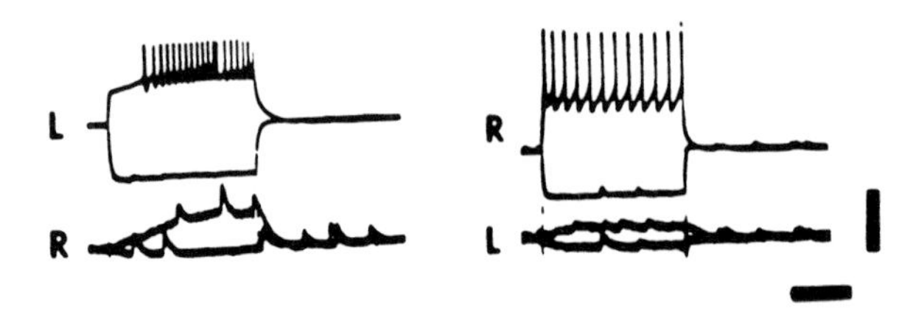

FIGURE 7. Apparent rectification at an electrotonic synapse between lateral (L) and rostral (R) motoneurons of leech ganglion 6, due to extrajunctional membrane properties. The rostral cell increases its input resistance when depolarized and decreases it when hyperpolarized. Therefore, polarizations of the lateral neuron more easily depolarize than hyperpolarize the rostral neuron (left). The input resistance of the lateral neuron is low and weakly voltage dependent over the range studied, so that it is polarized by the rostral neuron in a weak, symmetrical fashion (right). The junction itself is apparently linear. Calibrations: 1 sec; 10.4 mV. From Zipser (1979) with copyright permission of the American Physiological Society.

between pharyngeal expansion motoneurons of *Navanax,* which are well coupled under resting conditions (Spira and Bennett, 1972). When the pharyngeal nerves are stimulated, coupling is reduced because of the reduction in input resistance due to the activation of inhibitory chemical synapses. In principle, activation of an excitatory synapse could also decouple cells, but an inhibitory synapse has a smaller effect on the postsynaptic resting potential.

Synaptic modulation using an excitatory chemical synapse occurs between coupled inking control motoneurons in the abdominal ganglion of *Aplysia* (Carew and Kandel, 1976). Chemical synaptic inputs to the motoneurons evoke compound EPSPs that are characterized by an early decreased input resistance followed by a slow increased input resistance. The latter phase presumably results from a decrease of membrane potassium permeability and creates an augmentation of coupling between motoneurons over what would have been achieved under resting conditions.

At least two sites of functional decoupling have been proposed to occur in the mammalian central nervous system. One is in the cat inferior olive (Llinas *et al.*, 1974) in which nerve endings occur in clusters called glomeruli. Electrotonically coupled endings appear in the center of the glomeruli, surrounded by chemically coupled ones, suggesting that activation of the chemical synapses could decouple the junctions and thereby regulate the activity of the nucleus. The other example is in the hippocampus, where recurrent chemically inhibitory synapses are proposed to shunt the effects of electrotonic coupling between CA3 pyramidal cells (MacVicar and Dudek, 1981). This would explain the action of convulsive agents such as penicillin, which are believed to reduce recurrent inhibition, because such an agent would enhance coupling and therefore increase synchronous activity.

3.3. Frequency Modulation (Filtering)

Electrotonic synapses have the capability of acting as low-pass filters, omitting or summating rapidly changing signals (Bennett, 1966; Bennett and Pappas, 1983). The electrotonic synapse must charge the postsynaptic membrane capacitance by current flow through junctional channels. If the junctional resistance is high or the postsynaptic membrane time constant is long, then short-duration spikes can be significantly attenuated at the postsynaptic cell membrane. Membrane time constants of 100 msec or more have been measured with concomitant transmission failures in the leech (Zipser, 1979) and *Aplysia* (Tauc, 1969; Rayport and Kandel, 1980). Figure 3 illustrated an example of coupling between pleural ganglionic cells of *Tritonia,* showing how a fast presynaptic spike followed by a prolonged hyperpolarization is transformed into a weak EPSP followed by a relatively strong hyperpolarization (Getting, 1974).

Low-pass filtering can be an important way to influence postsynaptic excitability without actually generating a spike. The cells of the lobster cardiac ganglion, for example, are coupled for pacemaking of the heartbeat by electrotonic transmission of graded, subthreshold potentials while excluding the transmission of action potentials (Watanabe and Bullock, 1960).

3.4. Cytoplasmic and Neurohumoral Factors

Cytoplasmic constituents such as calcium and hydrogen ions are well-known uncouplers of gap junction pathways in nonnervous (Rose and Rick, 1978; Spray *et al.*, 1982) and nervous (Baux *et al.*, 1978; Giaume *et al.*, 1980) tissues. Calmodulin has also been suggested to cause uncoupling (Peracchia and Bernardini, 1984) and cAMP may decrease (De Mello, 1984) or increase (Teranishi *et al.*, 1983; Piccolino *et al.*, 1984) junctional resistance. Since these factors are all well regulated by cells, they could provide important *in vivo* junctional control mechanisms. Several common neurotransmitters that may activate some of these factors also influence coupling. Coupling between horizontal cells of the turtle retina is reduced by both GABA (Piccolino *et al.*, 1982) and dopamine (Teranishi *et al.*, 1983; Piccolino *et al.*, 1984). Coupling between pancreatic acinar cells is reduced by acetylcholine (Findlay and Petersen, 1982) and coupling between heart cells is increased by epinephrine (De Mello, 1982). These direct effects of neurotransmitters on coupling are to be distinguished from the actions of chemical synapses on the input resistance of the postsynaptic cell (see Section 3.2). Indeed, the effect of cytoplasmic factors on coupling may also be expressed indirectly as an effect on the resistance of the extrajunctional membrane.

Evidence for a hormonal effect on coupling in the nervous system has been obtained for magnocellular neuroendocrine cells of the rat hypothalamus (Cobbett and Hatton, 1984). Here, coupling is influenced by the hydration state of the tissue as signaled by a proposed androgenic hormone. Estrogen is postulated to control coupling between smooth muscle cells of rat myometrium (Garfield *et al.*, 1978).

One possible physiological example of a cytoplasmic effect is damage uncoupling: when one cell of a coupled pair is damaged, the pair spontaneously uncouple (Asada and Bennett, 1971), presumably to protect the undamaged cell from injury currents. The actual mechanism of damage uncoupling is unknown, but might be related to an influx of extracellular calcium.

3.5. Development and Regeneration

The modulation of electrotonic coupling is an important aspect of development and regeneration. At least up to the point of the establishment of functioning nerve cells, it is generally believed that the function of these junctions is primarily the transport of metabolites and developmental signals, though electrotonic potentials have been proposed as one possible developmental signal (Harris *et al.*, 1983).

All early blastula cells appear to be linked by gap junctions, the pattern of coupling changing as differentiation proceeds (Furshpan and Potter, 1968; Bennett *et al.*, 1981). In the amphibians *Rana* and *Ambystoma*, there is a large increase of coupling between cells of the neural tube during neurulation (Warner, 1973; Decker and Friend, 1974). Specific block of coupling at early embryonic stages by antibodies to gap junction protein results in developmental defects (Warner *et al.*, 1984).

Many early synaptic pathways appear to be transiently coupled by electrotonic synapses at early stages of connectivity, as demonstrated in *Xenopus* eye (Dixon and Cronly-Dillon, 1972) and grasshopper nerve cord (Goodman and Spitzer, 1979). Growth cones form gap junctions with their target cells in two insect nervous systems, grasshopper (Taghert *et al.*, 1982) and daphnia (LoPresti *et al.*, 1974). Dye and electrical coupling in rat neocortex are prevalent during the first 4 days after birth but then fall off dramatically by day 10, at which time chemical synapses rapidly begin to appear (Connors *et al.*, 1983). Electrotonic synapses do not necessarily precede chemical synapses. The adult avian ciliary ganglion contains mixed synapses in which the electrical component does not appear until after the chemical component is already functioning (Landmesser and Pilar, 1970, 1972).

Adult neurons can be stimulated to couple to each other by cell damage. During regeneration of axons in the ventral nerve cord of the leech, distal segments of severed s-cells do not degenerate but appear to heal over and reattach by electrotonic junctions to their original proximal segments (Carbonetto and Muller, 1977). This restores the electrical connectivity of the system long before the proximal segment completes the slower process of regeneration to its original length. In the buccal ganglion of *Helisoma*, axotomy results in increased strength of coupling at extant electrotonic synapses (Murphy *et al.*, 1983) as well as the formation of coupling between pairs of cells that are not coupled in the normal animal (Bulloch and Kater, 1982). New junctions seem to be able to form between any cells that are in a growing state (Hadley and Kater, 1983; Hadley *et al.*, 1985). Junctions formed between functionally unfavorable neuron pairs are transient (at least *in vivo*) as they are broken at the time of formation of stable junctions between functionally favorable pairs (Bulloch and Kater, 1982). Evidence for the formation of new electrotonic synapses as a result of dendrotomy has also been found in guinea pig neocortex (Gutnick *et al.*, 1985).

There are several possibilities for the fate of eliminated electrotonic synapses. Some may become functionally decoupled but remain physically in place as apparently occurs at the R_2–LPl_1 synapse in *Aplysia* (see Section 3.2). In other cases the junction may physically disappear. Damage to lateral giant axons of the crayfish nerve cord results in rapid uncoupling of adjoining septal junctions (Asada and Bennett, 1971) followed within 1 hr by separation of junctional membranes and interposition of Schwann cells (Pappas *et al.*, 1971; Bittner and Ballinger, 1980). This condition persists for a period of about 3 months, after which both electrical and ultrastructural coupling may be restored (Anderson and Bittner, 1980; Bittner and Ballinger, 1980). In the intervening period, intact junctional membranes have been noted to be internalized within the cytoplasm of the intact axon segment (Hanna *et al.*, 1984; see also Zampighi *et al.*, 1978). An alternative mechanism to internalization is the breakup and dispersion of junctions over the cell surface membrane (Lane and Swales, 1980; Hanna *et al.*, 1984). Either of these two mechanisms may lead to degradation of the channel protein, or else to storage for potential reuse.

4. EXPERIMENTAL DETERMINATION OF ELECTROTONIC COUPLING

4.1. Alternative Mechanisms of Intercellular Communication

Before discussing methods of determining electrotonic coupling, we must first consider alternative mechanisms of communication, namely, those other pathways we must eliminate once some sort of coupling has been established.

Obviously, it must be determined that the electrodes are not recording from the same cell, which can be especially misleading in the central nervous system where separated recording points can be connected by long but continuous neurites. Similarly, cells can be connected by true, membrane-bound cytoplasmic bridges (Fawcett, 1961). These have been noted to occur between dividing cells (Fawcett, 1961), between axons and Schwann cells of crayfish (Peracchia, 1981), and also between neurosecretory cells of goldfish preoptic nucleus (Gregory *et al.*, 1984), though others have noted bridges to occur as fixation artifacts (Bennett, 1973; Bennett *et al.*, 1973).

The unidirectional chemical synapse, of course, is the most common coupling pathway in the nervous system. More difficult to distinguish from electrotonic coupling, because of its bidirectionality, would be the reciprocal chemical synapse. This structure has both pre- and postsynaptic elements arranged on both sides of the synaptic cleft. These have been described between mitral and granule cells of the olfactory bulb (Rall *et al.*, 1966), between bipolar and amacrine cells of the retina (Dowling, 1968), possibly between photoreceptors and horizontal cells of the retina (see Stell, 1972), and between neurons of several invertebrates including in the tentacle ganglion of the terrestrial snail *Achatina fulica* (McCarrager and Chase, 1985). Also bidirectional, in the sense of transmitting both depolarizations and hyperpolarizations, are the tonically releasing synapses of some sensory receptors that can increase or decrease their rate of transmitter release (Obara and Oomura, 1973; Obara, 1974).

Any kind of ephaptic conduction, transmission by way of extracellular space, would mimic electrotonic coupling by gap junctions because of its high speed and bidirectionality. Ephaptic conduction seems to be a primary pathway for providing inhibition at the goldfish Mauthner cell (Furukawa and Furshpan, 1963; Korn and Faber, 1975), and excitation of synchronization in rat hippocampus (Taylor and Dudek, 1982b, 1984).

Another possible neuronal interaction is through potassium accumulation in a restricted extracellular space, from which impulse activity in one cell can cause long-lasting depolarization in another, closely apposed cell. There are two known examples of specific synaptic connections of this sort: between sensory cells of the circumesophageal nervous system of the nudibranch mollusc *Hermissenda* (Alkon and Grossman, 1978) and between giant interneurons of the cockroach metathoracic ganglion (Yarom and Spira, 1982). These are reciprocal and long-lasting, but insensitive to hyperpolarizations.

4.2. Morphology

Morphological identification of gap junctions by electron microscopy is the handiest method of demonstrating electrotonic synapses because virtually any part of the nervous system is accessible to histological analysis. The disadvantage of the technique, however, is that one cannot always confirm by other methods that the identified cells are indeed electrotonically coupled. Morphology is not sufficient proof. In the turtle retina, for example, there are anatomically defined gap junctions coupling all types of photoreceptors, yet only those with the same spectral sensitivity are coupled electrotonically (Raviola and Gilula, 1973; Detwiler and Hodgkin, 1979). We have already discussed the example of the R_2–LPl_1 synapse in *Aplysia,* which is present but apparently does not play a role in impulse transmission in the adult animal (Rayport and Kandel, 1980).

Similarly, there are reports of cells that are electrotonically coupled but with few or no anatomically defined gap junctions in smooth muscle (Daniel *et al.*, 1976), cultured tumor (Larsen *et al.*, 1977), cultured heart (Williams and DeHaan, 1981), and regenerating liver (Meyer *et al.*, 1981).

The ultrastructure of the gap junction is well known (Fig. 8) (Brightman and Reese, 1969; Gilula, 1974). In thin section with permanganate fixation the junction appears as a pentalaminar region of adjacent membrane fusion about 15 nm in total width (Fig. 8D,E). With glutaraldehyde or osmium fixation and heavy metal staining, this region appears septilaminar with the center layer representing an extracellular space 2–4 nm in width, reduced from the normal 10–20 nm (Fig. 8F). At high magnification the intercellular gap is seen to be bridged by thin processes about 10 nm apart, and may be filled by extracellular application of colloidal lanthanum (Fig. 8G). When viewed *en face*, either by thin section with lanthanum or by freeze-fracture, junctions often appear as hexagonal arrays of membrane particles with 9- to 10-nm center-to-center spacing, but can also appear in strands (Fig. 8H).

Proper histological procedures are extremely important for correct identification of gap junctions. In thin section, artifactual membrane appositions that may be indistinguishable from pentalaminar gap junctions form in fixatives that cause cell swelling (Brightman and Reese, 1969; Daniel *et al.*, 1976). Exposure to hyperosmotic sucrose solutions prior to fixation, on the other hand, results in reversible separation of junctional membranes which are indistinguishable from adjacent nonjunctional membranes (Barr *et al.*, 1965, 1968; Goodenough and Gilula, 1974). In the latter case, gap junctions still may be identified by freeze-fracture techniques, at least within a 30-sec exposure (Goodenough and Gilula, 1974). It should be noted that glutaraldehyde itself is a potent and rapid uncoupler of gap junctions (Bennett *et al.*, 1972), leaving open the question of how well even well-fixed tissues represent the physiological state.

Freeze-fracture is generally more useful than thin-section electron microscopy for locating gap junctions because the membrane-plane fracture face provides a wider expanse of surface for examination. This would be particularly important in the nervous system, where junctions often occur on fine, distal processes. For example, gap junctions have been demonstrated with the freeze-fracture technique, but not with thin-section, in retina (Raviola and Gilula, 1975), hippocampus (Schmalbruch and Jahnsen, 1981), and hypothalamus (Andrew *et al.*, 1981). These junctions tend to be small and rare, supporting the notion that failure to locate them in thin sections is a sampling problem. It may also be that, as with hyperosmotic sucrose uncoupling (Goodenough and Gilula, 1974), junctions seen only in freeze-fracture represent broken junctions resulting, perhaps, from cell damage during tissue preparation (Pappas *et al.*, 1971).

In the case of cells from which even freeze-fracture has failed to locate gap junctions, there might be another sort of sampling error. Gap junctions are generally defined as aggregates of at least a few membrane particles, but there is

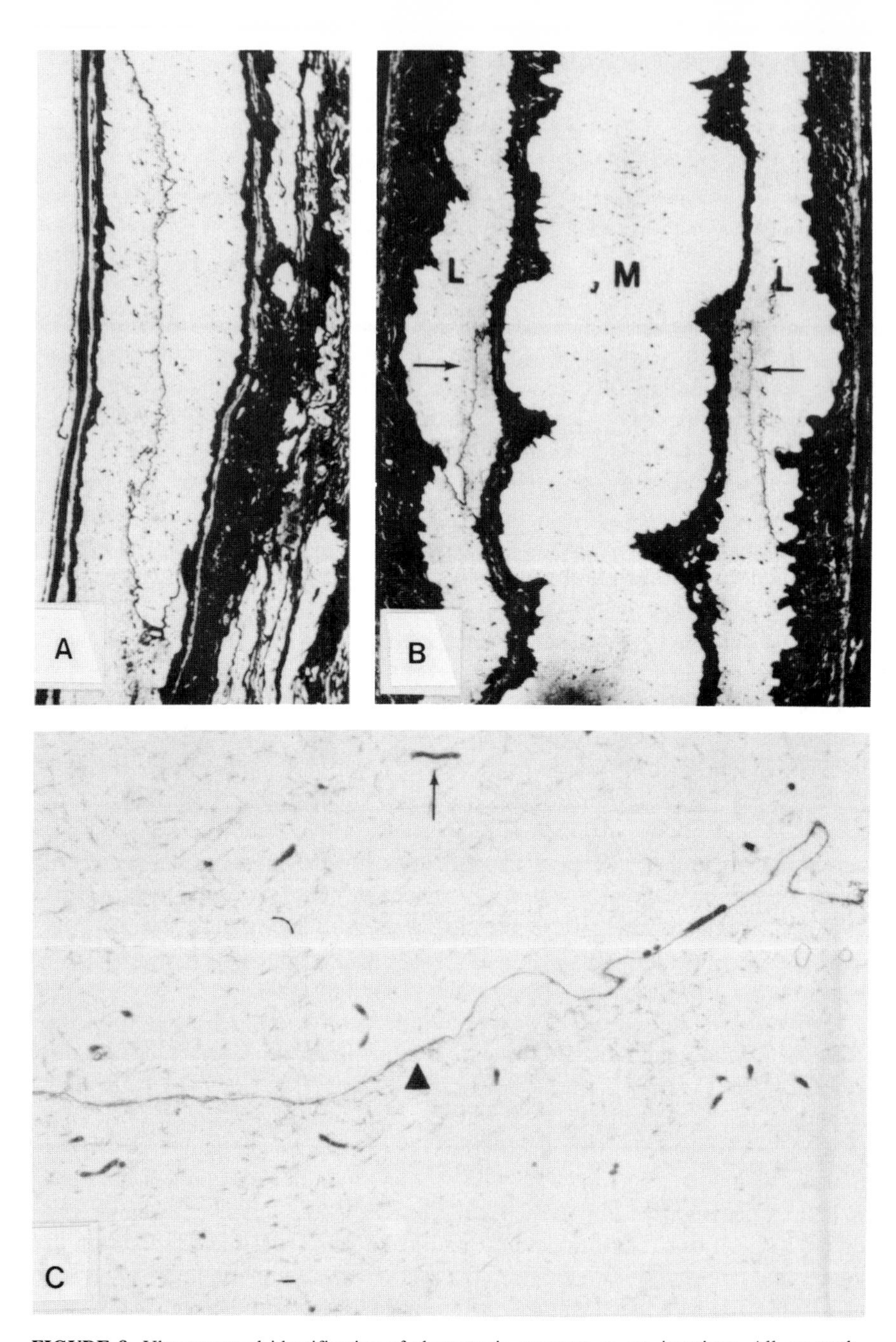

FIGURE 8. Ultrastructural identification of electrotonic synapses as gap junctions. All are earthworm lateral giant axon septal junctions except G, which is a *Mytilus* anterior byssus retractor muscle junction. (A) Light micrograph of a sagittal section through the nerve cord showing the oblique septum separating lateral axon segments. ×275. (B) Same as A but in dorsal section showing the paired lateral (L) and single median (M) giant axons. The septum of the median axon, which is not shown here, crosses transversely. ×275. (C) Low-magnification electron micrograph of the lateral axon septum. ×5000. (D) Higher-magnification electron micrograph of a septal junction fixed with

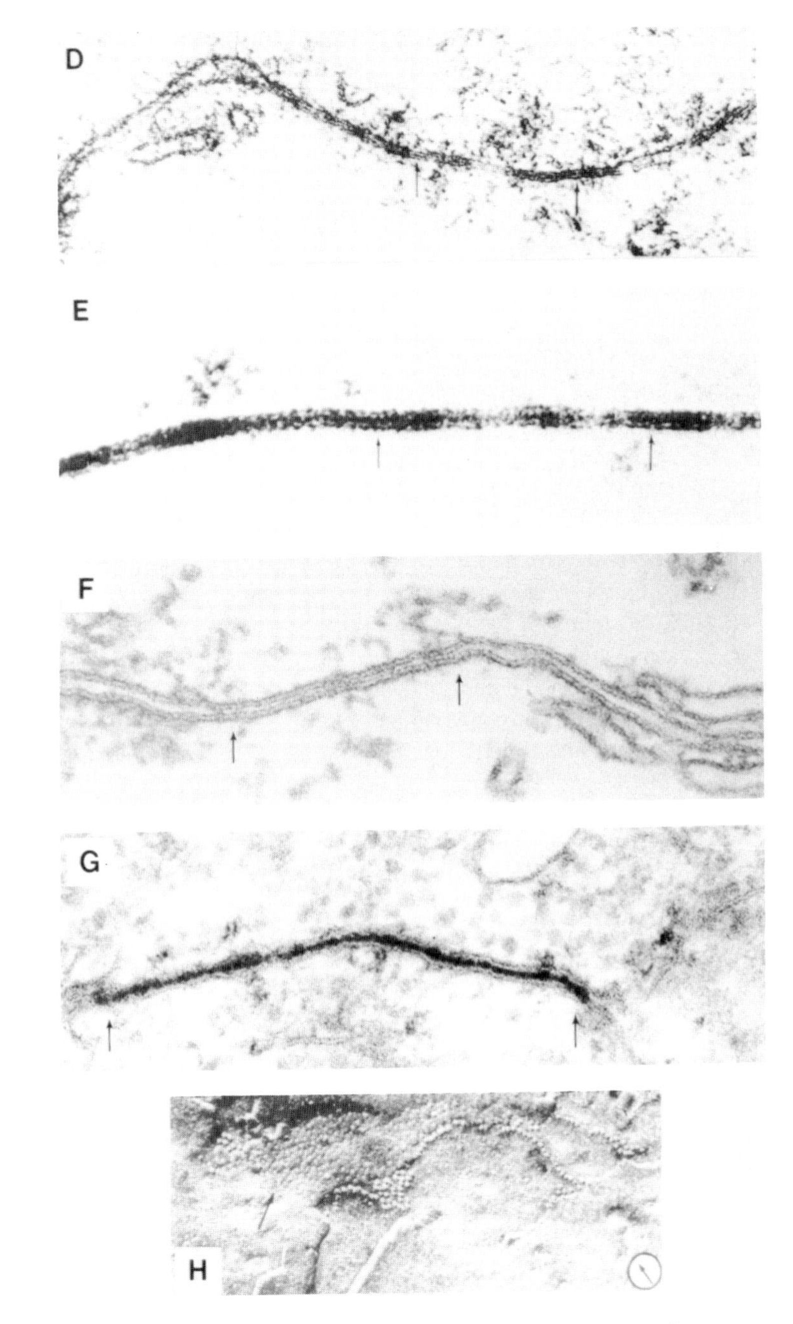

sodium permanganate, which results in a pentalaminar appearance. ×200,000. (E) Same as D. ×367,000. (F) Another septal junction fixed in glutaraldehyde and stained with uranyl acetate, resulting in a septilaminar appearance. ×162,500. (G) *Mytilus* junction with colloidal lanthanum precipitate filling the gap region of the junction. ×375,000. (H) Freeze-fracture of a septal junction showing pits and particles arranged in plaques (arrow) and strands. This A-type junction has pits on the E-face and particles on the P-face. ×84,000. All except G are reproduced with permission from Kensler (1978).

no reason at this time to believe that coupling cannot be subserved by nonaggregated junctional particles. In cultured heart cells treated with a 24- to 48-hr exposure to the protein synthesis inhibitor cycloheximide, there is a complete loss of aggregates of at least four particles (Williams and DeHaan, 1981). However, the electrotonic coupling remains strong and the total number of integral membrane particles in the nonjunctional membrane is unchanged by the treatment.

Freeze-fracture of hybrids of junctionally competent and incompetent tumor cells also show a loss of gap junctions while retaining electrotonic coupling (Larsen *et al.*, 1977). In this case, however. a new membrane structure is found that is fibrillar in appearance and may be a modified form of the gap junction protein of the cancerous cells.

One could also postulate several reasons why there would be gap junctions without electrotonic coupling. Not all junctions need be conducting. Pharmacologically uncoupled gap junctions can form dense crystalline plaques on the surface membrane, giving the appearance of well-coupled cells (Peracchia and Dulhunty, 1976). It cannot be excluded that some gap junctions, or gap junction-like structures, have a role other than cell-to-cell communication, like structural support, for example. Even if the junction is communicating, it may still be unable to support electrotonic coupling (see Section 3.2). Such junctions in the nervous system might function like junctions of other tissues in which metabolic coupling and molecular signaling are their apparent major roles (see Sections 3.5 and 4.5).

4.3. Electrophysiology

Electrophysiological testing can provide the most direct measure of electrotonic coupling. One simply measures current flow between cells (usually as displacement of membrane potential) and looks for rapid, bidirectional transmission of both depolarizations and hyperpolarizations (Fig. 2). Alternatively, one looks for some combination of the above that cannot be explained by any other type of coupling. Unfortunately, many preparations do not provide adequate access for proper electrophysiological measurement. Hence, there are two general approaches to the electrophysiology of electrotonic synapses. One, applicable mainly to the large neurons of invertebrates, involves placement of two electrodes into each cell of a coupled pair and provides for a direct measure of coupling. The second approach, utilized mainly in the vertebrate nervous system, involves placement of a single microelectrode in one cell and the use of indirect tests to demonstrate coupling. The following discussion will consider first four- and then single-microelectrode techniques.

It is possible to double the use of single electrodes by using a bridge circuit to simultaneously inject current and record voltage. It should be noted, however, that microelectrode tip resistance, and therefore bridge balance, is extremely sensitive to current intensity, polarity, and time, so that reliance on bridge circuits to measure coupling can lead to incorrect results (Martin and Pilar, 1963; Engberg and Marshall, 1979), as will be demonstrated at the end of this section.

A four-electrode system is the most accurate because of the separation of current and voltage electrodes and the ability to perform reciprocal tests (Fig. 9). The most common measure of coupling is the coupling coefficient, defined as the ratio of the change in membrane potential in each of two coupled cells for a current injected into one of them:

$$k_{12} = \Delta V_2/\Delta V_1 = 1/(1 + r_j/r_2) \qquad (1)$$
$$k_{21} = \Delta V_1/\Delta V_2 = 1/(1 + r_j/r_1)$$

The coefficients, k_{12} and k_{21}, are measured by injecting current into neurons 1 and 2, respectively; V_1 and V_2 are the membrane potential deviations, r_1 and r_2 are the nonjunctional membrane resistances, and r_j is the junctional resistance coupling the two cells (Bennett, 1966; Socolar, 1977). The coupling coefficient is usually measured under steady-state hyperpolarization. Since the postsynaptic voltage deviation can never be greater than that of the current-injected side, given passive extrajunctional resistances, coupling coefficients must be less than 1.

The coupling coefficient is a handy index of synaptic effectiveness. Giant neurons of invertebrates with short neurites often have large coefficients, provid-

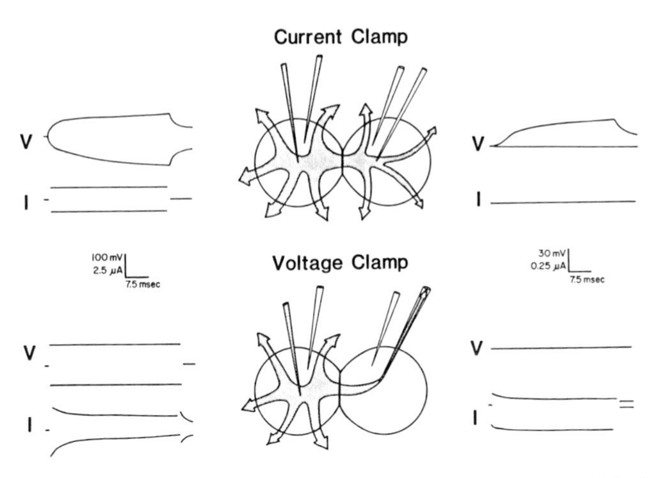

FIGURE 9. Diagram comparing four-microelectrode techniques for electrophysiological study of electrotonic synapses. (Top) Current clamp in which a constant-current pulse is injected into one cell and the resulting voltage change is measured in both cells and compared. The current across the junction is neither constant nor measured. (Bottom) Double voltage clamp in which both cells are individually clamped to bath ground. When one clamp is stepped, the other must respond with a current equal and opposite to the junctional current to maintain its membrane potential constant. Junctional voltage and current are both directly measured.

ing one-to-one spike transmission. For example, the crayfish septal junction has a coupling coefficient of about 0.45 (Watanabe and Grundfest, 1961; Asada and Bennett, 1971; Johnston and Ramon, 1981). One of the highest coefficients, 0.95, is found at the earthworm septal junction (Brink and Barr, 1977). This high value is due to myelination of the nonjunctional membrane, which results in very low transmembrane current loss. Vertebrate neurons tend to have lower coefficients. Hippocampal CA3 neurons, for example, are about 0.2 (MacVicar and Dudek, 1981) and pufferfish supermedullary neurons are about 0.05 (Bennett *et al.*, 1967a).

The coupling coefficient does not always accurately represent the actual degree of physiological spike coupling, which may be underestimated because of the passive cable properties or overestimated because of the regenerative properties of a neurite. For example, spike coupling in hippocampal CA3 neurons is about half that for hyperpolarizations, due to the low-pass filtering capacity of the system (MacVicar and Dudek, 1981; see Section 3.3). Spike coupling between pufferfish supermedullary neurons is slightly better than predicted by the measured coupling coefficient, because spikes can propagate directly to the synapse whereas hyperpolarizations are applied from a distance.

Coupling coefficients do not give detailed information about the properties of the junctional resistance itself because they are sensitive to postsynaptic nonjunctional membrane resistance, they cannot distinguish from among coupled networks, and they are insensitive to changes in junctional resistance as this value approaches that of the nonjunctional resistance (i.e., $k \rightarrow 0.5$) (Socolar, 1977). The measurement also includes the access resistance between the recording site and the junctions, which may be distant. Finally, the coupling coefficient depends upon which cell current is injected into, which must be specified.

A related parameter is the transfer resistance, defined as the voltage deviation in one cell for a current injected into the other:

$$R_t = \Delta V_1/\Delta I_2 = r_1 r_2/(r_1 + r_j + r_2) = \Delta V_2/\Delta I_1 \tag{2}$$

I_1 and I_2 represent the currents injected into neurons 1 and 2, respectively (Bennett, 1966). As with the coupling coefficient, transfer resistance is sensitive to nonjunctional membrane resistance and access resistance, but, for a linear system, it is insensitive to the direction of the test. Thus, the equality of forward and reverse transfer resistances is a quick and useful test for good microelectrode penetration. Transfer resistance can be used in the so-called pi-tee transform (Bennett, 1966) to calculate junctional resistance:

$$r_j = (R_1 R_2 - R_t^2)/R_t \tag{3}$$

R_1 and R_2 are the input resistances (junctional membrane in parallel with nonjunctional) of cells 1 and 2.

As an example of the use of these equations, we can analyze the records of Fig. 2, showing current injections of a crayfish septal junction. This is a high-conductance synapse with one-to-one spike transmission (top). The hyperpolarizing pulses (bottom) produce a presynaptic voltage change of ~ 43 mV and a postsynaptic change of ~ 23 mV, so the coupling coefficient is ~ 0.46 [equation (1)]. The coefficient is approximately the same measured in either direction because the two cells are nearly identical. For a 550-nA current, the input resistance is ~ 83 kohm for both sides and the transfer resistance is ~ 42 kohm from which a junctional resistance of ~ 122 kohm is calculated [equation (3)]. Previous work on this synapse has shown by extrapolation that the true junctional resistance is closer to 16 kohm, the difference being due to axoplasmic access resistance (Johnston and Ramon, 1981; see also Fig. 10). These equations are invalid if the membrane resistance changes during the test pulses, as in the depolarizing direction of these records.

A more powerful method of studying electrotonic synapses is by voltage clamp, which has been implemented in two ways. By the internal ground method, the junctional membrane is directly clamped, with the inside of one of the coupled cells defined as reference potential or ground (Johnston and Ramon, 1982; Verselis and Brink, 1984). With the double voltage clamp method (Fig. 9, bottom), two independent voltage clamps are utilized, each one clamping one of the coupled cells to external ground (Spray *et al.*, 1979; Jaslove and Brink, 1986). By default, the junctional membrane between the two must also be clamped. The latter method is easier to implement and has better current control capabilities.

The power of the voltage clamp may be understood by consideration of Fig. 9. With current clamp, a constant-current pulse is injected into one cell but divides between the junctional and nonjunctional membrane pathways, so that the actual junctional current is neither known nor constant. With voltage clamp, the voltage is the same everywhere within the space-clamped regions and therefore the actual voltage across the junction, V_j, is known and constant (Fig. 10). When one side of the junction is stepped, the voltage clamp on the other side must hold its cell potential constant by responding with exactly enough current to offset any junctional current. Therefore, junctional current is also known (Fig. 10) and junctional resistance is directly computable as:

$$r_j = \Delta V_j / \Delta I_j \tag{4}$$

This measurement is insensitive to both the direction of the test and the nonjunctional membrane resistance as long as the junctional membrane is within the

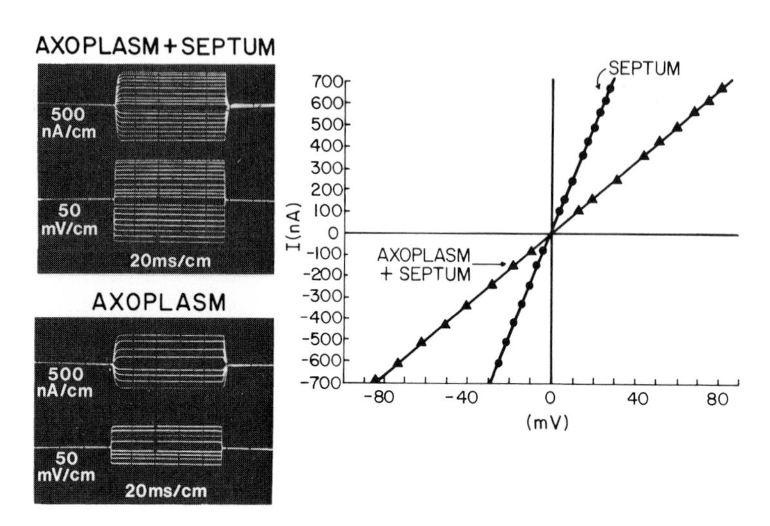

FIGURE 10. Voltage clamp records from a linear electrotonic synapse, the earthworm septal junction, recorded with an internal ground circuit. The symmetrical current and voltage records are shown on the left and the steady-state *I*/*V* curve is plotted on the right. These records also demonstrate the axoplasmic access resistance (bottom, left), which may be subtracted from the total to give the true septal resistance (average = 30–50 kohm). From Verselis and Brink (1984) with copyright permission of the Biophysical Society.

space-clamped region. All of the usual considerations for speed, accuracy, and stability of microelectrode voltage clamps apply (Finkel and Gage, 1985). In the case of crayfish lateral giant axons, lowered access resistance for current electrodes has been achieved by replacing the current microelectrodes with low-resistance cannulas (Johnston and Ramon, 1982; Jaslove and Brink, 1986). A similar approach using a double patch clamp has been applied to junctions between nonneuronal cells and has produced apparent single channel measurements (Neyton and Trautmann, 1985).

Indirect methods of studying electrotonic synapses are important for preparations such as the central nervous system in which only one recording point, a microelectrode in the postsynaptic cell, is usually available (Martin and Pilar, 1964; Bennett *et al.*, 1967b–d; Baker and Llinas, 1971; Korn *et al.*, 1973; Llinas *et al.*, 1974). These methods involve stimulation and analysis of electrotonic postsynaptic potentials (PSPs) generated by presynaptic spikes (Fig. 11). Electrotonic PSPs may occur singly or, in mixed synapses, as the early phase of a complex waveform (Fig. 18). If there is a low incidence of coupling, then presynaptic activation generates all-or-none PSPs of about 1-mV amplitude representing the all-or-none nature of the presynaptic spike (Fig. 11, left), as found in the rat mesencephalic nucleus (Baker and Llinas, 1971). However, if there is a high incidence of coupling, then graded stimulation recruits in a larger number of presynaptic neurons, which results in graded PSPs (Fig. 11, right). Such graded potentials can exceed 5 mV and evoke postsynaptic spikes in the rat lateral vestibular nucleus (Korn *et al.*, 1973) and cat inferior olive (Llinas *et al.*, 1974). Electrotonic PSPs have been referred to as short-latency depolarizations (SLD; Baker and Llinas, 1971), graded antidromic potentials (GAD; Korn *et al.*, 1973), and fast prepotentials (FPP; MacVicar and Dudek, 1981).

Indirect methods potentially provide three sources of stimulation: direct current injection through the recording microelectrode, orthodromic stimulation along the presynaptic pathway, and antidromic stimulation of the postsynaptic axon. Stimulation of the physiological presynaptic pathway is often impractical so antidromic stimulation of pools of coupled neurons is more commonly utilized. The strategy of this method is to record from a neuron with a high threshold to extracellular stimulation so that at low stimulus levels it is not directly activated but its coupled neighbors are.

Electrotonic PSPs are similar in shape to, and must be distinguished from, chemically mediated EPSPs and electrotonically decayed dendritic spikes (Wong *et al.*, 1979) and node of Ranvier spikes (Baker and Llinas, 1971). Tests include synaptic latency, frequency, polarization, and collision.

The latency test is based on the fact that the typical chemical synapse incurs a synaptic delay of up to several milliseconds associated with neurotransmitter release (see Silinsky, 1985). In principle, electrotonic synapses should have no delay, but, because of the charging time of the postsynaptic membrane, electrotonic delays are usually put at 0.2–0.4 msec, measured, say, as time to half peak (see Bennett, 1966). Longer delays would not necessarily prove chemical

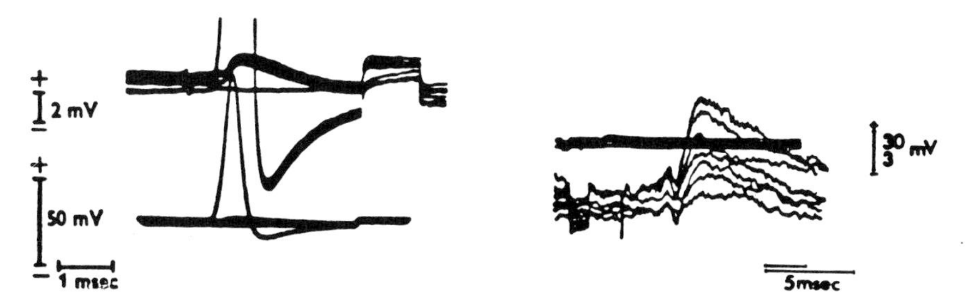

FIGURE 11. Superimposed electrotonic PSPs measured in the vertebrate central nervous system following antidromic stimulation. (Left) Rat mesencephalic neurons. Upper trace, high gain; lower trace, low gain. The low incidence of coupling in this nucleus leads to all-or-none PSPs reflecting the all-or-none nature of the presynaptic spike. One trace was suprathreshold for antidromic spike invasion of the recording soma. From Baker and Llinas (1971) with copyright permission of the Physiological Society. (Right) Cat inferior olive. Upper trace, low gain and slow sweep speed; lower trace, high gain and fast sweep speed. This highly coupled nucleus shows graded PSPs reflecting increasing recruitment of presynaptic fibers with increasing stimulus amplitude. From Llinas *et al.* (1974) with copyright permission of the American Physiological Society.

coupling, since there may simply be a very long membrane time constant (see Section 3.3). The measure of synaptic delay in a single-microelectrode system is ambiguous. Usually, time 0 is defined as the appearance of an extracellular field potential generated from what is assumed to be the activation of the presynaptic terminal (Fig. 12) (Korn *et al.*, 1973; Llinas *et al.*, 1974).

Frequency tests take advantage of the tendency of chemical synapses to become depressed or facilitated by repetitive stimulation, and to block completely at very high stimulation rates or long-lasting trains (see Silinsky, 1985). Electrotonic synapses are much less sensitive to frequency and duration of stimulation, although it is possible for pre- and postsynaptic capacitance to result in summation and facilitation of electrotonic PSPs, under the proper circumstances (Fig. 13) (Bennett and Pappas, 1983; see also Section 3.3). One possible source of error with high-frequency stimulation is spike propagation failure at axonal branch points or other regions of low safety factor. Such a block could prevent full activation of presynaptic terminals (Muller and Scott, 1981).

Polarization tests depend on the fact that chemical synaptic currents are driven by the difference between the postsynaptic membrane potential and the synaptic reversal potential. The latter is typically near 0 mV for an EPSP (Coombs *et al.*, 1955). Thus, hyperpolarization of the postsynaptic neuron should increase the height of the EPSP and depolarization should diminish or even reverse it. In contrast, electrotonic synapses (with the possible exception of the rare voltage-dependent rectifiers) should be insensitive to membrane poten-

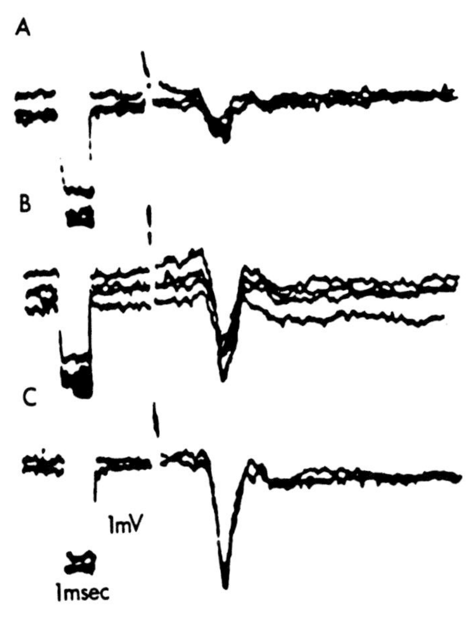

FIGURE 12. (A–C) Superimposed extracellular field potentials recorded in cat inferior olive for three different levels of stimulation of the contralateral cerebellar white matter. Upward deflection is the stimulus artifact. The field potential is often taken to represent time 0 for measurement of synaptic delay in systems for which there is no direct measurement of presynaptic depolarization. From Llinas *et al.* (1974) with copyright permission of the American Physiological Society.

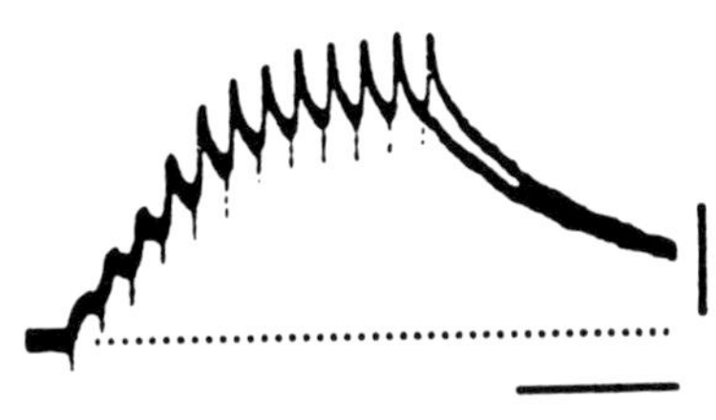

FIGURE 13. Summation and facilitation of an electrotonic PSP in the electric organ of the stargazer *Astroscopus* as a result of antidromic tetanic stimulation. This effect was ascribed mainly to presynaptic mechanisms: invasion of the presynaptic soma was facilitated because of incomplete time for repolarization between spikes, thus leading to a larger presynaptic potential reflected as a larger PSP. Calibrations: 5 mV; 50 msec. From Bennett (1968) with copyright permission of Prentice–Hall. See also Bennett and Pappas (1983).

tial and not be reversible. This has been demonstrated within a limited range of potentials by polarizing a postsynaptic electromotor neuron through the recording electrode during the arrival of a PSP (Fig. 14) (Bennett *et al.*, 1967c). However, incorrect results may occur using this method because of voltage-dependent changes in synaptic gating or nonjunctional membrane resistance which lead to extrapolation to unreasonable reversal potentials. An example of the former occurs at the rectifying electrotonic motor giant synapse of the hatchetfish (Auerbach and Bennett, 1969). Here, hyperpolarization of the postsynaptic axon increases the activation of the synapse and thus increases synaptic current and PSP amplitude, so that extrapolation results in an apparent reversal potential for this synapse of about 0 mV. Polarization of the postsynaptic side of the highly conducting crayfish septal junction affects the height of the presynaptic spike and therefore also the size of the PSP (Watanabe and Grundfest, 1961). Postsynaptic polarization should only be used in cells with low backwards coupling coefficients in order to avoid such interference with activation of presynaptic terminals.

Spike collision is an important control for tests involving antidromic stimulation, which eliminates the possibility of axonally generated currents entering the soma of the postsynaptic cell and generating PSP-like signals (Baker and Llinas, 1971; Korn *et al.*, 1973; Llinas *et al.*, 1974). The recording electrode is used to stimulate a postsynaptic orthodromic spike which, in the case of a

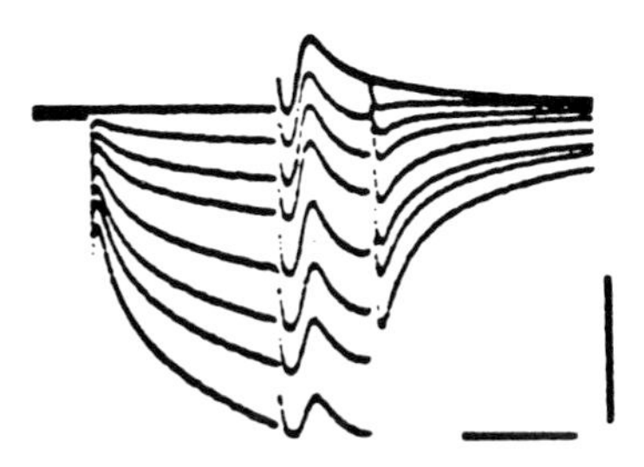

FIGURE 14. Demonstration of postsynaptic polarization of an electrotonic PSP recorded in the electric organ of the electric catfish *Malapterurus*. There should be little effect on the PSP so long as the reverse coupling coefficient is small. The small effect on PSP amplitude in this example was due to changes in presynaptic invasion or postsynaptic input resistance as a result of the hyperpolarizing pulse. There should be no reversal potential as would be found for a chemical synapse. Calibrations: 50 mV; 2 msec. From Bennett *et al.* (1967c) with copyright permission of the American Physiological Society.

chemical or weak electrotonic synapse, should not interfere with the generation of a subsequent PSP. The entire system is then fired antidromically but the propagating orthodromic spike blocks the arrival at the soma of an antidromic spike for a period of about twice the axonal propagation time plus the membrane refractory period. This procedure is also useful for blocking antidromic invasion of neurons that do not have the highest threshold in the stimulated trunk.

A possible source of error in all indirect tests is failure of antidromic invasion of the presynaptic terminal. This problem has been elucidated in an invertebrate model, the stomatogastric ganglion of the spiny lobster (Mulloney and Selverston, 1972). A motoneuron was studied that generated inhibitory chemical synaptic potentials when fired orthodromically, but failed to do so when fired antidromically. Antidromic spikes arriving at the soma were smaller and faster than soma-generated spikes due to geometrical factors that enhanced propagation in one direction and inhibited it in the other.

An illuminating example of the difficulties in using indirect methods to study electrotonic coupling in the central nervous system comes from work on the mechanism of the Ia EPSP of cat spinal motoneurons. This synapse is typically studied with a single microelectrode in the motoneuron soma and remote stimulation of afferent fibers. Early work utilizing a double-barreled microelectrode technique reported the Ia EPSP to be chemically mediated, based largely upon measurement of a reversal potential near 0 mV (Coombs *et al.*, 1955). Subsequent work confirmed the reversal of the late phase of the EPSP, but failed to produce a reversal of the early phase, suggesting that the synapse must be at a distant dendritic site (Smith *et al.*, 1967). Two later reports attempting to consider only proximally generated EPSPs recorded through a bridge circuit still failed to reverse the early phase of the EPSP and suggested a possible role of electrotonic coupling at this synapse (Edwards *et al.*, 1976; Werman and Carlen, 1976). The corresponding Ia synapse of the frog spinal cord is known to be mixed (Fig. 18), having early electrotonic and late chemical components (Shapovalov and Shiriaev, 1980). Ultimately, a unique system was developed that allowed penetration of two independent microelectrodes into a single spinal motoneuron in order to measure reversal potentials without the necessity of a bridge circuit (Engberg and Marshall, 1979). It was found that the early phase of the EPSP did reverse at a potential close to the 0 mV predicted for a chemical synapse, and it was suggested that the earlier, negative results were influenced by delayed rectification of the nonjunctional membrane, which drew too much injection current to maintain adequate bridge balance.

4.4. Dye Coupling

Gap junction channels are able to transport dyes and stains up to a molecular weight of about 1500–3000 daltons (Bennett, 1973; Simpson *et al.*, 1977; Schwarzmann *et al.*, 1981), so that complicated electrotonic networks can be traced with a single dye-injecting microelectrode penetration. The most well-known tracers are procion yellow (mol. wt. 500–600; Payton *et al.*, 1969), cobalt (atomic weight 58.93; Politoff *et al.*, 1974), fluorescein (mol. wt. 332.31; Kanno and Loewenstein, 1966; Furshpan and Potter, 1968), 6-carboxyfluorescein (mol. wt. 376.32; Weinstein *et al.*, 1977), and Lucifer Yellow (mol. wt. 457.25; Stewart, 1978, 1981). Horseradish peroxidase (HRP; mol. wt. 42,000) is excluded from gap junction channels (Bennett *et al.*, 1973) and so is a useful control.

Dye coupling has been used to demonstrate intercellular communication in hippocampus (MacVicar and Dudek, 1980), hypothalamus (Andrew *et al.*, 1981), neocortex (Gutnick and Prince, 1981), and frog spinal cord (Brenowitz *et al.*, 1983) as well as a variety of invertebrate preparations (Fig. 15A). Unfortunately, the technique has produced some anomalous results. In some systems that are known to be electrotonically coupled, dye coupling has not been found (Strausfeld and Obermayer, 1976; Goodman and Spitzer, 1979; Kaczmarek *et al.*, 1979; Audesirk *et al.*, 1982; Powell and Westerfield, 1984). Other systems, which are dye coupled, are not electrotonically coupled (Zieglgansberger and Reiter, 1974; Kuhnt *et al.*, 1979). Even within the same system there are sometimes conflicting results: frog spinal motoneurons are well coupled electrotonically (Grinnell, 1966; Sotelo and Taxi, 1970; Westerfield and Frank, 1982), yet dye coupling experiments have produced both positive (Brenowitz *et al.*, 1983) and negative (Powell and Westerfield, 1984) results. Cat spinal motoneurons do not appear to be coupled electrotonically (Engberg and Marshall, 1979), yet dye coupling has been reported (Zieglgansberger and Reiter, 1974). Dye coupling has been questioned as a possible artifact in both hippocampus (Knowles *et al.*, 1982) and neocortex (Gutnick *et al.*, 1985). In this section we will discuss possible reasons for obtaining false-positive or false-negative results from these experiments.

One obvious explanation for anomalous dye coupling is anomalous electrical coupling (Fig. 15B,C). If electrical coupling changes for any reason during tissue preparation, then dye flux should be similarly affected and not represent the *in vivo* state. There are a variety of ways in which this could occur. Both electrotonic and dye coupling increase between neurons of *Helisoma* after axotomy (Bulloch and Kater, 1982; Murphy *et al.*, 1983). Lucifer Yellow coupling in the guinea pig neocortical slice depends upon the plane of section of the slice: when cut radially, dye coupling increases, again suggesting increased coupling as a result of severed cell processes (Gutnick *et al.*, 1985). In crayfish, coupling decreases in strength between axons that have sustained membrane damage (Asada and Bennett, 1971) or lowered pH (Giaume *et al.*, 1980). Hypertonic medium decreases dye coupling at earthworm septa (Brink and Barr, 1977) and affects coupling between magnocellular neurons of the rat hypothalamus in a more complex way (Cobbett and Hatton, 1984).

A first consideration in the use of any tracer should be its nonjunctional membrane permeability. Dye coupling assumes that the stain moves between cells only by the gap junction pathway. Movement by way of nonjunctional

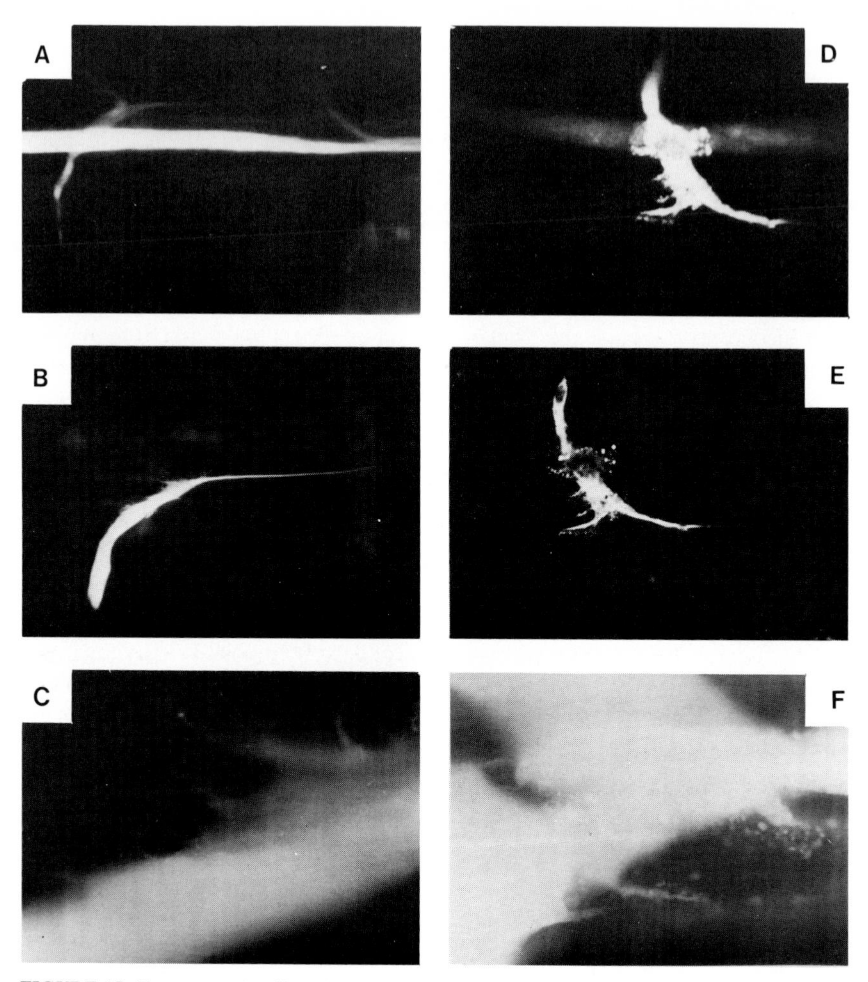

FIGURE 15. Fluorescent dye-fills of giant axons of the crayfish ventral nerve cord at the level of the second and third abdominal ganglia. All fills were microelectrode iontophoretic injections of 6-carboxyfluorescein except F, which was Lucifer Yellow. (A) A lateral axon was injected and dye crossed into both a motor axon (left) and a lateral axon of the next segment (right). (B) A motor axon was injected but no dye passed into the lateral axon because the junction was uncoupled, probably due to the voltage-dependent properties of the rectifying motor synapse. (C) Same specimen as B at higher magnification showing the fine synaptic processes of the motor axon. (D) Injection of another motor axon demonstrating blebbing at the site of penetration and possible artifactual filling of the lateral axon. (E) Same specimen as D after dehydration and clearing in alcohol and methyl salicylate without prior fixation. The heavy fill and blebbing of the motor axon was well preserved, but the light fill of the lateral axon was washed away. (F) High-magnification photo of another filled motor axon showing beading and disruption of the fine endings. For calibration, the diameter of the motor axon at its widest point in each photo was about 100 μm.

membranes would result in underestimation of coupling if most of the dye is lost to the extracellular solution, or overestimation if there is a high rate of transport between closely apposed membranes (Bennett, 1973; van Venrooij *et al.*, 1975; Brink and Ramanan, 1985). Of the dyes listed above, only fluorescein has significant membrane permeability, which led to the adoption of its more polar derivative, 6-carboxyfluorescein (Weinstein *et al.*, 1977).

A hole created artifactually between cells can also provide an adequate pathway for intercellular dye flux. Such holes have reportedly been created by the tips of microelectrodes resting between cells of carp retina (Kaneko *et al.*, 1981) and of rat hippocampus (Alger *et al.*, 1983). Dye coupling by rupture of closely apposed membranes can also occur as a result of overinjection. This error may be difficult to detect as the dye transfer can appear highly specific: there is often little leakage of dye apparent in extracellular space. We have found that high pressure can drive 6-carboxyfluorescein rapidly across uncoupled motor synapses of internally perfused crayfish axons (unpublished results). Procion Yellow similarly crosses the giant (chemical) synapse of the squid under too much pressure from microelectrode injection (Llinas, 1973, pp. 134 and 215). Rupturing may be related to blebbing of the membrane. The cell in Fig. 15D,E responded to overinjection of 6-carboxyfluorescein by growing membranous buds that sometimes pinched off. Similar examples of blebbing have been reported during injection of fluorescein into embryonic cells of *Fundulus* (Bennett *et al.*, 1978).

Damaged neuronal processes can take up dye from extracellular space. Cut crayfish lateral giant axon segments take up Procion Yellow, though their septal junctions are uncoupled (Payton *et al.*, 1969). In frog spinal cord, nonspecific staining of neurons by Lucifer Yellow has been found along the entire track of the withdrawn injection microelectrode (Powell and Westerfield, 1984). In guinea pig hippocampal slices, good Lucifer Yellow fills were obtained even when the injecting microelectrode was intentionally pushed through the target cell, as though dye leaked back up along the electrode shaft (Knowles *et al.*, 1982). In cat spinal cord, Procion Yellow produced large amounts of intracellular staining when injected by hypodermic needle, but not when applied electrophoretically (Zieglgansberger and Reiter, 1974). The large syringe needle apparently damaged numerous neuronal processes, which took up the stain.

Vesicularization may play a role in dye coupling. Schwann cells surrounding crayfish giant axons readily take up axonally injected Lucifer Yellow, though they do not appear to be electrotonically coupled to the axon itself (Viancour *et al.*, 1981). And severed crayfish giant axon segments remain viable for long periods with metabolic support from Schwann cells and other giant axons (Bittner, 1981). It has been suggested that these pathways rely upon coordinated exo-endocytosis (Fig. 16) capable of transferring large molecules between cells (Bittner, 1981). Electron microscopy has suggestively revealed apposed fusions of ~ 50-nm "synaptic vesicles" at adjacent membranes of the crayfish septal

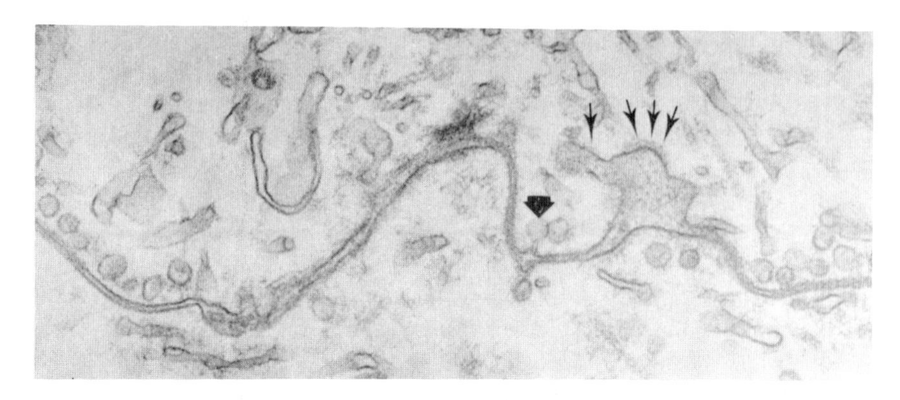

FIGURE 16. Electron micrograph of a crayfish lateral axon septal junction demonstrating possible coordinated exo-endocytosis (large arrow) in parallel with gap junctions. Such events have been suggested to be a mechanism for transfer of large molecules between cells. Other "synaptic vesicles" may be seen along the septal membranes. ×85,000. From Zampighi *et al.* (1978) with copyright permission of Longman Press.

junction (Zampighi *et al.*, 1978) and the vertebrate brain (Waxman *et al.*, 1980). A related process may have been visualized in one report using time-lapse photography through Nomarski optics to show "bubbling" of micrometer-sized vacuoles in parallel with electrotonic coupling at the crayfish septal junction (Politoff, 1977).

There have been several reports of dye transport along specific chemical synaptic pathways in the CNS including Procion Yellow in rat brain (Kuhnt *et al.*, 1979), cobalt in insect (Strausfeld and Obermayer, 1976), and even HRP in insect (Nassel, 1982) and vertebrate (Triller and Korn, 1981). In some cases, but not all, this transsynaptic dye transport has been correlated with the rate of stimulation of the chemical synapse (Kuhnt *et al.*, 1979). Again exo-endocytocis at the chemical synaptic region has been identified as the probable mechanism of dye transfer (Gomez-Ramos and Rodriguez-Echandia, 1981; Triller and Korn, 1981).

Connectivity by way of small processes in the nervous system can be especially troublesome for dye studies. Both electrical and dye coupling have been found to occur between pairs of *Aplysia* bag cells coupled near somata, but only electrical coupling was found between cells coupled by distal processes (Kaczmarek *et al.*, 1979). Intracellular dye diffusion is affected by cytoplasmic viscosity, extrajunctional membrane permeability, and binding to cytoplasmic proteins (Zimmerman and Rose, 1985; Brink and Ramanan, 1985). Given a long neuronal process, and particularly a thin one, these combined effects can significantly hinder tracer from reaching the synapse. Further, electrical coupling is related to the ratio of junctional membrane area to total postsynaptic membrane area, while dye coupling is related to the ratio of junctional area to the postsynaptic cell volume (Sheridan, 1973). Thus, for a constant injection current, dye coupling becomes relatively weaker than electrical coupling as postsynaptic cell size increases. Also, for a constant dye concentration, the fluorescence intensity becomes weaker as the cell size decreases (Fig. 17). It should be noted that failure of dye coupling between electrotonically coupled cells has also been reported to occur between nonneural cells without processes (Lo and Gilula, 1979, for example).

Another possible explanation for conflicting dye results is differential permeability of junctions to dye molecules. There are believed to be at least two classes of gap junctions. The "A" type, found in vertebrate, mollusk and annelid nervous systems, seems to be smaller and less permeable to dyes than the

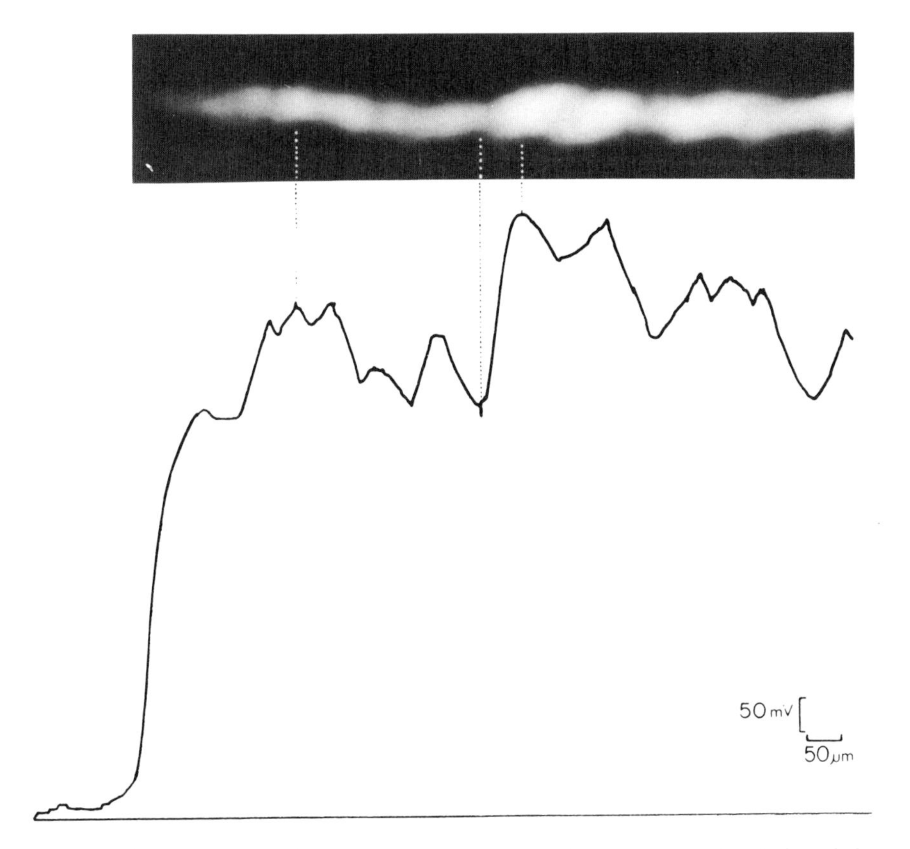

FIGURE 17. Demonstration of the influence of cell shape on fluorescence intensity. An irregularly shaped segment of an earthworm median giant axon was uniformly filled with 6-carboxyfluorescein by microelectrode iontophoresis (top). The fluorescence intensity profile, however, recorded with a scanning photomultiplier, varied strongly with axon diameter (bottom). From Brink and Ramanan (1985) with copyright permission of the Biophysical Society.

"B" type, found in arthropods (Gilula, 1974; Brink *et al.*, 1981). For example, Procion Yellow crosses septal junctions of the crayfish (Payton *et al.*, 1969) much more readily than septal junctions of the earthworm (Mulloney, 1970). There may also be differential permeability within these two categories. One report has suggested that intracellular calcium may close junctions by narrowing the channel diameter (Rose *et al.*, 1977), which would have the effect of increasing permselectivity, but more recent evidence is pointing toward all-or-none closure of junctional channels (Zimmerman and Rose, 1985). On the other hand, junctional walls appear to be negatively charged (Brink and Dewey, 1980) and dyes appear to permeate in charged, hydrated form (Brink, 1983; Brink and Ramanan, 1985) so that any treatment that would influence electrostatic fields, hydrophobicity, or viscosity within the channels (such as altering intracellular calcium, pH, or temperature) may indeed affect dye selectivity while not necessarily blocking small ion transport (i.e., electrotonic coupling).

Tracers can also have direct effects on junctional membranes. Cobalt chloride freely permeates crayfish septal junctions when present at low concentrations, but causes uncoupling at high concentrations (Politoff *et al.*, 1974). Many dyes are known to be toxic to nerve cells (Miller and Selverston, 1979). Lucifer Yellow is particularly toxic in intense blue light, a phenomenon that has been used to advantage in producing selective axotomy in lobster and crayfish (Miller and Selverston, 1979). Carboxyfluorescein has been utilized similarly in the *Helisoma* nervous system (Cohan *et al.*, 1983). Figure 15F illustrates an example of neurite damage to a crayfish motor axon in which the fine, feathery synaptic processes (Fig. 15C) have become beaded after Lucifer Yellow injection.

The final concern of this section is fixation. The common fixative glutaraldehyde has been reported to produce micrometer-sized holes through junctional membranes, resulting in subsequent intercellular HRP transport (Bennett *et al.*, 1973; Bennett, 1973). Postsynaptically transported Lucifer Yellow has been reported to wash out during histological preparation (Margiotta and Walcott, 1983). We have also found this, but only if the dye concentration was low. At high concentrations, crayfish giant axons bind intracellular Lucifer Yellow and 6-carboxyfluorescein without fixation, and retain them even after tissue dehydration and clearing (Fig. 15D,E). The usual procedure of dehydration and clearing by alcohol followed by xylene or methyl salicylate has been shown to produce significant morphological artifacts in guinea pig neocortex (Grace and Llinas, 1985). These include cell shrinkage with Lucifer Yellow and neurite deformation with HRP, either of which could lead to dye leakage or a misestimate of cell contact area. These artifacts were found to be eliminated by a new procedure for clearing using dimethyl sulfoxide (Grace and Llinas, 1985). Still, whenever possible, dye fills should be viewed without the aid of histological treatment.

In conclusion, dye coupling is a powerful but complex technique for tracing electrotonically coupled pathways. Several controls are essential. Whenever possible, simultaneous electrophysiological tests of coupling should be performed while injecting dyes. Extracellular application of dyes can test for finite membrane permeability or uptake through damaged processes. Injection of combinations of permeant and impermeant dyes can indicate whether the effective pore size is within a reasonable range, eliminating transport by fusion or exo-endocytosis. Finally, dye coupling can be compared before and after pharmacological uncoupling of the tissue as a test for alternate coupling routes, as will be described in Section 4.6.

4.5. Metabolic Coupling

The intercellular transfer of small molecules as metabolic agents or signals is of great importance in nonneural tissues (see Loewenstein, 1979) and probably also in the nervous system in development and regeneration (see Section 3.5). But the role of gap junctions in metabolic coupling in the adult nervous system has been overshadowed by their more obvious role in electrotonic coupling.

Tests utilizing metabolic coupling usually involve loading cells with a marker, such as labeled precursor, and then allowing these cells to couple to unloaded cells (Rieske *et al.*, 1975; see also Loewenstein, 1979). The latter group is later assayed for transfer of the marker, perhaps by autoradiography or simply by cell viability. Since this method involves a rather long time-scale and complex analysis, it is not generally used as a test for electrotonic coupling in the nervous system and will not be considered in detail. The considerations for metabolite transfer are the same as those previously discussed for similarly sized dye molecules. And, as with dye coupling, there is evidence that a significant amount of metabolic coupling in the nervous system proceeds by a process of coordinated exo-endocytosis (Bittner, 1981).

4.6. Pharmacology

Both chemical and electrotonic synapses have well-defined pharmacological interactions. Chemical synapses are readily blocked by replacement of extracellular calcium with magnesium or manganese (see Silinsky, 1985). Most electrotonic synapses are uncoupled by acidifying the extracellular solution with membrane-permeant agents (Giaume *et al.*, 1980; Spray *et al.*, 1982). When used in conjunction with electrophysiological or dye-coupling techniques, these agents provide a powerful means of distinguishing synaptic types. For example, the Ia synapse of frog spinal motoneurons is mixed. In the isolated, perfused cord, a low-Ca, high-Mn solution blocks the late chemical component of the EPSP, but not the early electrotonic component (Fig. 18) (Shapovalov and Shiriaev, 1980). Neurons of the neocortical slice are dye-coupled, and dye spread can be greatly reduced by bubbling CO_2 through the medium to lower its pH, suggesting that this coupling is by way of gap junctions and not cell fusions or holes (Gutnick and Lobel-Yaakov, 1983).

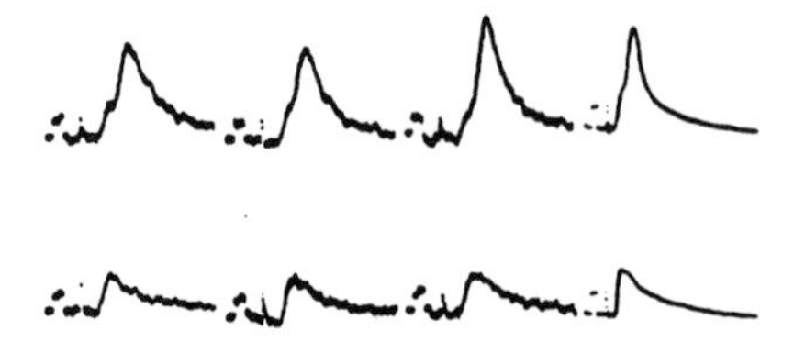

FIGURE 18. Demonstration of mixed Ia EPSPs recorded from motoneurons of an isolated, perfused frog spinal cord. The first three records in each row show individual EPSPs; the fourth record is an average of 200 individual EPSPs. Top row: Records from a cord in normal Ringer (including 1.8 mM Ca^{2+}, 0 mM Mn^{2+}) showing the fast electrotonic component on the rising phase. Bottom row: Records from the same preparation with modified Ringer (0 mM Ca^{2+}, 2 mM Mn^{2+}) which blocked the chemical component leaving a pure electrotonic PSP. Calibration pulses are 1 msec and 0.2 mV. From Shapovalov and Shiriaev (1980) with copyright permission of the Physiological Society.

Additional uncoupling agents may be found in a recent review of the pharmacology of gap junctions (Spray and Bennett, 1985). Care must be taken in manipulating extracellular calcium or pH because either can also influence nonjunctional membrane properties and other synapses in the system, resulting in misleading effects. One early study of coupling in avian ciliary ganglion utilized instead a specific chemical synaptic blocker, *d*-tubocurarine, to block the chemical component of a mixed synapse (Martin and Pilar, 1963). Antibodies to gap junction proteins are currently being developed and promise to provide a highly specific marker for electrotonic synapses. Intracellularly injected antibodies block both electrotonic and dye coupling (Warner *et al.*, 1984; Hertzberg *et al.*, 1985) and cause developmental errors when injected into the early embryo (Warner *et al.*, 1984). Antibodies can also be reacted with fluorescent label to allow visual identification of the synaptic site (Warner *et al.*, 1984; Hertzberg *et al.*, 1985).

5. CONCLUSION

Electrotonic coupling by way of gap junctions is a common and important mechanism of synaptic coupling in nervous systems from coelenterates to mammals. Its most important quality is the ability to provide for rapid and mutual excitation, and, as such, it has been found mainly in systems requiring high speed and synchronization. Rather than being purely passive circuit elements, electrotonic synapses are naturally modulated in a variety of ways, providing for the requisite plasticity of the nervous system. Demonstrating the existence of electrotonic coupling is often difficult, relying on morphological, electrophysiological, and dye-transfer techniques. Each of these has proven susceptible to numerous artifacts so that the strongest evidence of coupling is provided by a combination of all three, along with pharmacological controls.

Recent work has continued to demonstrate the importance of electrotonic coupling among the giant neurons of invertebrate nervous systems. Such preparations have provided most of the quantitative data regarding the structure and function of these synapses and have been fruitful substrates for studying the development and modulation of coupling. The intricacies of the vertebrate CNS have delayed equivalent progress in understanding coupling there, but improved techniques such as the brain slice and cell culture have become available for gaining access to coupled central neurons. Several groups have now succeeded in impaling coupled brain cells with pairs of electrodes for direct electrophysiological measurements (Engberg and Marshall, 1979; MacVicar and Dudek, 1981; Llinas and Yarom, 1981). Advances in methods of intracellular perfusion and voltage clamp are providing new data on the biophysics of electrotonic coupling (see Sections 3.1 and 4.3). Antibodies being developed to gap junctions can specifically block or label electrotonic synapses that may be inaccessible to other methods for studies of the function and distribution of intercellular junctions (see Section 4.6).

A paucity of data, thus far, has left the extent and role of coupling in the mammalian CNS largely unclear. While recent work has suggested that coupling in certain areas, such as cerebral cortex, may be less significant than at first it might have seemed (see Section 4.4), the importance of coupling in other areas, such as inferior olive, seems of little doubt. Coupling in the olivary nucleus appears to be associated with the normal spatial and temporal ($\sim$ 10/sec) synchronized activity patterns of cerebellar Purkinje cells (Llinas, 1985; Llinas and Yarom, 1986). Models of information processing in the brain have stressed the need to consider ensemble behavior of groups of neurons rather than the activity of individual neurons (e.g., Hopfield and Tank, 1986). Ensemble activity again implies electrotonic coupling as an underlying mechanism of synchronization (see Llinas and Yarom, 1986). On the pathological side of things, abnormally strong electrotonic coupling has been discussed as a basis for the characteristic oscillatory brain rhythms of at least two neurological diseases: epilepsy (Traub *et al.*, 1985) and Creutzfeldt–Jakob disease (Traub and Pedley, 1981) (though in the former case primarily by ephaptic field effects and in the latter by virus-induced membrane fusions). Thus, coupling would play an important role at the highest levels of brain function in both health and disease. The full extent and importance of electrotonic coupling in the mammalian CNS will ultimately be revealed as new techniques continue to be utilized to probe difficult-to-reach neurons and analyze complex neuronal circuits.

ACKNOWLEDGMENTS. We are grateful to Ms. Bonnie Bunch and Dr. Rodolfo Llinas for careful readings of the manuscript and to Ms. Lucille Betti and Mr. Steven Nash for their help with the illustrations. Work in our laboratory is supported by NIH Grants GM24905 and HL31299.

6. REFERENCES

Alger, B. E., McCarren, M., and Fisher, R. S., 1983, On the possibility of simultaneously recording from two cells with a single microelectrode in the hippocampal slice, *Brain Res.* **270:**137–141.

Alkon, D. L., and Grossman, L., 1978, Evidence for nonsynaptic neuronal interaction, *J. Neurophysiol.* **41:**640–653.

Anderson, T. E., and Bittner, G. D., 1980, Long term alteration of electrotonic synapses, *Brain Res.* **184:**13–36.

Andrew, R. D., MacVicar, B. A., Dudek, F. E., and Hatton, G. I., 1981, Dye transfer through gap junctions between neuroendocrine cells of rat hypothalamus, *Science* **211:**1187–1189.

Arvanitaki, A., and Chalazonitis, N., 1959, Interactions electriques entre le soma geant A et les somata immediatement contigus, *Bull. Inst. Oceanogr. Monaco* No. 1143, pp. 1–30.

Asada, Y., and Bennett, M. V. L., 1971, Experimental alteration of coupling resistance at an electrotonic synapse. *J. Cell Biol.* **49:**159–172.

Audesirk, G., Audesirk, T., and Bowsher, P., 1982, Variability and frequent failure of Lucifer Yellow to pass between two electrically coupled neurons in *Lymnaea stagnalis, J. Neurobiol.* **13:**369–375.

Auerbach, A. A., and Bennett, M. V. L., 1969, A rectifying electrotonic synapse in the central nervous system of a vertebrate, *J. Gen. Physiol.* **53:**211–237.

Baker, R., and Llinas, L., 1971, Electrotonic coupling between neurones in the rat mesencephalic nucleus, *J. Physiol. (London)* **212:**45–63.

Barr, L., Dewey, M. M., and Berger, W., 1965, Propagation of action potentials and the structure of the nexus in cardiac muscle, *J. Gen. Physiol.* **48:**797–823.

Barr, L., Berger, W., and Dewey, M. M., 1968, Electrical transmission at the nexus between smooth muscle cells, *J. Gen. Physiol.* **51:**347–368.

Baux, G., Simonneau, M., Tauc, L., and Segundo, J. P., 1978, Uncoupling of electrotonic synapses by calcium, *Proc. Natl. Acad. Sci. USA* **75:**4577–4581.

Baylor, D. A., and Nicholls, J. G., 1969, Chemical and electrical synaptic connexions between cutaneous mechanoreceptor neurones in the central nervous system of the leech, *J. Physiol. (London)* **203:**591–609.

Baylor, D. A., Fuortes, M. G. F., and O'Bryan, P. M., 1971, Receptive fields of cones in the retina of the turtle, *J. Physiol. (London)* **214:**265–294.

Bennett, M. V. L., 1966, Physiology of electrotonic junctions, *Ann. N.Y. Acad. Sci.* **137**(Art. 2):509–539.

Bennett, M. V. L., 1968, Similarities between chemically and electrically mediated transmission, in: *Physiological and Biochemical Aspects of Nervous Integration* (F. D. Carlson, ed.), pp. 73–128, Prentice–Hall, Englewood Cliffs, N.J.

Bennett, M. V. L., 1972, A comparison of electrically and chemically mediated transmission, in: *Structure and Function of Synapses* (G. D. Pappas and D. P. Purpura, eds.), pp. 221–256, Raven Press, New York.

Bennett, M. V. L., 1973, Permeability and structure of electrotonic junctions and intercellular movements of tracers, in: *Intracellular Staining in Neurobiology* (S. B. Kater and C. Nicholson, eds.), pp. 115–134, Springer-Verlag, Berlin.

Bennett, M. V. L., 1977, Electrical transmission: A functional analysis and comparison to chemical transmission, in: *Handbook of Physiology—The Nervous System I* (E. R. Kandel, ed.), pp. 357–416, Williams & Wilkins, Baltimore.

Bennett, M. V. L., and Pappas, G. D., 1983, The electromotor system of the stargazer: A model for integrative actions at electrotonic synapses, *J. Neurosci.* **3:**748–761.

Bennett, M. V. L., Aljure, E., Nakajima, Y., and Pappas, G. D., 1963, Electrotonic junctions between teleost spinal neurons: Electrophysiology and ultrastructure, *Science* **141:**262–264.

Bennett, M. V. L., Nakajima, Y., and Pappas, G. D., 1967a, Physiology and ultrastructure of electrotonic junctions. I. Supramedullary neurons, *J. Neurophysiol.* **30:**161–179.

Bennett, M. V. L., Pappas, G. D., Aljure, E., and Nakajima, Y., 1967b, Physiology and ultrastructure of electrotonic junctions. II. Spinal and medullary electromotor nuclei in mormyrid fish, *J. Neurophysiol.* **30:**180–208.

Bennett, M. V. L., Nakajima, Y., and Pappas, G. D., 1967c, Physiology and ultrastructure of electrotonic junctions, III. Giant electromotor neurons of *Malapterurus electricus, J. Neurophysiol.* **30:**209–235.

Bennett, M. V. L., Pappas, G. D., Gimenez, M., and Nakajima, Y., 1967d, Physiology and ultrastructure of electrotonic junctions. IV. Medullary electromotor nuclei in gymnotid fish, *J. Neurophysiol.* **30:**236–300.

Bennett, M. V. L., Spira, M. E., and Pappas, G. D., 1972, Properties of electrotonic junctions between embryonic cells of *Fundulus, Dev. Biol.* **29:**419–435.

Bennett, M. V. L., Feder, N., Reese, T. S., and Stewart, W., 1973, Movement during fixation of peroxidases injected into the crayfish septate axon, *J. Gen. Physiol.* **61:**254–255.

Bennett, M. V. L., Spira, M. E., and Spray, D. C., 1978, Permeability of gap junctions between embryonic cells of *Fundulus:* A reevaluation, *Dev. Biol.* **65:**114–125.

Bennett, M. V. L., Spray, D. C., and Harris, A. L., 1981, Electrical coupling in development, *Am. Zool.* **21:**413–427.

Bennett, M. V. L., Zimering, M. B., Spira, M. E., and Spray, D. C., 1985, Interaction of electrical and chemical synapses, in: *Gap Junctions* (M. V. L. Bennett and D. C. Spray, eds.), pp. 355–366, Cold Spring Harbor Laboratory, Cold Spring Harbor, N.Y.

Berry, M. S., 1972, Electrotonic coupling between identified large cells in the buccal ganglia of *Planorbis corneus, J. Exp. Biol.* **57:**173–185.

Bittner, G. D., 1981, Trophic interactions of CNS giant axons in crayfish, *Comp. Biochem. Physiol.* **68A:**299–306.

Bittner, G. D., and Ballinger, M. L., 1980, Ultrastructural changes at gap junctions between lesioned crayfish axons, *Cell Tissue Res.* **207:**143–153.

Brenowitz, G. L., Collins, W. F., III, and Erulkar, S. D., 1983, Dye and electrical coupling between frog motoneurons, *Brain Res.* **274:**371–375.

Brightman, M. W., and Reese, T. S., 1969, Junctions between intimately apposed cell membranes in the vertebrate brain, *J. Cell Biol.* **40:**648–677.

Brink, P. R., 1983, Effect of deuterium oxide on junctional membrane channel permeability, *J. Membr. Biol.* **71:**79–87.

Brink, P. R., and Barr, L., 1977, The resistance of the septum of the median giant axon of the earthworm, *J. Gen. Physiol.* **69:**517–536.

Brink, P. R., and Dewey, M. M., 1980, Evidence for fixed charge in the nexus, *Nature* **285:**101–102.

Brink, P. R., and Ramanan, S. V., 1985, A model for the diffusion of fluorescent probes in the septate giant axon of earthworm: Axoplasmic diffusion and junctional membrane permeability, *Biophys. J.* **48:**299–309.

Brink, P. R., Dewey, M. M., Colflesh, D. E., and Kensler, R. W., 1981, Polymorphic nexuses in the earthworm *Lumbricus terrestris, J. Ultrastruct. Res.* **77:**233–240.

Bulloch, A. G. M., and Kater, S. B., 1982, Neurite outgrowth and selection of new electrical connections by adult *Helisoma* neurons, *J. Neurophysiol.* **48:**569–583.

Carbonetto, S., and Muller, K. J., 1977, A regenerating neurone in the leech can form an electrical synapse on its severed axon segment, *Nature* **267:**450–452.

Carew, T. J., and Kandel, E. R., 1976, Two functional effects of decreased conductance EPSPs: Synaptic augmentation and increased electrotonic coupling, *Science* **192:**150–153.

Cobbett, P., and Hatton, G. I., 1984, Dye coupling in hypothalamic slices: Dependence on *in vitro* hydration state and osmolality of incubation medium, *J. Neurosci.* **4:**3034–3038.

Cohan, C. S., Hadley, R. D., and Kater, S. B., 1983, "Zap axotomy": Localized fluorescent excitation of single dye-filled neurons induces growth by selective axotomy, *Brain Res.* **270:** 93–101.

Connors, B. W., Bernardo, L. S., and Prince, D. A., 1983, Coupling between neurons of the developing rat neocortex, *J. Neurosci.* **3:**773–782.

Coombs, J. S., Eccles, J. C., and Fatt, P., 1955, Excitatory synaptic action in motoneurones, *J. Physiol. (London)* **130:**374–395.

Daniel, E. E., Daniel, V. P., Duchon, G., Garfield, R. E., Nichols, M., Malhotra, S. K., and Oki, M., 1976, Is the nexus necessary for cell-to-cell coupling of smooth muscle? *J. Membr. Biol.* **28:**207–239.

Decker, R. S., and Friend, D. S., 1974, Assembly of gap junctions during amphibian neurulation, *J. Cell Biol.* **62:**32–47.

De Mello, W. C., 1982, Changes in cell-to-cell coupling during the cardiac cycle, *Physiologist* **25:**197.

De Mello, W. C., 1984, Modulation of junctional permeability, *Fed. Proc.* **43:**2692–2696.

Detwiler, P. B., and Hodgkin, A. L., 1979, Electrical coupling between cones in turtle retina, *J. Physiol. (London)* **291:**75–100.

Dixon, J. S., and Cronly-Dillon, J. R., 1972, The fine structure of the developing retina in *Xenopus laevis, J. Embryol. Exp. Morphol.* **28:**659–666.

Dowling, J. E., 1968, Synaptic organization of the frog retina: An electron microscopic analysis comparing the retinas of frogs and primates, *Proc. R. Soc. London Ser. B* **170:**205–228.

Edwards, F. R., Redman, S. J., and Walmsley, B., 1976, The effect of polarizing currents on unitary Ia excitatory post-synaptic potentials evoked in spinal motoneurons, *J. Physiol. (London)* **259:**705–723.

Engberg, I., and Marshall, K. C., 1979, Reversal potential for Ia excitatory post synaptic potentials in spinal motoneurons of cats, *Neuroscience* **4:**1583–1591.

Fawcett, D. W., 1961, Intercellular bridges, *Exp. Cell Res. Suppl.* **8:**174–187.

Findlay, I., and Petersen, O. H., 1982, Acetylcholine-evoked uncoupling restricts the passage of Lucifer Yellow between pancreatic acinar cells, *Cell Tissue Res.* **225:**633–638.

Finkel, A. S., and Gage, P. W., 1985, Conventional voltage clamping with two intracellular microelectrodes, in: *Voltage and Patch Clamping with Microelectrodes* (T. G. Smith, H. Lecar, S. J. Redman, and P. W. Gage, eds.), pp. 47–94, American Physiological Society, Bethesda.

Fulton, B. P., Miledi, R., and Takahashi, T., 1980, Electrical synapses between motoneurons in the spinal cord of the newborn rat, *Proc. R. Soc. London Ser. B* **208:**115–120.

Furshpan, E. J., 1964, "Electrical transmission" at an excitatory synapse in a vertebrate brain, *Science* **144:**878–880.

Furshpan, E. J., and Potter, D. D., 1957, Mechanism of nerve-impulse transmission at a crayfish synapse, *Nature* **180:**342–343.

Furshpan, E. J., and Potter, D. D., 1959, Transmission at the giant motor synapses of the crayfish, *J. Physiol. (London)* **145:**289–325.

Furshpan, E. J., and Potter, D. D., 1968, Low-resistance junctions between cells in embryos and tissue culture, *Curr. Top. Dev. Biol.* **3:**95–127.

Furukawa, T., and Furshpan, E. J., 1963, Two inhibitory mechanisms in the Mauthner neurons of goldfish, *J. Neurophysiol.* **26:**140–176.

Garfield, R. E., Sims, S. M., Kannan, M. S., and Daniel, E. E., 1978, Possible role of gap junctions in activation of myometrium during parturition, *Am. J. Physiol.* **235:**C168–C179.

Getting, P. A., 1974, Modification of neuron properties by electrotonic synapses. I. Input resistance, time constant, and integration, *J. Neurophysiol.* **37:**846–857.

Getting, P. A., and Willows, A. O. D., 1974, Modification of neuron properties by electrotonic synapses. II. Burst formation by electrotonic synapses, *J. Neurophysiol.* **37:**858–868.

Giaume, C., and Korn, H., 1982, Ammonium sulfate induced uncouplings of crayfish septate axons with and without increased junctional resistance, *Neuroscience* **7**:1723–1730.

Giaume, C., and Korn, H., 1983, Biomechanical transmission at the rectifying electrotonic synapse: A voltage-dependent process, *Science* **220**:84–87.

Giaume, C., and Korn, H., 1984, Voltage-dependent dye coupling at a rectifying electrotonic synapse of the crayfish, *J. Physiol. (London)* **356**:151–167.

Giaume, C., Spira, M. E., and Korn, H., 1980, Uncoupling of invertebrate electrotonic synapses by carbon dioxide, *Neurosci. Lett.* **17**:197–202.

Gilula, N. B., 1974, Junctions between cells, in: *Cell Communication* (R. P. Cox, ed.), pp. 1–29, Wiley, New York.

Gogan, P., Gueritaud, J. P., Horcholle-Bossavit, G., and Tyc-Dumont, S., 1974, Electrotonic coupling between motoneurones in the abducens nucleus of the cat, *Exp. Brain Res.* **21**:139–154.

Gomez-Ramos, P., and Rodriguez-Echandia, E. L., 1981, Retrograde axonal transport and trans-neuronal transference of horseradish peroxidase in the rat ciliary ganglion, *Experientia* **37**:1337–1339.

Goodenough, O. A., and Gilula, N. B., 1974, The splitting of hepatocyte gap junctions and zonulae occludentes with hypertonic disaccharides, *J. Cell Biol.* **61**:575–590.

Goodman, C. S., and Spitzer, N. C., 1979, Embryonic development of identified neurones: Differentiation from neuroblast to neurone, *Nature* **280**:208–214.

Grace, A. A., and Llinas, R., 1985, Morphological artifacts induced in intracellularly stained neurons by dehydration: Circumvention using rapid dimethyl sulfoxide clearing, *Neurosciences* **16**:461–475.

Gregory, W. A., Hall, D. H., and Bennett, M. V. L., 1984, Ultrastructural studies of membrane specializations in the goldfish preoptic area, *Soc. Neurosci. Abstr.* **10**:1178.

Grinnell, A. D., 1966, A study of the interaction between motoneurons in the frog spinal cord, *J. Physiol. (London)* **182**:612–648.

Gutnick, M. J., and Lobel-Yaakov, R., 1983, Carbon dioxide uncouples dye-coupled neuronal aggregates in neocortical slices, *Neurosci. Lett.* **42**:197–200.

Gutnick, M. J., and Prince, D. A., 1981, Dye coupling and possible electrotonic coupling in the guinea pig neocortical slice, *Science* **211**:67–70.

Gutnick, M. J., Connors, B. W., and Ransom, B. R., 1981, Dye-coupling between glial cells in the guinea pig neocortical slice, *Brain Res.* **213**:486–492.

Gutnick, M. J., Lobel-Yaakov, R., and Rimon, G., 1985, Incidence of neuronal dye-coupling in neocortical slices depends on the plane of section, *Neuroscience* **15**:659–666.

Hadley, R. D., and Kater, S. B., 1983, Competence to form electrical connections is restricted to growing neurites in the snail, *Helisoma, J. Neurosci.* **3**:924–932.

Hadley, R. D., Bodnar, D. A., and Kater, S. B., 1985, Formation of electrical synapses between isolated, cultured *Helisoma* neurons requires mutual neurite elongation, *J. Neurosci.* **5**:3145–3153.

Hanna, R. B., Pappas, G. D., and Bennett, M. V. L., 1984, The fine structure of identified electrotonic synapses following increased coupling resistance, *Cell Tissue Res.* **235**:243–249.

Harris, A. L., Spray, D. C., and Bennett, M. V. L., 1983, Control of intercellular communication by voltage dependence of gap junctional conductance, *J. Neurosci.* **3**:79–100.

Hertzberg, E. L., Spray, D. C., and Bennett, M. V. L., 1985, Reduction of gap junctional conductance by microinjection of antibodies against the 27-kDa liver gap junction polypeptide, *Proc. Natl. Acad. Sci. USA* **82**:2412–2416.

Hopfield, J. J., and Tank, D. W., 1986, Computing with neural circuits: A model, *Science* **233**:625–633.

Jaslove, S. W., and Brink, P. R., 1986, The mechanism of rectification at the electrotonic motor giant synapse of the crayfish, *Nature* **323:**63–65.

Johnston, M. F., and Ramon, F., 1981, Electrotonic coupling in internally perfused crayfish segmented axons, *J. Physiol. (London)* **317:**509–518.

Johnston, M. F., and Ramon, F., 1982, Voltage independence of an electrotonic synapse, *Biophys. J.* **39:**115–117.

Kaczmarek, L. K., Finbow, M., Revel, J.-P., and Strumwasser, F., 1979, The morphology and coupling of *Aplysia* bag cells within the abdominal ganglion and in cell culture, *J. Neurobiol.* **1:**535–550.

Kaneko, A., Nishimura, Y., Tauchi, M., and Shimai, K., 1981, Morphological observation of retinal cells presumably made syncytial by an electrode penetration, *J. Neurosci. Methods* **4:**299–303.

Kanno, Y., and Loewenstein, W. R., 1966, Cell-to-cell passage of large molecules, *Nature* **212:**629–630.

Kater, S. B., 1974, Feeding in *Helisoma trivolvis:* The morphological and physiological bases of a fixed action pattern, *Am. Zool.* **14:**1017–1036.

Kensler, R. W., 1978, An untrastructural study of the nexuses in frog heart, earthworm lateral axon, and *Mytilus* ABRM, Doctoral dissertation, State University of New York, Stony Brook.

Knowles, W. D., Funch, P. G., and Schwartzkroin, P. A., 1982, Electrotonic and dye coupling in hippocampal CA1 pyramidal cells *in vitro, Neuroscience* **7:**1713–1722.

Korn, H., and Bennett, M. V. L., 1975, Vestibular nystagmus and teleost oculomotor neurons—Functions of electrotonic coupling and dendritic impulse initiation, *J. Neurophysiol.* **38:**430–451.

Korn, H., and Faber, D. S., 1975, An electrically mediated inhibition in goldfish medulla, *J. Neurophysiol.* **38:**452–471.

Korn, H., Sotelo, C., and Crepel, F., 1973, Electrotonic coupling between neurons in the rat lateral vestibular nucleus, *Exp. Brain Res.* **16:**255–275.

Korn, H., Sotelo, C., and Bennett, M. V. L., 1977, The lateral vestibular nucleus of the toadfish *Opanus tau:* Ultrastructural and electrophysiological observations with special reference to electrotonic transmission, *Neuroscience* **2:**851–884.

Kriebel, M. E., Bennett, M. V. L., Waxman, S. G., and Pappas, G. D., 1969, Oculomotor neurons in fish: Electrotonic coupling with multiple sites of impulse initiation, *Science* **166:**520–524.

Kuhnt, U., Kelly, M. J., and Schaumberg, R., 1979, Transsynaptic transport of Procion Yellow in the central nervous system, *Exp. Brain Res.* **35:**371–385.

Landmesser, L., and Pilar, G., 1970, Selective reinnervation of two cell populations in the adult pigeon ciliary ganglion, *J. Physiol. (London)* **211:**203–216.

Landmesser, L., and Pilar, G., 1972, The onset and development of transmission in the chick ciliary gangion, *J. Physiol. (London)* **222:**691–713.

Lane, J., and Swales, L. S., 1980, Dispersal of junctional particles, not internalization, during the *in vivo* disappearance of gap junctions, *Cell* **19:**579–586.

Larsen, W. J., Azarnia, R., and Loewenstein, W. R., 1977, Intercellular communication and tissue growth. IX. Junctional membrane structure of hybrids between communication-competent and communication-incompetent cells, *J. Membr. Biol.* **34:**39–54.

Levitan, H., Tauc, L., and Segundo, J. P., 1970, Electrical transmission among neurons in the buccal ganglion of a mollusc, *Navanax inermis, J. Gen. Physiol.* **55:**484–496.

Llinas, R. R., 1973, Discussion, in: *Intracellular Staining in Neurobiology* (S. B. Kater and C. Nicholson, eds.), pp. 134, 215, Springer-Verlag, Berlin.

Llinas, R. R., 1985, Electrotonic transmission in the mammalian central nervous system, in: *Gap Junctions* (M. V. L. Bennett and D. C. Spray, eds.), pp. 337–353, Cold Spring Harbor Laboratory, Cold Spring Harbor, N.Y.

Llinas, R., and Yarom, Y., 1981, Electrophysiology of mammalian inferior olivary neurones *in vitro:* Different types of voltage-dependent ionic conductances, *J. Physiol. (London)* **315:**549–567.

Llinas, R., and Yarom, Y., 1986, Oscillatory properties of guinea-pig inferior olivary neurones and their pharmacological modulation: An *in vitro* study, *J. Physiol. (London)* 376:163–182.

Llinas, R., Baker, R., and Sotelo, C., 1974, Electrotonic coupling between neurons in cat inferior olive, *J. Neurophysiol.* **37:**560–571.

Lo, C. W., and Gilula, N. B., 1979, Gap junctional communication in the post-implantation mouse embryo, *Cell* **18:**411–422.

Loewenstein, W. R., 1979, Junctional intercellular communication and the control of growth, *Biochim. Biophys. Acta* **560:**1–65.

LoPresti, V., Macagno, E. R., and Levinthal, C., 1974, Structure and development of neuronal connections in isogenic organisms: Transient gap junctions between growing optic axons and lamina neuroblasts, *Proc. Natl. Acad. Sci. USA* **71:**1098–1102.

McCarrager, G., and Chase, R., 1985, Quantification of ultrastructural symmetry at molluscan chemical synapses, *J. Neurobiol.* **16:**69–74.

MacVicar, B. A., and Dudek, F. E., 1980, Dye-coupling between CA3 pyramidal cells of the rat hippocampus, *Brain Res.* **196:**494–499.

MacVicar, B. A., and Dudek, F. E., 1981, Electrotonic coupling between pyramidal cells: A direct demonstration in rat hippocampal slices, *Science* **213:**782–785.

Margiotta, J. F., and Walcott, B., 1983, Conductance and dye permeability of a rectifying electrical synapse, *Nature* **305:** 52–55.

Martin, A. R., and Pilar, G., 1963, Dual mode of synaptic transmission in the avian ciliary ganglion, *J. Physiol. (London)* **168:**443–463.

Martin, A. R., and Pilar, G., 1964, An analysis of electrical coupling at synapses in the avian ciliary ganglion, *J. Physiol. (London)* **171:**454–475.

Meszler, R. M., Pappas, G. D., and Bennett, M. V. L., 1972, Morphological demonstration of electrotonic coupling of neurons by way of presynaptic fibers, *Brain Res.* **37:**412–415.

Meyer, D. J., Yancey, S. B., and Revel, J.-P., 1981, Intercellular communication in normal and regenerating rat liver: A quantitative analysis, *J. Cell Biol.***91:**505–523.

Miller, J. P., and Selverston, A. I., 1979, Rapid killing of single neurons by irradiation of intracellularly injected dye, *Science* **206:**702–704.

Muller, K. J., and Scott, S. A., 1981, Transmission at a "direct" electrical connexion mediated by an interneurone in the leech, *J. Physiol. (London)* **311:**565–583.

Mulloney, B., 1970, Structure of the giant fibers of earthworms, *Science* **168:**994–996.

Mulloney, B., and Selverston, A., 1972, Antidromic action potentials fail to demonstrate known interactions between neurons, *Science* **177:**69–72.

Murphy, A. D., Hadley, R. D., and Kater, S. B., 1983, Axotomy-induced parallel increases in electrical and dye coupling between identified neurons of *Helisoma, J. Neurosci.* **3:**1422–1429.

Nassel, D. R., 1982, Transneuronal uptake of horseradish peroxidase in the central nervous system of dipterous insects, *Cell Tissue Res.* **225:**639–662.

Neyton, J., and Trautmann, A., 1985, Single-channel currents of an intercellular junction, *Nature* **317:**331–335.

Nicholls, J. G., and Purves, D., 1970, Monosynaptic chemical and electrical connexions between sensory and motor cells in the central nervous system of the leech, *J. Physiol. (London)* **209:**647–667.

Obara, S., 1974, Receptor cell activity at "rest" with respect to the tonic operation of a specialized lateralis receptor, *Proc. J. Acad.* **50:**386–391.

Obara, S., and Oomura, Y., 1973, Disfacilitation as the basis for the sensory suppression in a specialized lateralist receptor of the marine catfish, *Proc. J. Acad.* **49:**213–217.

Pappas, G. D., Asada, Y., and Bennett, M. V. L., 1971, Morphological correlates of increased coupling resistance at an electrotonic synapse, *J. Cell Biol.* **49:**173–188.

Pappas, G. D., Waxman, S. G., and Bennett, M. V. L., 1975, Morphology of spinal electromotor neurons and presynaptic coupling pathways in the gymnotid *Sternarchus albifrons*, *J. Neurocytol.* **4:**469–478.

Payton, B. W., Bennett, M. V. L., and Pappas, G. D., 1969, Permeability and structure of junctional membranes at an electrotonic synapse, *Science* **166:**1641–1643.

Peracchia, C., 1981, Direct communication between axons and sheath glial cells in crayfish, *Nature* **290:**597–598.

Peracchia, C., and Bernardini, G., 1984, Gap junction structure and cell-to-cell coupling regulation: Is there a calmodulin involvement? *Fed. Proc.* **43:**2681–2691.

Peracchia, C., and Dulhunty, A. F., 1976, Low resistance junctions in crayfish, *J. Cell Biol.* **70:**419–439.

Piccolino, M., Neyton, J., Witkovsky, P. and Gerschenfeld, H. M., 1982, γ-Aminobutyric acid antagonists decrease junctional communication between L-horizontal cells of the retina, *Proc. Natl. Acad. Sci. USA* **79:**3671–3675.

Piccolino, M., Neyton, J., and Gerschenfeld, H. M., 1984, Decrease of gap junction permeability induced by dopamine and cyclic adenosine 3′:5′-monophosphate in horizontal cells of turtle retina, *J. Neurosci.* **4:**2477–2488.

Politoff, A. L., 1977, Protein semiconduction: An alternative explanation of electrical coupling, in: *Intercellular Communication* (W. C. De Mello, ed.), pp. 127–143, Plenum Press, New York.

Politoff, A., Pappas, G. D., and Bennett, M. V. L., 1974, Cobalt ions cross an electrotonic synapse if cytoplasmic concentration is low, *Brain Res.* **76:**343–346.

Powell, S. L., and Westerfield, M., 1984, The absence of specific dye-coupling among frog spinal neurons, *Brain Res.* **294:**9–14.

Rall, W. G., Shepherd, G. M., Reese, T. S., and Brightman, M. W., 1966, Dendrodendritic synaptic pathway for inhibition in the olfactory bulb, *Exp. Neurol.* **14:**44–56.

Raviola, E., and Gilula, N. B., 1973, Gap junctions between photoreceptor cells in the vertebrate retina, *Proc. Natl. Acad. Sci. USA* **70:**1677–1681.

Raviola, E., and Gilula, N. B., 1975, Intramembrane organization of specialized contacts in the outer plexiform layer of the retina: A freeze-fracture study in monkeys and rabbits, *J. Cell Biol.* **65:**192–222.

Rayport, S. G., and Kandel, E. R., 1980, Developmental modulation of an identified electrical synapse: Functional uncoupling, *J. Neurophysiol* **44:**555–567.

Rieske, E., Schubert, P., and Kreutzberg, G. W., 1975, Transfer of radioactive material between electrically coupled neurons of the leech central nervous system, *Brain Res.* **84:**365–382.

Ringham, G. L., 1975, Localization and electrical characteristics of a giant synapse in the spinal cord of the lamprey, *J. Physiol. (London)* **251:**395–407.

Rose, B., and Rick, R., 1978, Intracellular pH, intracellular free Ca, and junctional cell–cell coupling, *J. Membr. Biol.* **44:**377–415.

Rose, B., Simpson, I., and Loewenstein, W. R., 1977, Calcium ion produces graded changes in permeability of membrane channels in cell junctions, *Nature* **267:**625–627.

Schmalbruch, H., and Jahnsen, H., 1981, Gap junctions on CA3 pyramidal cells of guinea pig hippocampus shown by freeze-fracture, *Brain Res.* **217:**175–178.

Schwarzmann, G., Wiegandt, H., Rose, B., Zimmerman, A., Ben-Haim, D., and Loewenstein, W. R., 1981, Diameter of the cell-to-cell junctional membrane channels as probed with neutral molecules, *Science* **213:**551–553.

Shapovalov, A. I., 1980, Interneuronal synapses with electrical, dual and chemical mode of transmission in vertebrates, *Neuroscience* **5:**1113–1124.

Shapovalov, A. I., 1982, Evolution of the mechanisms of connection between neurons: Electrical, mixed and chemical synapses, *Neurosci. Behav. Physiol.* **12:**169–176.

Shapovalov, A. I., and Shiriaev, B. I., 1980, Dual mode of junctional transmission at synapses between single primary afferent fibres and motoneurones in the amphibian, *J. Physiol. (London)* **306:**1–15.

Sheridan, J. D., 1973, Functional evaluation of low resistance junctions: Influence of cell shape and size, *Am. Zool.* **13:**1119–1128.

Silinsky, E. M., 1985, The biophysical pharmacology of calcium-dependent acetylcholine secretion, *Pharmocol. Rev.* **37:**81–132.

Simpson, I., Rose, B., and Loewenstein, W. R., 1977, Size limit of molecules permeating the junctional membrane channels, *Science* **195:**294–296.

Smith, T. G., and Baumann, F., 1969, The functional organization within the ommatidium of the lateral eye of *Limulus, Prog. Brain Res.* **31:**313–349.

Smith, T. G., Baumann, F., and Fuortes, M. G. F., 1965, Electrical connections between visual cells in the ommatidium of *Limulus, Science* **147:**1446–1447.

Smith, T. G., Wuerker, R. B., and Frank, K., 1967, Membrane impedance changes during synaptic transmission in cat spinal motoneurons, *J. Neurophysiol.* **30:**1072–1096.

Socolar, S. J., 1977, The coupling coefficient as an index of junctional conductance, *J. Membr. Biol.* **34:**29–37.

Sotelo, C., and Korn, H., 1978, Morphological correlates of electrical and other interactions through low-resistance pathways between neurons of the vertebrate central nervous system, *Int. Rev. Cytol.* **55:**67–107.

Sotelo, C., and Taxi, J., 1970, Ultrastructural aspects of electrotonic junctions in the spinal cord of the frog, *Brain Res.* **17:**137–141.

Spencer, A. N., and Satterlie, R. A., 1980, Electrical and dye coupling in an identified group of neurons in a coelenterate, *J. Neurobiol.* **11:**13–19.

Spira, M. E., and Bennett, M. V. L., 1972, Synaptic control of electrotonic coupling between neurons, *Brain Res.* **37:**294–300.

Spira, M. E., Spray, D. C., and Bennett, M. V. L., 1976, Electrotonic coupling: Effective sign reversal by inhibitory neurons, *Science* **194:**1065–1067.

Spira, M. E., Spray, D. C., and Bennett, M. V. L., 1980, Synaptic organization of expansion motoneurons of *Navanax inermis, Brain Res.* **195:**241–269.

Spray, D. C., and Bennett, M. V. L., 1985, Physiology and pharmacology of gap junctions, *Annu. Rev. Physiol.* **47:**281–303.

Spray, D. C., Harris, A. L., and Bennett, M. V. L., 1979, Voltage dependence of junctional conductance in early amphibian embryos, *Science* **204:**432–434.

Spray, D. C., Stern, J. H., Harris, A. L., and Bennett, M. V. L., 1982, Gap junctional conductance: Comparison of sensitivities to H and Ca ions, *Proc. Natl. Acad. Sci. USA* **79:**441–445.

Stell, W. K., 1972, The morphological organization of the vertebrate retina, in: *Handbook of Sensory Physiology,* Vol. VII/2 (M. G. F. Fuortes, ed.), pp. 111–213, Springer-Verlag, Berlin.

Stewart, W. W., 1978, Functional connections between cells as revealed by dye-coupling with a highly fluorescent naphthalimide tracer, *Cell* **14:**741–759.

Stewart, W. W., 1981, Lucifer dyes—Highly fluorescent dyes for biological tracing, *Nature* **292:**17–21.

Strausfeld, N. J., and Obermayer, M., 1976, Resolution of intraneuronal and transynaptic migration of cobalt in the insect visual and central nervous systems, *J. Comp. Physiol.* **110:**1–12.

Taghert, P. H., Bastiani, M. J., Ho, R. K., and Goodman, C. S., 1982, Guidance of pioneer growth cones: Filopodial contacts and coupling revealed with an antibody to Lucifer Yellow, *Dev. Biol.* **94:**391–399.

Tauc, L., 1959, Interaction non synaptique entre deux neurones adjacents du ganglion abdominal de l'aplysie, *C.R. Acad. Sci.* **248:**1857–1859.

Tauc, L., 1969, Polyphasic synaptic activity, *Prog. Brain Res.* **31:**247–257.

Taylor, C. P., and Dudek, F. E., 1982a, A physiological test for electrotonic coupling between CA1 pyramidal cells in rat hippocampal slices, *Brain Res.* **235:**351–357.

Taylor, C. P., and Dudek, F. E., 1982b, Synchronous neural afterdischarges in rat hippocampal slices without active chemical synapses, *Science* **218:**810–812.

Taylor, C. P., and Dudek, F. E., 1984, Excitation of hippocampal pyramidal cells by an electrical field effect, *J. Neurophysiol.* **52:**126–142.

Teranishi, T., Negishi, K., and Kato, S., 1983, Dopamine modulates s-potential amplitude and dye-coupling between external horizontal cells in carp retina, *Nature* **301:**243–246.

Traub, R. D., and Pedley, T. A., 1981, Virus-induced electrotonic coupling: Hypothesis on the mechanism of periodic EEG discharges in Creutzfeldt–Jakob disease, *Ann. Neurol.* **10:**405–410.

Traub, R. D., Dudek, F. E., Taylor, C. P., and Knowles, W. D., 1985, Simulation of hippocampal after discharges synchronized by electrical interactions, *Neuroscience* **14:**1033–1038.

Triller, A., and Korn, H., 1981, Interneuronal transfer of horseradish peroxidase associated with exo/endocytotic activity in adjacent membranes, *Exp. Brain Res.* **43:**233–236.

van Venrooij, G. E. P. M., Hax, W. M. A., Schouten, V. J. A., Denier van der Gon, J. J., and van der Vorst, H. A., 1975, Absence of cell communication for fluorescein and dansylated amino acids in an electrotonic coupled cell system, *Biochim. Biophys. Acta* **394:**620–632.

Verselis, V., and Brink, P. R., 1984, Voltage clamp of the earthworm septum, *Biophys. J.* **45:**147–150.

Viancour, T. A., Bittner, G. D., and Ballinger, M. L., 1981, Selective transfer of Lucifer Yellow CH from axoplasm to adaxonal glia, *Nature* **293:**65–66.

Warner, A. E., 1973, The electrical properties of the ectoderm in the amphibian embryo during induction and early development of the nervous system, *J. Physiol. (London)* **235:**267–286.

Warner, A. E., Guthrie, S. C., and Gilula, N. B., 1984, Antibodies to gap-junctional protein selectively disrupt junctional communication in the early amphibian embryo, *Nature* **311:**127–131.

Watanabe, A., and Bullock, T. H., 1960, Modulation of activity of one neuron by subthreshold slow potentials in another in lobster cardiac ganglion, *J. Gen. Physiol.* **43:**1031–1045.

Watanabe, A., and Grundfest, H., 1961, Impulse propagation at the septal and commissural junctions of crayfish lateral giant axons, *J. Gen. Physiol.* **45:**267–308.

Waxman, S. G., Waxman, M., and Pappas, G. D., 1980, Coordinated micropinocytotic activity of adjacent neuronal membranes in mammalian central nervous system, *Neurosci. Lett.* **20:**141–146.

Weinstein, J. N., Yoshikami, S., Henkart, P., Blumenthal, R., and Hagins, W. A., 1977, Liposome–cell interactions: Transfer and intracellular release of a trapped fluorescent marker, *Science* **195:**489–491.

Werman, R., and Carlen, P. L., 1976, Unusual behavior of the Ia EPSP in cat spinal motoneurons, *Brain Res.* **112:**395–401.

Westerfield, M., and Frank, E., 1982, Specificity of electrical coupling among neurons innervating forelimb muscles of the adult bullfrog, *J. Neurophysiol.* **48:**904–913.

Williams, E. H., and DeHaan, R. L., 1981, Electrical coupling among heart cells in the absence of ultrastructurally defined gap junctions, *J. Membr. Biol.* **60:**237–248.

Wong, R. K. S., Prince, D. A., and Basbaum, A. I., 1979, Intradendritic recordings from hippocampal neurons, *Proc. Natl. Acad. Sci. USA* **76:**986–990.

Wylie, R. M., 1973, Evidence of electrotonic transmission in the vestibular nuclei of the rat, *Brain Res.* **50:**179–183.

Yarom, Y., and Spira, M. E., 1982, Extracellular potassium ions mediate specific neuronal interaction, *Science* **216:**80–82.

Zampighi, G., Ramon, F., and Duran, W., 1978, Fine structure of the electrotonic synapse of the lateral giant axons in a crayfish (*Procambarus clarkii*), *Tissue Cell* **10:**413–426.

Zieglgansberger, W., and Reiter, C., 1974, Interneuronal movement of Procion Yellow in cat spinal neurones, *Exp. Brain Res.* **20:**527–530.
Zimmerman, A. L., and Rose, B., 1985, Permeability properties of cell-to-cell channels: Kinetics of fluorescent tracer diffusion through a cell junction, *J. Membr. Biol.* **84:**269–283.
Zipser, B., 1979, Voltage-modulated membrane resistance in coupled leech neurons, *J. Neurophysiol.* **42:**465–475.

Chapter 5

Gap Junctions in Smooth Muscle

E. E. Daniel
Department of Neurosciences
McMaster University
Hamilton, Ontario L8N 3Z5, Canada

1. INTRODUCTION

This chapter will deal with control over cell-to-cell communication among smooth muscle cells. In particular, it will deal with the nature of the cell-to-cell junctions in smooth muscle, identification of those junctions (gap junctions and others) that may participate in cell-to-cell coupling, regulation of the formation of gap junctions and of the conductance of gap junctions in smooth muscle.

2. CELL-TO-CELL JUNCTIONS IN SMOOTH MUSCLE

A variety of cell-to-cell contacts have been described (Henderson *et al.*, 1971; Gabella, 1972, 1981; Henderson, 1975; Garfield, 1985) between smooth muscle cells. These include, besides gap junctions, which are usually considered the prime candidate for provision of cell-to-cell coupling, close appositions and intermediate contacts (Figs. 1–5). In addition, there are complex junctions that I will call cell-to-cell intrusions: often these involve long protrusions of portions of one cell into another with junctions of various sorts (gap junctions, close appositions, long intermediate contacts along the way). Figures 1 to 5 show examples of these various junctions and the captions describe them.

All smooth-muscle-containing tissues that I have examined contain all these types of junctions with the possible exception of gap junctions.

3. DO ALL SMOOTH MUSCLE CELLS HAVE GAP JUNCTIONS?

Many smooth muscle cells have obvious gap junctions but others do not, even after careful inspection. These are listed in Table I (with references) and

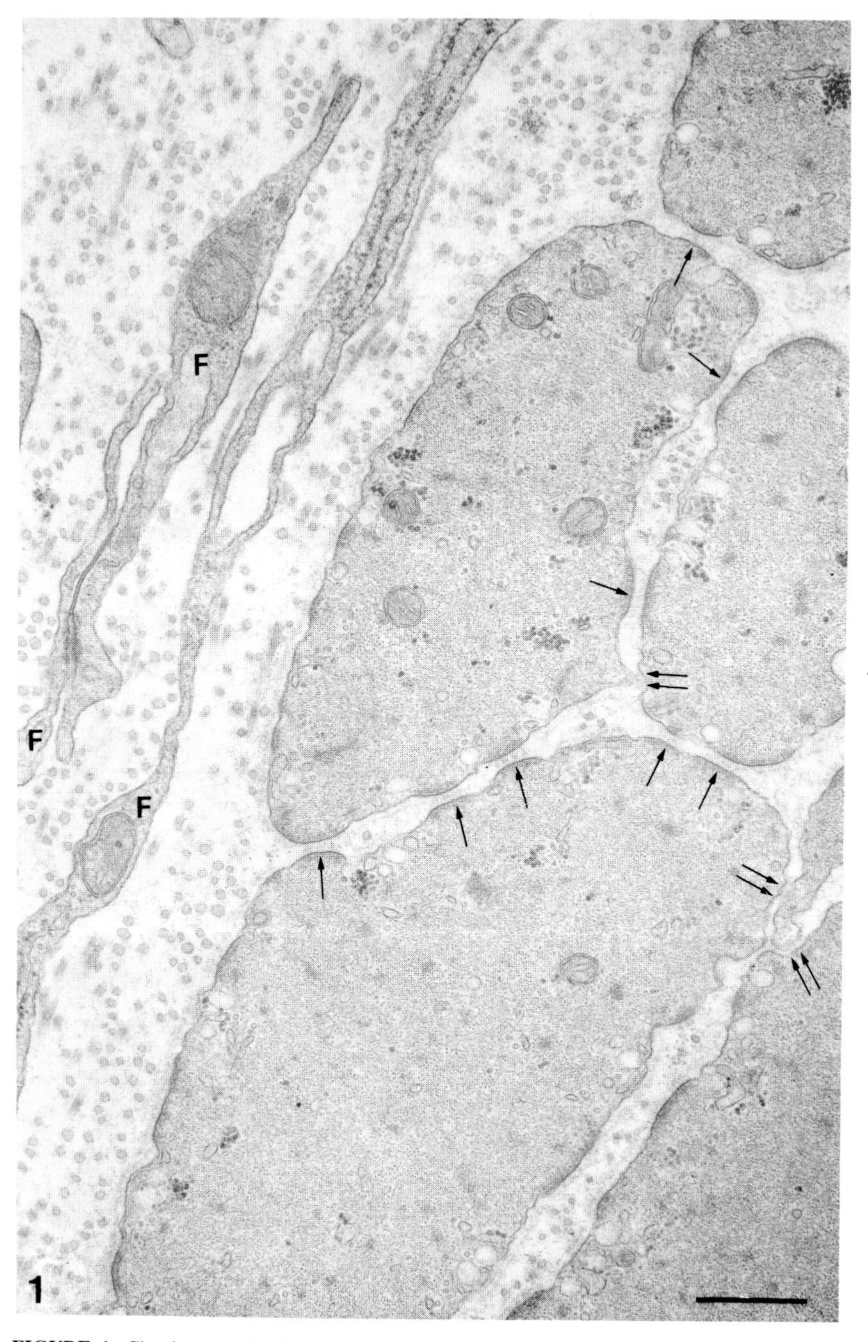

FIGURE 1. Circular muscle of nonpregnant rat myometrium. F, fibroblast. Various cell-to-cell contacts but no gap junctions are present; identified are intermediate contacts (arrows) and close apposition contacts (double arrows). Note gap junctions between fibroblasts.

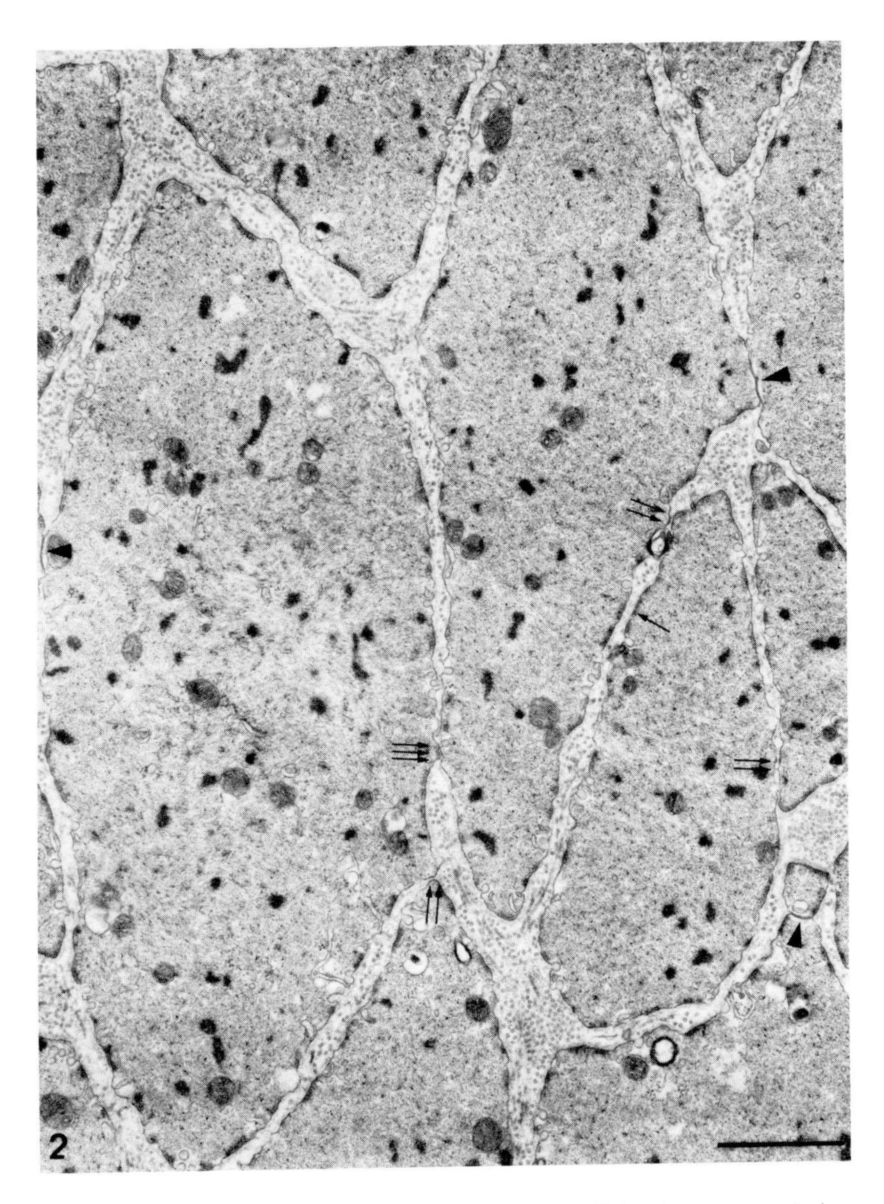

FIGURE 2. Circular muscle of dog ileum. Various cell-to-cell junctions are present. Arrows, intermediate contacts; double arrows, close apposition contacts; triple arrows, nexuslike contacts; arrowheads, gap junctions.

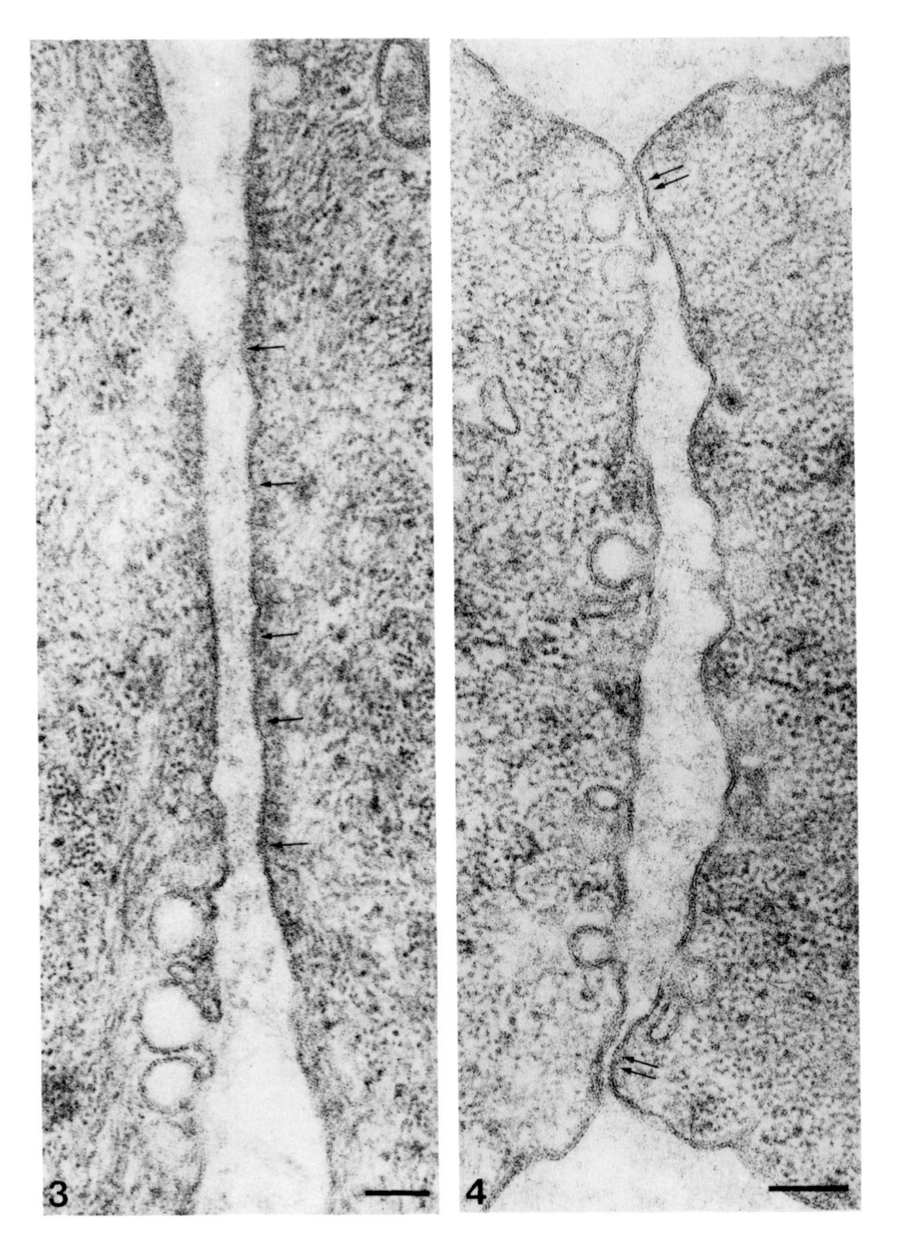

FIGURE 3. High magnification of intermediate contact (arrows). Longitudinal muscle of monkey colon. Scale bar = 0.1 μm.

FIGURE 4. High magnification of two close apposition contacts (double arrows). Human colon. Scale bar = 0.1 μm.

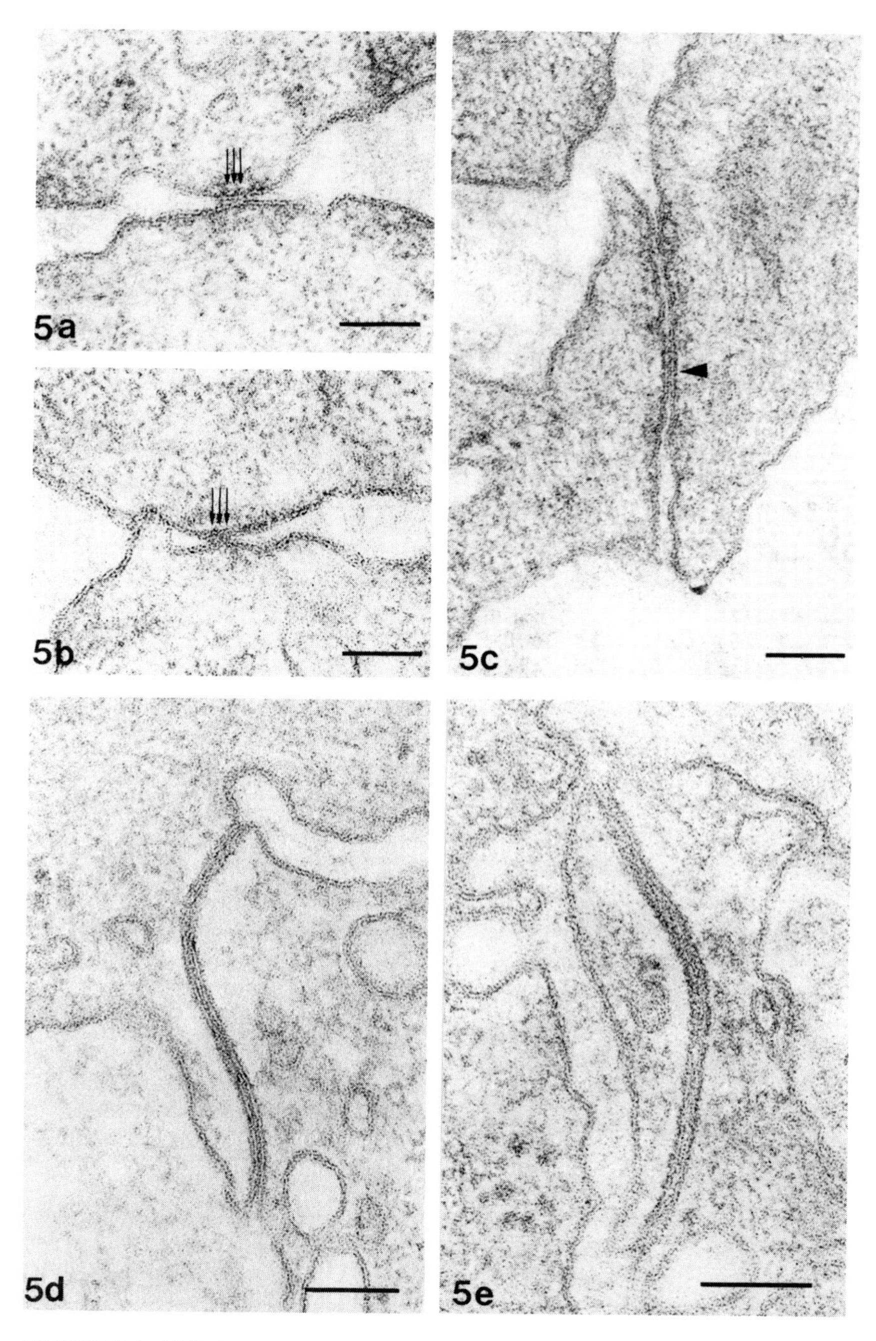

FIGURE 5. (a, b) High magnification of nexuslike contacts (triple arrows). Human colon. (c) Small gap junction (arrowhead). Circular muscle of 16-day-pregnant rat. (d, e) High-magnification micrographs of gap junction from circular muscle of rat myometrium fixed during delivery. Scale bars = 0.1 μm.

Table I
Smooth Muscle Distribution of Gap Junctions

A. Possessing obvious gap junctions

Tissue	Density of gap junctions: No./100 cross-sectional muscle cells	Density of gap junctions: No./mm plasma membrane[a]	Average profile length (nm)[b]	References
1. Small intestine (circular muscle)				
Dog	14.0	—		Daniel *et al.* (1971, 1972, 1976)
Guinea pig	NA	$48/mm^{2c}$	200–300	Gabella and Blundell (1979, 1981)
Cat	NA	Like guinea pig		Gabella and Blundell (1981)
Rabbit	NA	Like guinea pig		Gabella and Blundell (1981)
		Qualitative observations		Daniel (unpublished)
Human	NA	Qualitative observations		Daniel *et al.* (1975a)
2. Stomach (circular muscle)				
Dog—distal	18–35	NA	—	Oki and Daniel (1974), Daniel *et al.* (1971, 1972)
	10	NA	242	Daniel *et al.* (1984)
3. Esophagus (circular muscle)				
Opossum	4.1	NA	201	Daniel *et al.* (1984)
Human	2.1	NA	156	Fox *et al.* (1983)

Cat	NA	Qualitative observations		Daniel (unpublished)
Dog (LES)	14–59 (27)	37–55 (55)	149–210 (177)	Allescher *et al.* (submitted)
4. Myometrium (at term)				
Rat	NA	13.4–18	200	Garfield *et al.* (1978)
Sheep	NA	4	230	Garfield *et al.* (1979b)
Human	6	NA	NA	Garfield *et al.* (1979a)
Guinea pig	NA	3.6	205	Garfield *et al.* (1982b)
Rabbit	NA	7.8	185	Demianczuk *et al.* (1984)
Baboon	NA	4.4	280	Hayashi *et al.* (1986)
5. Trachea				
Dog	NA	6.7	150	Kannan and Daniel (1978)
		17.0–19.9	160–190	Agrawal and Daniel (1986)
Human	2.7	NA	115	Daniel *et al.* (1986a)
Bovine	8	20/mm^{2c}	NA	Cameron *et al.* (1982)
Guinea pig	1.2–2.1	NA	NA	Jones *et al.* (1980)
6. Bronchi				
Human (1st and 2nd order)		Like trachea		Daniel (unpublished)
Dog (at least to 4th order)		Like trachea		Daniel (unpublished)

[a]The figures in this column were obtained by counting each gap junction once instead of twice which is necessary when computing nm gap junction per mm plasma membrane. To obtain this value, the number of gap junctions per unit membrane length is multiplied by 2 and by the average length of gap junction profiles.
[b]Not corrected for decreases owing to sectioning through diameter less than maximum. This correction would increase values by 1.6-fold if gap junctions are circular.
[c]From freeze-fracture studies or using estimates from other freeze-fracture studies. This number is likely to be considerably greater than the number that would be obtained if data are expressed as No./1000 μm assuming circular shape of gap junction random location and equal likelihood of sectioning or fracturing through a gap junction of average diameter less than 300 nm.

(*continued*)

Table I (*Continued*)

B. Lacking gap junctions detectable on thin sectioning (and in some cases on freeze-fractures). Nexuslike contacts sometimes present

	Thin sections		Freeze-fracture		
	Absent	Rare	Absent	Rare	References
1. Small intestine					
Longitudinal muscle					
Dog	×		×		Henderson *et al*. (1971), Daniel *et al*. (1976)
Guinea pig	×		×		Gabella (1972, 1976), Gabella and Blundell (1981)
Human	×				Daniel *et al*. (1975a)
Rabbit		×		×	Gabella and Blundell (1981)
Cat		×		×	Gabella and Blundell (1981)
Mouse	×				Thuneberg (1982)
Musculoris mucosa	×				Daniel (unpublished)
2. Stomach					
Longitudinal muscle					
Dog	×				Oki and Daniel (1974), Daniel *et al*. (1971, 1972, 1974)
Human	×				Daniel *et al*. (1975a)
Circular muscle (proximal)					
Dog		×			Oki and Daniel (1974)
3. Colon					
Taenia coli					
Guinea pig		×		×	Gabella and Blundell (1981)
Longitudinal muscle					
Dog	×				Daniel and Berezin (unpublished)

Human	×		Daniel *et al*. (1975a)
Pig	×		Daniel and Berezin (unpublished)
Circular muscle			
Dog		×	Daniel and Berezin (unpublished)
Human		×	Daniel *et al*. (1975a), Daniel and Brezin (unpublished)
Pig		×	Daniel and Berezin (unpublished)
4. Esophagus			
Muscularis mucosa			
Opposum	×		Domoto *et al*. (1983)
5. Myometrium			
Nonpregnant and pregnant preterm			
Rat	×[a]	×[b]	Garfield *et al*. (1978, 1980a,b)
Sheep		×	Garfield *et al*. (1979b), Verhoeff *et al*. (1985)
Human		×	Garfield *et al*. (1979a)
Guinea pig		×	Garfield *et al*. (1982b)
Rabbit		×	Demianczuk *et al*. (1984)
Baboon	×		Hayashi *et al*. (1986)
Mouse	×		Dahl and Berger (1980)
6. Fallopian tube			
Human	×		Daniel *et al*. (1975b)
7. Vas deferens			
Rat	×		Paton *et al*. (1976)
8. Bronchi			
4th–7th order			
Human		×	Daniel *et al*. (1986a)

[a]Nonpregnant.
[b]Pregnant preterm.

include longitudinal muscle of the mammalian gastrointestinal tract, including the taenia coli, muscularis mucosa of the gut, vas deferens, oviduct, bladder, portal vein, and bronchial smooth muscle in some species. In general, circular muscle of the gut and large airways (trachea and first- and second-order bronchi) possess gap junctions; however, these are very scarce or absent in the circular muscle of the colon and sometimes absent in smaller airways. They are absent or rare in a number of tissues, such as arteries. This is also true of the myometrium at all times prior to and after term (Garfield and Daniel, 1974; Garfield *et al.*, 1977, 1978, 1979a,b, 1980a,b, 1982a,b). It is thus clear that gap junctions recognizable by thin-section electron microscopy are not necessarily present in all smooth muscle cells. Some of these muscles are well coupled electrically, e.g., taenia coli of guinea pig, longitudinal muscle of gut, myometrium (see Table I).

There are two relevant questions: Is it possible that the techniques for recognizing gap junctions may miss them? Are there alternate mechanisms that could account for coupling in the absence of gap junctions?

3.1. Recognition of Gap Junctions in Smooth Muscle

There are two main techniques for recognition of gap junctions. One involves examination of thin sections. When these are fixed properly and stained appropriately (McNutt and Weinstein, 1973; Garfield, 1985) and sectioned in cross section to the plane of the junction [best accomplished by cutting muscle cells in cross section since the oval or circular junctions are usually on protrusions from or on the body of the muscle cell (Henderson *et al.*, 1971; Henderson, 1975)], gap junctions can be recognized by their typical seven-lined structure (Fig. 5). Since tight junctions do not appear to exist in smooth muscle, most researchers also accept a five-lined structure (Fig. 2). Before the application of *en bloc* staining to delineate the 2-nm gap in the junctions, smooth muscle gap junctions were not distinguishable structurally from tight junctions and were often called nexuses (e.g., Barr *et al.*, 1968).

The problem is that some tissues may contain nexuslike contacts which are very small (Fig. 5a,b). In these structures seven-membered junctions are not observed and the question arises: are they small gap junctions in which the typical structures cannot be discerned because of their size in relation to section thickness? For example, if the section is 50 nm thick, a gap junction containing only a few connexons so that its diameter in sections is less than 50 nm, will not reveal a clearly defined characteristic structure. Some authorities have argued (see Fry *et al.*, 1977; Garfield, 1985) that, in the absence of obvious typical gap junctions, smooth muscle tissues that show electrical coupling are really coupled by tiny gap junctions, perhaps even by the presence of one functional connexon (Clapham *et al.*, 1980; Williams and DeHaan, 1981), unresolved in thin section.

There is no way to exclude this possibility at the present time using ultrastructural techniques. Successful application of immunocytochemical localization of gap junction protein at the ultrastructural level might help resolve this problem. The absence of such proteins in these nexuslike contacts would suggest they are not gap junctions; the presence of such proteins in them would suggest they may be gap junctions. To date, the antibodies to gap junction protein have not been successfully used in this way in smooth muscle.

Theoretically, small gap junctions might become identifiable if thin sections are made thinner. However, in a 40-nm section, the interspace between membranes becomes less discernible (Fig. 5d,e) because the connexons contribute a greater proportion to electron opacity of the stained material than in thicker sections in which the superposition of the outer membrane creates an apparent intermembrane gap of 2 to 4 nm. In thinner sections it may be difficult to identify the connexons in the interspace of small gap junctions (see also Forbes and Sperelakis, 1985, Figs. 29–31).

Additional approaches to clarify the composition of this interspace in very thin sections may be helpful; e.g., the use of colloidal lanthanum or tannic acid (see Garfield, 1985; Forbes and Sperelakis, 1985) to increase the electron density of the space not occupied by connexon protein.

The alternate accepted method to identify gap junctions is the use of freeze-fracture replicates (Daniel *et al.*, 1976; Peracchia, 1977, 1980; Gabella and Blundell, 1979, 1981) which reveals the presence of connexon proteins as assemblies of membrane particles usually protruding from the P face (protoplasmic face of the membrane) or as the presumed corresponding pits in the E face (extracellular membrane face). The gap junction particles are 7–9 nm in diameter and form an array which may be more or less regular with a particle spacing from 7 to 14 nm center-to-center.

However, there are difficulties with the application of this technique to identification of small nexuslike contacts as gap junctions. First, it has been established that gap junctions in smooth muscles are usually on cell protrusions (Henderson *et al.*, 1971; Daniel *et al.*, 1976; Garfield, 1985) and fracture planes often do not follow the contours of these protrusions. Thus, this method misses many gap junctions in smooth muscle as demonstrated by direct comparison (MacKenzie and Garfield, 1985). Second, there are many membrane particles not identifiable as related to connexons, but similar in size to gap junction particles. Thus, the occurrence of a small assembly of particles or pits in a freeze-fracture replica does not prove anything about their nature. Furthermore, even if the particles are gap junction particles, there needs to be proof that they are lined up with gap junction particles in the apposed cell membrane. This requires the uncommon occurrence of a fracture plane stepping from one cell to the next in the small junction. Obviously, this is highly unlikely in a very small gap junction, although it has been observed (e.g., Gabella and Blundell, 1981). Thus, this method has not answered the question as to whether there are small, unrecognized gap junctions present in tissues which are electrically coupled without classical gap junctions. In some tissues, such as longitudinal muscle of small intestine, there may be no evidence of small assemblies of particles on freeze-fracture (Gabella and Blundell, 1981).

It is worth noting that there are uncertainties about the minimum number of apposed gap junction particles required for electrical coupling of smooth muscle cells providing for the observed space constants of 1 to 2 mm. Studies of coupling between two cardiac myoballs have shown that coupling improves continuously with time and the estimate was made (Clapham *et al.*, 1980) that synchrony of the action potentials occurred when 500 functional connexons were found between 100 cells. In a further study, it was shown that all assemblies of particles recognizable as gap junctions by freeze-fracture disappeared 48 hr after inhibition of protein synthesis (Williams and DeHaan, 1981). The authors still found electrical coupling between cells and synchrony of action potentials; they suggested that disassembled single connexons were still functional. It is obvious that one such particle will provide some coupling and some authors contend that is sufficient (Clapham *et al.*, 1980), but until the conductance of the channel in such a particle is known with certainty, it is impossible to estimate whether a single or a few connexons could account for space constants of several millimeters when they are present in a three-dimensional array of smooth muscle cells. In addition, the topology of cell arrangements will have a major impact on the determination of the relation between connexons and measurements of coupling (Socolar, 1977); this is unknown.

Furthermore, as more and more smooth muscles with gap junctions are observed, it is becoming clear that their size is probably regulated to some degree such that in sections they appear to have an average length of between 150 and 200 nm (see references and Table I). Furthermore, even when gap junctions are changing in number (see Garfield, 1985; Agrawal and Daniel, 1986), they appear to achieve usually a similar average profile length of about 200 nm. If a single or very small number of gap junction particles can provide all the required coupling between smooth muscle cells, one wonders what the evolutionary and functional advantage would be to a system that often appears to operate to attain a much larger average gap junction size.

3.2. Necessity of Gap Junctions for Coupling in Smooth Muscle

The second major question is, do we need to conjure up undetectable gap junctions to account for coupling? The answer to that question is not agreed upon, but alternate modes of coupling have been proposed. For example, Sperelakis and colleagues (Sperelakis and Hoshiko, 1961; Sperelakis and Tarr, 1965; Sperelakis and MacDonald, 1974; Sperelakis and Mann, 1977; Sperelakis, 1979; Mann and Sperelakis, 1979; Sperelakis *et al.*, 1983, 1985) have argued over a number of years that low-resistance contacts (gap junctions) are not required for the spread of active or electrotonic currents between cardiac and smooth muscle cells. This review is not the appropriate place to deal with this hypothesis in detail; however, it contends that ephaptic transmission can occur across narrow clefts between muscle cells provided certain constraints are met (see below). This possibility has been supported by some experimental (Sperelakis and Tarr, 1965; but see Kobayashi *et al.*, 1967; Sperelakis and Hoshiko, 1961) and many theoretical studies (Sperelakis and Mann, 1977; Sperelakis *et al.*, 1983, 1985) using models of cell chains with reasonable values for active and passive electrical properties of membranes. More recently, a bundle of cells disconnected or connected by low-resistance junctions and an extracellular space and clefts between cells has been studied (Sperelakis and Picone, 1986). These studies suggest that, with clefts of a size likely to be present between muscle cells (e.g., 20 nm), propagation of action potentials can occur along a chain of cells provided that the action potential does not depolarize all membrane adjacent to the gap simultaneously. Furthermore, this propagation of action potentials was shown to be aided by the accumulation of K^+ which occurs in the cleft volume because of K^+ fluxes during the repolarization phase of the action potential (Sperelakis *et al.*, 1985). More recently, Sperelakis and colleagues (Sperelakis and Picone, 1986) have presented evidence that spread of electrotonically applied currents can occur via the cable formed by the anisotropic arrangement of muscle cells in bundles (Sperelakis and MacDonald, 1974), leading to a greater resistance to current flow transverse to the bundles than current flow in the long axis. Furthermore, as the result of these cablelike properties of the extracellular fluid, the use of various techniques in smooth muscle for measuring the space constant (Abe and Tomita, 1968; Tomita, 1969, 1975, Ohba *et al.*, 1976) and hence for estimating junctional resistance may be compromised, e.g., the Abe–Tomita partitioned bath. The occurrence of extracellular current flow also raises the question of whether the sucrose-gap methodology operates as is usually assumed—by current flow between cells via low-resistance pathways (see Sperelakis *et al.*, 1983; Daniel *et al.*, 1987).

Studies of cell-to-cell coupling and gap junctions in smooth muscles have provided further evidence about this problem. As mentioned above, there are numerous smooth muscle systems that have no, or very few, identifiable gap junctions. Three of these studied in some detail are the guinea pig taenia coli* and the nonpregnant, estrogen-dominated or 17-day-pregnant rat myometrium (Sims, 1982; Sims *et al.*, 1982). The space constants of each of these tissues have been measured in the Abe–Tomita partitioned bath: values range from 2.0 to 2.7 mm (Abe and Tomita, 1968; Tomita, 1969, 1975; Zelcer and Daniel,

*Early studies of this muscle reported that it had what were identified as gap junctions in thin sections after permanganate fixation (Dewey and Barr, 1962; Barr *et al.*, 1968); however, more recent studies using glutaraldehyde fixation and freeze-fracture (see Gabella and Blundell, 1981) do not confirm these findings. The early findings may be a consequence of swelling of cells after permanganate (Daniel *et al.*, 1976).

1979; Sims *et al.*, 1982), much longer than the length of cells [about 80 μm for nonpregnant myometrium (Kyozuka, Berezin, and Daniel, unpublished results), 438 μm for 17-day-pregnant rat myometrium (Cole *et al.*, 1985), and 575 μm for guinea pig taenia coli (Gabella, 1976)]. In addition, each has been studied by measurement of impedance to currents of various frequencies (Tomita, 1969; Ohba *et al.*, 1976; Sims *et al.*, 1982). These measurements are carried out in each case between nonpolarizing electrodes after sucrose perfusion has replaced (wholly or partly) perfusion with physiological saline in a snugly fitting glass tube. As washout of the low-impedance saline occurs, impedance at all frequencies rises, but the impedance at low frequencies rises to higher values. By plotting these changes against time, it is possible to distinguish a rapid and a slow phase of impedance increase, the former assumed to reflect washout of saline from the extracellular space, the latter assumed to reflect slow changes probably related to sucrose effects to shrink cells and to rupture gap junctions (see below). Assuming that the gap junction impedance can be represented by a resistance and capacitance in parallel (Fig. 6), the impedance at high (infinite) frequencies is assigned to the myoplasm and that at low frequencies to the sum of resistances of myoplasm and junctions (Figs. 6 and 7). Using this approach, the junctional resistances in these two smooth muscles were remarkably similar (about 300–400 Ωcm, respectively). Bortoff and Sillen (1986) reported a lower value (173 Ωcm) for circular muscle of cat small intestine, a tissue with many gap junctions (Taylor *et al.*, 1977; Gabella and Blundell, 1981). Similarly, Sims *et al.* (1982) reported a value of 134 Ωcm for myometrium when gap junctions were present. Table II summarizes these and related results including the finding by Bortoff and Sillen that after 100% CO_2 believed to completely uncouple gap junctions, a junctional impedance of 422 Ωcm was found.

When gap junctions appear and increase in number in the myometrium at term (see Garfield *et al.*, 1977, 1978, 1982; Puri and Garfield, 1982; Garfield, 1985), the junctional impedance declines (see Table II) and this is accompanied by an increase in the space constant from 2.6 to 3.5 mm. By making some assumptions, the two changes in myometrium at term can be suggested to be mutually compatible (Sims *et al.*, 1982). This suggests that the presence of gap junctions does affect the junctional properties of the smooth muscle tissues measured by the extracellular injection of current in the Abe–Tomita bath or in the impedance measurements. However, the low values for junctional resistance prior to the appearance of a significant number of detectable gap junctions (myometrium) and when they are absent or very rare (taenia coli) and the similar values obtained in a tissue that has its gap junctions uncoupled by 100% CO_2 (Bortoff and Sillen, 1986) all raise the question: what accounts for the measured values of junctional resistance? Are they really owing just to gap junctions?

There are further findings which raise questions. Myometrial cell-to-cell coupling was measured prior to term and at term by a technique in which [^{3}H]-2-deoxyglucose (2-DG) was taken up into cells within one perfused segment of

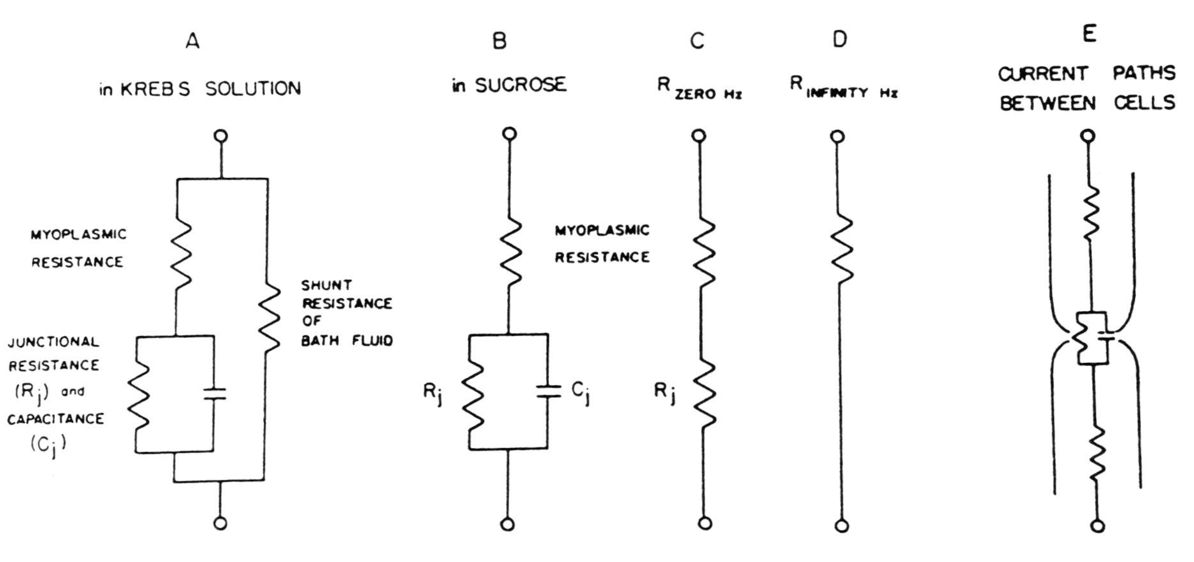

FIGURE 6. Model equivalent circuit of smooth muscle under various conditions [after Tomita (1969) and Ohba *et al.* (1976)]. (A) Total impedance of tissue with shunt resistance of extracellular medium. R_j and C_j are junctional resistance and capacitance. (B) Tissue impedance after washout of extracellular medium with sucrose. (C) Tissue resistance at 0 Hz is the sum of the myoplasmic and junctional components. (D) $R_{INFINITY}$ represents myoplasmic impedance. (E) Model of the current pathways between smooth muscle cells. From Sims (1982).

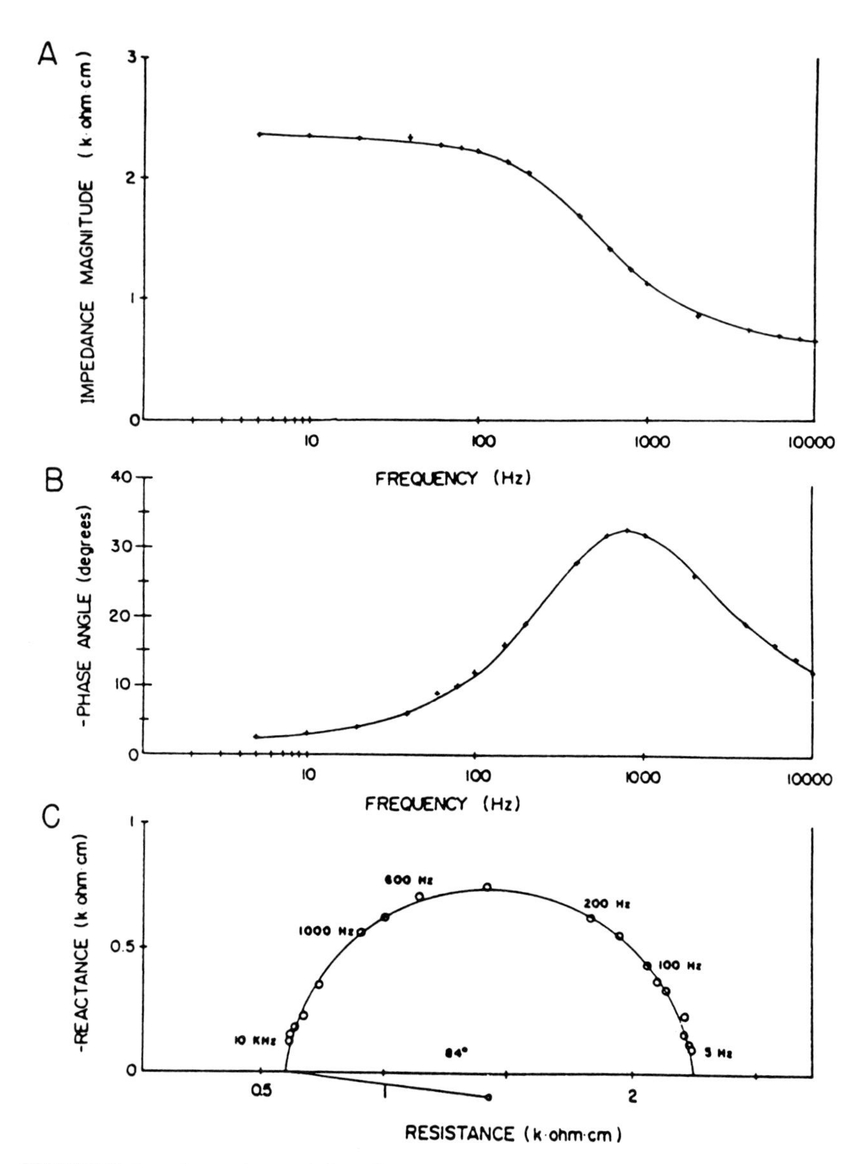

FIGURE 7. Impedance characteristics of nonpregnant myometrium after 5-min superfusion with sucrose. (A) Impedance magnitude decreases with increasing frequency of stimulation, but is independent of frequency at low and high extremes. (B) Phase angle between voltage and current shows a single peak. (C) Impedance locus, where resistance is the real part and reactance is the imaginary part of the impedance. Numbers near the circles indicate the measuring frequency. The center of the locus is depressed below the real axis with a phase angle of 84°. The characteristic frequency (f_0) occurs at 390 Hz. From Sims (1982).

Table II
Junctional Resistances in Smooth Muscle Calculated from Impedances

Tissue	Junctional resistance (ohm-cm)	Myoplasmic resistance (ohm-cm)	Gap junctions	References
Guinea pig taenia coli	180	190	Very rare	Tomita (1969)[a]
	372	214	Very rare	Ohba *et al.* (1976)
	419	233	Very rare	Sims *et al.* (1982)
Rat myometrium				
Nonpregnant	1333	595	None	Sims *et al.* (1982), Sims (1982)
17- to 22-day pregnant	323	319	Very rare	Sims *et al.* (1982), Sims (1982)
Delivery	134	340	Many	Sims *et al.* (1982), Sims (1982)
Postpartum	1358	756	Very rare	Sims *et al.* (1982), Sims (1982)
Rabbit myometrium				
Estrogen	92	191	Probably none	Bortoff and Gilloteaux (1980)[a]
Estrogen + progesterone	180	185	Probably none	Bortoff and Gilloteaux (1980)[a]
Cat intestinal circular muscle				
Control	173	134	Probably many	Bortoff and Sillen (1986)[a]
Hypertrophied[b]	96	128	(Assumed increased)	Bortoff and Sillen (1986)[a]
Atrophied[b]	340	151	(Assumed decreased)	Bortoff and Sillen (1986)[a]
In 100% CO_2				
Control	422	130	(Assumed conductance decreased)	Bortoff and Sillen (1986)[a]

[a]Technique used partial (50%) rather than complete displacement and Krebs solution with sucrose.
[b]Comparisons not made to control, but the two experimental groups had cells of different (~ 27%) lengths. Changes due to differential stretch were not ruled out.

strips of pregnant rat myometrium after gap junctions form and the spread of radioactivity into the tissue in a separately perfused compartment was measured (Cole *et al.*, 1985; Cole, 1985; Cole and Garfield, 1985). This glucose analogue is rapidly taken up into smooth muscle and other cells via the glucose transport system; then the molecule is phosphorylated and becomes no longer a substrate for the transport system. Since the phosphorylated analogue is a highly charged molecule, it does not diffuse out of cells; thus, aside from leakage via the extracellular space connecting the two tissue compartments, the diffusion of labeled 2-DG has to be within cells. Studies of cell size, 2-DG uptake, and diffusion of sucrose and mannitol in myometrium with or without gap junctions showed that any differences in cell coupling could not be attributed to cell geometry, glucose handling or diffusion in the extracellular space. There was a remarkable increase in the apparent diffusion coefficient (D_a) when gap junctions formed (Table III). Although not emphasized by those who carried out the study (Cole *et al.*, 1985), it is notable that the distinction between the extent of this diffusion without and with gap junctions was much greater than the extent of difference in the space constants and the junctional impedance (Table III summarizes the relevant comparisons). This suggests that the electrical measurements may reflect a cell-to-cell coupling mechanism which was not available to the diffusing glucose analogue. Moreover, there are examples in other cells in which electrical coupling could be demonstrated but diffusion of intracellular molecules could not (Larsen *et al.*, 1977; Loewenstein, 1979, 1980, 1981). However, there is no simple relationship between the values measured for electrical coupling and for diffusion in tissues (Socolar, 1977).

Other evidence suggests the possible existence of a mode of electrical cell-to-cell coupling other than low-resistance contacts (i.e., gap junctions). In a smooth muscle system (circular muscle of opossum esophagus) with many gap junctions (Daniel and Posey-Daniel, 1984a,b), a space constant of ~ 2 mm (Kannan *et al.*, 1985), and excellent performance in the single and double sucrose gap (Daniel *et al.*, 1983b, 1985; Jury *et al.*, 1985), we found using thin sections for quantitation of these structures that the gap junctions in the sucrose-containing compartment began to disappear in 15 min and seemed to disappear entirely by 2 hr. However, the sucrose gap performed well in terms of size of the inhibitory junction potential, transgap resistance, for up to 6 to 8 hr. Clearly the current flowing through the sucrose compartment after several hours may not

Table III
Comparison of Gap Junction Effects on Electrical Measurements and Diffusion of Intracellular Metabolites[a]

Tissue	λ (mm)	Impedance (ohm-cm)	D^a, 2-DG (cm^2/sec × 10^{-6})	% fractional area of gap junction
Pregnant, preparturient	2.6	323		0.005
	(1.4)	(2.4)		(6.52)
Delivery	3.7	134		0.24
Pregnant, preparturient			0.20	0.10
			(9.5)	(2.6)
Delivery			1.90	0.26

[a]Ratios for values for delivering compared to preparturient tissues or the inverse if appropriate are given in parentheses.

have been confined within cells; we suggest it was carried in an ion-exchange layer at the cell surface, which was resupplied for long periods by ions leaking from the cells. Since isosmotic sucrose solution was used to replace physiological saline wholly (Ohba *et al.*, 1976; Sims *et al.*, 1982) or in part (Tomita, 1969; Bortoff and Sillen, 1986) in the impedance measurements, it is possible that gap junctions rapidly became nonfunctional in those experiments as well. In the study of Sims *et al.*, (1982), this problem was partly obviated by extrapolating repeated impedance measurements back to the time of sucrose application.

Recently, alcohols such as heptanol and octanol have been found to uncouple conductance between cells in simple systems (Spray *et al.*, 1985). It is not yet clear whether they have observable effects at an ultrastructural level, but changes have been reported in connexon arrangements, which are partly reversible (Délèze and Herve, 1983).

From the above, we have to conclude that there is no straightforward way to measure junctional resistance in smooth muscle tissues which leads to an unambiguous interpretation. The best measure of cell-to-cell coupling appears to be diffusion of intracellularly confined metabolites. Therefore, neither structural nor functional studies provide an unequivocal affirmation that undetected gap junctions can account totally for electrical coupling between smooth muscle cells in which they cannot be demonstrated. An increasing number of contradictions or inconsistencies seem to exist for this hypothesis.

4. REGULATION OF GAP JUNCTIONS IN SMOOTH MUSCLE

4.1. Myometrium: Control of Synthesis of Gap Junctions

Early studies of nonpregnant rat myometrium failed to discover gap junctions between smooth muscle cells (Garfield and Daniel, 1974). However, when tissues from delivering rats were studied, numerous gap junctions were observed in both muscle layers (Garfield *et al.*, 1977, 1978, 1980a, 1982; Puri and Garfield, 1982); furthermore, it was found that gap junctions were very rare or absent until a few hours prior to delivery, appeared rapidly, and then disappeared a few hours after delivery. These basic findings have now been extended to a variety of mammalian species: sheep (Garfield *et al.*, 1979a,b), guinea pig (Garfield *et al.*, 1982b), mouse (Dahl and Berger, 1978; Dahl *et al.*, 1980; Verhoeff *et al.*, 1985), human (Garfield *et al.*, 1979a, 1980c; Garfield and Hayashi, 1980, 1981), rabbit (Demianczuk *et al.*, 1984) and baboon (Hayashi *et al.*, 1987). Figure 8 depicts some of these results diagrammatically. There are some quantitative differences in the proportion of gap junctional area to total area achieved at term but, in general, about 0.2 to 0.6% of the total (noncaveolar) membrane area becomes occupied by gap junctions. Garfield (1985) has estimated that there are about 1000 gap junctions per cell at term in rat myometrium.

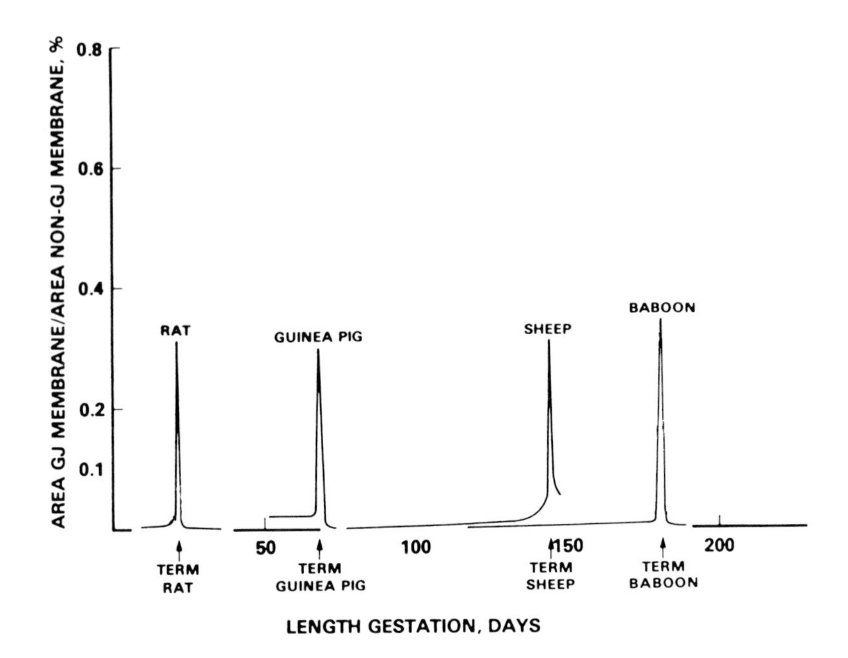

FIGURE 8. Diagram of gap junction membrane content (as a proportion of total sarcolemmal membrane in myometrium) of several species during pregnancy (modified from Garfield, 1985). Gap junction membrane proportions are given in relation to length of gestation. Proportions of gap junction membrane are shown for rat, guinea pig, sheep, and baboon. Note the sharp increase in gap junction membrane at term in all species, achieving similar proportions of junctional membrane in several species. Note, also, that guinea pig and sheep have a small amount of gap junction membrane prior to term while rat and baboon have almost none.

There are some differences among species with regard to the frequency of gap junctions prior to term; a few are found in sheep and guinea pig (Garfield *et al.*, 1979b, 1982b), almost none in rat and baboon (Garfield *et al.*, 1978, 1980a,b; Sims *et al.*, 1982; Garfield, 1985; Hayashi *et al.*, 1987). However, part of the difference may arise from the diligence of the search for gap junctions and other questions of technique.

The first question is: how is this precise control over gap junction number executed?—it prevents or limits gap junction formation prior to term, causes rapid formation just at term, and degrades gap junctions rapidly after term. There is evidence that protein synthesis is required for gap junction formation. The initiation of gap junction formation at term can be mimicked if rat myometrium from pregnant or nonpregnant animals exposed to estrogen is organ-cultured (Garfield *et al.*, 1980a,b); in such cases, gap junction formation is prevented if protein synthesis is inhibited. Furthermore, gap junctions are produced if rats are given repeated large doses of estrogens (see below) and this, too, was markedly inhibited if cycloheximide was given at a critical time during the estrogen treatment (MacKenzie and Garfield, 1986b).

4.1.1. Estrogens

Estrogens (estradiol and estrone) appear to promote gap junction formation at term in the sense that gap junction formation correlated with an increase in circulating estrogen levels in sheep (Garfield *et al.*, 1979b; Verhoeff *et al.*, 1985) and rat (Puri and Garfield, 1982). Furthermore, high pharmacological doses of estrogen continued over several days cause gap junction formation in nonpregnant animals (Dahl and Berger, 1978; MacKenzie *et al.*, 1983; Burghardt *et al.*, 1984a,b) and premature formation of gap junctions in preterm pregnant animals (MacKenzie and Garfield, 1985, 1986a). As hypophysectomy does not prevent estrogen-induced gap junction formation (Merk *et al.*, 1980), pituitary interactions are not required. Thus, estrogen appears to promote the synthesis of gap junction protein under the control of the myometrial cell genome. How this is controlled in terms of DNA and RNA systems is not known. However, an antiestrogen, tamoxifen, which antagonizes estrogen effects by binding to the estrogen receptor and interfering with its action in the nucleus, prevents gap junction formation by estrogens (MacKenzie and Garfield, 1986b). In addition, estrogen promotes the synthesis of cytosolic receptors for progesterone (see Fuchs, 1973, 1978; Nathanielsz, 1978) which appear to be necessary if progesterone is to exert effects on gap junction formation (Garfield *et al.*, 1980a).

4.1.2. Progesterone

Progesterone levels in plasma and uterine tissue are inversely correlated with gap junction formation in rats (Garfield *et al.*, 1982a; Puri and Garfield, 1982), sheep (Garfield *et al.*, 1979b; Verhoeff *et al.*, 1985) and baboons (Garfield and Hayashi, 1981; Hayashi *et al.*, 1981); i.e., the lower the progesterone level the greater the number of gap junctions. Progesterone given prior to term in rats delays myometrial gap junction formation (Garfield *et al.*, 1980a) as it does in cultures of rat myometrium provided that the tissues have been exposed to estrogen *in vivo* prior to tissue isolation or *in vitro* after isolation (Garfield *et al.*, 1980a,b). Apparently, estrogen is needed to promote synthesis of progesterone receptors. If progesterone is withdrawn as a consequence of ovariectomy of pregnant rats on day 16, gap junctions form prematurely and this can be prevented by administration of progesterone (Garfield *et al.*, 1980a, 1982a; Puri and Garfield, 1982). In general, the data support the hypothesis that progesterone inhibits the ability of estrogens to initiate gap junction formation and that in those species in which estrogen levels rise and progesterone levels fall at term (see Fuchs, 1978; Thorburn and Challis, 1979), the altered ratio of these steroid hormones promotes gap junction formation at that time.

4.1.3. Prostaglandins

In some species in which progesterone levels do not fall at term and progesterone administration cannot delay delivery, prostaglandins may play a significant role. Even in species with evidence of progesterone inhibition of delivery such as rats and sheep, prostaglandins may play a role in regulation of gap junctions. So far, it is unclear whether there are direct actions of prostanoids to promote gap junction formation and whether such actions depend upon protein synthesis. Since prostaglandins also alter the synthesis of ovarian steroid hormones, their actions may be indirect by way of actions of these agents. In organ cultures *in vitro* (Garfield *et al.*, 1980a,b) and in myometria of pregnant rats *in vivo* (Garfield *et al.*, 1980a,b), indomethacin (20 μg/ml) delayed gap junction formation. *In vitro,* this usually begins after 2 hr and reaches a maximum at 48 hr. This effect was overcome by a thromboxane analogue, but not by PGE_2, PGI_2, or $PGF_{2\alpha}$. In another *in vitro* study (Garfield *et al.*, 1980b), an inhibitor of thromboxane synthesis and a thromboxane antagonist as well as a stable prostacyclin analogue (carbacyclin) delayed gap junction formation. However, there was no attempt to establish the selectivity of the action of these agents under the experimental conditions used (usually 10 μg/ml or 10^{-5} M was applied), so the relevant pathway of prostanoid metabolism involved may still require elucidation. It is also unclear whether they act directly on smooth muscle to produce these effects. In guinea pigs near term, a prostanoid (16,16-dimethyl PGE_2) which initiated premature delivery, also elicited premature gap junction formation (Garfield *et al.*, 1982b). Surprisingly, indomethacin administration *in vivo* in rats enhanced the effect of pharmacological doses of estradiol to cause gap junction formation in nonpregnant rat myometrium (MacKenzie *et al.*, 1983; MacKenzie and Garfield, 1985). The mechanisms of action as well as the particular prostanoids involved in regulation of gap junctions in myometrium are obscure: among the possibilities proposed (Garfield, 1985) are (1) increased cross-linking of gap junction proteins; (2) regulation of steroid receptor number or binding capacity or function; (3) effects on gap junction degradation. There is, at present, no evidence for any of these hypotheses.

Other studies have shown that pelvic neurectomy had little effect on gap junction formation at term, although it blocked parturition (Burden *et al.*, 1979). Also, uterine distension together with estradiol (neither alone was very effective) restored gap junctions in postpartum rats (Wathes and Porter, 1982). The mechanisms involved in these effects are unknown, but distension effects were postulated to act through prostaglandins.

4.1.4. Control of Degradation of Gap Junctions

Degradation of gap junctions after parturition is much less well understood than their formation at term. Berezin *et al.* (1982) showed that ovariectomy of pregnant rats just prior to term did not significantly affect their disappearance after delivery. In another study (MacKenzie and Garfield, 1986a), continued daily administration of various doses of estrogen from day 21 onward to day 23 to pregnant rats, did not inhibit the disappearance of gap junctions which occurred normally after delivery on day 22. Thus, estrogen withdrawal is not the cause of gap junction disappearance after delivery, consistent with the results of ovariectomy. However, distension of the uterus with balloons after delivery and supplemented by estrogen treatment was reported to restore gap junctions in the rat myometrium (Wathes and Porter, 1982); for how long was not determined. Clearly the control events determining that gap junctions will disappear after delivery are not known, but the role of loss of stretch requires further study. It alone is probably an insufficient explanation for loss of gap junctions since Garfield *et al.* (1978) showed that a smaller but clear-cut increase in gap junctions occurred at term when one uterine horn had no fetus. Neither is the mechanism of their degradation clear; Garfield (1985) has postulated that they are rapidly internalized.

4.1.5. Role of Gap Junctions in Myometrium

A variety of evidence suggests that gap junctions regulate uterine motor function in pregnancy: (1) premature formation of gap junctions after ovariectomy (Garfield *et al.*, 1980a,b, 1982a), induced by abortifacient prostaglandins (Garfield *et al.*, 1982b), or high-dose, prolonged administration of estrogen (MacKenzie and Garfield, 1986a) all cause premature delivery. There was, however, a delay between gap junction formation and delivery after estrogen. (2) Delayed formation of gap junctions caused by pelvic neurectomy (Burden *et al.*, 1979), by indomethacin (Garfield *et al.*, 1980a) or administration of progesterone following ovariectomy (Garfield *et al.*, 1980a) was associated with delayed delivery. In all species studied, gap junction formation precedes delivery. A reasonable hypothesis is that gap junction absence or scarcity is necessary to allow pregnancy to proceed to term and that gap junction presence is necessary to allow normal labor and delivery. Their presence may be necessary but not sufficient for delivery since pelvic neurectomy delays or prevents parturition without preventing gap junction formation (Burden *et al.*, 1979) and pharmacological doses of estrogen induced premature gap junction formation in pregnant rats without causing immediate delivery (MacKenzie and Garfield, 1986a).

The effects that might underlie the function of gap junctions at term are usually considered to be improved spread of action potential bursts and individual action potentials (Verhoeff *et al.*, 1985; S. Miller, unpublished). Recently, evidence has been found (S. Miller, unpublished) that gap junction formation (along with changes in electrical properties of circular muscle) may allow coupling of longitudinal and circular muscle layers at term so that contractions are simultaneous around the uterine circumference as well as expulsive.

There are alternative or complementary explanations of the control of labor;

e.g., the control over oxytocin receptor number which is reported to increase and decline in some species over the same time scale in relation to pregnancy as the gap junctions (Soloff *et al.*, 1979; see also Fuchs, 1978). Thus, the improved spread of electrical events mediated by gap junctions could be complemented by an enhanced excitation of uterine electrical events by increased sensitivity of oxytoxin. So far, no tests have been made of the relative importance of these two mechanisms or whether both are required for normal delivery.

4.2. Trachea

The tracheae of most species studied to date [dog, guinea pig, and human (Kannan and Daniel, 1978, 1980; Daniel *et al.*, 1980; Jones *et al.*, 1982; Cameron *et al.*, 1982; Agrawal and Daniel, 1986; Daniel *et al.*, 1986a)] have gap junctions. Moreover, their numbers can be rapidly modulated *in vitro*, at least in canine trachea (Kannan and Daniel, 1978; Agrawal and Daniel, 1986). This was first noted when the effects of tetraethylammonium ions (TEA) and 4-aminopyridine (4-AP) were examined (Kannan and Daniel, 1978). Within 5–10 min after exposure to concentrations assumed to block K^+ channels, there was an increase in the number of gap junctions. The increase often amounted to a doubling of the proportion of gap junction membrane and was complete within 20 min. This increase occurred in the absence of detectable protein synthesis (Kannan and Daniel, 1978); thus, it was postulated to result from an enhanced assembly or a decreased rate of degradation of preformed gap junction protein.

The original mechanism was thought to involve K^+ channels since TEA and 4-AP were assumed to block these channels. However, it was observed that the excitatory effects of 4-AP on mechanical function were blocked by atropine but gap junction formation was not (Kannan and Daniel, 1978). Later, detailed electrophysiological studies (Kannan *et al.*, 1984) revealed that 4-AP acted via cholinergic mechanisms to affect electrical and mechanical events and did not affect space constants or rectification of voltage changes by depolarizing currents as expected of K^+ channel blockade either before or after atropine. Thus, its effects on gap junctions were not likely to result from a mechanism involving K^+ channels.

Next, the hypothesis was tested that arachidonic acid release might be involved (Agrawal and Daniel, 1986). 4-AP or arachidonic acid enhanced the proportion of gap junction membrane but this change was reversed by a selective (indomethacin) inhibitor of cyclooxygenase in concentrations which prevented the relaxation caused by the transformation of arachidonic acid (presumably to PGE_2 and PGI_2). Thus, both 4-AP and arachidonic acid decreased gap junctions when cyclooxygenase was inhibited. However, all changes from 4-AP were abolished by nonselective [5,8,11,14-eicosa-tetraynoic acid (ETYA) and nordihydroguariatic acid (NDGA)] inhibitors of cyclooxygenase plus lipoxygenase.

These results suggested that formation of PGE_2 and/or PGI_2 was responsible for the increase in gap junctions of 4-AP. In support of this idea, PGE_2 and

PGI_2, applied exogenously, increased the proportion of gap junction membrane. Only PGE_2, of all the procedures increasing gap junctions, caused an increase in the average size as well as the number of gap junction profiles. This suggested that within limits the size of gap junctions was more closely regulated than their number. These findings supported a role of prostanoids released endogenously or added exogenously to affect gap junction turnover and increase gap junction membrane.

The decrease in gap junctions caused by 4-AP and arachidonate after indomethacin suggested that arachidonate metabolites via the lipoxygenase pathway might decrease gap junctions. After ETYA or NDGA, which probably inhibited both cyclooxygenase and lipoxygenase, 4-AP caused neither an increase nor a decrease in tracheal gap junctions. This suggested a role for the lipoxygenase pathway perhaps to inhibit gap junction formation when the cyclooxygenase pathway is selectively blocked. This effect, if it exists, does not appear to be mediated by LTC_4 or LTD_4 since FPL-55217 was unable to prevent the decrease in gap junctions caused by 4-AP after indomethacin. Further study of this suggestion seems warranted in view of the recent observation (Mansour and Daniel, unpublished) that gap junction membrane proportions in trachea are reduced after acute sensitization of guinea pigs with ovalbumin.

Consequences of altered gap junction membranes on tracheal function have not been fully analyzed in terms of effects of space constant, junctional resistance, or responsiveness of tracheal muscle. One difficulty is that many of the agents enhancing gap junction formation also affect membrane resistance, thereby complicating interpretation of space constants and impedance changes. Also, the presumed increases in coupling suggested by increases in gap junctions, may not lead to the same consequence as in myometrium. This is because the trachea is normally incapable of producing action potentials owing to strong rectification via voltage-sensitive K^+ channels of effects of depolarizing currents. Thus, action potentials are not normally produced by depolarization and 4-AP does not alter this situation (after atropine); increased synchronization of contraction because of improved spread of action potentials, therefore, seems unlikely. Conceivably, improved spread of the excitatory junction potential and its contractile effects from the local release of acetylcholine could result; however, it is also possible that improved coupling would increase the amount of membrane capacity which would have to be depolarized by local excitatory junction potentials and the resulting electrotonic currents. This might inhibit the size of those potentials which result from electrotonic propagation. The effect of increased gap junction membrane on electrotonic spread of electrical activities requires further analysis.

4.2.1. Functional Modulation of Smooth Muscle Gap Junctions

So far, there have been no reports of studies of cell-to-cell coupling between pairs or between linear arrays of smooth muscle cells. Thus, evidence to date derives from studies of spread of currents or diffusion of intracellular molecules as already discussed. In only two of these studies were correlations made with the proportion of gap junctions at term between myometrial cells. One study (Sims *et al.*, 1982) compared tissues without (prior to term) or with (at term) numerous gap junctions. Thus, this provided no insight into modulation of gap junction conductance. The other study of Cole (1985; Cole *et al.*, 1985) involved diffusion of intracellular 2-DG (see Section 3.2). This was shown to be enhanced when gap junctions appeared at term but was markedly reduced (D_a from 1.72 to 0.18×10^{-6} cm^2/sec) when intracellular Ca^{2+} was presumably elevated by the ionophore A23187 (1 μM). This change in apparent diffusion coefficient (D_a) was abolished when Ca^{2+} was omitted from the external medium, a procedure which by itself had no effect on D_a. Concentrations of calmodazolium from 0.01 to 10 μM or of chlorpromazine from 0.1 to 100 μM prevented in a concentration-dependent manner the effects of A23187 to lower D_a from 2-DG in the presence of external calcium. These changes in D_a were attributed to the calmodulin-antagonist effect of the two agents and the suggestion made that calcium entry initiated a calmodulin-mediated effect on gap junction conductance. In agreement with this suggestion, there was no change in the proportion of plasma membrane occupied by gap junctions or in the leakage of 2-DG from the cells. Cole also reported that $Ca^{2+}{}_i$ elevation by ordinary contractile doses of oxytocin did not reduce D_a but that very high concentrations of this peptide (1 μM) did. Therefore, variation over the usual range of $Ca^{2+}{}_i$ would not cause uncoupling. Studies in other tissues suggest that physiological variation in $Ca^{2+}{}_i$ does not, but such variation in pH_i does affect gap junction conductance (see Gilula and Epstein, 1976; De Mello, 1980; Turin and Warner, 1980; Spray *et al.*, 1982, 1985; Spray and Bennett, 1985). However, $Ca^{2+}{}_i$ was not assessed in any of these experiments and it is likely that the levels of $Ca^{2+}{}_i$ near the cell membrane through which calcium enters or from which it is released may differ from the average level of $Ca^{2+}{}_i$. Nevertheless, this conclusion is consistent with findings made in other tissues and discussed elsewhere (see Chapter 3, this volume) as is the postulated role of calmodulin (Peracchia, 1984; Peracchia *et al.*, 1983; Peracchia and Girsch, 1985).

Cole (1985), measuring D_a for intracellular diffusion of 2-DG, also reported the surprising finding that elevation of cAMP by dibutyryl cAMP (0.5 or 1 mM), by 8-Br cAMP (0.2 mM), by forskolin (1 mM), or theophylline (1 mM) reduced D_a (e.g., 0.1 mM 8-Br cAMP reduced D_a from 1.58 to 0.3×10^{-6} cm^2/sec). 5′-AMP and butyrate had no effect on D_a. Gap junction area was insignificantly increased owing to larger junction size in these experiments, and efflux of 2-DG was unchanged. In the experiments in which cAMP was presumed to be elevated because adenylate cyclase was stimulated, but not in those in which cAMP was presumed to be elevated by its diffusion down a concentration gradient, there were electron-opaque deposits (possibly containing Ca^{2+} or $PO_4{}^{3-}$) on the cytosolic surfaces of gap junctions. These findings were assumed to indicate that cAMP elevation by any means would inhibit gap junction conductance, but only when adenylate cyclase was activated were there deposits in junctions. If correct, this is in contrast to the effect claimed for cAMP to enhance gap junction conductance of heart muscle (De Mello, 1983, 1984) and several other tissues (for review see Flagg-Newton *et al.*, 1981; Spray and Bennett, 1985).

Cole (1985) further reported decreases in D_a by agents which increase cAMP levels in uterus; e.g., by relaxin (o.1 μg/ml), by 10^{-6} M isoprenaline, by 10^{-6} M carbacyclin (which inhibited gap junction formation *in vitro;* see above), and by 10^{-6} M PGE_2, but not by 10^{-6} M $PGF_{2\alpha}$ or 10^{-7} M oxytocin which cause contraction but may not elevate cAMP. As already discussed, 10^{-6} M oxytocin, which caused supramaximal contraction and $Ca^{2+}{}_i$ elevation, did decrease D_a. Indomethacin (5×10^{-6} M) did not decrease D_a but 30 μM ETYA did. No changes in relative area of gap junctions were found. Electron-opaque deposits were found in tissues treated with isoprenaline, relaxin, carbacyclin, and ETYA. The mode of action of ETYA was not investigated further. Since cAMP levels were not measured in these studies or in those involving adenylate cyclase stimulation of diffusion of cAMP or its analogues, no direct evidence was available to relate cAMP to gap junction permeability. Neither was reversibility of these effects demonstrated or shown to relate to cAMP levels. Thus, the proposed role of cAMP to decrease junctional conductance across myometrial gap junctions remains an interesting hypothesis.

Some studies of impedance by the methods of Tomita (1969) have been carried out in various smooth muscles by Bortoff and colleagues. Unfortunately, he has never correlated any of his findings with structural analysis of the tissues studied. One study found, for example, that progesterone inhibited longitudinal impedance in the myometrium of estrogen-treated nonpregnant rabbits (Bortoff and Gilloteaux, 1980). However, this tissue probably had no detectable gap junctions (Demianczuk *et al.*, 1984; Gilloteaux, personal communication).

Recently, Bortoff and Sillen (1986) reported that circular muscle strips of cat jejunum had a specific junctional resistance (R_j) of 134 ohm-cm in control strips and that hypertrophy (produced by taking gut from residual areas after bypass) was associated with a decrease in R_j to 97 ohm-cm while atrophy (taking gut from the bypassed segment) was associated with an increase in R_j to 340 ohm-cm. There was an increase in about 20 or 30% in cell length in two samples studied from the "hypertrophied" compared to the "atrophied" muscle, but cell volume, surface area, and extracellular space were not measured and neither were gap junction areas. Previous studies of Gabella (1979) in guinea pig intestine were assumed to apply. Thus, it is difficult to assess the mechanism(s) of these changes in R_j obtained by the method of Tomita. More interesting was the observation that perfusion with 100% CO_2 increased R_j reversibly and the pH-sensitive component was estimated to be 290, 134, and 1038 ohm-cm for control, hypertrophied, and atrophied circular muscle, respectively. This treatment is reported to uncouple gap junctions (Iwatsuki and Petersen, 1979; Délèze and Herve, 1983; Majet *et al.*, 1985). In these tissues 100% CO_2 caused a definite, but not remarkable increase in R_j; e.g., to 422 ohm-cm and the residual R_j presumably represented a shunt resistance in parallel with the junctional capacitance. The authors argue that this shunt must be located at the junction, but this argument depends upon the correctness of the simple equivalent circuit of the muscle used; i.e., a junctional impedance consisting of parallel resistance and capacitance in series with a myoplasmic resistance. As discussed above, there is no direct evidence for a junctional capacitance and the occurrence of most gap junctions at abutments between cells at regions other than their ends strongly suggests that the actual current pathway may not be adequately represented by the placing of myoplasmic resistance and junctional impedance in series. It is not excluded that the extracellular current flow via the field coupling mechanism postulated by Sperelakis (see Section 3.2) of the flow of current using ion-exchange systems at the cell surface (see Section 3.2) accounts for the failure of 100% CO_2 and resultant lowered pH_i to increase total longitudinal and "specific junctional resistance" to the expected degree (to infinite values assuming complete uncoupling) in the experiments of Bortoff and Sillen (1986).

4.2.2. Gap Junctions between Smooth Muscles and Interstitial Cells of Cajal

Gap junctions have been found consistently between smooth muscle cells of the circular muscle layers of various parts of the gastrointestinal tract and interstitial cells of Cajal from a variety of species [lovebird intestine (Imaizumi and Hama, 1969); canine intestine (Henderson *et al.*, 1971; Daniel *et al.*, 1971, 1972, 1978; Duchon *et al.*, 1974; Daniel, 1977); stomach and lower esophageal sphincter (Oki and Daniel, 1974; Daniel *et al.*, 1977, 1984; Allescher *et al.*, submitted); mouse intestine (Yamamoto, 1977; Thuneberg, 1982; Rumessen and Thuneberg, 1982; Rumessen *et al.*, 1982; Pellegrini-Faussone, 1984); opossum esophagus (Daniel *et al.*, 1979, 1983b, 1985; Daniel and Posey-Daniel, 1984a,b; Daniel, 1985); human esophagus (Fox *et al.*, 1983; for early studies see Taxi, 1959, 1965)]. Gap junctions are also common between interstitial cells. Interstitial cells of Cajal make close nexuslike contact with circular muscle of the colon in several species (Daniel, unpublished), which have very few gap junctions between muscle cells. Similar contacts are made between a different type of interstitial cell of Cajal, located in the myenteric plexus, and longitudinal and circular muscle as well as between these interstitial cells (Thuneberg, 1982). One report in cat intestine claims that these cells make gap junctions with both layers (Taylor *et al.*, 1977). The cat may be unusual in possessing some small gap junctions in longitudinal muscle (Gabella and Blundell, 1981). It is of interest that small nexuslike junctions occur between interstitial cells of Cajal in the deep muscular plexus and the innermost layer of circular muscle which lacks gap junctions even though many obvious gap junctions are made between these cells and the other circular muscle cells which have gap junctions (Duchon *et al.*, 1974; Daniel, 1977) and between the interstitial cells themselves. It appears probable that the ability of the interstitial cells of the circular muscle to make classical gap or nexuslike junctions depends on the junction-forming properties of the smooth muscle cells they contact, while the inability of interstitial cells of the myenteric plexus to make classical gap junctions may determine their failure to form such junctions with circular muscles, which clearly have this ability.

There are two theories about the function of these connections, i.e., as pacemakers or as sites of action of nonadrenergic, noncholinergic nerve mediators. In the rabbit intestine, there is evidence of poor electrotonic coupling between the two muscle layers (Cheung and Daniel, 1980); yet they oscillated together producing simultaneous slow waves. One suggestion is that slow waves are initiated by a pacemaker located in the myenteric plexus and that this pacemaker is the plexus of interconnected interstitial cells found there. The current belief is that slow waves are myogenic in origin (Daniel and Sarna, 1978). Alternately, the interstitial cell may provide a pathway for coupling between the two layers: however, only in cat intestine has a morphological basis for this been reported (Taylor *et al.*, 1977). Some suggestive evidence supports the pacemaker hypothesis; e.g., high-amplitude slow waves near this plexus (Taylor *et al.*, 1976) and the inability of muscle presumably lacking this plexus to oscillate (Hara *et al.*, 1986; Hara and Szurszewski, 1986; Suzuki *et al.*, 1986). The interstitial cell plexus of the deep muscular plexus may also function as a pacemaker since its presence seems to be sufficient to allow slow waves to be initiated when the myenteric plexus and associated interstitial cells are absent (Hara *et al.*, 1986; Hara and Szurszewski, 1986). If both sets of interstitial cells are capable of functioning as pacemakers, then there must be a mechanism for their coordination, since both muscle layers oscillate together. It must be admitted that more direct evidence about the pacemaker hypothesis is needed; e.g., evidence that these cells have oscillating membrane potentials and that they can drive associated smooth muscle cells.

The other theory, which is not completely incompatible with the first, is that these cells are the sites of innervation by varicosities of nonadrenergic, noncholinergic inhibitory nerves of the circular muscle of various parts of the gastrointestinal tract. Morphological studies have shown (Daniel *et al.*, 1979; Daniel and Posey-Daniel, 1984a,b) that there are many close nerve varicosity–interstitial cell contacts in this layer and that there are relatively few or no close nerve–smooth muscle contacts except perhaps near the deep muscular plexus; it was also argued that the ability to initiate inhibitory junction potentials in the intestine is correlated to some degree with the presence of such relationships (Daniel, 1985). Further, it was shown that there is an inhibitory response which is elicited by longer ($\geq$ 2 msec) pulses of field stimulation and is not abolished by tetrodotoxin or scorpion venom toxin (Daniel *et al.*, 1979; Daniel and Posey-Daniel, 1984b). Also, the inhibitory junction potentials in some cells show a very short delay after field stimulation which might occur because these recordings are taken from interstitial cells or smooth muscles in close gap junction contact with them (Kannan *et al.*, 1985). Furthermore, the proposed mediator of the inhibitory junction potential, vasoactive intestinal polypeptide (VIP), has electrophysiological effects (small hyperpolarization, decreased membrane conductance, reversal potential about -35 mV) different from those of the NANC inhibitory mediator (large hyperpolarization, increased membrane conductance, reversal potential about -90 mV), yet the effect of massive release of nerve mediator by scorpion venom resembles the effect of VIP (Daniel *et al.*, 1983b, 1985, 1986b). Again, direct evidence is required to evaluate this hypothesis critically; e.g., evidence that interstitial cells can initiate inhibitory junction potentials or other responses on neural release of mediator, that there is a special close relationship between VIP-containing nerves and interstitial cells, that such responses can be transmitted electrotonically to smooth muscle or can activate some alternate mechanism such as release of intracellular calcium in smooth muscle to cause the K^+ conductance increase.

5. CONCLUSIONS

The existence of gap junctions in many smooth muscles is established. Their absence in some smooth muscles is strongly indicated. Certainly there are smooth muscles in which classical gap junctions do not exist but which are electrically coupled and a major question is what the mechanism is. Small, undetected gap junctions and coupling by field effects not requiring low-resistance contacts have been suggested. An increasing body of evidence suggests that the latter plays some role, although most authorities favor the former.

Gap junctions between myometrial cells are physiologically controlled: absent or very rare prior to term in pregnant (as well as nonpregnant uteri), produced in large numbers at term, rapidly degraded after delivery. This series of events is dependent on protein synthesis controlled by ovarian hormones, by prostanoids, and the gap junctions formed seem to be necessary for normal parturition. They are modulated in conductance by $Ca^{2+}{}_i$, by cAMP, and possibly by other factors.

Gap junctions between tracheal cells are subject to rapid increases in number, not requiring protein synthesis, which appear to be caused by release of prostanoids, and may be opposed by release of lipoxygenase products. The function of these changes requires further study.

Gap junctions or nexuslike contacts between smooth muscle cells and interstitial cells of Cajal occur in the gastrointestinal tract and may play a role in pacemaker activity or in neurotransmission.

Further advances in our understanding of events will depend upon the introduction of new techniques to evaluate cell-to-cell coupling unequivocally and to record from simpler systems which enable direct tests of current hypotheses.

ACKNOWLEDGMENTS. The skilled assistance of Dr. I. Berezin in preparing figures for this chapter is acknowledged. Our work is supported by the Medical Research Council of Canada.

6. REFERENCES

Abe, Y., and Tomita, T., 1968, Cable properties of smooth muscle, *J. Physiol. (London)* **196:**87–100.

Agrawal, R., and Daniel, E. E., 1986, Control of gap junction formation in canine trachea by arachidonic acid metabolites, *Am. J. Physiol.* **250:**C495–C505.

Allescher, H. D., Berezin, I., Jury, J., and Daniel, E. E., Development of a model for electrophysiological evaluation of nonadrenergic, noncholinergic neurotransmission in lower esophageal sphincter, *Am. J. Physiol.* (submitted).

Barr, L., Berger, W., and Dewey, M. M., 1968, Electrical transmission at the nexus between smooth muscle cells, *J. Gen. Physiol.* **48:**797–823.

Berezin, I., Daniel, E. E., and Garfield, R. E., 1982, Ovarian hormones are not necessary for postpartum regression of gap junctions, *Can. J. Physiol. Pharmacol.* **60:**1567–1572.

Bortoff, A., and Gilloteaux, J., 1980, Specific tissue impedances of estrogen- and progesterone-treated rabbit myometrium, *Am. J. Physiol.* **238:**C34–C42.

Bortoff, A., and Sillen, L. F., 1986, Changes in intercellular electrical coupling of smooth muscle accompanying atrophy and hypertrophy, *Am. J. Physiol.* **250:**C292–C298.

Burden, H. W., Capps, M. L., and Lawrence, I. E., 1979, Gap junctions in the myometrium of pelvic-neurectomized rats with blocked parturition, *Am. J. Anat.* **156:**105–111.

Burghardt, R. C., Matheson, R. L., and Gaddy, D., 1984a, Gap junction modulation in rat uterus. I. Effects of estrogen on myometrial and serosal cells, *Biol. Reprod.* **30:**239–248.

Burghardt, R. C., Mitchell, P. A., and Kurten, R., 1984b, Gap junctional modulation in rat uterus. II. Effects of antiestrogens on myometrial and serosal cells, *Biol. Reprod.* **30:**249–255.

Cameron, A. R., Bullock, C. G., and Kirkpatrick, C. T., 1982, The ultrastructure of bovine tracheal muscle, *J. Ultrastruct. Res.* **81:**290–305.

Cheung, D. W., and Daniel, E. E., 1980, Comparative study of the smooth muscle layers of the rabbit duodenum, *J. Physiol. (London)* **309:**13–27.

Clapham, D. E., Schrier, A., and DeHaan, R. L., 1980, Junctional resistance and action potential delay between embryonic heart cell aggregates, *J. Gen. Physiol.* **75:**633–654.

Cole, W. C., 1985, Gap junctions in uterine smooth muscle, Ph.D. thesis, McMaster University, Hamilton, Ontario, Canada.

Cole, W. C., and Garfield, R. E., 1985, Alterations in coupling in uterine smooth muscle, in: *Gap Junctions* (M. V. L. Bennett and D. C. Spray, eds.), pp. 215–230, Cold Spring Harbor Laboratory, Cold Spring Harbor, N.Y.

Cole, W. C., Garfield, R. E., and Kirkaldy, S. J., 1985, Gap junctions and direct intercellular communication between rat uterine smooth muscle cells, *Am. J. Physiol.* **249:**C20–C31.

Dahl, G., and Berger, W., 1980, Nexus formation in the myometrium during parturition and induced by estrogen, *Cell Biol. Int. Rep.* **2:**381–387.

Dahl, G., Azarnia, B., and Werner, R., 1980, *De novo* construction of cell-to-cell channels, *In Vitro* **16:**1068–1075.

Daniel, E. E., 1977, Nerves and motor activity of the gut, in: *Nerves and the Gut* (F. P. Brooks and P. W. Evers, eds.), pp. 154–196, Slack, Thorofare, N.J.

Daniel, E. E., 1985, Non-adrenergic, non-cholinergic (NANC) neuronal inhibitory interactions with smooth muscle, in: *Calcium and Contractility* (A. K. Grover and E. E. Daniel, eds.) pp. 385–426, Humana Press, Clifton, N.J.

Daniel, E. E., and Posey-Daniel, V., 1984a, Neuromuscular structures in opossum esophagus: Role of interstitial cells of Cajal, *Am. J. Physiol.* **246:**G305–G315.

Daniel, E. E., and Posey-Daniel, V., 1984b, Effects of scorpion venom on structure and function of esophageal lower sphincter (LES) and body circular muscle (BCM) from opossum, *Can. J. Physiol. Pharmacol.* **62:**360–373.

Daniel, E. E., and Sarna, S. K., 1978, The generation and conduction of activity in smooth muscle, *Annu. Rev. Pharmacol. Toxicol.* **18:**145–166.

Daniel, E. E., Robinson, K., Duchon, G., and Henderson, R. M., 1971, The possible role of close contacts (nexuses) in the propagation of control electrical activity in the stomach and small intestine, *Am. J. Dig. Dis.* **16:**(7):611–622.

Daniel, E. E., Duchon, G., and Henderson, R. M., 1972, The ultrastructural bases for coordination of intestinal motility, *Am. J. Dig. Dis. N. S.* **17**(4)**:**289–298.

Daniel, E. E., Duchon, G., and Bowes, K. L., 1975a, The structural bases for control of human gastrointestinal motility, in: *Proc. Vth Int. Symp. GI Motility* (G. Van Trappen, ed.), pp. 142–151, Typoff Press, Louvain, Belgium.

Daniel, E. E., Posey, V. A., and Paton, D. M., 1975b, A structural analysis of the myogenic control systems of the human fallopian tube, *Am. J. Obstet. Gynecol.* **121:**1054–1066.

Daniel, E. E., Daniel, V. P., Duchon, G., Garfield, R. E., Nichols, M., Malhotra, K., and Oki, M., 1976, Is the nexus necessary for cell-to-cell coupling of smooth muscle? *J. Membr. Biol.* **28:**207–239.

Daniel, E. E., Sarna, S., and Crankshaw, J., 1977, Mechanism of tetrodotoxin insensitive relaxation of opossum lower esophageal sphincter, *Proc. VIth Int. Symp. GI Motility*, Edinburgh, pp. 523–533.

Daniel, E. E., Garfield, R. E., Kannan, M. S., Zelcer, E., and Sims, S., 1978, The nature of control over coupling between smooth muscle cells and layers; its contribution to the synchrony of smooth muscle contraction, *Jpn. J. Smooth Muscle Res.* **14**(Suppl.)**:**37–38.

Daniel, E. E., Crankshaw, J., and Sarna, S., 1979, Prostaglandins and tetrodotoxin-insensitive relaxation of opossum lower esophageal sphincter, *Am. J. Physiol.* **236:**E153–E172.

Daniel, E. E., Davis, C., Jones, T. R., and Kannan, M. S., 1980, Control of airway smooth muscle, in: *Airway Reactivity* (F. Hargreave, ed.), pp. 80–107, Astra Pharmaceutical, Mississauga, Ontario, Canada.

Daniel, E. E., Cowan, W., and Daniel, V. P., 1983a, Structural bases for neural and myogenic control of human detrusor muscle, *Can. J. Physiol. Pharmacol.* **61:**1247–1273.

Daniel, E. E., Helmy-Elkholy, A., Jager, L. P., and Kannan, M. S., 1983b, Neither a purine nor VIP is the mediator of inhibitory nerves of opossum esophageal smooth muscle, *J. Physiol. (London)* **336:**243–260.

Daniel, E. E., Sakai, C. Y., Fox, J. E. T., and Posey-Daniel, V., 1984, Structural bases for function of circular muscle of canine corpus, *Can. J. Physiol. Pharmacol.* **62:**1304–1314.

Daniel, E. E., Jager, L. P., Jury, J., Helmy-Elkholy, A., Kannan, M. S., and Posey-Daniel, V., 1985, The mediators and mechanisms causing the non-adrenergic, non-cholinergic nerve responses in opossum esophagus: Role of interstitial cells of Cajal, *Biomed. Res.* **5:**67–84.

Daniel, E. E., Kannan, M., Davis, C., and Posey-Daniel, V., 1986a, Ultrastructural studies on the neuromuscular control of human tracheal and bronchial smooth muscle, *Respir. Physiol.* **63:**109–128.

Daniel, E. E., Jager, L. P., and Jury, J., 1986b, Catecholamines release mediators in the opossum oesophageal circular smooth muscle, *J. Physiol. (London)* (in press).

Daniel, E. E., Posey-Daniel, V., Jager, L. P., Berezin, I., and Jury, J., 1987, Structural effects of exposure in the sucrose gap apparatus—How does the gap work? *Am. J. Physiol.* (accepted for publication).

Délèze, J., and Herve, J. C., 1983, Effect of several uncouplers of cell-to-cell communication on gap junction morphology in mammalian heart, *J. Membr. Biol.* **74:**203–215.

De Mello, W. C., 1980, Influence of intracellular injection of H^+ on the electrical coupling in cardiac Purkinje fibers, *Cell Biol. Int. Rep.* **4:**51.

De Mello, W. C., 1983, The role of cAMP and Ca on the modulation of junctional conductance: An integrated hypothesis, *Cell Biol. Int. Rep.* **7:**1033–1040.

De Mello, W. C., 1984, Effect of intracellular injection of cAMP on the electrical coupling of mammalian cardiac cells, *Biochem. Biophys. Res. Commun.* **119:**1001–1007.

Demianczuk, N., Lowel, M. E., and Garfield, R. E., 1984, Myometrial electrophysiological activity and gap junctions in the pregnant rabbit, *Am. J. Obstet. Gynecol.* **149:**485–491.

Dewey, M. M., and Barr, L., 1962, Intercellular connection between smooth muscle cells: The nexus, *Science* **137:**670–672.

Domoto, T., Jury, J., Berezin, I., Fox, J. E. T., and Daniel, E. E., 1983, Does substance P comediate with acetylcholine in nerves of opossum esophageal muscularis mucosa? *Am. J. Physiol.* **245:**G19–G28.

Duchon, G., Henderson, R., and Daniel, E. E., 1974, Circular muscle layers in the small intestine, *Proc. Fourth Int. Symp. GI Motility,* pp. 635–646, Mitchell Press, Vancouver.

Flagg-Newton, J. L., Dahl, G., and Loewenstein, W. R., 1981, Cell junctions and cyclic AMP: Upregulation of junctional membrane permeability and junctional membrane particles by administration of cyclic nucleotide or phosphodiesterase inhibitor, *J. Membr. Biol.* **63:**105–121.

Forbes, M. S., and Sperelakis, N., 1985, Intercalated discs of mammalian heart: A review of structure and function, *Tissue Cell* **17:**605–648.

Fox, J. E. T., Daniel, E. E., deFaria, C. R., deRezende, J. M., Rassi, L., deRezende, J., Jr., and Posey-Daniel, V., 1983, Relationship of functional changes to structural changes in megaesophagus of Chaga's disease, in: *Gastrointestinal Motility* (C. Roman, ed.), pp. 51–58, MTP Press.

Fry, G. N., Devine, C. E., and Burnstock, G., 1977, Freeze-fracture studies of nexuses between smooth muscles, *J. Cell Biol.* **72:**26–34.

Fuchs, A. R., 1973, Parturition in rabbits and rats, *Mem. Soc. Endocrinol.* **20:**163–185.

Fuchs, A. R., 1978, Hormonal control of myometrial function during pregnancy and parturition, *Acta Endocrinol. (Copenhagen) Suppl.* **221:**1–70.

Gabella, G., 1972, Intercellular junctions between circular and longitudinal intestinal muscle layers, *Z. Zellforsch. Mikrosk. Anat.* **125:**191–199.

Gabella, G., 1976, Quantitative morphological study of smooth muscle cells of the guinea pig taenia coli, *Cell Tissue Res.* **170:**168–186.

Gabella, G., 1979, Hypertrophic smooth muscle. III. Increase in number and size of gap junctions, *Cell Tissue Res.* **201:**263–276.

Gabella, G., 1981, Structure of smooth muscles, in: *Smooth Muscle* (E. Bulbring, A. F. Brading, A. W. Jones, and T. Tomita, eds.), pp. 1–46, Arnold, London.

Gabella, G., and Blundell, D., 1979, Nexuses between smooth muscle cells of the guinea pig ileum, *J. Cell Biol.* **82:**239–247.

Gabella, G., and Blundell, D., 1981, Gap junctions of the muscles of the small and large intestine, *Cell Tissue Res.* **219:**469–488.

Garfield, R. E., 1985, Cell-to-cell communication in smooth muscle, in: *Calcium and Contractility* (A. K. Grover and E. E. Daniel, eds.), pp. 143–174, Humana Press, Clifton, N. J.

Garfield, R. E., and Daniel, E. E., 1974, The structural basis of electrical coupling (cell-to-cell contacts) in rat myometrium, *Gynecol. Invest.* **5:**284–300.

Garfield, R. E., and Hayashi, R. H., 1980, Presence of gap junctions in the myometrium of women during various stages of menstruation, *Am. J. Obstet. Gynecol.* **138:**569–574.

Garfield, R. E., and Hayashi, R. H., 1981, Appearance of gap junctions in the myometrium of women during labor, *Am. J. Obstet. Gynecol.* **140:**254–260.

Garfield, R. E., Sims, S., and Daniel, E. E., 1977, Gap junctions: Their presence and necessity in myometrium during parturition, *Science* **198:**958–960.

Garfield, R. E., Sims, S. M., Kannan, M. S., and Daniel, E. E., 1978, The possible role of gap junctions in activation of the myometrium during parturition, *Am. J. Physiol.* **235:**C168–C179.

Garfield, R. E., Rabideau, S., Challis, J. R. G., and Daniel, E. E., 1979a, Ultrastructural basis for maintenance and termination of pregnancy, *Am. J. Obstet. Gynecol.* **133:**308–315.

Garfield, R. E., Rabideau, S., Challis, J. R. G., and Daniel, E. E., 1979b, Hormonal control of gap junctions in sheep myometrium, *Biol. Reprod.* **21:**999–1007.

Garfield, R. E., Kannan, M. S., and Daniel, E. E., 1980a, Gap junction formation in myometrium: Control of estrogens, progesterone and prostaglandins, *Am. J. Physiol.* **238:**C81–C89.

Garfield, R. E., Merrett, D., and Grover, A. K., 1980b, Gap junction formation and regulation in myometrium, *Am. J. Physiol.* **239:**C217–C228.

Garfield, R. E., Merrett, D., and Grover, A. K., 1980c, Studies on gap junctions in the myometrium of women during various stages of menstruation, *Am. J. Obstet. Gynecol.* **138:**569–574.

Garfield, R. E., Puri, C. P., and Csapo, A. I., 1982a, Endocrine, structural and functional changes in the uterus during premature labor, *Am. J. Obstet. Gynecol.* **142:**21–27.

Garfield, R. E., Daniel, E. E., Dukes, M., and Fitzgerald, J. D., 1982b, Changes in gap junctions in myometrium of guinea pig at parturition and abortion, *Can. J. Physiol. Pharmacol.* **60:**335–341.

Gilula, N. B., and Epstein, M. L., 1976, Cell-to-cell communication, gap junctions and calcium, *Symp. Soc. Exp. Biol.* **30:**257–272.

Hara, Y., and Szurszewski, J. H., 1986, Effect of potassium on canine intestinal smooth muscle, *J. Physiol. (London)* **372:**521–538.

Hara, Y., Kubota, M., and Szurszewski, J. H., 1986, Electrophysiology of smooth muscle of the small intestine of some mammals, *J. Physiol. (London)* **372:**501–520.

Hayashi, R. H., Garfield, R. E., and Kuehl, T. J., 1987, Gap junction formation in the myometrium of pregnant baboon: A possible model, *Am. J. Obstet. Gynecol.* (submitted).

Henderson, R., 1975, Cell-to-cell contacts, in: *Methods in Pharmacology,* Vol. 3 (E. E. Daniel and D. M. Paton, eds.), pp. 47–77, Plenum Press, New York.

Henderson, R. M., Duchon, G., and Daniel, E. E., 1971, Cell contacts in duodenal smooth muscle layers, *Am. J. Physiol.* **221:**564–574.

Imaizumi, M., and Hama, K., 1969, An electron microscopic study on the interstitial cells of the gizzard of the love bird (Uronloncha domestica), *Z. Zellforsch. Mikrosk. Anat.* **97:**351–357.

Iwatsuki, N., and Petersen, O. H., 1979, Pancreatic acinar cells: The effects of CO_2, NH_4Cl and acetylcholine on intercellular communication, *J. Physiol. (London)* **291:**317–326.

Jones, T. R., Kannan, M. S., and Daniel, E. E., 1980. Ultrastructural study of guinea pig tracheal smooth muscle and its innervation, *Can. J. Physiol. Pharmacol* **58:**974–983.

Jury, J., Jager, L. P., and Daniel, E. E., 1985, Unusual potassium channels mediate NANC-nerve mediated inhibition in opossum esophagus, *Can. J. Physiol. Pharmacol.* **63:**107–112.

Kannan, M. S., and Daniel, E. E., 1978, Formation of gap junctions by treatment *in vitro* with potassium conductance blockers, *J. Cell Biol.* **78:**338–348.

Kannan, M. S., and Daniel, E. E., 1980, Structural and functional study of control of canine tracheal smooth muscle, *Am. J. Physiol.* **238:**C27–C33.

Kannan, M. S., Jager, L. P., Daniel, E. E., and Garfield, R. E., 1984, Effects of 4-aminopyridine and tetraethylammonium chloride on the electrical activity and cable properties of canine tracheal smooth muscle, *J. Physiol. Exp. Ther.* **227:**706–715.

Kannan, M. S., Jager, L. P., and Daniel, E. E., 1985, Electrical properties of smooth muscle cell membrane of opossum esophagus, *Am. J. Physiol.* **248:**G342–G346.

Kobayashi, M., Prosser, C. L., and Nagai, T., 1967, Electrical properties of intestinal muscle as measured intracellularly, *Am. J. Physiol.* **213:**275–286.

Larsen, W. J., Azarnia, R., and Loewenstein, W. R., 1977, Junctional membrane structure of hybrids between communication-competent and communication-incompetent cells, *J. Membr. Biol.* **34:**39–54.

Loewenstein, W. R., 1979, Junctional intercellular communication and the control of growth, *Biochim. Biophys. Acta* **560:**1–65.

Loewenstein, W. R., 1980, Junctional cell-to-cell communication and growth control, *Ann. N.Y. Acad. Sci.* **339:**39–45.

Loewenstein, W. R., 1981, Junctional intercellular communication: The cell-to-cell membrane channel, *Physiol. Rev.* **61:**829–913.

MacKenzie, L. W., and Garfield, R. E., 1985, Hormonal control of gap junctions in the myometrium, *Am. J. Physiol.* **248:**C296–C308.

MacKenzie, L. W., and Garfield, R. E., 1986a, Effects of 17β-estradiol on myometrial gap junctions and pregnancy in the rat, *Can. J. Physiol. Pharmacol.* **64:**462–466.

MacKenzie, L. W., and Garfield, R. E., 1986b, Effects of tamoxifen citrate and cycloheximide on estradiol induction of rat myometrial gap junctions, *Can. J. Physiol. Pharmacol.* **64:**703–706.

MacKenzie, L. W., Puri, C. P., and Garfield, R. E., 1983, Effects of estradiol-17β and prostaglandins on rat myometrial gap junctions, *Prostaglandins* **26:**925–941.

McNutt, N., and Weinstein, R. S., 1973, Membrane ultrastructure at mammalian intercellular junctions, *Prog. Biophys. Mol. Biol.* **26:**45–101.

Majet, F., Dunia, I., Vassort, G., and Majet, J. L., 1985, Ultrastructure changes in gap junctions associated with CO_2 uncoupling in frog atrial fibres, *J. Cell Sci.* **74:**51–63.

Mann, J. E., and Sperelakis, N., 1979, Further development of a model for electrical transmission between myocardial cells not connected by low resistance pathways, *J. Electrocardiol.* **12:**23–33.

Merk, R. B., Kwan, P. W. L., and Leav, I., 1980, Gap junctions in the myometrium of hypophysectomized estrogen-treated rats, *Cell Biol. Int. Rep.* **4:**287–294.

Nathanielsz, P. W., 1978, Endocrine mechanisms of parturition, *Annu. Rev. Physiol.* **40:**411–437.

Ohba, M. Y., Sakamoto, H., Tokuno, H., and Tomita, T., 1976, Impedance components in longitudinal direction in the guinea-pig taenia coli, *J. Physiol. (London)* **256:**527–540.

Oki, M., and Daniel, E. E., 1974, Ultrastructural basis for electrical coupling in the dog stomach, *Proc. Fourth Int. Symp GI Motility*, pp. 85–95, Mitchell Press, Vancouver.

Paton, D. M., Bucklund-Nicks, J., and Johns, A., 1976, Postjunctional supersensitivity of the rat vas deferens and gap junctions, *Can. J. Physiol. Pharmacol.* **54:**412–416.

Pellegrini-Faussone, M. S., 1984, Morphogenesis of the special circular layer of the interstitial cells of Cajal related to the plexus profundus of mouse intestinal muscle coat, *Anat. Embryol.* **169:**151–158.

Peracchia, C., 1977, Gap junctions: Structural changes after uncoupling procedures, *J. Cell Biol.* **72:**628–641.

Peracchia, C., 1980, Structural correlates of gap junction permeation, *Int. Rev. Cytol.* **66:**81.

Peracchia, C., 1984, Communicating junctions and calmodulin: Inhibition of electrical uncoupling in Xenopus embryo by calmodazolium, *J. Membr. Biol.* **81:**49–58.

Peracchia, C., and Girsch, S. J., 1985, Functional modulation of cell coupling: Evidence for a calmodulin-driven channel gate, *Am. J. Physiol.* **248:**H765–H782.

Peracchia, C., Bernardine, G., and Peracchia, L. L., 1983, Is calmodulin involved in the regulation of gap junction permeability? *Pfluegers Arch.* **399:**152–154.

Puri, C., and Garfield, R. E., 1982, Changes in hormone levels and gap junction during pregnancy, labor and parturition in the rat, *Biol. Reprod.* **27:**967–975.

Rumessen, J. J., and Thuneberg, L., 1982, Plexus muscularis profundus and associated interstitial cells. I. Light microscopical studies of mouse small intestine, *Anat. Rec.* **203:**115–129.

Rumessen, J. J., Thuneberg, L., and Mikkelsen, H. B., 1982, Plexus muscularis profundus and associated interstitial cells. II. Ultrastructural studies, *Anat. Rec.* **203:**130–146.

Sims, S. M., 1982, Gap junction formation in uterine smooth muscle at parturition and accompanying changes in the electrical properties, Ph.D. thesis, McMaster University.

Sims, S. M., Daniel, E. E., and Garfield, R. E., 1982, Improved electrical coupling in uterine smooth muscle is associated with increased numbers of gap junctions at parturition, *J. Gen. Physiol.* **80:**353–375.

Socolar, S. J., 1977, The coupling coefficient as an index of junctional conductance, *J. Membr. Biol.* **34:**29–37.

Soloff, M. S., Alexandrova, M., and Fernstrom, M. J., 1979, Oxytocin receptors: Triggers for parturition and lactation, *Science* **204:**1313.

Sperelakis, N., 1979, Propagation mechanisms in heart, *Annu. Rev. Physiol.* **41:**441–457.

Sperelakis, N., and Hoshiko, T., 1961, Electrical impedance of cardiac tissue, *Circ. Res.* **9:**1280–1283.

Sperelakis, N., and MacDonald, R. L., 1974, Ratio of transverse to longitudinal resistivities of isolated cardiac muscle fiber bundles, *J. Electrocardiol.* **7:**301–314.

Sperelakis, N., and Mann, J. E., Jr., 1977, Evaluation of electric field changes in the cleft between excitable cells, *J. Theor. Biol.* **64:**71–96.

Sperelakis, N., and Picone, J., 1986, Cable analysis is relevant to cell coupling in cardiac muscle and smooth muscle bundles, *Inter. Tech. Biol. Med.* **7:**433–457.

Sperelakis, N., and Tarr, M., 1965, Weak electronic interaction between neighboring visceral smooth muscle cells, *Am. J. Physiol.* **208:**737–747.

Sperelakis, N., Marschall, R. A., and Mann, J. E., 1983, Propagation down a chain of excitable cells by electric field interactions in the junctional clefts: Effect of variation in extracellular resistance, including a ''sucrose gap'' simulation, *IEEE Trans. Biomed. Eng.* **30:**658–664.

Sperelakis, N., Lobracco, B., Jr., Mann, J. E., and Marschall, R., 1985, Potassium accumulation in inter-cellular junctions combined with electrical field interactions for propagation in cardiac muscle, *Inter. Tech. Biol. Med.* **6:**24–43.

Spray, D. C., and Bennett, M. V. L., 1985, Physiology and pharmacology of gap junctions, *Annu. Rev. Physiol.* **47:**281–303.

Spray, D. C., Harris, A. L., and Bennett, M. V. L., 1982, Comparison of pH and calcium dependence of gap junctional conductance, *Intercellular pH*, pp. 445–461, Liss, New York.

Spray, D. C., White, R. L., Mazet, F., and Bennett, M. V. L., 1985, Regulation of gap junctional conductance, *Am. J. Physiol.* **248:**H753–H764.

Suzuki, N., Prosser, N. L., and Dahms, V., 1986, Boundary cells between longitudinal and circular layers: Essential for electrical slow waves in cat intestine, *Am. J. Physiol.* **250:**G287–G294.

Taxi, J., 1959, Sur la structure des travees du plexus d'Auerbath: confrontation des donnees fournies par le microscope ordinaire et par le microscope electronique, *Ann. Sci. Nat. Zool.* **12:**571–592.

Taxi, J., 1965, Contribution a l'etude des connexions des neurones moteurs du systeme nerveux autonome, *Ann. Sci. Nat. Zool. Biol. Anim.* **7:**413–674.

Taylor, A. B., Kruelen, D., and Prosser, C. L., 1977, Electron microscopy of the connective tissues between longitudinal and circular muscle of the small intestine of the cat, *Am. J. Anat.* **150:**427–442.

Taylor, G. S., Daniel, E. E., and Tomita, T., 1976, Origin and mechanism of intestinal slow waves, in: *Proceedings of the Fifth International Symposium on Gastrointestinal Motility* (G. Van Trappen, ed.), pp. 142–151, Typoff Press, Herentals, Belgium.

Thorburn, G. D., and Challis, J. R. G., 1979, Endocrine controls of parturition, *Physiol. Rev.* **59:**863.

Thuneberg, L., 1982, Interstitial cells of Cajal, intestinal pacemaker cells? *Adv. Anat. Embryol. Cell Biol.* **71:**1–100.

Tomita, T., 1969, The longitudinal tissue impedance of the smooth muscle of guinea pig taenia coli, *J. Physiol. (London)* **201:**145–149.

Tomita, T., 1975, Electrophysiology of mammalian smooth muscle, *Prog. Biophys. Mol. Biol.* **30:**185–203.

Turin, L., and Warner, A. E., 1980, Intracellular pH in early Xenopus embryos: Its effects on current flow between blastomeres, *J. Physiol. (London)* **300:**489–534.

Verhoeff, A., Garfield, R. E., Ramoudt, J., and Wallenburg, H. D. D., 1985, Electrical and mechanical uterine activity and gap Junctions in peripartal sheep, *Am. J. Obstet. Gynecol.* **153:**447–454.

Wathes, D. C., and Porter, D. G., 1982, Effects of uterine distention and estrogen treatment on gap junction formation in the myometrium of the rat, *J. Reprod. Fertil.* **65:**497–505.
Williams, E. H., and DeHaan, R. L., 1981, Electrical coupling among heart cells in the absence of ultrastructurally defined gap junctions, *J. Membr. Biol.* **60:**237–248.
Yamamoto, M., 1977, Electron microscopic studies on the innervation of the smooth muscle and interstitial cell of Cajal in the small intestine of the mouse and rat, *Arch. Histol. Jpn.* **40:**171–201.
Zelcer, E., and Daniel, E. E., 1979a, Electrical coupling in rat myometrium during pregnancy, *Can. J. Physiol. Pharmacol.* **57:**490–495.

Chapter 6

Cell Communication and Growth

Judson D. Sheridan
Department of Cell Biology and Neuroanatomy
University of Minnesota Medical School
Minneapolis, Minnesota 55455

1. INTRODUCTION

The possibility that gap junctions play an important role in the control of cell proliferation (Furshpan and Potter, 1968; Loewenstein, 1968) has probably stimulated more interest and generated more controversy than any of their other suggested functions. Much of the attention given this idea has in fact been elicited by the corollary notion that defects in the junctions may contribute to the aberrant control of cell division in cancer (Loewenstein, 1968). Numerous comprehensive reviews, appearing in the last several years, have summarized the rather checkered history of these two related ideas (e.g., Loewenstein, 1979, 1981; Trosko *et al.*, 1983; Sheridan and Atkinson, 1985; Vitkauskas and Canellakis, 1985). Rather than take a similar, and arguably redundant, approach, this chapter will focus on underlying concepts, using several more recent and selected research findings from various laboratories as illustrative examples.

2. BASIC PRINCIPLES

2.1. Gap Junctions

As described in other chapters, the basic structural unit of the gap junction is the connexon, which is comprised of a hexameric, intramembranous protein and which is connected to another connexon in the adjacent cell membrane (see Peracchia, 1980, for general structural description). The pairs of apposed connexons, which are typically arranged in plaques or, on occasion, in other patterns, are also the functional units of the gap junction, with each pair forming a hydrophilic channel directly linking the cytoplasms of the adjacent cells (Loewenstein, 1981).

Several lines of evidence indicate that gap junctional channels have effective diameters of about 1.6 to 2.0 nm in nonarthropods and a bit larger in arthropods (Loewenstein, 1981). Consequently, the channels act as molecular sieves that allow the passage of small inorganic ions and molecules the size of small cellular metabolites and restrict the passage of macromolecules. Charge, molecular shape, and chemical affinity with the connexons themselves may also modify the permeability for certain molecules (Loewenstein, 1981).

Because gap junctions are formed by aggregates of connexons, it has been commonly assumed that the permeability, or, for current flow, the conductance, of a gap junction is proportional to the number of connexons (or, more precisely, connexon pairs) comprising it, which in turn is proportional to junctional area (see e.g., Sheridan, 1973; Sheridan *et al.*, 1978; Brink and Dewey, 1978). These notions, of course, require that all of the channels, or some fixed proportion, be open, but we now know that channels can be closed under different physiological or pathological states (De Mello, this volume; Peracchia, this volume). Moreover, there is some evidence, reviewed below, that even in a single cell type there may be a spectrum of single channel conductances (Neyton and Trautmann, 1985) or permeabilities (Biegon *et al.*, 1986). Thus, it is more accurate to consider that the number of connexons sets the upper limit of permeability.

All experiments indicate that the net transfer of molecules or ions through junctional channels is a passive process, depending only on electrochemical gradients. Although the general effect of junctional transfer is to reduce these gradients, the outcome in terms of any particular cell can be to raise or to lower its electrical potential and/or its concentration of critical ions or molecules. That is, any individual cell can in effect serve as a "sink" or a "source" during junctional transfer. This rather simple distinction can have important physiological consequences in a cell population, for example whether junctional transfer leads to propagation of a signal, emphasizing the cellular "sources" of the signal, or to dampening of a response, emphasizing the cellular "sinks." Moreover, at different times or under different conditions, the behavior of a single population of junctionally interconnected cells can be dominated by either the "sources" or the "sinks" even for a particular signal molecule.

The concept of "source" and "sink" can be illustrated by appropriate junctionally dependent systems. The most obvious example where the "source" is predominant is in the propagation of action potentials throughout heart muscle (De Mello, this volume). An important factor in this system is the cellular reinforcement of the transmitted signal. Examples in which the "sink" predominates are more difficult to identify. One system in which a mixture of "source" and "sink" effects appears to operate involves cocultures of ouabain-sensitive and -insensitive cells (Corsaro and Migeon, 1977a; discussed further below). Ouabain kills isolated sensitive cells by inhibiting the Na^+/K^+ ATPase and thereby allowing intracellular Na^+ to rise and K^+ to fall following passive membrane leakage. When the sensitive cells form gap junctions with ouabain-insensitive cells, however, the insensitive cells ("source") are apparently able to supply the needed K^+ to and remove the excess Na^+ from the sensitive cells ("sink"), thus "rescuing" them from ouabain-induced death.

These principles dictate the essential features of any physiological process involving transfer by way of gap junctions: (1) The process must depend on or be modulated by small ions or molecules and/or electrical potentials; (2) the cells must have gap junctions whose channels are open; (3) the junctionally connected cells must have temporal or spatial variations in the critical substance concentrations and/or electrical potentials.

The presence of these features ensures that the process *must* be affected by junctional transfer provided that the number of open channels is sufficiently large. This "junctional imperative" is often overlooked in discussions about the "possible" biological role(s) of gap junctions. Yet it should be emphasized that this statement does not say that the involvement of gap junctions in a cell process excludes other forms of communication. It is quite possible, if not likely, gap junctions act in concert with other regulatory mechanisms. Conversely, identification of nonjunctional factors in a particular process does not rule out gap junctional contributions.

2.2. Cell Proliferation

The unit of cell proliferation is the cell cycle (Mitchison, 1971). In its most general form, the cell cycle is conventionally divided into a series of stages through which a nascent cell must proceed before it divides into two daughter cells and starts the process all over again. Most dividing cells have four stages: G1, S (during which most of the DNA is synthesized), G2, and M (for mitosis). The lengths of the stages, and of the overall cycle, vary from cell type to cell type, and even within a single cell population. The two stages flanking S-phase show considerable temporal variability and in fact account for most of the variability in average cycle duration in different cell types. G1 is a particularly interesting stage for our current purposes because it is during this stage (or its more temporally extended counterpart, G0, or resting phase) that cells make the "decision" to continue (or, in the case of G0 cells, to "reenter") the cycle. While there is no reason to exclude a role for gap junctions in other stages of the cycle (some possibilities are discussed below), the most interesting possibility is some influence on the decision process itself.

One attractive model of this process was developed some time ago by Smith and Martin (1973). Their "transition probability" model proposes that sometime during G1 each cell enters a state of temporary arrest which it leaves on a random, probabilistic basis. The variability in the time spent in this so-called "A-state" accounts for the major variability in the individual cycle durations in the population. The remainder of the "cycle," which they call the "B-state," is essentially of constant duration for any particular cell type, and includes part of G1 and all of S, G2, and M.

The primary support for this model was derived from time-lapse cinematographic analysis of the intermitotic times for cultured cells. When the proportion of cells not yet divided was plotted versus time after mitosis, there was a nearly constant period during which no cells divided followed by a time over which the proportion fell off exponentially. These results were consistent with a nearly fixed duration of the B-state and a constant probability of a cell leaving the A-state.

Subsequent work has provided support for this model (e.g., Minor and Smith, 1974; Shields and Smith, 1977; but see Murphy *et al.*, 1984), although an alternative, which has more mechanistic flavor, has gained particular attention. In this alternative, which can be called the "unstable protein" model, it is proposed that during the equivalent of the A-state, each cell is synthesizing a critical protein which must reach a threshold concentration (or activity) for the cell to begin synthesizing DNA (i.e., to enter the S-phase) (Schneiderman *et al.*, 1971; Rossow *et al.*, 1979). This protein is very unstable and must be synthesized very rapidly in order for threshold to be exceeded. According to this model, the apparently indeterminate nature of the A-state is actually a reflection of variability in the synthetic rates for the unstable protein in different cells.

Although on the surface these models appear to be different, they are not totally exclusive. In their original article, Smith and Martin (1973) suggest that the probabilistic nature of the A-state could result from an oscillation in a rare control molecule. If the amount of the molecule were "subject to considerable biochemical noise," randomness could result. Similarly, with the "unstable protein" model, oscillations in the synthetic rate, owing to biochemical feedback, could be affected by "biochemical noise" with the consequence that the net rate of accumulation of the protein could vary randomly within the population.

3. POSSIBLE EFFECTS OF GAP JUNCTIONS ON DIFFERENT PHASES OF THE CYCLE

3.1. G1 (G0)-Phase

3.1.1. General Considerations

Hypotheses for the involvement of gap junctions in the control of cell proliferation have most often focused either explicitly or implicitly on this stage. With a few exceptions, the arguments have remained rather general (Loewenstein, 1968, 1979; Burton, 1971; Burton and Canham, 1973). Progress through the cell cycle is postulated to depend on the intracellular concentration of a small molecular signal that is produced asynchronously by cells in the population. As in the "unstable protein" model discussed above, there is a critical concentration threshold that must be exceeded, but the difference is that the signal molecule is small enough to pass through gap junctional channels. As the cell population grows, cells pack more tightly and are able to form more extensive and more frequent gap junctions. Provided the new gap junctions have open channels, as the signal molecules are produced they diffuse to other cells, dampening out the stimulatory effect. This is obviously an example of the "sink" effect.

A variant on this theme proposes that there is an inhibitory small molecule, which is high in concentration during most of the cycle, but which transiently decreases at the critical "decision point" during G1. This "disinhibition," it is suggested, can be reduced or eliminated if the inhibitory signal is junctionally transmitted by neighboring cells acting as "sources" (see, e.g., Mehta *et al.*, 1986). As in the case of a stimulatory event, the signal will be more effectively transmitted when the cell population increases in density and forms more open gap junctional channels. In both, the result is a density-dependent arrest of cells at the same critical decision point.

3.1.2. Relationship to Cell Cycle Models

Either of the two models for cell cycle control discussed above, the "transition probability" model and the "unstable protein" model, can be readily modified to include effects of gap junctions. As already mentioned, Smith and Martin (1973) suggested that the probabilistic nature of the A-state was due to random noise-like fluctuations in the concentration of a critical, positive control molecule or molecules, while the cell was poised just below threshold. If the control molecules are junctionally permeant, the implications of junctional transfer follow rather directly from the more general ideas discussed above. With cells that are isolated, either because they are not in contact or because they do not have patent junctional channels, the probability that threshold would be exceeded with a "pulse" of control molecules would be determined by intrinsic properties of the cell itself, e.g., how close to threshold it was during the A-state, what sort of amplitude distribution the concentration pulses had, what the frequency distribution of the pulses was, and so on. As two cells became interconnected via patent junctions, however, two factors would operate in addition. First, it would be unlikely that both cells would be in the A-state at the same time, and therefore one cell would probably act simply as a passive sink dampening the amplitude distribution of signal pulses in the other cell. Second, even if both cells were in the A-state, they would be unlikely to generate signal pulses at the same time, and each would dampen the other cell's signals. Obviously, as long as the signals continued to be generated asynchronously, larger groups of interconnected cells would produce greater dampening because of the larger effective volume. This effect probably would predominate even if the larger number of cells would slightly increase the chances that more than one of the connected cells would be in the A-state at the same time. [Burton and Canham (1973) have discussed such ideas in more quantitative terms.]

For the "unstable protein" model to be influenced by gap junctions, one additional step would have to be postulated because the control protein itself would not be able to pass through the junctions. Here the level of junctional control would have to be proximal to the fluctuations in the protein, perhaps operating on either its synthesis or degradation. Once again, small molecules would have to be involved, and their concentrations would have to differ in cells at, or away from, the "decision point." Junctional transfer would average out concentrations, thereby maintaining, for example, the synthetic rate below that needed to exceed threshold.

3.1.3. Dependence of the Decision Process on Small Molecules

Of course it is one thing to show that the earlier, more general concepts of junctional effects of G1 can be fitted into the two models of the "decision" process in G1 and quite another to provide more specific evidence to support the association. The pivotal question is whether or not junctionally permeant molecules act at the "decision" point, and, as we will see, there is no shortage of possibilities.

In an earlier review (Sheridan, 1976) I discussed a model based on a speculated role of two molecules, cAMP and cGMP, as junctionally transmitted messengers in growth control. This model shared major general features with earlier models (e.g., Loewenstein, 1968; Burton, 1971; Burton and Canham, 1973; as detailed by Loewenstein, 1979), but it was independently derived on the basis of the small size, reported antagonistic actions, and cell cycle-related concentration changes of the cyclic nucleotides. As described further below, subsequent work has provided some support for the involvement of cAMP as growth inhibitory messenger (e.g., Mehta *et al.*, 1986) (although some workers have argued for its growth stimulatory action *in vivo;* e.g., MacManus *et al.*, 1975), but cGMP as a stimulatory messenger (Seifert and Rudland, 1974; Goldberg *et al.*, 1975) has been largely supplanted by a more complex sequence of events leading to increased cytoplasmic Ca^{2+} and K^{+} (see below).

In addition to cyclic nucleotides, several other small molecules have surfaced as potential contributors to the regulatory events in G1. Some of the earliest evidence for a critical "decision" point in G1 (called the "restriction point") came from experiments in which cells were arrested in G1 by a reduction in certain essential amino acids, e.g., isoleucine (Pardee, 1974). The interesting feature of the arrested cells is that when the amino acid was restored, the entire cell population took nearly the same amount of time to begin synthesizing DNA, i.e., to enter S-phase. On the surface this evidence seems to be difficult to reconcile with the Smith and Martin model. In fact, it appears to fit best with the "unstable protein" model: it is easy to think of a decrease in amino acids as decreasing synthesis of proteins generally and of the unstable protein specifically. However, as Smith and Martin (1973) explain, their model could accommodate such evidence as well if the effect of the amino acid deficiency were to greatly lower the probability of leaving the A-state. Restoration of the amino acid would substantially raise the probability so that most of the cells would enter S within a relatively short period of time. When compared to the long time in the A-state during amino acid deprivation, the variation in time taken for different cells to enter S would be minimal and appear relatively synchronous. Irrespective of which model ultimately applies, variations in the amount of amino acids in cells in the A-state and non-A-state would be averaged out by junctional transfer which could substantially influence the distributions of cycle times in the population.

These arguments, however, are derived from rather artificial experimental conditions. Extreme fluctuations in amino acid concentrations are not likely to be normally involved in growth control, although some variability would be expected in the cell populations. Therefore, it is reasonable to look for other small molecules whose G1-related changes and effects point toward a regulatory role.

Several lines of evidence have suggested the involvement of inorganic ions in the control of G1. Lubin (1967), for example, provided evidence for the importance of maintaining high intracellular K^+ in order for cells to continue to synthesize macromolecules and to proliferate. Moreover, based on this evidence, he even implicated cell coupling as a possible mechanism by which cells could integrate their ionic control of cell growth. He also suggested that brief or prolonged uncoupling could allow cells to grow independently during embryogenesis. He did not try to identify a particular K^+-sensitive stage of the cell cycle, but the importance of continued protein synthesis in at least one of the models for G1 control makes this an appropriate point to discuss K^+ effects further.

These early suggestions were made initially with little direct evidence for the transfer of significant quantities of K^+ (or other specific ions, e.g., Na^+) from cell to cell via permeable junctions, at least for cells lacking electrical excitability. The discovery that induced electrotonic potentials could be transferred among nonexcitable cells ("electrical coupling") (Kuffler and Potter, 1964; Kanno and Loewenstein, 1964; Potter *et al.*, 1966), made it likely that some transfer of K^+, along with other small ions, would occur with appropriate electrochemical potential differences. Moreover, for cardiac muscle, Weidmann (1966) demonstrated the movement of radiolabeled K^+ along the length of cardiac muscle fiber bundles and concluded that the K^+ followed the same low-resistance pathways from cell to cell taken by electrical current during propagation of the cardiac action potential. Nevertheless, it remained unclear, particularly for nonexcitable cells, whether there could be enough transfer to significantly change intracellular ionic concentrations.

An important group of experiments, utilizing ouabain to lower intracellular

K^+ (and raise intracellular Na^+), helped to show, at least indirectly, that such bulk transfer via gap junctions not only could occur but could influence cell growth. The paradigm, which has already been mentioned briefly as an example of the "sink" and "source" concept, involves cocultures of ouabain-sensitive and -insensitive cells (Corsaro and Migeon, 1977a). When grown alone in the presence of ouabain, the ouabain-sensitive cells stop growing as they progressively lose K^+, and gain Na^+. The ouabain-insensitive cells show neither the growth nor the ionic changes in the presence of ouabain. When small numbers of the ouabain-sensitive cells are cocultured with the insensitive cells, however, they continue to grow even when ouabain is present, provided that both cell types are capable of forming permeable junctions with each other. This "rescue" of the sensitive by the insensitive cells is most likely due to the transfer of K^+ from the insensitive to the sensitive cells and of Na^+ in the reverse direction and to the ability of the unaffected Na^+/K^+ ATPase in the insensitive cells to keep all the cells in ionic homeostasis.

Evidence for the K^+ side of this postulated two-way ionic traffic was provided by Ledbetter and Lubin (1979) who analyzed the contents of ^{86}Rb, as a K^+ analogue, in ouabain-treated and -untreated cocultures. When the cocultured sensitive and insensitive cells were "junctionally competent," the amount of ^{86}Rb was essentially unaffected by the presence of ouabain. When either cell type was replaced with the appropriate junctionally incompetent cell, the amount of ^{86}Rb fell in accord with the number of sensitive cells.

In association with the changes in ^{86}Rb transfer, Ledbetter and Lubin (1979) monitored protein synthesis using incorporation of labeled amino acids which they detected autoradiographically and measured by scintillation counting of cell extracts. As anticipated from the growth studies, protein synthesis was maintained in ouabain-treated cocultures of sensitive and insensitive cells, provided that they were junctionally competent, and thus paralleled the maintenance of ^{86}Rb levels.

There have been no comparable studies of the Na^+ exchange in such cocultures although the elevation of cellular Na^+ in the presence of ouabain for sensitive cells may be of equal or even greater importance in reducing growth. High Na^+ would be expected to have several effects, e.g., release of intracellular stores of Ca^{2+}, acidification due to reversal of the Na^+/H^+ exchange, that, if excessive, could disrupt many cellular processes including those critical for continued growth.

Although these experiments demonstrate that gap junctions can reduce the growth-inhibiting effects of K^+ loss (and/or Na^+ gain) in these model cocultures, what role if any do changes in K^+ or Na^+ play in the normal control of cell growth, particularly in relation to the decision process in G1?

To get at the answer to this question, we must consider the complex series of events that have been shown to occur in several different cell types as they are stimulated to leave G1. The following events have been defined by experiments from many laboratories and have recently been summarized by Macara (1985) (for other recent reviews see Michell, 1982; Anderson *et al.*, 1985; Nishizuka, 1986): (1) A mitogen, such as PDGF, first interacts with a membrane receptor. (2) This interaction leads to activation of phospholipase C which rapidly breaks down phosphatidylinositol biphosphate into inositol triphosphate (IP_3) plus diacylglycerol (DAG). (3) The IP_3 causes the release of Ca^{2+} from intracellular stores whereas the DAG, in the presence of appropriate Ca^{2+} levels, activates a specialized kinase, protein kinase C. (4) By action of protein kinase C and/or Ca^{2+} and calmodulin, there is a stimulation of the Na^+/H^+ exchange mechanism, leading to alkalinization of the cytoplasm as H^+ is pumped out of the cell and Na^+ is pumped in. Additional Na^+ may accumulate as it enters in exchange for Ca^{2+}. (5) The increase in intracellular Na^+ activates the Na^+/K^+ pump, which raises the intracellular K^+ content while removing the excess Na^+. It is reasonable to suggest further that the increase in K^+ leads to progression of the cells from G1 to the G1/S boundary and into S phase, perhaps by promoting the synthesis of the critical unstable protein discussed earlier.

This cascade of events is typically elicited by first arresting (or, from Smith and Martin's perspective, greatly slowing) a cell population, for example by reducing serum, and then stimulating the cells with the mitogen. The inference is that single cells in an asynchronously dividing population experience a similar set of changes of ions and small molecules as each cell proceeds through the critical part of G1.

Thus, there is evidence not only for K^+ and Na^+ changes during G1 but also for changes in several other small ions and molecules. Consequently, the presence of gap junctions could in principle affect many of the steps in this cascade, providing a built-in safety factor. A single cell in G1, tightly coupled to its neighbors, would have considerable difficulty in lowering its H^+ concentration (because of transfer from its neighbors) or raising its Na^+ or K^+ concentrations (because of transfer to its neighbors). Whether junctional transfer could affect the accumulation of IP_3, DAG, or Ca^{2+} is more questionable, even though they are small enough to be junctionally permeant. IP_3 and DAG are apparently quite labile (Nishizuka, 1986) and may act rather locally. Also, DAG is quite lipid soluble and may exert most of its effects within membrane domains (Bell, 1986). The question of Ca^{2+} transfer is somewhat different. Although Ca^{2+} ions are small enough to pass through junctional channels, they have been shown to greatly decrease junctional permeability at the appropriate concentration (Loewenstein, 1981; Spray *et al.*, 1982). The key here is that the level needed to close channels is reported to be in the range of 5×10^{-6} M and above and it has been argued that this range is above that typically needed for Ca^{2+} effects on most intracellular processes (Spray *et al.*, 1982). Therefore, if the level of Ca^{2+} resulting from the mitogenic cascade is below the junctionally inhibitory value, Ca^{2+} might well be transferred to neighbors from the cell responding to the mitogen.

In addition to the small molecules directly implicated in the decision process, there are other mitogen-induced changes that could be affected by gap junctional transfer. As discussed some time ago by Socolar (1973), if such changes as increased membrane transport of sugars or amino acids occurred in a single cell connected with gap junctions to its neighbors, there would be a minimal increase in the intracellular concentration of the transported molecules (or their immediate derivatives) because of junctional transfer.

3.2. S-Phase

Although control of cell proliferation is not normally exerted on cells already synthesizing DNA, there still are important possible influences of junctional transfer. The nucleotides from which DNA is synthesized in most cells are derived from a combination of *de novo* synthesis and reutilization of degraded nucleic acids via the so-called scavenger pathways. It has been known for some time that cells that lack thymidine kinase (TK^-) can incorporate labeled thymidine into their DNA when they are fed labeled nucleotides via their junctions with TK^+ cells (Pitts, 1971). Because the thymidine nucleotide polls are relatively short-lived (Pitts and Simms, 1977), the critical transfer must occur while the recipient cells are in S-phase. If the TK^- and TK^+ cells are prevented from making TTP by *de novo* synthesis (e.g., by treating with aminopterin), the growth of the TK^- becomes strictly dependent on having thymidine in the medium, and having adequate gap junctions with TK^+ cells (Pitts, 1971).

Whether such extreme differences in TK activity ever occur in normal cell populations is unclear, but certainly similar "metabolic cooperation" could balance out even smaller differences in enzyme activity. Here it should be noted that non-S-phase cells would have to have appreciable TK activity in order to contribute the necessary amount of TTP to the TK^- cells.

This conclusion suggests another, perhaps even more interesting, possibility for cooperation during S-phase. Single S-phase cells could actually depend on their neighbors for "help" in obtaining exogenous thymidine. In essence, the DNA synthetic machinery of a single cell would function as a sink for several contributing thymidine sources, increasing the efficiency with which the cell population as a whole could utilize exogenous nucleoside.

3.3. G2-Phase

This phase, like S-phase and M-phase, is not usually affected by growth-controlling mechanisms. Therefore, in the absence of evidence for well-defined alterations in small molecules or ions, there is little basis for suggesting any particular junctional effect on progress through G2.

3.4. M-Phase

Although the mitotic, or M, phase, is also not involved in normal growth regulation, it is reasonable to ask whether any of the complex mechanisms of mitosis, i.e., chromosome condensation, alignment, and separation, and/or of cytokinesis might be influenced by gap junctions. There are no clear answers, but a few interesting points can be made: (1) Many of the events of M phase are controlled by intracellular Ca^{2+} levels (Berridge, 1975), which in turn can be influenced by several other small molecules, Na^{+} (which participates in Na^{+}/Ca^{2+} exchange), cAMP (which tends to promote Ca^{2+} reuptake into cellular storage sites; Berridge, 1975), and IP_3 (which promotes Ca^{2+} release from cellular stores; Burgess *et al.*, 1984). (2) cAMP levels are minimal during mitosis and are severalfold higher throughout most of the rest of the cycle (Friedman *et al.*, 1976). The reduction could be important for allowing cytoplasmic Ca^{2+} to rise. (3) If nonmitotic cells remain well-coupled to a cell entering mitosis, they could transfer cAMP in amounts sufficient to offset the normal cAMP reduction. (4) Thus, some decrease in coupling might be advantageous for the mitotic process.

There is little quantitative information on this point. Several workers have reported that mitotic cells remain coupled to their neighbors as judged by the retention of gap junctions (Merk and McNutt, 1972) and of ionic (O'Lague *et al.*, 1970) and dye transfer (Atkinson, personal communication). However, none of these studies directly measured junctional permeability. One recent study on early mammalian embryos demonstrated a pronounced reduction in dye spread between mitotic and nonmitotic cells (Goodall and Maro, 1986). Whether this result is peculiar to this preparation or reflects an extreme case of events occurring during mitosis in general remains to be determined.

4. EVIDENCE FOR GAP JUNCTION INVOLVEMENT IN CONTROL OF PROLIFERATION

With the exception of the experiments on cocultures of ouabain-sensitive and -insensitive cells, the preceding arguments for a role of gap junctions in cellular growth control are rather speculative. They are based on the existence of gradients of ions and other small molecules in adjacent cells in different stages of the cell cycle and on the general property of gap junctions to reduce such gradients. More direct evidence is needed to implicate junctional transfer and such evidence has been obtained from experimental manipulations of normal cells and from analysis of gap junctions and junctional transfer in cells deficient in growth control.

4.1. Normal Cells

Most of the emphasis in the preceding section was on the expected inhibitory effect of gap junctions on cell growth, in particular as potential contributors to density-dependent or contact inhibition of cell proliferation. If gap junctions play this role, removal of the junctions or reduction in their permeability to the critical regulatory molecules should reverse, or prevent the establishment of, the inhibitory effects (Potter *et al.*, 1966; Loewenstein, 1966, 1968). This line of reasoning has led to numerous experiments analyzing the effects on gap junctions and/or junctional transfer of agents or conditions known to prevent or reverse density-dependent arrest (or slowdown) of cell proliferation. Some selected examples are considered in the following discussion.

It has long been known that some cells "arrested" at confluency can be stimulated to undergo one or more rounds of division by an abrupt increase in serum concentration (Holley and Kiernan, 1968). Although there are many possible explanations for this effect (e.g., Stoker, 1973; Folkman and Moscona, 1978; Wieser and Oesch, 1986), it has been shown that some cells, at least, respond to such a serum step by a decrease in junctionally mediated transfer of fluorescent dye (Flagg-Newton and Loewenstein, 1981). Moreover, one particular cell line, CL-1D cells, which at high density and normal serum lack gap junctions, will form functional junctions when the serum is decreased (Azarnia *et al.*, 1981). (These cells are "transformed," but their junctional response to serum resembles that of "normal" cells.)

It has been suggested that the decrease in junctional transfer induced by the serum step is a result of a decrease in intracellular cAMP levels which promotes junction formation. This suggestion is based on several findings: (1) cAMP levels are lower in the various cells in the presence of serum than in its absence (Azarnia *et al.*, 1981; Flagg-Newton and Loewenstein, 1981); (2) treatment of several cultured mammalian cell lines with dibutyryl cAMP leads to increased junction formation (Flagg-Newton *et al.*, 1981); (3) treatment of CL-1D cells with dibutyryl cAMP also leads to junctional development even at high cell density and normal serum (Azarnia *et al.*, 1981); (4) several other cells increase junctional transfer when treated with hormones which raise intracellular cAMP (Radu *et al.*, 1982).

One problem with relating these junctional changes to the serum-induced mitogenesis is that the serum effects on junctional transfer may take as long as 24 hr or more to develop. However, growth stimulation by an increase in serum, as already pointed out, is quite rapid, with cells entering S-phase within a few hours (the actual stimulation event probably occurs even sooner). It is not clear from the published work whether cAMP level or junctional transfer was studied within very short times after serum decrease. Because cAMP can influence junctional transfer after only a few hours (Flagg-Newton *et al.*, 1981), the kinetics of the junctional changes following serum steps (up and down) should be reexamined.

Another traditional method for stimulating division by density-inhibited cells is to remove some of the cells by making a "wound" (Todaro *et al.*, 1965). The typical result is migration and proliferation by the cells within a few cell diameters of the wound, leading ultimately to wound closure. It has been suggested that the cells which proliferate do so because they have lost their junctional coupling to their neighbors (Loewenstein, 1979). Obviously the cells directly adjacent to the wound have lost coupling to the neighbors that were removed, but what about the stimulated cells farther from the wound edge? In an attempt to evaluate coupling by these more distant cells, Stoker (1975) seeded TK^- cells (TG1) on a wounded culture of the 3T3 cells and determined the effectiveness with which the 3T3 cells transferred thymidine-derived nucleotides to the TK^- cells various distances from the wound. He found no quantitative changes in the extent of "metabolic cooperation" near the wound. However, this protocol actually tested the formation of nucleotide-permeable junctions, not the permeability of junctions existing at the time of the wound.

In another system, a decrease in amount of gap junctions in the region of a wound has been reported by two groups (Schwartz *et al.*. 1975; Spangoli *et al.*, 1982). The intimal surfaces of rabbit aortae were wounded with a balloon catheter and gap junctions near and away from the wound measured at different times after wounding. In the more quantitative study with freeze-fracture methods (Spangoli *et al.*, 1982), there was an 87% decrease in amount of gap junctions between endothelial cells in the region near the wound. [Qualitatively similar changes have been reported in endothelial monolayers wounded in cell culture (Ryan *et al.*, 1982; Brown *et al.*, 1982).] These studies all suggest that a decrease in junctional transfer might coincide with wound-induced endothelial cell proliferation.

However, recent studies by Jackman indicate that the story might not be quite so simple, at least in the intact vessel (Jackman *et al.*, 1981; Jackman, 1986). He has developed a method for culturing rat aortic explants that retain their ability to undergo complete endothelial regeneration following *in vitro* wounding of the intimal surface (Jackman, 1982, 1986). In order to compare the ability of endothelial cells in the regenerating and intact regions of the intima to carry out junctional transfer, he seeded donor cells (typically bovine aortic endothelial cells), prelabeled with [^{3}H]uridine, on the explant. The donors were generally more adherent to the regenerating cells than to those farther away from the wound edge, but instances of nucleotide transfer were seen in both regions. Unexpectedly, the extent of recipient labeling was greater in the regenerating region, a result with several possible explanations, including greater amount of gap junctions in that region (in conflict with the earlier reports), greater rate of RNA synthesis by the regenerating cells, or lower rate of junction formation between the donor and recipients in the intact region, allowing less time for nucleotides to be transferred. Whatever the explanation, the result was not straightforward.

Another type of wound, which has provided a useful model for growth regulation, is partial hepatectomy. Removal of part of the liver results in a generalized stimulation of DNA synthesis and cell division leading to regeneration of the organ essentially to its original size. Revel's group has studied several features of the gap junctions in the liver during the regeneration process and has demonstrated the following: (1) there is a precipitous decrease in the size and frequency of gap junctions, reaching a minimum about 29–35 hr after surgery (Yee and Revel, 1978; Yancey *et al.*, 1979); (2) electrical coupling and dye transfer among hepatocytes are reduced, but not eliminated at this same time (Meyer *et al.*, 1981); (3) as the liver regains its original size and proliferation ceases, however, the gap junctions are morphologically restored (Yee and Revel, 1978). These changes are generally consistent with the hypothesized role for gap junctions in control of cell proliferation.

Besides wounding, there is another method for rapidly decreasing cell density, i.e., trypsinization and replating of dispersed cells. When this procedure is carried out with confluent, density-inhibited cells, it always leads to stimulation of growth, which continues until the cells again reach confluency. In this case there is no doubt that junctional communication is interrupted. The questionable issue is whether the cessation of division is associated with reestablishment of junctional communication. According to the general models of gap junction involvement in growth control, we would expect that each cell in the growing population would form more gap junctions as the cell density increased and, as a consequence, the cell came into contact with greater numbers of other cells. Tests of this expectation, however, have given rather mixed results. Yamasaki *et al.* (1985) used dye injection to study two subclones of Balb/c 3T3 and demonstrated that both clones showed an increase in junctional transfer as they approached confluence. However, once confluence was established, and growth was inhibited, the two clones behaved differently: one maintained its increased junctional capability whereas the other lost most of the capability over the next few days. In a dye injection study of a different Balb/c 3T3 line, Flagg-Newton and Loewenstein (1981) showed that the proportion of cell interfaces having dye-permeable junctions actually decreased as the cells grew toward (or were plated at) progressively higher densities. The lowest incidence of junctional transfer occurred at the highest cell densities. The decreased incidence was correlated with a decrease in cAMP levels in the cells.

Although some of these results seem to conflict with the basic model of gap junction involvement in growth control, there are some complicating factors. First, in the Yamasaki *et al.* study, the cells that decreased their junctional communication as they became confluent actually grew to a higher density before growth ceased than did the cells whose junctional communication remained elevated. This result, therefore, does fit with the general model. Second, in the Flagg-Newton and Loewenstein study, it is not clear that the confluent cells had in fact stopped growing. Because the medium was changed every day, the cells were being continually exposed to serum growth factors which may have influenced the junctions. Therefore, it is difficult to relate this study to the general model, although it does support the hypothesized role for cAMP in maintaining junctional communication.

Before shifting to a discussion of experiments involving cells deficient in growth control, a final example involving normal cells is worth a few comments. By the time that mammalian oocytes reach the stage of development that directly precedes ovulation, they have gone into meiotic arrest. When ovulation occurs, the oocyte is released from the arrest, and completes meiosis as it proceeds down the fallopian tube in preparation for fertilization. When it was shown that the oocyte in the follicle formed permeable gap junctions with surrounding cumulus cells (Anderson and Albertini, 1976; Gilula *et al.*, 1978), which in turn formed gap junctions with other granulosa cells (Albertini and Anderson, 1974) and that these gap junctions as well as junctional transfer disappeared shortly after ovulation (Dekel *et al.*, 1981), there was speculation that the transfer of some signal through the junctions kept the preovulatory oocyte in meiotic arrest (Dekel and Beers, 1980). Moreover, there was a prime candidate for the inhibitory signal, cAMP; exogenous cAMP derivatives prevented the resumption of meiosis by the postovulatory oocyte, cAMP was high in the cumulus and granulosa cells in the presence of FSH, and the oocyte had no FSH sensitivity and rather low adenylate cyclase activity (Schultz *et al.*, 1983).

There were two problems with these suggestions, however. First, the decrease in junctions and junctional transfer between oocyte and immediate cumulus neighbors did not precede germinal vesicle breakdown (the initial event in the release from meiotic arrest) (Eppig, 1982). Second, oocytes separated from cumulus cells after FSH treatment of the cumulus–oocyte complex did not have higher levels of cAMP than oocytes treated with FSH after isolation (Schultz *et al.*, 1983; Beers and Olsiewski, 1985).

Recent studies have now resolved these two problems. First, quantitative freeze-fracture studies have indicated that there is a breakdown of gap junctions between the cumulus cells just preceding ovulation (Larsen *et al.*, 1986). Since the major binding sites for FSH are on the more peripheral granulosa cells, it is likely that this loss of junctions could separate the oocyte from the major source of FSH-induced cAMP. Second, experiments in which the degradation of cAMP in the oocyte is minimized by high levels of phosphodiesterase inhibitors during its isolation from cumulus cells show that the oocytes do in fact have elevated cAMP levels after FSH, forskolin, or cholera toxin treatment of the cumulus–oocyte complex (Bornslaeger and Schultz, 1985). Thus, the most recent evidence is fully consistent with the original ideas concerning the role of gap junctions and cAMP transfer in meiotic arrest.

4.2. Cells Deficient in Growth Control

The speculation that defective regulation of cell proliferation might follow when junctional areas or permeability is reduced has also spawned numerous searches for junctional abnormalities in cancerous, or transformed, cell systems (Loewenstein, 1979). Many of the earlier studies focused on all-or-none changes, in part because of the limited availability of methods for detecting more subtle, quantitative changes in the junctions, and in part because of the scarcity of systems in which transformation and/or junctional alterations could be manipulated reversibly. The results from the earlier studies indicated that pronounced junctional defects were evident in only a subset of transformed cells, making it difficult to assess the etiological importance of the junctional changes.

Over the last several years there have been a number of new research findings and directions that point to a more general relationship between junctional defects and cell transformation than was previously anticipated. These new efforts can be conveniently divided into two groups, those involving viral transformation and those involving tumor promoters and related molecules. As will become evident, however, this division is rather arbitrary and at the mechanistic level there are several interesting points of convergence.

4.2.1. Viral Transformation

4.2.1a. Junctional Permeability. Until recently, cells infected by transforming viruses were given little attention by researchers interested in gap junctions, chiefly because in a few early studies, a number of lines of such cells had been shown to lack obvious defects in gap junctions or in junctional transfer of small molecules (Potter *et al.*, 1966; Furshpan and Potter, 1968; Pinto da Silva and Gilula, 1972; O'Lague and Dalen, 1974). An important clue to the possibility of more subtle defects came from the studies of Corsaro and Migeon (1977b) who showed that several lines of SV40-transformed cells, as well as lines transformed by unknown or other means, were deficient in "metabolic cooperation" as assayed by a "kiss of death" protocol (Goldfarb *et al.*, 1974). For this procedure, which as described below has also been useful in detecting the junctional effects of tumor promoters, $HGPRT^+$ (hypoxanthine-guanine phosphoribosyl transferase) and $HGPRT^-$ cells are cocultured at different ratios in the presence of 6-thioguanine. This purine analogue is readily incorporated by the $HGPRT^+$ cells into their nucleic acids, leading to cell death. $HGPRT^-$ cells are resistant unless they form permeable junctions with the $HGPRT^+$ cells, in which case they too are killed because of the junctional transfer of the 6-TG-derived nucleotides. Cells that survive to form colonies in the 6-TG-treated cocultures are $HGPRT^-$ cells that have failed to develop functional gap junctions with the $HGPRT^+$ cells (or for some other reason have been unaffected by the transferred 6-TG metabolites; see below). The number of surviving colonies can easily be counted and, with the caveat mentioned above, provides an inverse measure of the junctional capabilities of the coculture pair. Thus, in Corsaro and Migeon's study, many of the transformed cells left significant numbers of surviving colonies in the cocultures.

The difficulty with these studies, however, and with many of the previous comparisons of transformed and nontransformed cells was the possible contribution of other differences in the cells such as length of time in culture, fundamental differences in nucleotide metabolism (other than the scavenger pathway), and so on. That is, the changes in "metabolic cooperation" could have been due to nonjunctional factors, or, if junctionally based, could have come from some intrinsic differences in the cells other than the presence or absence of transformation.

Here an important approach has involved the use of temperature-sensitive viral mutants, allowing the cells to be shifted reversibly between the transformed and nontransformed states by a simple change in incubation temperature. Atkinson *et al.*, (1981) first utilized such a system, involving normal rat kidney (NRK) cells infected with LA-25, a Rous sarcoma virus (RSV) mutant, to study the changes in junctional transfer of a fluorescent dye by cells grown at the transforming and nontransforming temperatures. For these initial studies, the quantitative measure of junctional transfer was the "transfer time interval" defined as the shortest time before dye injected into one cell was first detected in an adjacent cell. The results confirmed earlier reports that virus-transformed cells were capable of junctional transfer, but they demonstrated that the rate of dye transfer was substantially slower than for nontransformed cells. Moreover, the change in transfer rate was detectable within 15 min of temperature downshift (i.e., to the transforming temperature) or temperature upshift (i.e., back to the normal temperature), making the junctional change one of the earlier transformation-related events.

We have recently developed a more quantitative method for assessing junctional permeability to fluorescent dyes utilizing computer-assisted video analysis (Liu *et al.*, 1982; Biegon *et al.*, 1987) and have applied this method to LA-25 cells and to other cell lines infected with different retroviruses (Atkinson and Sheridan, 1985). In this method the measure of junctional transfer is more direct, providing an estimate of junctional permeance (the product of the permeability coefficient and the junctional area). Our results with the LA-25 cells confirmed our earlier findings but in addition indicated that the decrease in junctional permeability upon shifting to the transforming temperature was even more dramatic than originally detected. Moreover, the average junctional permeance 1 hr after temperature shift was even lower than that found after 1 day's growth at the new temperature (Atkinson, unpublished). One possible implication of this finding is considered further below.

Our finding of reduced junctional permeance in RSV-transformed cells and the earlier results with SV40-infected cells by Corsaro and Migeon (1977b), both of which were recently confirmed (Azarnia and Loewenstein, 1984a,b; Chang *et al.*, 1985) using several different cell lines, raised the immediate question of whether other types of transforming viruses would produce similar effects. Using our video method, we have found alterations in junctional permeability in two other virus-infected NRK cell lines (Atkinson and Sheridan, 1985). The first, a clonal line (6m2) infected with a mutant of Moloney murine sarcoma virus (ts110) which is temperature sensitive for transformation, has substantially reduced junctional permeability when grown at the transforming temperature, although the kinetics of the changes following temperature shift have not yet been worked out. Kirsten virus-infected NRK, expressing a k-ras oncogene, have reduced permeability as compared to uninfected cells (these are not temperature sensitive).

Thus, these studies have shown that three different types of retrovirus, expressing different oncogenes, and one type of DNA virus, have the common effect of apparently reducing, but not eliminating, the capability for junctional transfer. In addition, where it has been possible to study the kinetics, the apparent junctional change is associated with very early events in the transformation process. Together the results strengthen the argument for an important role for junctional alteration in cell transformation.

4.2.1b. Possible Mechanisms. With all of these studies it was first important to ensure that the changes in dye transfer were directly related to changes in the junctions. As already implied, the interpretation of changes in "metabolic cooperation" is problematic because several factors, such as differences in nucleic acid synthesis or in *de novo* nucleotide synthesis and nucleotide pools, could influence the amount of incorporation of the transferred molecules. On the other hand, the dye injection and transfer experiments are more straightforward, particularly when a membrane-impermeant dye such as Lucifer Yellow CH is used. For such dye methods the only nonjunctional factors of potential concern are differences in cell volume or in extent of cell damage produced by the impalement. Yet in the typical case, transformed cells tend to be smaller, the opposite direction to explain an apparent decrease in transfer by difference in cell volume. The leakage effect of cell damage is easily ruled out when the injected cell retains a high concentration of dye and any effect of cell damage on junctional permeability is unlikely when the junctional transfer rate follows simple diffusion kinetics for some time following removal of the injecting pipette.

The absence of nonjunctional artifacts indicated that the changes in dye transfer in the different virus-infected cell systems were junctional in origin and, in principle, could have resulted from changes in the number of gap junction channels, in the proportion of open channels, or in the diameter of the open channels. In order to examine the first possibility for our LA-25 NRK cells, we used freeze-fracture to analyze the amount of gap junction (frequency and areas) linking cells under one of four different conditions (Atkinson *et al.*, 1986): (1) cells grown for more than 24 hr at the transforming or (2) nontransforming temperature; or (3) cells shifted for an hour to the transforming or (4) nontransforming temperature. For additional comparison, uninfected NRK cells were studied under these conditions as well. The results indicated a decrease in junctional area in the cells grown for a day at the transforming temperature but a similar change was seen in the noninfected cells. However, by 1 hr after temperature shift in either direction, at a time when the changes in junctional permeability were fully expressed, there was no significant change in the junctional areas. Yet there were changes in particle arrangements (though not in uninfected cells), with a more regular arrangement being associated with the nontransforming temperature and a more random arrangement being associated with the transforming temperature. Thus, from these and other data, it is apparent in the LA-25 system at least that the permeability changes do not reflect changes in total numbers of channels but rather in proportion of open channels or changes in channel diameter. The alterations in particle arrangements are more difficult to interpret because the cells were fixed. However, their close correlation with the temperature-induced changes in junctional permeability makes them quite interesting. They may reflect some transformation-related change in membrane properties and, if so, similar changes might be seen in other viral transformation systems.

As yet there are no quantitative structural data on gap junctions in the other virus-transformed cell systems and, therefore, it remains possible that changes in junctional areas or frequency may contribute to the changes in dye transfer in those cases. In the absence of appropriate studies, it seems more productive at this point to consider the possible factors that might cause the changes in the permeability of individual junctional channels.

As already noted, in principle the permeability of single channels can be reduced by totally closing them or by reducing their open diameters. Distinguishing between the two possibilities requires a very precise comparison of the number of connexons and total permeability, a detailed analysis of the distribution of single channel ionic conductances, or a study of the relative permeability of single junctions to probes of different sizes. Studies of the first type, with the necessary resolution, are not practical. Of the few studies of the second type, as yet carried out only with normal cells, the results are mixed, ranging from a preponderance of a single, unitary channel conductance (or unit multiples) (Spray *et al.*, 1986; Veenstra and DeHaan, 1986) to a substantial proportion of channels with intermediate conductances or substates (Neyton and Trautmann, 1985; Zampighi *et al.*, 1985; see also Loewenstein *et al.*, 1978). Studies of the third type have also given mixed results. In some studies of the third type, applied to certain normal cells, the channels appear to be either totally open or totally closed (or the proportion of channels of different diameters remains constant as the overall permeability is altered) (Zimmerman and Rose, 1985). We have recent evidence, however, that Novikoff hepatoma cells may have more than one channel diameter, with the proportion varying between different pairs of cells (Biegon *et al.*, 1986). In this system, the ratio of junctional permeance to Lucifer Yellow CH and Lissamine Rhodamine B (actually the hydrolyzed, hydrophilic product) varies as much as eightfold from cell pair to cell pair. Hülser and Brümmer (1982) have also concluded that junctional channels might gradually change their diameters in a system involving "spheroids" of cultured mammary tumor cells, although differential sensitivity of the electrophysiological and due transfer methods might have contributed to the findings. Therefore, based on current evidence, there is no reason to exclude the possibility that the decrease of junctional permeability associated with viral transformation involves altered channel diameter as well as (or instead of) complete closure of channels.

The control of the gating of junctional channels is addressed in more general terms by De Mello and Peracchia in their chapters. As they indicate, in a variety of cell systems, junctional channels can be induced to close by increased cytoplasmic Ca^{2+}, decreased cytoplasmic pH (i.e., increased H^{+}), and/or membrane potential changes. However, rather than begin with these possibilities, it is useful to consider first the nature of the transforming gene product, pp60v-src, coded by the v-src gene, which a recent elegant experiment indicates is responsible for the change in junctional permeability (Chang *et al.*, 1985).

pp60v-src is a protein kinase that catalyzes the phosphorylation of a number of cellular proteins at tyrosine residues (Collett and Erikson, 1978; Hunter and Sefton, 1980). The LA-25 mutant codes for a temperature-sensitive enzyme that is active at the transforming temperature and inactive at the nontransforming temperature (Wyke, 1973; Wang and Goldberg, 1979). The kinetics of the temperature-dependent activation of the pp60v-src in a related mutant, LA-23, is very similar to that of the decrease in junctional permeability (Goldberg, personal communication, mentioned in Atkinson *et al.*, 1981). Therefore, it is reasonable to suggest that there may be only a few steps between phosphorylation of a critical target protein and closing of junctional channels.

A rather direct effect would be phosphorylation of tyrosines on the junctional protein itself. Although there is as yet no evidence that this occurs, pp60v-src has been localized to the membrane at regions of cell contact (perhaps involving gap junctions) (Willingham *et al.*, 1979). Another effect, which would be slightly less direct, which would also be consistent with the membrane localization of the enzyme, and which might implicate one of the agents previously associated with junctional changes, is suggested by evidence that pp60v-src enhances the turnover of phosphatidylinositol (Macara, 1985; Sugimoto and Erikson, 1985), which, as noted above, may play an intermediary role in growth stimulation by mitogens. Breakdown of PI leads to the release of IP_3 which in turn promotes Ca^{2+} movement into the cytoplasm from cellular stores. Whether or not the Ca^{2+} levels are elevated sufficiently to close junctional channels is unclear, but at least the direction of the change is appropriate to explain the junctional effect of pp60v-src.

PI breakdown also releases DAG, which, in the presence of Ca^{2+}, activates another kinase, protein kinase C. This enzyme promotes phosphorylation of serines and threonines on several proteins, including the cellular form of the src gene product, pp60c-src (Gould *et al.*, 1985). pp60-src has a constitutively low tyrosine protein kinase activity (Iba *et al.*, 1985) that, it has been reported (Pietropaolo *et al.*, 1981; but see Goldberg *et al.*, 1980), may be increased after it is phosphorylated by protein kinase C. It is reasonable to suggest that the increased kinase activity of pp60c-src would stimulate PI turnover, as does pp60v-src. As discussed below, there is independent evidence that activation of protein kinase C leads to decreased junctional permeability (e.g., Yotti *et al.*, 1979; Enomoto and Yamasaki, 1985a) (and, in one system, even decreased junctional area; Yancey *et al.*, 1982). Taken together, all of these various suggestions lead to the possibility that the initial activation of pp60v-src might induce a cascade of self-amplifying events leading to decreased junctional transfer: (1) pp60v-src causes PI turnover, releasing IP_3 and DAG; (2) the IP_3 leads to increased cytoplasmic Ca^{2+}; (3) the Ca^{2+} in the presence of DAG activates protein kinase C; (4) protein kinase C phosphorylates two proteins, one which causes junctional channel modification (perhaps the junctional protein itself) and the other, pp60c-src, which becomes more active; (5) the activated pp60c-src further increases PI turnover, reinforcing the initial effect of the pp60v-src.

The evidence for many of the preceding comments about the interrelationships among pp60v-src, pp60c-src, protein kinase C, PI turnover, and altered junctional permeability has come from studies of the effects of tumor promoters, chiefly TPA, on these various proteins and activities. Thus, it is appropriate to consider the tumor promoter effects before completing the discussion of possible mechanisms and turning finally to implications for normal growth control.

4.2.2. Tumor Promoters

Whereas certain of the transforming viruses we have been discussing may have the full capability for converting normal cells into cancer cells, carcinogenesis in many experimental systems, and probably in the more common human cases, involves more than one agent. A simplified model for mouse skin carcinogenesis depends on the serial action of two agents, an initiator and a promoter (Berenblum and Shubik, 1947), but even this system has proved too complex to allow close study of the mechanisms by which the promoters act. Consequently, the action of promoters has been carefully investigated in cell culture, and the results have provided considerable insight into possible mechanisms for promotion of transformation. Moreover, they have also given clues to the process of normal growth regulation and the possible role of gap junctions (Trosko *et al.*, 1983).

4.2.2a. Junctional Permeability. Given the earlier suggestions of a possible role for decreased junctional transfer by transformed cells, there was considerable interest in the first reports that treatment of certain normal cells by a particular tumor promoter, 12-*O*-tetradecanoylphorbol 13-acetate (TPA), reduced their ability to carry out metabolic cooperation (Murray and Fitzgerald, 1979; Yotti *et al.*, 1979). The method used in the studies by Yotti *et al.* (1979) has already been described above and involved cocultures of HGPRT$^+$ and HGPRT$^-$ cells in the presence of 6-TG with or without TPA or a related compound lacking promoter activity. TPA, but not the nonpromoting analogue, led to the survival of numerous HGPRT$^-$ colonies, an effect which apparently reflected the reduced junctional transfer of the toxic, 6-TG-derived, metabolites from the HGPRT$^+$ to the HGPRT$^-$ cells.

Murray and Fitzgerald (1979), on the other hand, monitored the transfer of uridine-derived nucleotides as a measure of junctional permeability and demonstrated that the effects of TPA and a related promoter were evident within a 4-hr treatment period.

Shortly after the initial reports appeared, it was shown by other workers (Enomoto *et al.*, 1981) that TPA produced an even more rapid inhibition of electrical coupling (partial effect detected within 1 hr; full inhibition in about 5 hr versus 16 hr), which, in the absence of nonjunctional resistance changes, indicated that the effect was on the junctions themselves and not on one of the several other factors influencing the incorporation of nucleotides by the recipient cells in the earlier experiments. A particularly critical finding was the equally rapid reversibility of the TPA effect.

Still more recently, there have been reports from a number of laboratories of rapid, and reversible, inhibition of dye transfer in cultures treated by TPA or by any one of many different promoters. The extensive series of papers by Yamasaki and colleagues has revealed a number of particularly interesting features of the junctional changes induced by promoters (chiefly TPA and related compounds): (1) the effects have their most rapid onset when the cells are preconfluent and still growing (Enomoto and Yamasaki, 1985b); (2) the effects on growth-phase cells reverse spontaneously despite the continued presence of the agent (Enomoto and Yamasaki, 1985b; see also Fitzgerald *et al.*, 1983); (3) once the cells become confluent and reach a saturation density (which is higher for the promoter-treated cells), there is a progressive, but slower, decrease in junctional transfer, stabilizing at low levels within a few days (Enomoto and Yamasaki, 1985b); (4) cell lines which are resistant to the promoting effect of TPA also lack changes in junctional transfer (Rivedal *et al.*, 1985); (5) the synthetic diacylglycerol, OAG (1-oleoyl-2-acylglycerol), produces a decrease in dye transfer comparable in time course and magnitude to that produced by TPA (Enomoto and Yamasaki, 1985a) (a result also demonstrated independently by Yada *et al.*, 1986); (6) db-cAMP + caffeine reverses the TPA effect in Balb/c 3T3 cells, but not in V79 cells, an antagonism that requires protein synthesis (Enomoto *et al.*, 1984).

Despite the variability in the sensitivity of different cell lines to TPA action (either on the junctions or on promotion), the effects on the sensitive lines are so reproducible that it may be tempting to suggest that study of junctional transfer would provide a general means for identifying promoters. This extrapolation of the results with known promoters may be a bit premature (Trosko *et al.*, 1983). Certainly the interpretation of metabolic cooperation assays should be made guardedly in light of the many nonjunctional factors that could be influenced by any promoter candidate. Although dye injection methods should give more unambiguous information about junctional changes, the question remains whether all agents leading to decreased junctional transfer ought to be considered potential promoters.

4.2.2b. Possible Mechanisms. The fact that three different methods, metabolic cooperation, electrical coupling, and dye injection, all indicated a reduction of junctional transfer after cells were treated with tumor promoter strongly argues for an effect on the junctions rather than on nonjunctional processes. As with the viral effects discussed above, there are three possible basic mechanisms by which the junctional changes could have been induced, i.e., a decrease in number of junctional channels, a decrease in the proportion of open channels, or a decrease in channel diameter.

There has been only one freeze-fracture study of gap junctions in cultures treated with promoters (Yancey *et al.*, 1982), and, unfortunately, only one cell system, hamster V79 cells, and one promoter, TPA, were tested along with a nonpromoting analogue. After 16 hr of exposure to TPA, at a level shown in other studies to effectively inhibit metabolic cooperation, the V79 cells had substantially fewer gap junctions with smaller total area than seen in cultures with either control or nonpromoting analogue treatment.

Although this study provided definitive evidence for a decrease in gap junctions, several important questions remain. Is the decrease in gap junctions a general feature of the promoter effect or is it specific to the cell system or the particular promoter, TPA, used? Does the junctional change even in this system occur as rapidly as the promoter-induced decrease in junctional transfer, which can be detected within less than 1 hr? If in fact a decrease in gap junctions proves to be the general consequence of promoter action, is the junctional change secondary to cell shape change or is it due to some more fundamental alteration in the junction formation or degradation processes? Although we do not have definitive answers to these questions, they are sufficiently fundamental to the whole issue of junctional involvement in growth control that they warrant more detailed consideration.

To date there have been no other structural studies of promoter-treated cells. Therefore, the generality of the junctional changes observed with TPA and V79 cells remains unclear.

There are also uncertainties regarding the temporal relation between the structural changes in the junctions in the V79 and the alterations in junctional transfer. Junctional formation between cultured cells is a very dynamic process (Johnson *et al.*, 1974; Sheridan, 1978; Loewenstein, 1981), with kinetics fully compatible with the rapid changes in junctional transfer following removal of

TPA from an inhibited culture. Less is known, however, about the rates of junctional loss, except after cell dissociation where the process may be quite specific for the particular cell type (Preus *et al.*, 1981a). In regenerating liver the loss of junctions occurs over a few hours, while, for example, that in the myometrium following parturition occurs much more slowly (Garfield *et al.*, 1977). The decrease in gap junctions in cultures treated with TPA would have to be more rapid, at least to account for the rate of onset of the TPA effect on growth-phase cells (Enomoto and Yamasaki, 1985b).

If alterations in formation and/or degradation of gap junctions contribute to changes in junctional transfer, what are the possible underlying mechanisms? A few interesting possibilities are suggested by the change in cell shape, which usually occurs in cells treated with tumor promoters. Typically, cells become more rounded and smaller, and, in the more extreme cases, the cells actually retract from each other and reduce their apparent areas of contact. This retraction would appear to guarantee a smaller amount of gap junctions, but in fact, upon closer inspection, this conclusion is not as simple as it first seems. If our interest is in the possible role of a decrease in junctional transfer in the initial stimulation of cells to reenter the cycle or to make a "decision" to move toward DNA synthesis, then we need to look at relatively early changes in the amount of gap junction. That is, the problem focuses on the loss of existing gap junctions rather than the lack of formation of new gap junctions. Here the effects of cell retraction and consequent decrease in contact area are far from clear. Gap junctions do not cover a large proportion of the area of cell contact, particularly in cell culture. For example, the ratio of gap junction area to contact area may be as little as a fraction of a percent (Revel *et al.*, 1971) (even in liver the ratio is less than 10%; Yancey *et al.*, 1979). Moreover, gap junctions are extremely effective in holding cell membranes together. Therefore, when cells retract from each other, the separation most likely occurs in nonjunctional regions while the gap junctions probably remain intact in the residual areas of contact, much like the situation occurring when embryonic blastomeres are physically pulled apart, but remain junctionally coupled (Ito *et al.*, 1974). While these arguments do not rule out a rapid loss of gap junctions with promoter treatment, any such change would most likely be due to something more complicated than the change in total contact area resulting from retraction.

While there appears to be a short-term action of promoters on cell growth and junctional transfer, the longer-term effects may be more relevant to the overall process of tumor promotion. If so, retraction and the resulting change in contact area may have a direct effect on junctional area. Once cells have begun to divide, new junctions have to be formed between daughter cells. It is likely that formation is dependent on the area of cell contact and the number of junctional precursors per unit area. Therefore, a decrease in contact area between daughter cells might reduce the amount of newly formed junctions, leading, over a few cell divisions, to a population of cells with substantially reduced junctional areas.

Once again, however, the actual situation is not so simple because the rate at which junctional communication recovers following removal of TPA is so great that total restoration of normal shape and contact area seems precluded.

Here it is interesting to note that in temperature shift to the transforming temperature for LA-25 NRK cells, discussed above, there is a partial "recovery" of junctional permeability at a time that is associated temporally with the appearance of smaller junctions (Atkinson *et al.*, 1986). Although uninfected cells also have smaller junctions at the lower temperature, the development of smaller junctions in the transformed cells could reflect some attempt of the cells to form new junctions in response to the closing of existing junctional channels.

In those promoter-treated (or frankly transformed) cells which undergo a more modest shape change, e.g., simply a decrease in size and an increase in thickness, there may actually be an increase in apparent cell contact provided at the same time the cells pack together more tightly as their density decreases. Based on the previous arguments, the result should be an increase in junctional area if only contact area is important. On the other hand, if a decrease in junctional area is seen, or even if the junctional area remains constant, other factors must be having an effect. An interesting possibility is suggested if we consider the initial pretreatment condition where the cells are still rather flat. The junctions between the cells occur predominantly on lateral surfaces, which, because of the flatness of the cells, have limited areas, or at regions of cell overlap, which, with contact-inhibited cells, are also very limited in area. For these cells it seems quite possible that the junctional precursors as well are essentially restricted to the lateral or overlapping membrane regions, in order to concentrate them where junction formation occurs. The restriction could be maintained by anchoring of the precursors with cytoskeletal elements or by inserting new precursors into those areas and restricting extensive precursor diffusion to the apical and basal membranes by having a generally decreased membrane fluidity. It has been shown that membrane proteins in certain transformed cells have larger diffusion coefficients in the membrane plane, a possible consequence of loss of cytoskeletal anchoring and/or an increase in membrane fluidity. One consequence of these changes, therefore, would be to remove the postulated restrictions on junctional precursors. As they diffused throughout the entire membrane, they would become substantially more diffuse. Given the likely dependence of gap junction formation rate and extent on density of precursors (Sheridan, 1978; Preus *et al.*, 1981b), junctional size could easily fall or remain constant despite an increase in contact area.

Despite these arguments, the similarities between the effects of tumor promoters and RSV-induced transformation, and the lack of early changes in gap junctions in the viral systems, make it likely that channel gating is also influenced by promoters. As mentioned earlier, increased intracellular Ca^{2+}, decreased intracellular pH, and changes in membrane potential have each been shown to promote channel closure in several different systems (see chapters by

De Mello and Peracchia). It has been suggested that elevated Ca^{2+} might mediate the junctional effects of tumor promoters (e.g., Enomoto *et al.*, 1981; Yada *et al.*, 1986), but it is not clear that the level of Ca^{2+} attained with the promoters ever reaches the magnitude generally believed necessary to exert the junctional effect.

Within the past few years, it has become clear that TPA and other related tumor promoters activate protein kinase C, one of the critical elements in the pathway influenced by the breakdown products of PI (Nishizuka, 1986). Thus, the activation of protein kinase C is a common feature of stimulation by certain mitogens and of tumor promoter action. (The possible role of this pathway is the stimulation of cell proliferation by certain mitogens, such as PDGF, was discussed earlier). Moreover, even pp60v-src, the transforming gene product of RSV, is believed to stimulate this same kinase, but indirectly by leading to the phosphorylation of PI and facilitating its subsequent breakdown (Sugimoto and Erikson, 1985). The convergence of these various processes on protein kinase C suggests the intriguing possibility that this enzyme acts as the same final common pathway for stimulating cell growth by certain mitogens, promoters, and pp60v-src (Yamasaki and Enomoto, 1985). If so, we might ask whether the reduction in junctional permeability by pp60v-src and by TPA could possibly both be exerted by protein kinase C rather than by two independent mechanisms (Yamasaki and Enomoto, 1985).

Support for this idea has come from experiments, already mentioned, in which junctional dye transfer was tested in cells treated with OAG (Enomoto and Yamasaki, 1985a; Yada *et al.*, 1986). OAG produced a prompt and profound decrease in dye transfer, similar to that produced by TPA in parallel experiments. Since the only known action of OAG is activation of protein kinase C, it is clear that this enzyme can lead to junctional alterations, a result that brings us back to the possibility that the key event is phosphorylation of junctional protein, a possibility that bears further comment.

There have been only a few reports of junctional protein phosphorylation, one of them involving a protein, MP26, from calf lens (Johnson *et al.*, 1986), whose role in junctional transfer is somewhat controversial, and one indicating *in vitro* phosphorylation by cAMP protein kinase, by Ca^{2+}–calmodulin-dependent protein kinase, and by protein kinase C (Hertzberg and Spray, 1985; see also Saez *et al.*, 1986), and another simply indicating that phosphorylation occurred without any evidence for the specific enzymatic mechanism (Willecke *et al.*, 1985). Although cAMP protein kinases catalyze the phosphorylation of serine or theonine residues, as does protein kinase C, the residues involved would have to be different since cAMP enhances while protein kinase C inhibits junctional transfer. Clearly, further investigation of the extent and modulation of phosphorylation of junctional protein is necessary and may be critical for our understanding of the action of promoters and transforming agents such as oncogenes (Loewenstein, 1985).

4.2.3. Implications for Normal Growth Control

There are two major features of these studies of viral transformation and tumor promoters that provide some insight into the possible role of gap junctions in the regulation of cell proliferation. First, there is the close correlation between increased cell growth, or in the case of certain tumor promoter effects, increased saturation density, and diminished junctional transfer. Second, there is the evidence that the mechanism by which certain, if not many, tumor promoters act involves a natural regulator of cell proliferation, protein kinase C, and that this enzyme may even play a part in the action of at least one viral oncogene product.

4.2.3a. Correlation of Junctional Deficiency and Increased Cell Growth. Whereas this correlation and its implications are generally rather obvious from the preceding sections, two issues are worthy of elaboration. The first arises from a closer analysis of the long-term effects of TPA on the growth of Balb/c 3T3 cells. As previously mentioned in passing, TPA treatment of these cells in their growth phase transiently inhibits junctional transfer which then gradually decreases to and stabilizes at a lower level after the cells have reached saturation (Enomoto and Yamasaki, 1985b). This result, superficially, seems to contradict the general hypothesis that junctional inhibition promotes cell growth. However, the key may lie in the actual density at which the cells cease growing, which is nearly two times that for untreated cells, and in the fact that the junctional transfer, though reduced, is still present (Enomoto and Yamasaki, 1985b). These two facts suggest the following possibility: TPA may reduce the proportion of cell interfaces having detectable junctional transfer, but, because of the increased cell density, each cell would have more contacting neighbors and thus could be connected to enough cells to become inhibited by one of the various mechanisms discussed earlier.

This possibility is in essence a specific application of the volume dilution model proposed by Loewenstein (1968, 1979). Provided the signals for growth stimulation (or inhibition) and the reception mechanisms are unaffected as the cells grow, the population will ultimately arrest when the effective volume of junctionally connected cells reaches an appropriate size. If some cells are totally disconnected from the others, the cell density will have to increase to achieve the stable volume. If, on the other hand, all cells are interconnected, but many have severely reduced junctions, the ultimate cell density will reflect the time required to raise (or lower) the signal levels and, for relatively short-lasting signals, the cell density will again have to increase.

The second issue regards the fact that promoters need be present for only a brief time before full transformation of an "initiated" cell is expressed and a tumor cell colony is produced. Because the effect of promoters on junctional transfer is reversible, the continued growth of the initiated cell (i.e., transformation) must either occur independent of continued inhibition of junctional transfer or continue the junctional inhibition via some other mechanism (see Yamasaki and Enomoto, 1985, for further discussion). The first of these alternatives may apply to those transformed cells that show no obvious alteration in junctional transfer (although, as discussed above, small decreases in junctional capability have not been ruled out in most of these instances). The second alternative itself can occur in two different forms, inhibition of junctions among the transformed cells and inhibition of junctions only between the transformed cells and their nontransformed neighbors. This latter form has been demonstrated by Enomoto and Yamasaki (1984), who stressed the potential importance of this junctional isolation for the continued expansion of the transformed cell colonies. Their arguments are in part based on the prediction that maintained junctional communication between an "initiated" cell and its neighbors would act to prevent the "initiated" cell from growing further. Recently, this prediction has been tested by Mehta *et al.* (1986) who reported a positive correlation between the extent of junctional transfer between transformed and normal cells, increased with db-cAMP or decreased with retinoids, and the reduction in the ability of the transformed cells to produce colonies.

4.2.3b. A Final Common Pathway for Junctional Modulation during Normal Cell Growth. As we have seen, a single enzyme, protein kinase C, appears to be the focus of a remarkable convergence of various growth-promoting and junction-altering agents and conditions. First, the stimulation of quiescent cells to proliferate often begins with PI turnover, with subsequent increase in cytoplasmic IP_3, Ca^{2+}, and DAG which in turn activate protein kinase C and lead ultimately to elevated cytoplasmic Na^+ and K^+. The latter may promote synthesis of a critical "unstable" protein that moves the cell toward S-phase. Second, protein kinase C also decreases junctional transfer, perhaps directly. Third, protein kinase C may also activate the protooncogene, pp60c-src, whose effect could resemble that of the viral form, pp6v-src, which has been shown to decrease junctional permeability. Even this effect of pp60v-src may occur via protein kinase C, following stimulated PI turnover.

If protein kinase C is the final common pathway for both the reduction in junctional permeability and the stimulation of proliferation by exogenous agents, we immediately run into a logical problem with our earlier explanations of density-dependent inhibition of cell growth. We have suggested that density-inhibited cultures resist the effects of growth-promoting agents, e.g., in serum, by allowing stimulatory signal molecules, e.g., K^+, to leak away to neighboring cells via junctions. How could this transfer ever occur if the stimulatory signals are being produced at the same time the junctional permeability is being reduced? One possibility, of course, is that the two events have slightly different kinetic or dose–response characteristics, with the junctional change being the slower and/or less sensitive. A second possibility is that IP_3, DAG, or Ca^{2+}, which are increased prior to the activation of protein kinase C, are the molecules diluted out by junctional transfer. A third possibility is that density-dependent inhibition does in fact result from transfer of an inhibitory molecule, one that antagonizes both the growth-promoting and the junction-inhibiting effects of protein kinase C. In this regard, it is quite interesting that, for some cells at least, cAMP has these very effects (Slaga *et al.*, 1982; Enomoto *et al.*, 1984). Moreover, cAMP is elevated except during M-phase and early G1 (Seifert and Rudland, 1974; Friedman *et al.*, 1976) and, as suggested some time ago (Crick, 1970; Sheridan, 1971, 1974; Pitts, 1972) and as indicated by recent studies (e.g., Murray and Fletcher, 1984), probably is able to move through gap junctions. This intriguing idea brings us nearly full circle to the earlier suggestion that junctional transfer of cAMP may be a critical component of density-dependent inhibition of growth (Sheridan, 1974, 1976).

5. CONCLUSIONS AND FUTURE CONSIDERATIONS

Studies over the past several years have added support to the idea, nearly two decades old, that gap junctions may be involved in the control of cell proliferation. The heuristic attractiveness of earlier models for this involvement has been complemented by evidence for many possible cell-to-cell signal molecules, e.g., not only cAMP, suggested some time ago, but IP_3, DAG, Ca^{2+}, Na^+, and K^+, for several regulatory agents, e.g., protein kinase C and cAMP-dependent protein kinase, and for new potential mechanisms, e.g., phosphorylation of junctional protein. More sensitive methods for studying junctional permeability have provided us new evidence of an association between subtle junctional deficiencies and cell transformation by viruses on the one hand and tumor promotion by various chemical agents on the other hand.

Despite these new findings, we still do not have definitive evidence for junctional transfer of any specific proliferation signals. Nor is it clear that a decrease in junctional communication is a characteristic shared by all transformed cells rather than by a large, but select, subset. We have numerous clues as to the possible mechanisms by which junctional alterations are produced, but none has yet been worked out in detail.

Nevertheless, the convergence of some of the conclusions from studies of the effects of tumor promoters and viral oncogenes on cell growth and on gap junction structure and permeability provides some optimism that we are proceeding in the right direction in our pursuit of a role for junctional communication in the regulation of cell proliferation.

ACKNOWLEDGMENTS. The author wishes to thank Dr. Michael Atkinson for many helpful criticisms of earlier versions of the manuscript and Ms. Becca

Vance for assistance with the manuscript preparation. Work from this laboratory was supported in part by NIH Grants CA 16335, CA 30129, AM 30519, and AM 33655.

6. REFERENCES

Albertini, D. F., and Anderson, E., 1974, The appearance and structure of intercellular connections during ontogeny of the rabbit ovarian follicle with particular reference to gap junctions, *J. Cell Biol.* **63:**234–250.

Anderson, E., and Albertini, D. F., 1976, Gap junctions between the oocyte and companion follicle cells in the mammalian ovary, *J. Cell Biol.* **71:**680–686.

Anderson, W. B., Estival, A., Tapiovaara, H., and Gopalakrishna, R., 1985, Altered subcellular distribution of protein kinase C (a phorbol ester receptor). Possible role in tumor promotion and the regulation of cell growth: Relationship to changes in adenylate cyclase activity, in: *Advances in Cyclic Nucleotide and Protein Phosphorylation Research,* Vol. 19 (D. M. F. Cooper and K. B. Seamon, eds.), pp. 287–306, Raven Press, New York.

Atkinson, M. M., and Sheridan, J. D., 1985, Reduced junctional permeability in cells transformed by different viral oncogenes, in: *Gap Junctions* (M. V. L. Bennett and D. C. Spray, eds.), pp. 205–213, Cold Spring Harbor Laboratory, Cold Spring Harbor, N.Y.

Atkinson, M. M., Menko, A. S., Johnson, R. G., Sheppard, J. R., and Sheridan, J. D., 1981, Rapid and reversible reduction of junctional permeability in cells infected with a temperature sensitive mutant of avian sarcoma virus, *J. Cell Biol.* **91:**573–578.

Atkinson, M. M., Anderson, S. K., and Sheridan, J. D., 1986, Modification of gap junctions in cells transformed by a temperature-sensitive mutant of Rous sarcoma virus, *J. Membr. Biol.* **91:**53–64.

Azarnia, R., and Loewenstein, W. R., 1984a, Intercellular communication and the control of growth. X. Alteration of junctional permeability by the src gene. A study with temperature-sensitive mutant Rous sarcoma virus, *J. Membr. Biol.* **82:**191–205.

Azarnia, R., and Loewenstein, W. R., 1984b, Intercellular communication and the control of growth. XII. Alteration of junctional permeability by simian virus 40, roles of the large and small T antigens, *J. Membr. Biol.* **82:**213–220.

Azarnia, R., Dahl, G., and Loewenstein, W. R., 1981, Cell junction and cyclic AMP. III. Promotion of junctional membrane permeability and junctional membrane particles in a junction-deficient cell type, *J. Membr. Biol.* **63:**133–146.

Beers, W. H., and Olsiewski, P. J., 1985, Junctional communication and oocyte maturation, in: *Gap Junctions* (M. V. L. Bennett and D. C. Spray, eds.), pp. 307–314, Cold Spring Harbor Laboratory, Cold Spring Harbor, N.Y.

Bell, R. M., 1986, Protein kinase C activation by diacylglycerol second messengers, *Cell* **45:**631–632.

Berenblum, I., and Shubik, P., 1947, A new, quantitative approach to the study of the stages of chemical carcinogenesis in the mouse's skin, *Br. J. Cancer* **1:**383–391.

Berridge, M. J., 1975, The interaction of cyclic nucleotides and calcium in the control of cellular activity, *Adv. Cyclic Nucleotide Res.* **6:**1–98.

Biegon, R. P., Atkinson, M. M., and Sheridan, J. D., 1986, Comparison of two fluorescent dye transfer rates through gap junctions in Novikoff hepatoma cells, *Biophys. J.* **49:**342a.

Biegon, R. P., Atkinson, M. M., Liu, T. F., Kam, E., and Sheridan, J. D., 1987, Permeance of Novikoff Lepatoma gap junctions: Quantitative video analysis of dye transfer, *J. Membr. Biol.* (in press).

Bornslaeger, E. A., and Schultz, R. M., 1985, Regulation of mouse oocyte maturation: Effect of elevating cumulus cell cAMP on oocyte cAMP levels, *Biol. Reprod.* **33:**698–704.

Brink, P. R., and Dewey, M. M., 1978, Nexal membrane permeability to anions, *J. Gen. Physiol.* **72:**67–86.

Brown, L. M., Ryan, U. S., Absher, M., and Olzabal, B. M., 1982, Mathematical analysis of endothelial sibling pair cell–cell interactions using time-lapse cinematography data, *Tissue Cell* **14:**651–655.

Burgess, G. M., Godfrey, P. P., McKinney, J. S., Berridge, M. J., Irvine, R. F., and Putney, J. W., 1984, The second messenger linking receptor activation to internal Ca release in liver, *Nature* **309:**63–66.

Burton, A. C., 1971, Cellular communication, contact inhibition, cell clocks, and cancer, *Perspect. Biol. Med.* **14:**301–318.

Burton, A. C., and Canham, P. B., 1973, The behaviour of coupled biochemical oscillators as a model of contact inhibition of cellular division, *J. Theor. Biol.* **39:**555–580.

Chang, C., Trosco, J. E., Kung, H., Bombick, D., and Matsumura, F., 1985, Potential role of the src gene product in inhibition of gap-junctional communication in NIH/3T3 cells, *Proc. Natl. Acad. Sci. USA* **82:**5360–5364.

Collett, M. S., and Erikson, R. L., 1978, Protein kinase activity associated with the avian sarcoma virus src gene product, *Proc. Natl. Acad. Sci USA* **75:**2021–2024.

Corsaro, C. M., and Migeon, B. R., 1977a, Contact-mediated communication of ouabain resistance in mammalian cells in culture, *Nature* **268:**737–739.

Corsaro, C. M., and Migeon, B. R., 1977b, Comparison of contact-mediated communication in normal and transformed human cells in culture, *Proc. Natl. Acad. Sci. USA* **74:**4476–4480.

Crick, F., 1970, Diffusion in embryogenesis, *Nature* **225:**420–422.

Dekel, N., and Beers, W. H., 1980, Development of the rat oocyte in vitro: Inhibition and induction of maturation in the presence or absence of the cumulus oophorus, *Dev. Biol.* **75:**247–257.

Dekel, N., Lawrence, T. S., Gilula, N. B., and Beers, W. H., 1981, Modulation of cell-to-cell communication in the cumulus–oocyte complex and the regulation of oocyte maturation by LH, *Dev. Biol.* **86:**356–362.

Enomoto, T., and Yamasaki, H., 1984, Lack of intercellular communication between chemically transformed and surrounding nontransformed BALB/c 3T3 cells, *Cancer Res.* **44:**5200–5203.

Enomoto, T., and Yamasaki, H., 1985a, Rapid inhibition of intercellular communication between BALB/c 3T3 cells by diacylglycerol, a possible endogenous functional analogue of phorbol esters, *Cancer Res.* **45:**3706–3710.

Enomoto, T., and Yamasaki, H., 1985b, Phorbol ester-mediated inhibition of intercellular communication in BALB/c 3T3 cells: Relationship to enhancement of cell transformation, *Cancer Res.* **45:**2681–2688.

Enomoto, T., Sasaki, Y., Shiba, Y., Kanno, Y., and Yamasaki, H., 1981, Tumor promoters cause a rapid and reversible inhibition of the formation and maintenance of electrical cell coupling in culture, *Proc. Natl. Acad. Sci. USA* **78:**5628–5632.

Enomoto, T., Martel, N., Kanno, Y., and Yamasaki, H., 1984, Inhibition of cell communication between BALB/c 3T3 cells by tumor promoters and protection by cAMP, *J. Cell. Physiol.* **121:**323–333.

Eppig, J. J., 1982, The relationship between cumulus cell–oocyte coupling, oocyte meiotic maturation and cumulus expansion, *Dev. Biol.* **89:**268–272.

Fitzgerald, D. J., Knowles, S. E., Ballard, F. J., and Murray, A. W., 1983, Rapid and reversible inhibition of junctional communication by tumor promoters, *Cancer Res.* **43:**3614–3618.

Flagg-Newton, J. L., and Loewenstein, W. R., 1981, Cell junction and cyclic AMP. II. Modulations of junctional membrane permeability, dependent on serum and cell density, *J. Membr. Biol.* **63:**123–131.

Flagg-Newton, J. L., Dahl, G., and Loewenstein, W. R., 1981, Cell junction and cyclic AMP. 1. Upregulation of junctional membrane permeability and junctional membrane particles by administration of cyclic nucleotide or phosphodiesterase inhibitor, *J. Membr. Biol.* **63:**105–121.

Folkman, J., and Moscona, A., 1978, Role of cell shape in growth control, *Nature* **273**:345–349.

Friedman, D. L., Johnson, R. A., and Zeilig, C. E., 1976, The role of cyclic nucleotides in the cell cycle, *Cyclic Nucleotide Res.* **7**:69–111.

Furshpan, E. J., and Potter, D. D., 1986, Low-resistance junctions between cells in embryos and tissue culture, *Curr. Top. Dev. Biol.* **3**:95–127.

Garfield, R. E., Sims, S., and Daniel, E. E., 1977, Gap junctions: Their presence and necessity in myometrium during parturition, *Science* **198**:58–960.

Gilula, N. B., Epstein, M. L., and Beers, W. H., 1978, Cell-to-cell communication and ovulation: A study of cumulus–oocyte complex, *J. Cell Biol.* **78**:58–75.

Goldberg, A. R., Delclos, K. B., and Blumberg, P. M., 1980, Phorbol ester action is independent of viral and cellular src kinase levels, *Science* **208**:191–193.

Goldberg, N. G., Haddox, M. K., Nicol, S. E., Glass, D. B., Sanford, C. H., Kuehl, F. A., Jr., and Estensen, R., 1975, Biologic regulation through opposing influences of cyclic GMP and cyclic AMP: The Yin–Yang hypothesis, *Adv. Cyclic Nucleotide Res.* **5**:307–330.

Goldfarb, P. S. G., and Slack, C., Subak-Sharpe, J. H., and Wright, E. D., 1974, Metabolic cooperation between cells in tissue culture, *Symp. Soc. Exp. Biol.* **28**:463–484.

Goodall, H., and Maro, B., 1986, Major loss of junctional coupling during mitosis in early mouse embryos, *J. Cell Biol.* **102**:568–575.

Gould, K. L., Woodgett, J. R., Cooper, J. A., Buss, J. E., Shalloway, D., and Hunter, T., 1985, Protein kinase C phosphorylates pp60src at a novel site, *Cell* **42**:849–857.

Hertzberg, E. L., and Spray, D. C., 1985, Studies of gap junctions: Biochemical analysis and use of antibody probes, in: *Gap Junctions* (M. V. L. Bennett and D. C. Spray, eds.), pp. 57–65, Cold Spring Harbor Laboratory, Cold Spring Harbor, N. Y.

Holley, R. W., and Kiernan, J. A., 1968, "Contact inhibition" of cell division in 3T3 cells, *Proc. Natl. Acad. Sci. USA* **60**:300–304.

Hülser, D. F., and Brümmer, F., 1982, Closing and opening of gap junction pores between two- and three-dimensionally cultured tumor cells, *Biophys. Struct. Mech.* **9**:83–88.

Hunter, T., and Sefton, B. M., 1980, The transforming gene product of Rous sarcoma virus phosphorylates tyrosine, *Proc. Natl. Acad. Sci. USA* **77**:1311–1315.

Iba, H., Cross, F. R., Garber, E. A., and Hanafusa, H., 1985, Low level of cellular protein phosphorylation by nontransforming overproduced pp60c-src, *Mol. Cell. Biol.* **5**:1058–1066.

Ito, S., Sato, E., and Loewenstein, W. R., 1974, Studies on the formation of a permeable cell membrane junction. I. Coupling under various conditions of membrane contact. Effects of colchicine, cytochalasin B, dinitrophenol, *J. Membr. Biol.* **19**:305–338.

Jackman, R. W., 1982, Persistence of axial orientation cues in regenerating intima of cultured aortic explants, *Nature* **296**:80–83.

Jackman R. W., 1986, Regenerating endothelium in vascular organ culture, Ph.D. thesis, University of Minnesota, Minneapolis.

Jackman, R. W., Anderson, S. K., and Sheridan, J. D., 1981, Regeneration of vascular endothelium on cultured aorta explants following in vitro injury, *J. Cell Biol.* **91**:40a.

Johnson, K. R., Panter, S. S., and Johnson, R. G., 1986, Phosphorylation of lens membranes with a cyclic AMP-dependent protein kinase purified from the bovine lens, *Biochim. Biophys. Acta* **844**:367–376.

Johnson, R., Hammer, M., Sheridan, J., and Revel, J.-P., 1974, Gap junction formation between reaggregated Novikoff hepatoma cells, *Proc. Natl. Acad. Sci. USA* **71**:4536–4543.

Kanno, Y., and Loewenstein, W. R., 1964, Intercellular diffusion, *Science* 143:959–960.

Kuffler, S. W., and Potter, D. D., 1964, Glia in the leech central nervous system: Physiological properties and neuron–glia relationships, *J. Neurophysiol.* **27**:290–320.

Larsen, W. J., Wert, S. E., and Brunner, G. D., 1986, A dramatic loss of cumulus cell gap junctions is correlated with germinal vesicle breakdown in rat oocytes, *Dev. Biol.* **113**:517–521.

Ledbetter, M. L. S., and Lubin, M., 1979, Transfer of potassium: A new measure of cell–cell coupling, *J. Cell Biol.* **80:**150–165.
Liu, T.-F., Kam, E. Y., and Sheridan, J. D., 1982, Dye transfer through permeable junctions between cultured mammalian cells: Quantitative video analysis, *J. Cell Biol.* **95:**107a.
Loewenstein, W. R., 1966, Permeability of membrane junctions, *Ann. N.Y. Acad. Sci.* **137:**441–472.
Loewenstein, W. R., 1968, Communication through cell junctions: Implications in growth control and differentiation, *Dev. Biol.* **19**(Suppl. 2)**:**151–183.
Loewenstein, W. R., 1979, Junctional intercellular communication and the control of growth, *Biochim. Biophys. Acta Cancer Rev.* **560:**1–65.
Loewenstein, W. R., 1981, Junctional intercellular communication: The cell-to-cell membrane channel, *Physiol. Rev.,* **61:**829–913.
Loewenstein, W. R., 1985, Regulation of cell-to-cell communication by phosphorylation. *Biochem. Soc. Symp.* **50:**43–58.
Lowenstein, W. R., Kanno, Y., and Socolar, S. J., 1978, Quantum jumps of conductance during formation of membrane channels at cell–cell junction, *Nature* 274:133–136.
Lubin, M., 1967, Intracellular potassium and macromolecular synthesis in mammalian cells, *Nature* **213:**451–453.
Macara, I. G., 1985, Oncogenes, ions, and phospholipids, *Am. J. Physiol.* **248:**C3–C11.
MacManus, J. P., Whitfield, J. F., Boynton, A. L., and Rixon, R. H., 1975, Role of cyclic nucleotides and calcium in the positive control of cell proliferation, *Adv. Cyclic Nucleotide Res.* **5:**719–734.
Mehta, P. P., Bertram, J. S., and Loewenstein, W. R., 1986, Growth inhibition of transformed cells correlates with their junctional communication with normal cells, *Cell* **44:**187–196.
Merk, F. B., and McNutt, N. S., 1972, Nexus junctions between dividing and interphase granulosa cells of the rat ovary, *J. Cell Biol.* **55:**511–519.
Meyer, D. J., Yancey, B., and Revel, J.-P., 1981, Intercellular communication in normal and regenerating rat liver: A quantitative analysis, *J. Cell Biol.* **91:**505–523.
Michell, R. H., 1982, Inositol lipid metabolism in dividing and differentiating cells, *Cell Calcium* **3:**429–440.
Minor, P. D., and Smith, J. A., 1974, Explanation of degree of correlation of sibling generation times in animal cells, *Nature* **248:**241–243.
Mitchison, J. M., 1971, *The Biology of the Cell Cycle,* Cambridge University Press, London.
Murphy, J. S., Landsberger, F. R., Kikuchi, T., and Tamm, I., 1984, Occurrence of cell division is not exponentially distributed: Differences in the generation times of sister cells can be derived from the theory of survival of populations, *Proc. Natl. Acad. Sci. USA* **81:**2379–2383.
Murray, A. W., and Fitzgerald, D. J., 1979, Tumor promoters inhibit metabolic cooperation in cocultures of epidermal and 3T3 cells, *Biochem. Biophys. Res. Commun.* **91:**395–401.
Murray, S. A., and Fletcher, W. H., 1984, Hormone-induced intercellular signal transfer dissociates cyclic AMP-dependent protein kinase, *J. Cell Biol.* **98:**1710–1719.
Neyton, J., and Trautmann, A., 1985, Single-channel currents of an intercellular junctions, *Nature* **317:**331–335.
Nishizuka, Y., 1986, Studies and perspectives of protein kinase C, *Science* **233:**305–312.
O'Lague, P., and Dalen, H., 1974, Low resistance junction between normal and between virus transformed fibroblasts in tissue culture, *Exp. Cell Res.* **86:**374–382.
O'Lague, P., Dalen, H., Rubin, H., and Tobias, C., 1970, Electrical coupling: Low resistance junctions between mitotic and interphase fibroblasts in tissue culture, *Science* **170:**464–466.
Pardee, A. B., 1974, A restriction point for control of normal animal cell proliferation, *Proc. Natl. Acad. Sci. USA* **71:**1286–1290.

Peracchia, C., 1980, Structural correlates of gap junction permeation, *Int. Rev. Cytol.* **66:**81–146.
Pietropaolo, C., Laskin, J. D., and Weinstein, I. B., 1981, Effect of tumor promoters on arc gene expression in normal and transformed chick embryo fibroblasts, *Cancer Res.* **41:**1565–1571.
Pinto da Silva, P., and Gilula, N. B., 1972, Gap junctions in normal and transformed fibroblasts in culture, *Exp. Cell Res.* **71:**393–401.
Pitts, J. D., 1971, Molecular exchange and growth control in tissue culture, in: *Ciba Symposium on Growth Control in Cell Cultures* (G. E. W. Wolstenholme and J. Knight, eds.), pp. 89–105. Churchill Livingston, London.
Pitts, J. D., 1972, Direct interaction between animal cells, in: *Cell Interactions, Third Lepetit Colloquium* (L. G. Silvestri, ed.), pp. 277–285, North-Holland, Amsterdam.
Pitts, J. D., and Simms, J. W., 1977, Permeability of junctions between animal cells: Intercellular transfer of nucleotides but not macromolecules, *Exp. Cell Res.* **123:**153–163.
Potter, D. D., Furshpan, E. J., and Lennox, E. S., 1966, Connections between cells of the developing squid as revealed by electrophysiological methods, *Proc. Natl. Acad. Sci. USA* **55:**328–335.
Preus, D., Johnson, R. G., and Sheridan, J. D., 1981a, Gap junctions between Novikoff hepatoma cells following dissociation and recovery in the absence of cell contact, *J. Ultrastruct. Res.* **77:**248–262.
Preus, D., Johnson, R. G., Sheridan, J. D., and Meyer, R. M., 1981b, Analysis of gap junctions and formation plaques between reaggregating Novikoff hepatoma cells, *J. Ultrastruct. Res.* **77:**263–276.
Radu, A., Dahl, G., and Loewenstein, W. R., 1982, Hormonal regulation of cell junction permeability: Upregulation by catecholamine and prostaglandin E1, *J. Membr. Biol.* **70:**239–251.
Revel, J.-P., Yee, A. G., and Hudspeth, A. J., 1971, Gap junctions between electrotonically coupled cells in tissue culture and in brown fat, *Proc. Natl. Acad. Sci. USA* **68:**2924–2927.
Rivedal, E., Sanner, T., Enomoto, T., and Yamasaki, H., 1985, Inhibition of intercellular communication and enhancement of morphological transformation of Syrian hamster embryo cells by TPA: Use of TPA-sensitive and TPA-resistant cell lines, *Carcinogenesis* **6:**899–902.
Rossow, P. W., Riddle, V. G., and Pardee, A. B., 1979, Synthesis of labile, serum-dependent protein in early G1 controls animal cell growth, *Proc. Natl. Acad. Sci. USA* **76:**4446–4450.
Ryan, U. S., Absher, M., Olazabal, B. M., Brown, L. M., and Ryan, J. W., 1982, Proliferation of pulmonary endothelial cells: Time-lapse cinematography of growth to confluence and restitution of monolayer after wounding, *Tissue Cell* **14:**637–649.
Saez, J. C., Spray, D. C., Hertzberg, E. L., Nairu, A. C., Greenyard, P., and Bennett, M. V. L., 1986, cAMP increases junctional conductance and stimulates phosphorylation of the 27KDa principal gap junction polypeptide, *Proc. Natl. Acad. Sci. USA* **83:**2473–2477.
Schneiderman, M. H., Dewey, W. C., and Highfield, D. P., 1971, Inhibition of DNA synthesis in synchronized Chinese hamster cells treated in G1 with cycloheximide, *Exp. Cell Res.* **67:**147–155.
Schultz, R. M., Montgomery, R. R., Ward-Bailey, P. F., and Eppig, J. J., 1983, Regulation of oocyte maturation in the mouse: Possible roles of intercellular communication, cAMP, and testosterone, *Dev. Biol.* **95:**294–304.
Schwartz, S. M., Stemerman, M. B., and Benditt, E. P., 1975, The aortic intima. II. Repair of the aortic lining after mechanical denudation, *Am. J. Pathol.* **81:**15–42.
Seifert, W. E., and Rudland, P. S., 1974, Possible involvement of cyclic GMP in growth control of cultured mouse cells, *Nature* **248:**138–140.
Sheridan, J. D., 1971, Dye movement and low resistance junctions between reaggregated embryonic cells, *Dev. Biol.* **26:**627–636.
Sheridan, J. D., 1973, Functional evaluation of low resistance junctions: Influence of cell shape and size, *Am. Zool.* **13:**1119–1128.

Sheridan, J. D., 1974, Low resistance junctions: Some functional considerations, in: *The Cell Surface in Development* (A. A. Moscona, ed.), pp. 187–206, Wiley, New York.

Sheridan, J. D., 1976, Cell coupling and cell communication during embryogenesis, in: *The Cell Surface in Animal Embryogenesis and Development* (G. Poste and G. L. Nicolson, eds.), pp. 409–447, Elsevier/North-Holland, Amsterdam.

Sheridan, J. D., 1978, Junction formation and experimental modification, in: *Receptors and Recognition, Series B,* Vol. 2 (J. Feldman, N. B. Gilula, and J. D. Pitts, eds.), pp. 39–59, Halsted, New York.

Sheridan, J. D., and Atkinson, M. M., 1985, Physiological roles of permeable junctions: Some possibilities, *Annu. Rev. Physiol.* **47**:337–353.

Sheridan, J. D., Hammer-Wilson, M., Preus, D., and Johnson R. G., 1978, Quantitative analysis of low-resistance junctions between cultured cells and correlation with gap-junctional areas, *J. Cell Biol.* **76**:532–544.

Shields, R., and Smith, J. A., 1977, Cells regulate their proliferation through alterations in transition probability, *J. Cell. Physiol.* **91**:345–355.

Slaga, T. J., Fischer, S. M., Weeks, C. E., Nelson, K., Mamrack, M., and Klein-Szanto, A. J. P., 1982, Specificity and mechanisms of promoter inhibitors in multistage promotion in carcinogenesis, in: *Carcinogenesis—A Comprehensive Survey,* Vol. 7 (E. Hecker, N. E. Fusenig, W. Kunz, F. Marks, and H. W. Thielmann, eds.), pp. 19–34, Raven Press, New York.

Smith, J. A., and Martin, L., 1973, Do cells cycle? *Proc. Natl. Acad. Sci USA* **70**:1263–1267.

Socolar, S. J., 1973, Cell coupling in epithelia, *Exp. Eye Res.* **15**:693–698.

Spangoli, L. G., Pietra, G. G., Villaschi, S., and Johns, L. W., 1982, Morphometric analysis of gap junctions in regenerating arterial endothelium, *Lab Invest.* **46**:139–148.

Spray, D. C., Stern, J., Harris, A. L., and Bennett, M. V. L., 1982, Gap junctional conductance: Comparison of sensitivities to H and Ca ions, *Proc. Natl. Acad. Sci. USA* **79**:441–445.

Spray, D. C., Saez, J. C., Brosius, D., Bennett, M. V. L., and Hertzberg, E. L., 1986, Isolated liver gap junctions: Gating of transjunctional currents is similar to that in intact pairs of rat hepatocytes, *Proc. Natl. Acad. Sci. USA* **83**:5494–5497.

Stoker, M. P., 1973, Role of diffusion boundary layer in contact inhibition of growth, *Nature* **246**:200–203.

Stoker, M. P., 1975, The effects of topoinhibition and cytochalasin B on metabolic cooperation, *Cell* **6**:253–257.

Sugimoto, Y., and Erikson, R. L., 1985, Phosphatidylinositol kinase activities in normal and Rous sarcoma virus-transformed cells, *Mol. Cell. Biol.* **5**:3194–3198.

Todaro, G. J., Lazar, G. K., and Green, H., 1965, The initiation of cell division in a contact-inhibited mammalian cell line, *J. Cell. Comp. Physiol.* **66**:325.

Trosko, J. E., Chang, C.-C., and Medcalf, A., 1983, Mechanisms of tumor promotion: Potential role of intercellular communication, *Cancer Invest.* **1**(6):511–526.

Veenstra, R. D., and DeHaan, R. L., 1986, Measurement of single channel currents from cardiac gap junctions, *Science* **233**:972–974.

Vitkauskas, G. V., and Canellakis, E. S., 1985, Intercellular communication and cancer chemotherapy, *Biochim. Biophys. Acta* **823**:19–34.

Wang, E., and Goldberg, A. R., 1979, Effects of the src gene product on microfilament and microtubule organization in avian and mammalian cells infected with the same temperature-sensitive mutant of Rous sarcoma virus, *Virology* **92**:201–210.

Weidmann, S., 1966, The diffusion of radiopotassium across intercalated disc of mammalian cardiac muscle, *J. Physiol. (London)* **187**:323–342.

Wieser, R. J., and Oesch, F., 1986, Contact inhibition of growth of human diploid fibroblasts by immobilized plasma membrane glycoproteins, *J. Cell Biol.* **103**:361–367.

Willecke, K., Traub, O., Janssen-Timmen, U., Rixen, U., Dermietzel, R., Leibstein, A., Paul, D., and Rabes, H., 1985, Immunochemical investigations of gap junction protein in different mammalian tissues. in: *Gap Junctions* (M. V. L. Bennett and D. C. Spray, eds.), pp. 67–76, Cold Spring Harbor Laboratory, Cold Spring Harbor, N. Y.

Willingham, M. C., Jay, G., and Pastan, I., 1979, Localization of the ASV src gene product to the plasma membrane of transformed cells by electron microscopic immunocytochemistry, *Cell* **18:**125–134.

Wyke, J. A., 1973, The selective isolation of temperature-sensitive mutants of Rous sarcoma virus, *Virology* **52:**587–590.

Yada, T., Rose, B., and Loewenstein, W. R., 1986, Diacylglycerol downregulates junctional membrane permeability: TMB-8 blocks this effect, *J. Membr. Biol.* **88:**217–232.

Yamasaki, H., and Enomoto, T., 1985, Role of intercellular communication in BALB/c 3T3 cell transformation, in: *Carcinogenesis,* Vol. 9 (J. C. Barrett and R. W. Tennant, eds.), pp. 179–194, Raven Press, New York.

Yamasaki, H., Enomoto, T, Shiba, Y., Kanno, Y., and Kakunaga, T., 1985, Intercellular communication capacity as a possible determinant of transformation sensitivity of BALB/c 3T3 clonal cells, *Cancer Res.* **45:**637–641.

Yancey, S. B., Easter, D., and Revel, J.-P., 1979, Cytological changes in gap junctions during liver regeneration, *J. Ultrastruct. Res.* **67:**229–242.

Yancey, S. B., Edens, J. E., Trosko, J. E., Chang, C.-C., and Revel, J.-P., 1982, Decreased incidence of gap junctions between Chinese hamster V-79 cells upon exposure to the tumor promoter 12-O-tetradecanoyl phorbol-13-acetate, *Exp. Cell Res.* **139:**329–340.

Yee., A., and Revel, J.-P., 1978, Loss and reappearance of gap junctions in regenerating liver, *J. Cell Biol.* **78:**554–564.

Yotti, L. P., Chang, C.-C., and Trosko, J. E., 1979, Elimination of metabolic cooperation in Chinese hamster cells by a tumor promoter, *Science* **206:**1089–1091.

Zampighi, G., Hall, J., and Kreman, M., 1985, Purified lens junctional protein forms channels in planar lipid films, *Proc. Natl. Acad. Sci. USA* **82:**8468–8472.

Zimmerman, A. L., and Rose, B., 1985, Permeability properties of cell-to-cell channels: Kinetics of fluorescent tracer diffusion through a cell junction, *J. Membr. Biol.* **84:**269–283.

Chapter 7

Intercellular Communication in Embryos

Sarah C. Guthrie
Department of Anatomy
University College London
London WC1E 6BT, United Kingdom

1. INTRODUCTION

Where cells appose one another a number of membrane specializations can be found, one of the most interesting of which is the gap junction. Here, membranes come close together, reducing the intercellular space to 2–4 nm. Structures known as connexons lie in clusters within each membrane, uniting in pairs across the gap. Since each connexon is annular, enclosing a channel, the junction appears to provide a direct, low-resistance pathway between cell interiors. There is much evidence to show that the presence of gap junctions is correlated with electrical coupling and the ability to transfer small molecules (e.g., Gilula *et al.*, 1972). *In vivo* it is thought that ions, metabolites, and perhaps signals can pass through gap junctions. Recently, much energy has been expended in minute examination of the structure and composition of gap junctions. Despite the insights thus afforded, a question mark hangs over the role of the gap junction. In development the part played by junctions has proved particularly elusive.

Much interest has centered on a putative role for gap junctions in development. This is because the embryological literature furnishes abundant examples of cell interactions, some of which might involve junctions. Experimentation has built up a picture of many embryos as "regulative," able to replace missing parts to generate a normal embryo. Regulation implies recognition of injury, and communication in some manner between cells of the remaining portion of the embryo, allowing it to restore normal development. The capability for cell interaction thus revealed is also a feature of the intact embryo. Numerous specific cases of cell interactions have been described. One example is neural induction,

Present address for SCG: Department of Molecular Biology, Research Institute of Scripps Clinic, La Jolla, California 92037.

in which one population of cells, the mesoderm, influences another population of cells, the ectoderm, to form nervous tissue.

In the quest to match phenomenon to mechanism it would be wrong to overemphasize the role of intercellular communication at the expense of extracellular pathways. But empirically, gap junctions have the credentials to allow developmental signaling by small molecules. Tracer studies have set the upper size limit of their permeability at approximately 1000 daltons in mammalian cells and 2500 daltons in insect cells (Furshpan and Potter, 1969; Flagg-Newton *et al.*, 1979; Schwarzmann *et al.*, 1981). In culture, the passage of metabolites such as adenine nucleotides between cells has been demonstrated (Burk *et al.*, 1968) giving rise to the idea that cells engage in "metabolic cooperation," or sharing of resources. Moreover, myocytes and granulosa cells cultured together are capable of exchanging AMP, a molecule implicated as a messenger in many systems (Lawrence *et al.*, 1978). Unfortunately, data have not been forthcoming on the nature or action of substances that might embody signals in development *in vivo*.

Following the pioneering work of Potter *et al.* (1966) showing electrical coupling in the squid embryo, gap junctional coupling has been widely demonstrated in embryos. The temporal and spatial occurrence of gap junctions varies, however, and has provided some interesting correlations. There is a striking tendency for gap junctional coupling to be first detected at times when cells are deciding their fates. Further into development, junctions may display reduced permeability or even disappear between cells that pursue divergent paths of differentiation. Potter *et al.* proposed that pattern formation could depend on "interruptions in the current-carrying continuum." To evaluate the worth of this and other theories requires a close look at the available evidence relating to gap junctions and development.

2. THE OOCYTE

2.1. Oocyte Gap Junctions

My survey begins with the oocyte. Within the unfertilized egg, some degree of preparative pattern formation is known to take place, for example the laying down of cytoplasmic determinants. In *Drosophila,* maternal information specifies dorsoventral polarity in the egg (Anderson and Nusslein-Volhard, 1984). Several mutants have been isolated showing patterning defects caused by a deficiency in components of the cytoplasm. Mutant embryos can be rescued to varying degrees by injection of maternal wild-type poly(A) mRNA, showing that mRNA provided by the mother is necessary for patterning.

There has been speculation that developmental functions such as the specification of polarity might be fulfilled by oocyte gap junctions. While the mammalian oocyte lies dormant in the ovary, it is ensheathed by an acellular layer, the zona pellucida, beyond which is an array of follicle cells, the cumulus oophorus. At intervals, protrusions from the innermost cumulus cells penetrate the zona and contact the oocyte. Gap junctions are present at this interface (Albertini and Anderson, 1974). In amphibia and insects, the detailed anatomy differs from the mammal, but gap junctions are also found between oocyte and follicle cells (Vandenhoef *et al.*, 1984; Huebner, 1981).

2.2. The Control of Oocyte Maturation

Oocytes awaiting maturation are maintained in a state of meiotic arrest. In mammals, oocyte nutrition during this period is provided by the follicle cells (Heller *et al.*, 1981). Since oocyte denuded of their follicle cells mature spontaneously (Dekel and Beers, 1980), follicle cells are also implicated in maintaining meiotic arrest. In rat ovarian follicles, junctional communication was found to cease near the time of loss of follicle cells and shedding of the oocyte (Gilula *et al.*, 1978). Some authors therefore proposed that junctional uncoupling was more than an incidental feature of maturation, and constituted a regulatory step in the process (Gilula *et al.*, 1978; Dekel *et al.*, 1981). It is necessary to assess this possibility in the light of other data on factors that influence maturation.

The spontaneous maturation of denuded oocytes *in vitro* could be prevented by the addition of membrane-permeable derivatives of cAMP (Nekola and Moor-Smith, 1975), suggesting that cAMP can maintain meiotic arrest. Dekel *et al.* (1981) showed that artificially elevated levels of cAMP caused by addition of cyclic nucleotide phosphodiesterase inhibitors could interrupt communication between cumulus and oocyte in culture. Thus, cAMP had a second role, to uncouple gap junctions. Meiotic arrest could also be relieved by gonadotropins (Dekel and Beers, 1978). This led to the following hypothesis. A tonic level of cAMP communicated from follicle to oocyte was supposed to sustain meiotic arrest. At the preovulatory surge of gonadotropin, junctional shutdown ensued, and transfer of cAMP to the oocyte was halted, whereupon the oocyte resumed development. Whether cAMP can augment the effect of gonadotropin by closing junctions *in vivo* is not clear.

But details of the precise timing of events may give cause to doubt this hypothesis. Firstly, meiotic arrest generally continues even after intercellular coupling has begun to vanish, both *in vitro* (Dekel *et al.*, 1981) and *in vivo* (Eppig, 1982), suggesting that coupling loss does not itself cause maturation. Indeed, Eppig postulated that cessation of communication might relate more closely to physical loss of cohesion than to important biochemical events. Secondly, it appears that significant quantities of cAMP do not cross from follicle cells to occyte. Olsiewski and Beers (1983) examined synthesis of cAMP in oocytes and cumulus–oocyte complexes under the influence of various stimulatory agents. Initially, they examined the oocyte for endogenous cAMP. They found that even when induced to do so, the oocyte was only capable of synthesizing tiny amounts of cAMP (~ 1%) relative to the cumulus. To discover whether cAMP could be transferred from follicle cells to oocyte, follicles were incubated in radioactive adenine (Olsiewski and Beers, 1985). There was negligible transfer. Further experiments to determine levels of cAMP in the oocyte before and after maturation showed no evidence of a drop in oocyte cAMP before germinal vesicle breakdown had been accomplished. All this makes it unlikely that cAMP is the signal causing meiotic arrest, or that junctional closure permits maturation. Current opinion favors the idea that maturation depends on a fall in concentration of a maturation inhibitory factor, produced by the cumulus (Eppig *et al.*, 1983; Freter and Schultz, 1984). However, Larsen and co-workers pursued the idea that junctions might play a role, and found a close correlation between loss of gap junctions from the oocyte interface and maturation (Larsen *et al.*, 1984). Breakdown of the germinal vesicle corresponded to a reduction both in the total area of the gap junctions, and in the mean area of individual gap junctions. So far though, a direct role for gap junctions in formative events in the mammalian oocyte has not been established.

In the amphibian there is similar loss of junctional coupling between oocyte and follicle before the oocyte matures (Vilain *et al.*, 1980). In contrast to the mammal, however, gonadotropins enhance junctional coupling, which may be involved in transferring a signal that enhances vitellogenin uptake (Brown *et al.*, 1979). Meiotic maturation relies on progesterone binding directly to the oocyte surface, without involving the follicle; nevertheless, the effect may be mediated by a fall in oocyte cAMP (Masui and Clarke, 1979). Insects show temporal regulation of junctions similar to that in amphibia and mammals. Coupling is lost in the final phase of development, chorion formation (Huebner, 1981). Junctions may also facilitate the generation of electrical fields around follicles in insects. In *Cecropia* there is an internal current loop, entering the anterior end of the egg, and possibly dictating the early polarization (Jaffe and Woodruff, 1979). There is evidence for the localization of a morphogenetic head factor in *Cecropia,* and the authors hypothesize that a similar posterior loop might help activate or organize the posterior germ plasm.

3. EARLY CLEAVAGE STAGES

In the embryo, gap junctions typically appear at early cleavage stages. Many workers have made surveys of gap junctional occurrence and coupling in early embryos. The problem with drawing conclusions from these data is that the morphological identification of junctions does not necessarily reflect function, since junctions show diverse configurations (reviewed by Larsen, 1985). Furthermore, the characteristics of channel regulation may change with development, complicating analysis. In the amphibian embryo, for example, junctions can be found from the 8-cell stage onwards (Sanders and Dicaprio, 1976), but do not become sensitive to closure by low pH until the 64-cell stage (Turin and Warner, 1977). Conversely, in the chick lens junctions become less sensitive to uncoupling with carbon dioxide as development proceeds (Schuetze and Goodenough, 1982).

Research into gap junctions in embryos has concentrated principally on three aspects: observations of junctions by electron microscopy, electrical coupling, and dye coupling assays. But it is known that the presence of recognizable junctions is not necessarily synonymous with electrical coupling, nor electrical coupling with dye coupling. For instance, in *Patella*, junctions are present from the 2-cell stage (see Table I) but dye coupling is not present until the 32-cell stage (Dorresteijn *et al.*, 1983), when an increase in the frequency and size of gap junctional plaques was observed. Dye coupling studies in embryos have also suffered from a tendency to treat all probes as equivalent, as long as they are below the assumed channel size limit of 1000 daltons. In fact, channel permeability may depend on other variables such as diameter and charge of the probe, position of the cell assayed, and time in development. Attempts to distinguish embryonic junctions as a class from adult junctions must take account of these various caveats. In view of this, the great heterogeneity of results obtained from different species is not so surprising. The following section describes the evidence in more detail, leaving the amphibian embryo until last to illustrate some particularly diverse results.

Table I
Summary of Transfer of Small Molecules in Early Embryos of Various Species

Molecule	Species	Stage	Transfer	References
Lucifer Yellow	*Arbacia*	2 cells	No	Pochapin *et al.* (1983)
Fluorescein, Procion Yellow	*Asterias*	32 cells	No	Tupper and Saunders (1972)
Lucifer Yellow	*Patella*	32 cells	Yes	Dorresteijn *et al.* (1983)
Fluorescein	*Xenopus*	64 cells	No	Slack and Palmer (1969)
Lead EDTA	*Xenopus*	64 cells	No	Turin (1978)
Potassium argentocyanate	*Xenopus*	64 cells	Yes	Turin (1978)
Lucifer Yellow	*Xenopus*	32 cells	Yes	Guthrie (1984)
Carboxyfluorescein, Lucifer Yellow	*Mus*	8 cells	Yes	Goodall and Johnson (1982, 1984)

3.1. The Occurrence of Gap Junctions, Electrical Coupling, and Dye Coupling

Investigation of the sea urchin embryo did not succeed in finding gap junctions, though there were other junctional types (Spiegel and Howard, 1983). In the sea urchin at the 2-cell stage, Pochapin *et al.* (1983) found that electrical coupling was absent, and Lucifer Yellow (M_r 457) did not pass between blastomeres. In another echinoderm, the starfish, widespread electrical coupling but no transfer of the dyes fluorescein (M_r 332) or Procion Yellow (M_r 500) was reported at the 32-cell stage or at earlier stages (Tupper and Saunders, 1972). Using electron microscopy, junctions were first seen in the mollusk *Patella* at the 2-cell stage (Dorresteijn *et al.*, 1982) and in *Lymnea* at the 4-cell stage (Dorresteijn *et al.*, 1981). Dye transfer was not detected in *Patella* until the 32-cell stage (Dorresteijn *et al.*, 1983). Studies on the embryo of the teleost fish *Fundulus* showed electrical coupling (Bennett and Trinkaus, 1970) but fluorescein failed to transfer between isolated blastomeres (Bennett *et al.*, 1972), until a reassessment some years later yielded exactly the opposite result (Bennett *et al.*, 1978), when passage of both fluorescein and Lucifer Yellow was reported. In the mouse embryo, gap junctions were first detected by electron microscopy at the 8-cell stage (Ducibella *et al.*, 1975). Until then, neither carboxyfluorescein (M_r 376) nor Lucifer Yellow could fill more than two sister cells, but after cleavage to 8 cells, both dyes spread thoughout the embryo (Goodall and Johnson, 1982). Horseradish peroxidase (M_r 10,000), which can move only via cytoplasmic bridges and not gap junctions, could fill up to 4 cells (Goodall and Johnson, 1984), hinting that in this embryo, cytoplasmic connections are more extensive than between sister cells, and may contribute significantly to the permeability pattern.

The diffusion of dyes larger than M_r 500 has seldom been tested; there are few available probes. In the original work on the squid embryo by Potter *et al.* (1966), electrical coupling was recorded between cells distant from each other, but the dye Niagara Sky Blue (M_r 992) only transferred very infrequently. Similarly, in the chick embryo, electrical coupling was not concomitant with movement of this dye (Sheridan, 1968). In these two embryos, junctions apparently echo the size limit of adult mammalian channels, which is around M_r 1000.

3.2. Patterns of Communication in the Early Amphibian Embryo

Electrical coupling was detected in the newt (Ito and Hori, 1966) and *Xenopus* (Slack and Palmer, 1969), where gap junctions were first observed at the 8-cell stage, though they were small and sparse (Sanders and Dicaprio, 1976). But in *Xenopus* injections of fluorescein failed to reveal any dye communication at the 64-cell stage (Slack and Palmer, 1969). When Turin (1978) extended these studies using other tracers, he found that lead EDTA (mol. wt.

387) did not traverse cell borders at the 64-cell stage while smaller ion complexes such as potassium argentocyanate did. These compounds require chemical reaction to form a precipitate followed by sectioning to assess transfer, perhaps providing more reliable data than subjective judgments of fluorescence in whole embryos. Recently, however, injections of Lucifer Yellow into 32-cell *Xenopus* embryos demonstrated unambiguously that dye frequently moved between cells (Guthrie, 1984). There was a pronounced asymmetry in the tendency of cells to transfer dye; cells fated to give rise to dorsal tissue transferred with much higher frequency than those in ventral regions. At the 64-cell stage, a similar pattern existed, though the total amount of transfer was much less (Guthrie and Warner, manuscript in preparation).

These data suggest that in the amphibian and perhaps in other embryos, permeability may not be spatially uniform. Contradictions in the literature might be reconciled by assuming that many authors performed their experiments without taking account of cell position. The observations of Slack and Palmer might be explained if these authors had by chance injected cells lying at the ventral side of the embryo, where cells have low permeability. Since even dorsal cells do not invariably transfer dye, a small sample size of injections could have given negative results. Whether the pattern depends on differential distribution or permeability of gap junctions is unknown, and might be difficult to discover, due to the paucity of junctions in the *Xenopus* embryo at this stage. What is clear is that it is dangerous to seek to discover general rules about junctional permeability by injecting blastomeres that are positionally anonymous.

A consensus has thus not been forthcoming on the permeability of embryonic channels. It is striking, however, that electrical coupling seems to be ubiquitous among embryos, whereas the frequency of movement of small dyes varies. Thus, the idea of Potter *et al.* that pattern formation may depend on "interruptions in the current-carrying continuum," could be modified to consider the "dye-carrying continuum."

4. EARLY EVENTS IN DEVELOPMENT

The next step in evaluating the role of junctions is to set their static properties in the temporal context of development. Early in development the equivalence of cell types may require free communication, while later on differentiation may call for selective communication. Support for this idea would depend on discovering correlations between patterns of junctional permeability and development. Indeed, there are many examples of permeability patterns that change in developmental time, or coincide with important events. For example in the squid embryo, localized uncoupling of blastoderm cells from the giant yolk cells occurs at gastrulation (Potter *et al.*, 1966), a phenomenon also observed in the teleost fish, *Fundulus* (Kimmel *et al.*, 1984). There are many other examples of changes in patterns of communication, which will be described under separate themes such as compartments in development, and uncoupling of cell types during differentiation. In this section I shall focus on two specific examples of early cell interactions which have been subjects for investigating the role of gap junctions: cell determination in the mollusk, and compaction in the mouse embryo.

4.1. Communication and Cell Determination in the Molluscan Embryo

In the embryo of the mollusk *Patella,* dye coupling is universal between cells from the 32-cell stage onward, but subsequently becomes restricted to regions in the larva corresponding to presumptive tissue compartments (Dorresteijn *et al.*, 1983; Serras *et al.*, 1985). A decisive event at the 32-cell stage is the commitment of one of the vegetal macromeres to form the mother cell or "D" cell of the mesodermal lineage. One cell is induced to assume this fate by intruding into the center of the embryo and touching the animal quartet of micromeres (van den Biggelaar, 1977). Some authors entertained the idea that the interaction might be mediated by gap junctions (van den Biggelaar and Guerrier, 1979). In *Lymnea,* surveys of gap junctions showed a higher concentration between the presumptive D cell and the micromeres (Dorresteijn *et al.*, 1981). In *Patella* it was initially observed that dye injected into the D macromeres passed into the animal pole cells (de Laat *et al.*, 1980). But sections did not confirm the dye transfer that had been observed in whole embryos, nor could gap junctions be found at the interface in this species (Dorresteijn *et al.*, 1983), implying that mesodermal determination took place without junctional involvement.

4.2. Communication and Compaction in the Early Mouse Embryo

A central problem of early mammalian embryogenesis is how cells become determined to embark on one of two paths of differentiation. Preceding this first developmental decision is the process of compaction, when at the 8-cell stage cells lose their outlines and flatten onto one another. Simultaneously, they start to polarize, manifested by the appearance of a cap of microvilli and the aggregation of lectin-binding sites at the part of the cell farthest from the site of apposition with other cells. Asymmetric cell division from the 8- to 16-cell stage then creates two distinct populations of cells: the large outer polar cells possess villi while the small inner apolar cells do not (Johnson and Ziomek, 1982). Polar cells will contribute predominantly to the trophectodermal outer protective layer, and apolar cells will constitute the inner cell mass, from which develops the embryo proper and other derivatives. During this phase of development. cells are amenable to change mediated by cell contact. Polarization occurs in response to asymmetric cell contact (Ziomek and Johnson, 1980) at the 8-cell stage and cannot develop when a cell is completely surrounded by other cells. Investigations of cell behavior *in vitro* showed that cells can change their fate depending on their surroundings. Polar cells can generate inner cell mass in the absence of asymmetric cell contacts while apolar cells can make trophectoderm by polarizing in response to asymmetric contacts (Johnson and Ziomek, 1983).

Attempts have been made to distinguish the part played in these processes by cell contact at a physical level and junctional communication. Bearing in mind that cells begin to communicate at the late 8-cell stage, a role for gap junctions is feasible. By experimental recombination of isolated blastomeres, Johnson and Ziomek (1981) looked at the ability of blastomeres of various ages to induce polarity in blastomeres from 8-cell embryos. Polarization was assessed on the presence of fluorescent Con A binding to the characteristic cap of lectin receptors. The results obtained showed that the older the inducing cell, the more likely polarity was to develop in the 8-cell-stage blastomere. Inducing ability developed at the 2-cell stage, and was present at the 4-, 8-, and 16-cell stages. This induction is thus unlikely to take place via gap junctions, since they do not function until the 8-cell stage. Indeed, polarization appears to be remarkably resistant to intervention of all kinds. One set of experiments (Pratt *et al.*, 1982) used a range of treatments such as cytochalasin D, colcemid, and calcium-free medium to interfere with cell adhesion and the cytoskeleton. Polarization continued regardless, even when cell flattening and apposition had been almost completely prevented.

The least injurious treatment used by Pratt and colleagues was low-calcium medium, reinvestigated by Goodall (1986). In addition, he used an antiserum to an embryonal carcinoma line (anti-EC) and a monoclonal antibody recognizing a specific part of the calcium adhesion system. The monoclonal antibody, ECCD-1, did not prevent communication. Low-calcium medium and anti-EC both prevented ionic and dye coupling, but polarity ensued as previously described. This seems to rule out a role for junctions in polarization, especially since the lack of ionic coupling infers that not even a few residual junctions near the center of the embryo remained coupled.

The accumulated evidence suggests that polarization is a robust event that is able to occur in cells deprived of their normal associations with neighbors. Yet it may be that polarization as defined by increased ligand-binding capacity is not the only or the best criterion of development. Little data exist on the capability of embryos whose junctions have been disrupted to continue development. Furthermore, neither of the treatments efficacious in preventing the onset of junctional coupling could reverse coupling once established. The maintenance of the compacted state and early pattern formation may owe more to junctions than do the early differentiative events of polarization. Correlations between patterns of coupling and pattern formation in the mouse will be discussed later under the theme of developmental compartmentalization.

5. COMPARTMENTS, GRADIENTS, AND GAP JUNCTIONS

Orderly patterns of coupling or uncoupling at early stages of embryogenesis support the idea that spatial variation in junctional communication might be correlated with the division of the embryo into developmental territories. In many organisms such as the mouse embryo, the edges of most presumptive developmental units are unknown. By contrast, extensive clonal analysis in the insect has shown that the entire embryonic body plan is mapped out in a number of areas, the boundaries of which are frequently visible. Garcia-Bellido and colleagues (1973) described how in *Drosophila* groups of presumptive epidermal cells are set aside at the early blastoderm stage, and are thereafter unable to stray outside a well-defined developmental area or "compartment." Determination then ensues according to the compartmental address of the cell, since each compartment has its own combination of active genes (Crick and Lawrence, 1975).

The contribution of gap junctions to compartmentalization has often been proposed as one of separation. Regions of restricted permeability, for instance, might separate units that develop into different structures, giving them a degree of independence. Within the compartment's perimeters, morphogenetic gradients could exist. Gradients were strongly implied by the experiments of Stumpf (1966) on the moth *Manduca* and of Locke (1959) on *Rhodnius* which implied a gradient of morphogenetic information within the segment. In *Rhodnius,* pieces of cuticle were transplanted with varying degrees of rotation, their ripple pattern showing how they behaved in the novel environment. The result was that they remembered their original orientation to some extent, but also adjusted to their new milieu, as if they were in a gradient field of continuous positional values.

Gradients have thus been popular candidates for the bases of developmental information, perhaps specifying a series of positional values (Wolpert, 1971). Wolpert has postulated that pattern formation could be a two-step process: positional values are first assigned to parts of the embryo, and cells respond accordingly to pursue various paths of differentiation. A gradient of some substance could provide these positional values; a source at one end and a sink at the other would produce a smooth gradient. Where structures are repeated such as the insect larval segments, gradients could form a sawtooth array. Apart from the indirect evidence, though, there are little data on the nature of morphogens. Certainly the only ones so far identified, in *Hydra,* conform to a monotonic gradient (Schaller *et al.*, 1979).

Compartmentalization in the insect has been investigated with respect to a possible contribution in the process by junctions. Here I shall review the evidence relating to this question, an issue bound up with the generation of morphogenetic gradients. As a postscript I shall also look at intercellular communication and developmental divisions in the mouse embryo.

5.1. Communication across Insect Compartment Borders

Another insect, the milkweed bug *Oncopeltus,* provided the opportunity to investigate whether compartment borders had special junctional properties. In the larva, compartment borders are also segment borders, clearly marked by a change in pigmentation (Lawrence, 1973). At the border are found the same range of intercellular associations—desmosomes, septate junctions, and gap junctions—common throughout the segment (Lawrence and Green, 1975). Experiments on other insects—*Rhodnius* (Warner and Lawrence, 1973) and *Tenebrio* (Caveney, 1974)—showed no departure from uniform electrical coupling within the segment and across the segment boundary. In one respect, however, junctions at the border did appear atypical, being selective in the size of molecules afforded passage. This was investigated in *Calliphora* using the tracer lead EDTA (M_r 387) and Lucifer Yellow (M_r 457) and in *Oncopeltus* using Lucifer Yellow (Warner and Lawrence, 1982). In both insects, Lucifer Yellow was transferred freely between cells within the segment, but transfer was frequently restricted at the border. In *Calliphora* Lucifer Yellow never traversed the border, and in *Oncopeltus* it was restricted in 90% of cases. The slightly smaller and more compact tracer lead EDTA could always cross the border in *Calliphora.* Gap junctions at the border thus appear to have properties different from those within the segment. A one- to three-cell-wide strip of particularly impermeable cells at the margin may be responsible (Blennerhassett and Caveney, 1984).

5.2. Communication in Insect Imaginal Disks

Drosophila has been a major subject of clonal analysis and has given revolutionary insights into strategies of development. In this organism also, evidence was obtained showing that dye restriction coincides with developmental compartments. Weir and Lo (1982) injected Lucifer Yellow into the imaginal wing disk, and observed that dye did not cross the anterior/posterior boundary. Further experiments revealed a number of additional restriction lines that fitted with the clonal map (Weir and Lo, 1984). The question arises of whether these communication domains arise as a precursor to the partitioning into developmental regions, or are simply a result of compartmentalization. To try to resolve this, Weir and Lo turned their attention to the wing disk in flies mutant for *engrailed,* a deficiency causing posterior compartments to resemble anterior ones. Perhaps surprisingly, dye transfer was still absent at the border in mutant disks, which comprise two identical anterior compartments (Weir and Lo, 1985). However, variations in contour and thickness of the imaginal disk have raised the question of whether dye transfer could be limited by factors other than junctional occlusion. In later experiments, dye-fills with Lucifer Yellow have shown a contradictory pattern where small clusters of up to 20 cells receive dye, with no apparent reproducibility of pattern between disks (Fraser and Bryant, 1985).

If the presence of communication domains is to be believed, the interpretation is difficult. Weir and Lo contend that the presence of restriction boundaries accords with the notion that communication domains are significant for patterning. Yet what is their function?—presumably to prevent the movement of morphogenetic molecules between compartments, molecules that provoke the expression of different genes in adjacent compartments. The presence of the communication border in the *engrailed* disk may be compatible with this idea, since a border might be formed irrespective of whether the two compartments it divides are similar or not. Nevertheless, it detracts from the idea that junctional differences are instrumental in setting up the spheres of influence of the primary homoeotic genes such as the *Bithorax* Complex, *Antennapedia* Complex, and *engrailed*. Rather, the argument ventures into realms that are not well understood, farther down the genetic hierarchy. Once the homoeotic or selector genes have operated to create a series of compartments with positional values within the insect, realizator genes must come into action, to translate the positional instructions. Compartmentalization at the junctional level might be important in maintaining the positional system so that the subsequent genetic steps can be executed.

5.3. Communication in the Early Mouse Embryo

In the mouse embryo, where compartments have not been defined so strictly as in the insect, zones of communication have also been reported. Lo and Gilula (1979a) studied patterns of dye passage in the mouse preimplantation blastocyst and observed that Lucifer Yellow or fluorescein injected into the inner cell mass of trophectoderm could spread to the entire embryo. A little later, at postimplantation stages, Lucifer Yellow did not move between the outer trophectoderm cells and the inner cell mass (Lo and Gilula, 1979b). Thus, lack of communication may reflect the divergence of cell types at the first developmental decision. Still later, groups of cells within the inner cell mass ceased to communicate, a property Lo and Gilula speculated might be related to the pursuit of different fates.

5.4. Gap Junctions and Gradients

Morphogenetic gradients have been easier to conceive theoretically than to demonstrate practically. Is it plausible that gap junctions could make a contribution to maintaining such gradients? At the junctional level this might require widespread permeability within the segment, allowing the establishment of a smooth gradient, with restricted permeability at the segment border, which constitutes the interface between maxima and minima of adjacent gradient fields. Both these conditions may be judged to be fulfilled on the basis of the findings of

Warner and Lawrence (1982) showing the restriction of Lucifer Yellow transfer across the insect segment border.

Theoretical models for the diffusion of morphogens in tissues require that molecules diffuse faster the larger the developmental field. Work on *Tenebrio* showed that junctional communication increases with age of the insect and size of the compartment (Safranyos and Caveney, 1985). The diffusion rates of small dyes such as carboxyfluorescein increase in older segments in a manner not related to obvious changes in parameters of the segment. Another piece of evidence supporting the notion that morphogens move via gap junctions is that some of the few known morphogens, in *Hydra,* have sizes (500–1100 kD) compatible with junctional transmission (Schaller *et al.*, 1979; Schaller and Bodenmuller, 1981).

At the time of establishment, gradient fields are probably small, about 100 cells long or 1 mm. Calculations by Crick (1970) showed that with an effective diffusion constant of 2×10^{-7} cm^2/sec, a small molecule (say M_r 300) could form a gradient in a few hours. Recent measurements on tracer diffusion in embryonic cells show adequate rapidity. Lucifer Yellow (M_r 457) can diffuse throughout the 8-cell mouse embryo in less than 20 min (McClachlin *et al.*, 1983). Estimates for the rate of diffusion in insect epidermis (Safranyos and Caveney, 1985) agree well with Crick's prediction, giving a diffusion constant of $1\text{–}4 \times 10^{-7}$ cm^2/sec.

6. RESTRICTION OF COMMUNICATION LATER IN DEVELOPMENT

During later phases of development, when cells finish acquiring differentiated status, junctional coupling is sometimes lost between individual cells or tissue compartments. In the history of the apical ectodermal ridge (AER), the structure dictating proximodistal organization in limb development, gap junctions are lost as development proceeds (Kelley and Fallon, 1976). The loss of junctions from the edge of the AER coincides with the decline in size of the ridge from its maximum size to its disappearance. Other examples of developmental uncoupling abound particularly in nerve and muscle; these will be described next.

6.1. Gap Junctional Uncoupling in Neural Tissues

In the axolotl, *Ambystoma,* Warner (1973) monitored electrical coupling at the stage of neural fold closure. Though electrical coupling was widespread between cells in the neural plate, and between those in the epidermis, electrical coupling was lost between the two tissue compartments as the folds closed. Simultaneously, low-resistance contacts were formed across the midline, by cells previously separated by the neural plate. In another amphibian, *Xenopus,* stage-dependent uncoupling of Rohon–Beard neurons occurs as the immature calcium action potential is replaced by the sodium-dependent adult action potential (Spitzer, 1982). Another developing neural tissue is the retina, where at early stages junctions connect retinal cells, and join them to cells of the pigmented epithelium. Heterotypic junctions may mediate control of retinal growth by the epithelium, and enable retinal cells to inhibit the epithelium from forming ectopic neuroretina (Fujisawa *et al.*, 1976). The retina grows radially; cells being added on at the periphery. Loss of junctions between cells in the middle of the amphibian retina (Dixon and Cronly-Dillon, 1972) and the retina and epithelium (Dixon and Cronly-Dillon, 1974) is concomitant with cell fate determination, at a time when the axial polarity of the retina is being specified. In the embryo of the locust, *Schistocerca,* development is punctuated by the appearance of a well defined sequence of junctions. Gap junctions are expressed transiently between undifferentiated cells, but disappear when cells begin to cluster into the presumptive visual units, the ommatidia (Eley and Shelton, 1976).

In the visual system of *Daphnia,* transient gap junctions have been detected between the optic axons growing from an ommatidium and the neuroblasts of the optic lamina. Eight optic fibers grow out in sequence, each in turn making contact with an undifferentiated neuroblast. Each neuroblast wraps around the fiber and gap junctions are formed (Lopresti *et al.*, 1974). A similar short-lived interaction is encountered in the navigation of grasshopper "pioneer" neurons. While crossing the virgin territory of the limb bud en route to the central nervous system, these axons contact a few strategically placed "guidepost" cells, whose ablation causes the pioneers to lose their way (Bentley and Caudy, 1984). As they come into apposition with the guideposts, the neurons form dye-passing junctions (Taghert *et al.*, 1982). Both these examples make it tempting to believe that the junctional association is important for subsequent development.

6.2. Gap Junctional Uncoupling in Muscle Development

In myogenesis, junctional coupling has been described between muscle cells both *in vivo* and *in vitro* at early stages in various organisms (Rash and Fambrough, 1973; Blackshaw and Warner, 1976; Kalderon *et al.*, 1977; Chow and Poo, 1984). Between mature muscle fibers, coupling is generally absent. When gap junctions, ionic coupling, and metabolic coupling were detected between chick embryonic myoblasts in culture (Kalderon *et al.*, 1977), this led to the suspicion that junctions might be involved in myoblast fusion. However, when fusion was inhibited, the inhibition did not appear to operate by breaking gap junction contacts. Possibly, junctions have a more indirect influence, conferring on cells the competency to respond to a signal regulating fusion. The time course of uncoupling of muscle cells *in vivo* varies very much according to species. In rats, where cells fuse early in development, coupling may persist up until birth (Schmalbruch, 1982). Experiments on *Xenopus* myotomes *in vivo* showed that junctional coupling is lost some hours before fusion (Armstrong *et al.*, 1983). But in this species, myotome development is unusual, with cells developing to a fully functional state while remaining mononucleate. Cell fusion is a late event, and takes place in a manner different from that in other vertebrate species.

In amphibians, gap junctions may have a different role in myogenesis. The purpose of junctional coupling at early stages may be to allow current spread and muscle activation at a time before innervation is complete. In three amphibian species, *Xenopus*, *Bombina*, and *Ambystoma*, the delineation of the somites from the unsegmented mesoderm is accompanied by loss of electrical coupling between presumptive somite blocks (Blackshaw and Warner, 1976). In *Xenopus* and *Bombina*, coupling is resumed when segmentation is complete, a process that may be necessitated by their mode of development. These two species hatch early and show spontaneous swimming and avoidance reactions; this may be achieved by the propagation of current from somite to somite in the incompletely innervated musculature. *Ambystoma* does not exhibit this early behavior since it remains encased in a tough capsule for longer. By the time of hatching, innervation has already proceeded far enough to allow motor nerve activation of the muscle without recourse to a junctional route. The somites remain electrically insulated from one another after uncoupling. In *Xenopus*, cells then become uncoupled after the maturation of neuromuscular transmission (Armstrong *et al.*, 1983).

Gap junctions have occasionally been observed between nerve and muscle at an early stage during innervation (Fischbach, 1972). That the cells were dye coupled was shown by Allen (1986), who observed transfer of Lucifer Yellow from muscle to nerve in *Xenopus* cultures, presumably by gap junctions. When the muscle cells were loaded with gap junction antibody in addition to Lucifer Yellow, transfer was blocked. By searching for structural or functional effects of the antibody on neuromuscular connection, the importance of this interaction might be determined.

7. USING ANTIBODIES TO EXPLORE THE ROLE OF JUNCTIONS IN DEVELOPMENT

For some while it has been highly desirable to test directly the significance of junctions in development. But attempts have been frustrated because agents that can interfere with permeability, such as high calcium and low pH, often have deleterious effects on aspects of cell viability. The situation has changed, however, with the advent of antibodies to gap junction proteins which have been raised by a number of laboratories. These antibodies were raised against the 27k gap junctional protein of rat liver. Two studies have now shown that intercellular communication can be effectively interrupted by the injection of antibodies into cells. These experiments were performed in the early *Xenopus* embryo (Warner *et al.*, 1984) and in cultures of hepatocytes, myocardial cells, and superior cervical ganglion neurons (Hertzberg *et al.*, 1985). In both studies, electrical and dye coupling were abolished by the antibody injection. The study of Warner *et al.* went on to examine development.

In the *Xenopus* embryo at the 32-cell stage, a cell which normally transfers dye at high frequency was chosen to assay dye coupling after injection of antibody into its precursor at the 8-cell stage. Dye transfer was found to be inhibited in the antibody-containing region in 70–80% of injected embryos. Nor were cells in injected regions electrically coupled, though cells in uninjected regions of the same embryo retained strong electrical coupling. If the antibody could so completely block intercellular communication, what might be the consequences for later development? When antibody-injected embryos were left to develop into tadpoles, they exhibited a characteristic range of developmental abnormalities. These defects were confined to regions containing progeny cells of the original injected blastomere, in the ectoderm and mesoderm of the right-hand anterior part of the tadpole. Often, differentiation had taken place relatively normally, while pattern formation had been perturbed. The disruption of pattern formation as distinct from differentiation might imply distortion of normal positional values, which, simplistically, could be caused by discontinuities in gradients specifying this parameter. Normal patterns of gap junctional communication at early stages may thus be important for development.

Future Prospects

A number of avenues of investigation arise from these findings. Firstly, the anti-27k protein antibody cross-reacts with material from a wide range of species in both vertebrate and invertebrate kingdoms (Green *et al.*, manuscript in preparation). Such conservation of composition may also indicate conservation of structure and function. Thus, the antibody may be used to explore developmental events in many different organisms. In particular, injection into embryos whose character is more mosaic might yield specific defects with discrete absence of structures. There is scope also for examining other more specific cell interactions in embryogenesis. The embryological literature gives examples such as neural induction, where specific cell populations are known to interact. Mesodermal cells induce overlying ectoderm to form nervous tissue, a process that can be simulated in sandwiches of the two cell types (Saxen, 1961). Experiments interposing filters between the cell layers have given equivocal results as to whether cell–cell contact is required for induction. Examination of filters did not at first reveal any processes (Toivonen *et al.*, 1975), while in later experiments areas of membrane apposition were found (Toivonen and Wartiovaara, 1976), which may have contained gap junctions. The mechanism of neural induction remains uncertain. The data leave room to suppose that the inductive signal comprises several elements of which junctional communication is just one, an issue which could be explored using antibodies, either in sandwiches or in culture of the two cell types.

8. CONCLUSIONS

A wealth of evidence links gap junctions with events in development. Gap junctions are present in the embryos of almost all phyla. They often appear at times when cells are beginning to decide their fates, and disappear when that fate has been sealed. It has been theoretically difficult to conceive of a specific role for gap junctions, which are not thought to be highly discriminatory in terms of what passes through them, and act as a common thoroughfare for many molecules. But it is conceivable that in some embryos, cells are in many respects equivalent. Equipped with a diversity of requirements laid down in the egg, they may not need junctions to perform metabolic exchange. Thus, gap junctions may be able to fulfill a role in developmental signaling. Gap junctions might thus have one function early in development, and another later on, when they take up their normal disposition in the tissues of the mature organism. Parallels are found in embryos where genes are expressed in a lineage-specific manner early on, while their later patterns of expression may be quite different. Antibodies and other molecular probes may now be used to capitalize on existing knowledge from correlative studies and help define the role of intercellular communication in development.

ACKNOWLEDGMENTS. I am indebted to Anne Warner for her valuable comments and help with the manuscript. I also thank Colin Green and Francesca Allen for encouraging conversation.

9. REFERENCES

Albertini, D. F., and Anderson, E., 1974, The appearance and structure of intercellular connections during the ontogeny of the rabbit ovarian follicle with particular reference to gap junctions, *J. Cell Biol.* **63:**234.

Allen, F., 1986, Gap junctional communication during neuromuscular junction formation in *Xenopus* cultures, *J. Physiol.* **377:**77P.

Anderson, K. V., and Nusslein-Volhard, C., 1984, Information for the dorsal–ventral pattern of the *Drosophila* embryo is stored as maternal mRNA, *Nature* **311:**223.

Armstrong, D. L., Turin, L., and Warner, A. E., 1983, Muscle activity and loss of electrical coupling between striated muscle cells in *Xenopus* embryos, *J. Neurosci.* **3:**1414.

Bennett, M. V. L., and Trinkaus, J. P., 1970, Electrical coupling between embryonic cells by way of extracellular space and specialised junctions, *J. Cell Biol.* **44:**3592.

Bennett, M. V. L., Spira, M., and Pappas, G. D., 1972, Properties of electrotonic junctions between embryonic cells of *Fundulus*, *Dev. Biol.* **29:**419.

Bennett, M. V. L., Spira, M., and Spray, D. C., 1978, Permeability of gap junctions between embryonic cells of *Fundulus:* A re-evaluation, *Dev. Biol.* **65:**114.

Bentley, D., and Caudy, M., 1984, Pioneer axons lose directional growth after selective killing of guidepost cells, *Nature* **304:**63.

Blackshaw, S. E., and Warner, A. E., 1976, Low resistance junctions between mesoderm cells during development of the trunk muscles, *J. Physiol. (London)* **255:**209.

Blennerhassett, M. G., and Caveney, S., 1984, Separation of developmental compartments by a cell type with reduced junctional permeability, *Nature* **309:**361.

Brown, C. L., Wiley, S. H., and Dumont, J. N., 1979, Oocyte–follicle cell gap junctions in *Xenopus laevis* and the effects of gonadotropin on their permeability, *Science* **203:**182.

Burk, R. R., Pitts, J. D., and Subak-Sharpe, J., 1968, Exchange between hamster cells in culture, *Exp. Cell Res.* **53:**297.

Caveney, S., 1974, Intercellular communication in a positional field: Movement of small ions between insect epidermal cells, *Dev. Biol.* **40:**311.

Chow, I., and Poo, M. M., 1984, Formation of electrical coupling between embryonic *Xenopus* muscle cells in culture, *J. Physiol. (London)* **346:**181.

Crick, F., 1970, Diffusion in embryogenesis, *Nature* **225:**420.

Crick, F. H. C., and Lawrence, P. A., 1975, Compartments and polyclones in insect development, *Science* **189:**340.

Dekel, N., and Beers, W. H., 1978, Rat oocyte maturation *in vitro:* Relief of cyclic AMP inhibition by gonadotropins, *Proc. Natl. Acad. Sci. USA* **75:**4369.

Dekel, N., and Beers, W. H., 1980, Development of the rat oocyte *in vitro:* Inhibition and induction of maturation in the presence and absence of the cumulus oophorus, *Dev. Biol.* **75:**247.

Dekel, N., Lawrence, T. S., Gilula, N. B., and Beers, W. H., 1981, Modulation of cell-to-cell communication in the cumulus–oocyte complex and the regulation of oocyte maturation by LH, *Dev. Biol.* **86:**356.

de Laat, S. W., Tertoolen, L. G. J., Dorresteijn, A. W. C., and van den Biggelaar, J. A. M., 1980, Intercellular communication patterns are involved in cell determination in early molluscan development, *Nature* **287:**546.

Dixon, J. S., and Cronly-Dillon, J. R., 1972, The fine structure of the developing retina in *Xenopus laevis, J. Embryol. Exp. Morphol.* **28:**659.

Dixon, J. S., and Cronly-Dillon, J. R., 1974, Intercellular gap junctions in pigment epithelium cells during retinal specification in *Xenopus laevis, Nature* **251:**505.

Dorresteijn, A. W. C., van den Biggelaar, J. A. M., Bluemink, J. G., and Hage, W. J., 1981, Electron microscopical investigation of the intercellular contacts during the early cleavage stages of *Lymnea stagnalis* (Mollusca, Gastropoda), *Wilhelm Roux Arch. Dev. Biol.* **190:**215.

Dorresteijn, A. W. C., Bilinski, S. M.. van den Biggelaar, J. A. M., and Bluemink, J. G., 1982, The presence of gap junctions during early *Patella* embryogenesis: An electron microscopical study, *Dev. Biol.* **91:**397.

Dorresteijn, A. W. C., Wagemaker, H. A., de Laat, S. W., and van den Biggelaar, J. A. M., 1983, Dye-coupling between blastomeres in early embryos of *Patella vulgata;* Its relevance for cell determination, *Wilhelm Roux Arch. Dev. Biol.* **192:**262.

Ducibella, T., Albertini, D. F., Anderson, E., and Biggers, J. D., 1975, The preimplantation mammalian embryo: Characterisation of intercellular junctions and their appearance during development, *Dev. Biol.* **45:**231.

Eley, S., and Shelton, P. M. J., 1976, Cell junctions in the developing compound eye of the desert locust *Schistocerca gregaria, J. Embryol. Exp. Morphol.* **36:**2409.

Eppig, J. J., 1982, The relationship between cumulus cell–oocyte coupling, oocyte meiotic maturation and cumulus expansion, *Dev. Biol.* **86:**268.

Eppig, J. J., Freter, R. R., Ward-Bailey, P. F., and Schultz, R. M., 1983, Inhibition of oocyte maturation in the mouse: Participation of cAMP, steroids, and a putative maturation inhibitory factor, *Dev. Biol.* **100:**39.

Fischbach, G. D., 1972, Synapse formation between dissociated nerve and muscle cells in low density cultures, *Dev. Biol.* **28:**407.

Flagg-Newton, J. L., Simpson, I., and Loewenstein, W. R., 1979, Permeability of the cell to cell membrane channels in mammalian cell junctions, *Science* **205:**404.

Fraser, S., and Bryant, P., 1985, Patterns of dye coupling in the imaginal wing disc of *Drosophila melanogaster, Nature* **317:**533.

Freter, R. R., and Schultz, R. M., 1984, Microinjection of murine oocytes with a low molecular weight fraction of bovine follicular fluid inhibits meiosis, *J. Cell Biol.* **99:**389a.

Fujisawa, H., Morioka, H., Watanabe, K., and Nakamura, H., 1976, A decay of gap junctions in association with cell differentiation of neural retina in chick embryonic development, *J. Cell Sci.* **22:**585.

Furshpan, E. J., and Potter, D. D., 1969, Low-resistance junctions between cells in embryos and tissue culture, *Curr. Top. Dev. Biol.* **3:**95.

Garcia-Bellido, A., Ripoll, P., and Morata, G., 1973, Developmental compartmentalisation of the wing disc of *Drosophila, Nature New Biol.* **245:**251.

Gilula, N. B., Reeves, O. R., and Steinbach, A., 1972, Metabolic coupling, ionic coupling and cell contact, *Nature* **235:**262.

Gilula, N. B., Epstein, M. L., and Beers, W. H., 1978, Cell to cell communication and ovulation; a study of the cumulus–ooctye complex, *J. Cell Biol.* **78:**58.

Goodall, H., 1986, Manipulation of gap junctional communication during compaction of the early mouse embryo, *J. Embryol. Exp. Morphol.* **91:**283.

Goodall, H., and Johnson, M. H., 1982, Use of carboxyfluorescein diacetate to study formation of permeable channels between mouse blastomeres, *Nature* **295:**524.

Goodall, H., and Johnson, M. H., 1984, The nature of intercellular coupling in the pre-implantation mouse embryo, *J. Embyrol. Exp. Morphol.* **79:**53.

Guthrie, S. C., 1984, Patterns of junctional permeability in the early amphibian embryo, *Nature* **311:**149.

Heller, D. T., Cahill, D. M., and Schultz, R. M., 1981, Biochemical studies of mammalian oogenesis: Metabolic cooperativity between granulosa cells and growing mouse oocytes, *Dev. Biol.* **84:**455.

Hertzberg, E. L., Spray, D. C., and Bennett, M. V. L., 1985, Reduction of gap junctional conductance by microinjection of antibodies against the 27 kD rat liver gap junction polypeptide, *Proc. Natl. Acad. Sci. USA* **82:**2412.

Huebner, E., 1981, Oocyte–follicle cell interaction during normal oogenesis and atresia in an insect, *J. Ultrastruct. Res.* **74:**95.

Ito, S., and Hori, N., 1966, Electrical characteristics of *Triturus* eggs during cleavage, *J. Gen. Physiol.* **49:**1019.

Jaffe, L. F., and Woodruff, R. I., 1979, Large electrical currents traverse developing *Cecropia* follicles, *Proc. Natl. Acad. Sci. USA* **76:**1328.

Johnson, M. H., and Ziomek, C. A., 1981, Induction of polarity in mouse 8-cell blastomeres, *J. Cell Biol.* **91:**303.

Johnson, M. H., and Ziomek, C. A., 1982, Cell subpopulations in the late morula and early blastocyst of the mouse, *Dev. Biol.* **91:**431.

Johnson, M. H., and Ziomek, C. A., 1983, Cell interactions influence the fate of mouse blastomeres undergoing the transition from the 16 to the 32-cell stage, *Dev. Biol.* **95:**211.

Kalderon, N., Epstein, M. L., and Gilula, N. B., 1977, Cell to cell communication and myogenesis, *J. Cell Biol.* **75:**788.

Kelley, R. O., and Fallon, J. F., 1976, Ultrastructural analysis of the apical ectodermal ridge during vertebrate limb morphogenesis. I. The human forelimb with special reference to gap junctions, *Dev. Biol.* **51:**241.

Kimmel, C. B., Spray, D. C., and Bennett, M. V. L., 1984, Developmental uncoupling between blastoderm and yolk cell in the embryo of the teleost *Fundulus, Dev. Biol.* **102:**483.

Larsen, W. J., 1985, Relating the population dynamics of gap junctions to cellular function, in: *Gap Junctions* (M. V. L. Bennett and D. C. Spray, eds.), p. 289, Cold Spring Harbor Laboratory, Cold Spring Harbor, N.Y.

Larsen, W. J., Wert, S. E., and Brunner, G. D., 1984, The disruption of rat cumulus cell gap junctions could provide a signal to the egg to resume meiotic maturation, *J. Cell Biol.* **99:**345a.

Lawrence, P. A., 1973, A clonal analysis of segment development in *Oncopeltus, J. Embryol. Exp. Morphol.* **30:**3681.

Lawrence, P. A., and Green, S. M., 1975, The anatomy of a compartment border: The intersegmental boundary in *Oncopeltus, J. Cell Biol.* **65:**373.

Lawrence, T. S., Beers, W. H., and Gilula, N. B., 1978, Transmission of hormonal stimulation by cell-to-cell communication, *Nature* **272:**501.

Lo, C. W., and Gilula, N. B., 1979a, Gap junctional communication in the pre-implantation mouse embryo, *Cell* **18:**399.

Lo, C. W., and Gilula, N. B., 1979b, Gap junctional communication in the post-implantation mouse embryo, *Cell* **18:**411.

Locke, M., 1959, The cuticular pattern in an insect, *Rhodnius prolixus, J. Exp. Biol.* **39:**459.

Lopresti, V., Macagno, E. R., and Levinthal, C., 1974, Structure and development of neural connections in isogenic organisms: Transient gap junctions between growing optic neurons and lamina neuroblasts, *Proc. Natl. Acad. Sci. USA* **71:**1098.

McClachlin, J. R., Caveney, S., and Kidder, G. M., 1983, Control of gap junction formation in early mouse embryos, *Dev. Biol.* **98:**155.

Masui, Y., and Clarke, H. J., 1979, Oocyte maturation, *Int. Rev. Cytol.* **57:**186.

Nekola, M. V., and Moor-Smith, D. M., 1975, Failure of gonadotropins to induce *in vitro* maturation of oocytes treated with dibutyryl cyclic AMP, *J. Exp. Zool.* **194:**529.

Olsiewski, P. J., and Beers, W. H., 1983, cAMP synthesis in the rat oocyte, *Dev. Biol.* **100:**287.

Olsiewski, P. J., and Beers, W. H., 1985, Junctional communication and oocyte maturation, in: *Gap Junctions* (M. V. L. Bennett and D. C. Spray, eds.), p. 307, Cold Spring Harbor Laboratory, Cold Spring Harbor, N. Y.

Pochapin, M. B., Sanger, J. M., and Sanger, J. W., 1983, Microinjection of Lucifer Yellow CH into sea urchin eggs and embryos, *Cell Tissue Res.* **234:**309.

Potter, D. D., Furshpan. E. J., and Lennox, E. S., 1966, Connections between cells of the developing squid as revealed by electrophysiological methods, *Proc. Natl. Acad. Sci. USA* **55:**328.

Pratt, H. P. M., Ziomek, C. A., Reeve, W. J. D., and Johnson, M. H., 1982, Compaction of the mouse embryo: An analysis of its components, *J. Embryol. Exp. Morphol.* **70:**113.

Rash, J. E., and Fambrough, D., 1973, Ultrastructural and electrophysiological correlates of cell coupling and cytoplasmic fusion during myogenesis *in vitro, Dev. Biol.* **30:**166.

Safranyos, R. G. A., and Caveney, S., 1985, Rates of diffusion of fluorescent molecules via cell-to-cell membrane channels in a developing tissue, *J. Cell Biol.* **100:**736.

Sanders, E. J., and Dicaprio, R. A., 1976, Intercellular junctions in the *Xenopus* embryo prior to gastrulation, *J. Exp. Zool.* **197:**415.

Saxen, L., 1961, Transfilter neural induction of amphibian ectoderm, *Dev. Biol.* **3:**140.

Schaller, H. C., and Bodenmuller, H., 1981, Isolation and amino acid sequence of a morphogenetic peptide from *Hydra, Proc. Natl. Acad. Sci. USA* **78:**7000.

Schaller, H. C., Schmidt, T., and Grimmelikhuijzen, C. J. P., 1979, Separation and specificity of action of four morphogens from *Hydra, Wilhelm Roux Arch.* **186:**139.

Schmalbruch, H., 1982, Skeletal muscle fibres of newborn rats are coupled by gap junctions, *Dev. Biol.* **91:**485.

Schuetze, S. M., and Goodenough, D. A., 1982, Dye transfer between cells of the embryonic chick lens becomes less sensitive to CO_2 treatment with development, *J. Cell Biol.* **92:**694.

Schwarzmann, G., Wiegandt, H., Rose, B., Zimmerman, A., Ben-Haim, D., and Loewenstein, W. R., 1981, Diameter of the cell to cell junctional membrane channels as probed with neutral molecules, *Science* **213:**551.

Serras, F., Kuhtreiber, W. M., Krul, M. R. L., and van den Biggelaar, J. A. M., 1985, Communication compartments in molluscan embryos, *Cell Biol. Int. Rep.* **9:**8.

Sheridan, J. D., 1968, Electrophysiological evidence for low resistance intercellular junctions in the early chicken embryo, *J. Cell Biol.* **37:**650.

Slack, C., and Palmer, J. F., 1969, The permeability of intercellular junctions in the early embryo of *Xenopus laevis,* studied with a fluorescent tracer, *Exp. Cell Res.* **55:**416.

Spitzer, N. C., 1982, Voltage and stage-dependent uncoupling of Rohon–Beard neurons during embryonic development of *Xenopus* tadpoles, *J. Physiol.* **330:**145.

Spiegel, E., and Howard, L., 1983, Development of cell junctions in sea urchin embryos, *J. Cell Sci.* **62:**27.

Stumpf, H., 1966, Mechanism by which cells estimate their location in the body, *Nature* **212:** 431.

Taghert, P. H., Bastiani, M. J., Ho, R. K., and Goodman, C. S., 1982, Guidance of pioneer growth cones; filopodial contacts and coupling revealed with an antibody to Lucifer Yellow, *Dev. Biol.* **94:**391.

Toivonen, S., and Wartiovaara, J., 1976, Mechanism of cell interaction during primary embryonic induction studied in transfilter experiments, *Differentiation* **5:**61.

Toivonen, S., Tarin, D., Saxen, L., Tarin, P. J., and Wartiovaara, J., 1975, Transfilter studies on neural induction in the newt, *Differentiation* **4:**1.

Tupper, J. T., and Saunders, J. W., 1972, Intercellular permeability in the early *Asterias* embryo, *Dev. Biol.* **27:**546.

Turin, L., 1978, The physiology of intercellular communication in the early amphibian embryo, Ph.D. thesis, University of London.

Turin, L., and Warner, A. E., 1977, Carbon dioxide reversibly abolishes ionic communication between cells of the early amphibian embryo, *Nature* **270:**56.

van den Biggelaar, J. A. M., 1977, Development of dorsoventral polarity and mesoblast determination in *Patella vulgata, J. Morphol.* **154:**157.

van den Biggelaar, J. A. M., and Guerrier, P., 1979, Dorsoventral polarity and mesentoblast determination as concomitant results of cellular interactions in the mollusc *Patella vulgata, Dev. Biol.* **68:**462.

Vandenhoef, M. H. F., Dictus, W. J. A. G., Hage, W. J., and Bluemink, J. G., 1984, The ultrastructural organisation of gap junctions between follicle cells and the ooctye in *Xenopus laevis, Eur. J. Cell Biol.* **33:**242.

Vilain, J.-P., Moreau, M., and Guerrier, P., 1980, Uncoupling of oocyte–follicle cells triggers reinitiation of meiosis in amphibian oocytes, *Dev. Growth Differ.* **22:**687.

Warner, A. E., 1973, The electrical properties of the ectoderm in the amphibian embryo during induction and early development of the nervous system, *J. Physiol (London)* **235:**267.

Warner, A. E., and Lawrence, P. A., 1973, Electrical coupling across developmental boundaries in insect epidermis, *Nature* **245:**47.

Warner, A. E., and Lawrence, P. A., 1982, Permeability of gap junctions at the segmental border in insect epidermis, *Cell* **28:**243.

Warner, A. E., Guthrie, S. C., and Gilula, N. B., 1984, Antibodies to gap-junctional protein selectively disrupt junctional communication in the early amphibian embryo, *Nature* **311:**127.

Weir, M. P., and Lo, C. W., 1982, Gap junction communication compartments in the *Drosophila* wing disc, *Proc. Natl. Acad. Sci. USA* **79:**3232.

Weir, M. P., and Lo, C. W., 1984, Gap junction communication compartments in the *Drosophila* wing imaginal disc, *Dev. Biol.* **102:**130.

Weir, M. P., and Lo, C. W., 1985, An anterior/posterior communication compartment border in

engrailed wing discs: Possible implications for *Drosophila* pattern formation, *Dev. Biol.* **110**:84.
Wolpert, L., 1971, Positional information and pattern formation, *Curr. Top. Dev. Biol.* **6**:183.
Ziomek, C. A., and Johnson, M. H., 1980, Cell surface interaction induces polarization of mouse and 8-cell blastomeres at compaction, *Cell* **21**:935.

Chapter 8

Mechanisms of Cell-to-Cell Communication Not Involving Gap Junctions

Walmor C. De Mello
Department of Pharmacology
Medical Sciences Campus
University of Puerto Rico
San Juan, Puerto Rico 00936

1. CHEMICAL COMMUNICATION BETWEEN CELLS

The nervous system is certainly one of the most important systems of cell communication in the body.

The initial morphological concept of the nervous system was centralized in the idea of a complex netlike structure (Gerlach, 1871) in which the neurons are fused together—a concept challenged by His (1886) and Forel (1887) who visualized the nervous system as composed of isolated neurons.

Ramón y Cajal (1909) utilizing the Golgi technique provided strong evidence that the nervous system is not a reticulum but consists of isolated neurons. The studies of Koelliker (1890), van Geruchten (1891), and Waldeyer (1891) supported the view that neurons are functionally connected not by continuity but by contiguity. This theory, called the "neuron theory," proposed that the site of cell contiguity was the "terminal brushes" Ramón y Cajal (1909) also called "Faserkorbe" (Held, 1891).

It is interesting to add that despite the overwhelming evidence in favor of the neuron theory, some distinguished morphologists like Held (1905) still believed, at the beginning of the century, that neurofibrils provided a continuity between neurons. It was only in 1934 that Hinsey, and later on Nonidez (1944), demonstrated that these structures were nonnervous.

The contiguity between neurons and its implications to nervous function received the attention of morphologists and physiologists. Sherrington (1897, 1906) describing the characteristics of the reflex arc assumed that the same specific properties of cell membranes that separated two apposing neurons (called a synapse) were responsible for the one-way conduction in the reflex arc and explained the delay of impulse conduction found in addition to the conduction time typical of the nervous pathway.

The mechanism involved in the conduction of the electrical impulse across the "synapse" was a controversial issue for many years.

Du Bois-Reymond (1877) was probably the first to propose that the junctional transmission could be either electrical or chemical. In 1904, Elliot suggested that the influence of sympathetic nerves on smooth muscle was related to the release of adrenaline at the junctional zones. Certainly, the observations of Loewi (1933, 1945) showing that the vagus nerve inhibits the heart by means of the release of acetylcholine at the nerve endings, represent the most decisive evidence in favor of the chemical theory of synaptic transmission.

At this time, some electrophysiologists, still reluctant to accept this view, were finally convinced when Fatt and Katz (1950, 1952) demonstrated, with the use of intracellular microelectrodes, that the end-plate potential was due to the removal of charge from the postsynaptic membrane and that this charge was a thousand times larger than any removal that can be produced by presynaptic action currents *per se*.

As emphasized by Katz (1966) the impedance of the nerve ending and skeletal muscle fiber are mismatched, supporting the view that the electrical mechanism of synaptic transmission is not correct.

At the moment when the chemical theory of synaptic transmission was unquestionably accepted, a pure electrical excitatory synapse was described by Furshpan and Potter (1959) in crayfish abdominal nerve cord (see Fig. 1).

Chemical synapses are of two types—excitatory and inhibitory—and the inhibitory process can be localized in the presynaptic or postsynaptic areas. The events that occur at the chemical synapse constitute a chain that is characterized by: (1) depolarization of the presynaptic ending and release of the transmitter; (2) diffusion of the transmitter through the synaptic cleft; (3) interaction of the transmitter with specific receptors located at the surface of the postsynaptic membrane; (4) change in permeability of the postsynaptic membrane and production of a postsynaptic potential which evokes an action potential if the postsynaptic potential reaches threshold; (5) removal of the transmitter from the synaptic region by diffusion, enzymatic breakdown, or uptake.

In the vertebrate neuromuscular junction (an example of excitatory synapse) the excitatory postsynaptic potential (EPP) recorded with intracellular microelectrodes at different points along a curarized skeletal muscle fiber, is reduced in amplitude at zones distant from the end plate. This decrement of the EPP indicates that the latter is a localized event at the motor end plate that propagates electronically along the muscle fiber. The interaction of the transmitter with the receptor opens ionic channels in the postsynaptic membrane and enhances the membrane conductance. If the postsynaptic membrane is only permeable to Na^+

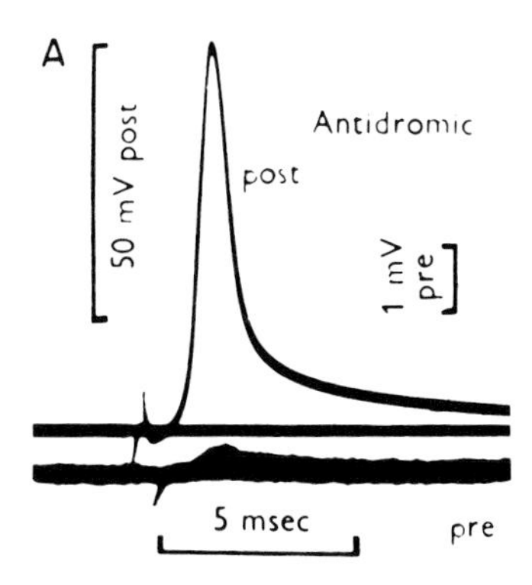

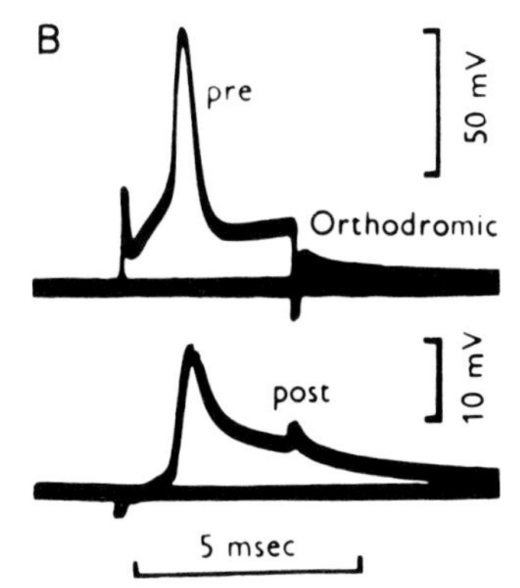

FIGURE 1. Transmission of electrical impulses through an electrotonic synapse of the crayfish. (A) Antidromic impulse in the postsynaptic motor axon produces only a small change in potential in the presynaptic fiber. (B) A directly evoked impulse in the presynaptic fiber causes a large PSP in the motor axon. From Furshpan and Potter (1969) with permission.

ions, the membrane potential will be displayed in the direction of the sodium equilibrium potential which is given by the following equation:

$$E_{Na} = \frac{RT}{F} \ln \frac{[Na]_o}{[Na]_i}$$

when R, T, and F are the gas constant, absolute temperature, and Faraday's constant, respectively. $[Na]_o$ and $[Na]_i$ are the extracellular and intracellular sodium concentration. Usually, the permeability to K^+ ions is also increased by the interaction of the transmitter with the receptor located at the end plate and the current flowing through the membrane (I_S) is equal to $I_{Na} + I_K$. The increased permeability to K^+ prevents the end-plate potential from moving beyond the zero potential.

The release of acetylcholine, which is the neurotransmitter at the frog or mammalian end plate, represents a "short circuit" placed across the membrane depolarizing it toward a value which corresponds to the diffusion potential one

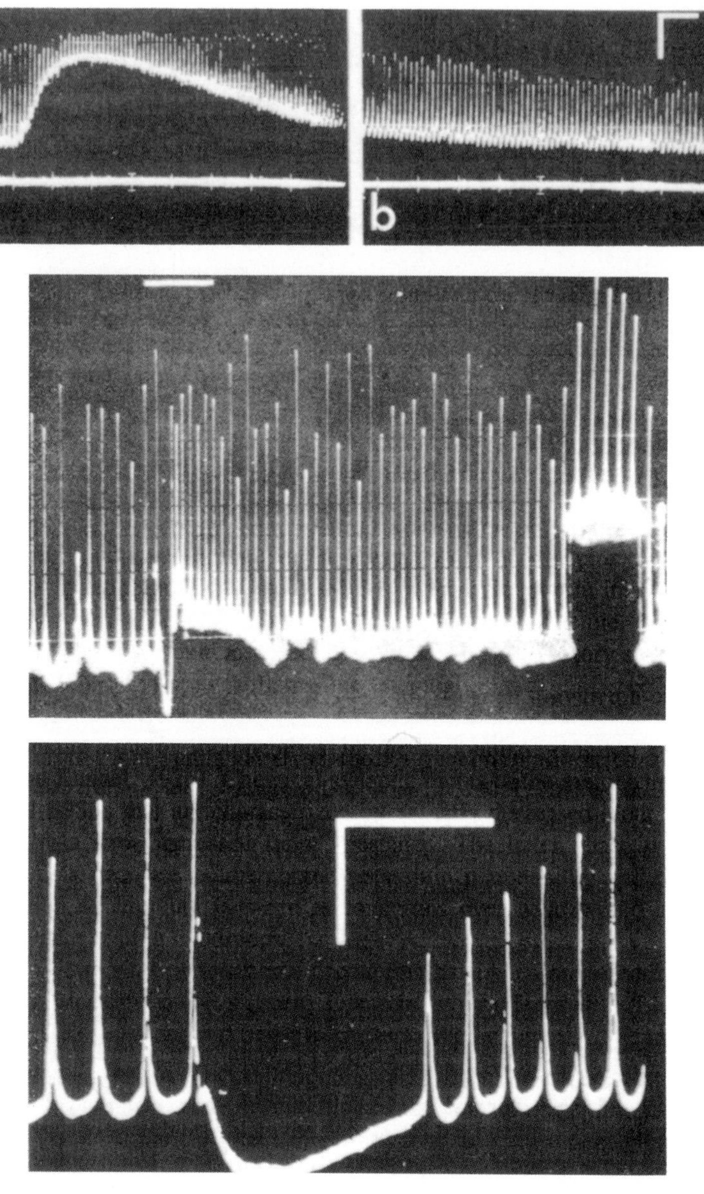

FIGURE 2. (Top) Records showing phasic depolarizations produced by electrophoretic application of ACh to the nerve cord region of an *Ascaris* muscle preparation. Two seconds. From del Castillo *et al.* (1964) with permission. (Center and bottom) Spontaneous electrical activity recorded from a somatic muscle cell of *Ascaris*. (Center) Stimulation of excitatory nerve fibers; (bottom) inhibition of spontaneous activity elicited by stimulation of inhibitory nerve fibers. From De Mello and Maldonado (1985) with permission.

might obtain if the membrane suffered a microlesion (see Katz, 1966). In a more primitive excitatory synapse like that located at the neuromuscular junction of *Ascaris lumbricoides,* the endogenous release of ACh or its electrophoretic application to the surface cell membrane also depolarizes the muscle cell (del Castillo *et al.*, 1964; see Fig. 2).

There is now general agreement that the synaptic excitation of all nerve cells produces a depolarization of the postsynaptic membrane.

The inhibitory postsynaptic potential (IPSP) (Fatt and Katz, 1953) is due to inhibitory postsynaptic current which flows outwardly when the IPSP is hyperpolarizing and inwardly when it is depolarizing. In crustacean neuromuscular junction (Otsuka, 1972) as well as in the somatic musculature of *Ascaris* (del Castillo *et al.*, 1964), GABA has been found to be the inhibitory transmitter (see Fig. 3).

In many tissues the release of the inhibitory neurotransmitter results in an increase in membrane permeability to K^+ or Cl^- ions or both. In cardiac pacemaker cells, for instance, the release of ACh by vagus nerve endings increases the membrane potential through an increase in K^+ permeability. Burgen and Terroux (1953) as well as Trautwein and Dudel (1958) found that the reversal potential for the action of ACh on the dog atrium fits well with the potassium equilibrium potential $\{E_K = (RT/F) \ln [K]_o/[K]_i\}$. Moreover, the rate of loss of ^{42}K in the frog sinus venosus is increased by nerve stimulation (Harris and Hutter, 1956). The increase in potassium conductance of cardiac pacemaker cells produced by ACh seems to be the major cause of cell decoupling seen with the neurotransmitter (see De Mello, 1980).

In crustacean neuromuscular junction, however, the postsynaptic inhibition is interpreted as due to a specific increase of Cl^- conductance (Boistel and Fatt, 1958). In the giant somatic muscle cells of *Ascaris,* GABA, which mimicks the stimulation of inhibitory nerves (see Fig. 3), enhances the membrane potential

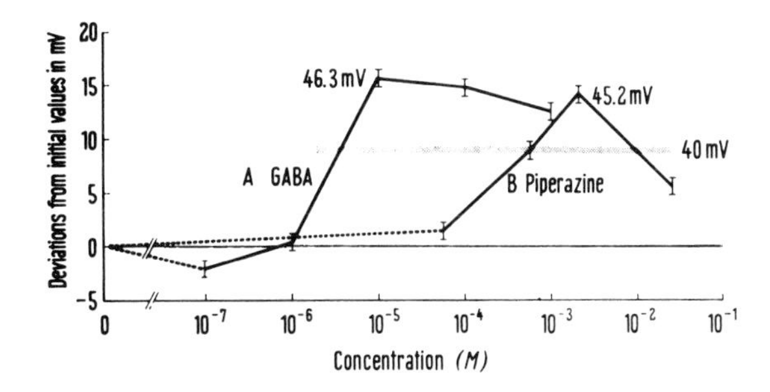

FIGURE 3. Graph showing the effect of GABA and piperazine on the membrane potential of *Ascaris* muscle cells. Ordinate: average membrane potential in millivolts, expressed as the difference from the initial values in absence of drugs. Abscissa: concentration of the drugs. From del Castillo *et al.* (1964) with permission.

and abolishes the spike activity by increasing the Cl^- conductance of cell membranes (del Castillo *et al.*, 1964). Indeed, when GABA was administered to somatic muscle cells of *Ascaris* exposed to Cl^--free solutions, no change in membrane potential was produced (del Castillo *et al.*, 1964). The specific change in Cl^- conductance produced by GABA is represented in the equivalent circuit for inhibitory responses in *Ascaris* shown in Fig. 4 (De Mello and Maldonado, 1985). An increase in Cl^- conductance is also involved in the process of synaptic inhibition in crayfish giant motor axons (Ochi, 1969), crayfish stretch receptor neurons (Ozawa and Tsuda, 1973), as well as in the Mauthner cells (Furukawa and Furshpan, 1963).

The communication between nerve and muscle can also be suppressed by "presynaptic inhibition." In 1961, Dudel and Kuffler discovered that the stimulation of inhibitory nerve fibers did not alter the ionic permeability of the cell membrane in crustacean muscle but interfered with the release of the excitatory neurotransmitter (see Fig. 5). This type of presynaptic interaction, which occurs through the establishment of a synaptic contact between the excitatory and inhibitory presynaptic nerve endings, exists in the autonomic nervous system involved in the control of caliber of peripheral vessels in mammals and also in the central nervous system (see Eccles, 1961).

Evidence is available that in the peripheral as well as in the central nervous system, the release of norepinephrine is modulated through presynaptic receptors. This important regulatory process has also been described for dopaminergic, GABAergic, and cholinergic neurons. In the release of norepinephrine, angiotensin (Ackerly *et al.*, 1976) and β agonists (Langer, 1976) enhance release

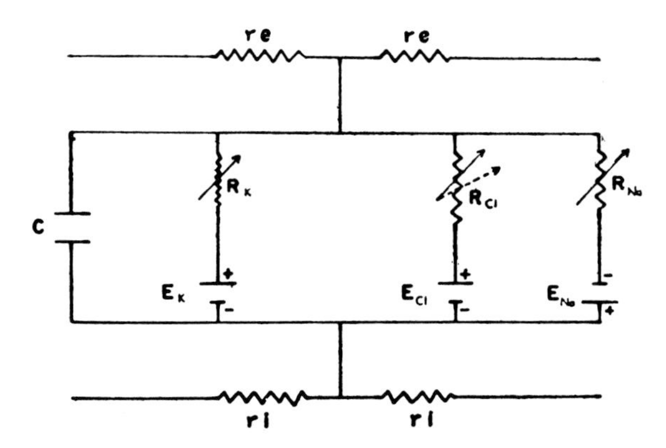

FIGURE 4. Schematic diagram of the nonjunctional membrane of the somatic muscle cell of *Ascaris lumbricoides* including the electromotive forces for potassium (E_K), sodium (E_{Na}), and chloride (E_{Cl}), as well as corresponding resistances for each channel. Change in R_{Cl} represents the change in Cl^- conductance produced by GABA; r_i, internal resistance. From De Mello and Maldonado (1985) with permission.

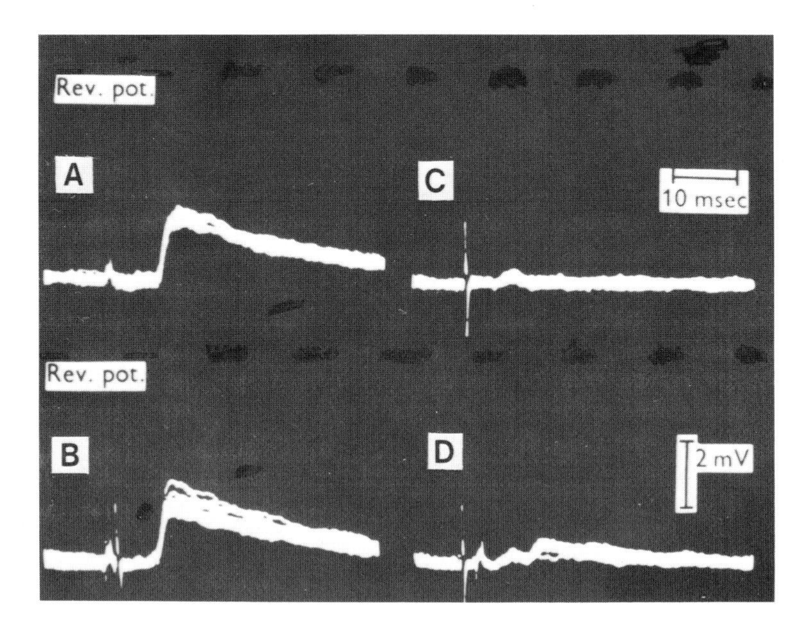

FIGURE 5. Intracellular records from crayfish neuromuscular junction. (A) EJP caused by stimulation of excitatory nerve. (C) IJP produced by inhibitory nerve stimulation. Rev. pot., reversal potential for IJP. (B) IJP are superimposed, producing small changes in amplitude of EJP. (D) EJP is greatly suppressed when inhibitory stimulation precedes excitatory stimulation. From Dudel and Kuffler (1961) with permission.

of the transmitter while prostaglandins (Starke, 1977) and cholinergic muscarinic agonists (Löffelholz and Muscholl, 1970) inhibit its release.

The transmission of the impulse in chemical synapses requires the synthesis and storage of chemical transmitters at the nerve endings. In cholinergic neurons the synthesis of the transmitter seems to occur in the neuroplasm and then the molecule is transferred into vesicles (Potter, 1970). Estimates of the average content of ACh in these vesicles vary between 2000 and 50,000 molecules (Katz, 1969; Potter, 1970; Whittaker, 1965).

Depolarization is required for the secretion of the neurotransmitter and Ca^{2+} is essential for the excitation–secretion coupling (Douglas and Poisner, 1964). Katz and Miledi (1967) investigated the influence of the extracellular Ca^{2+} concentration on the release of ACh. In these studies, they measured the membrane potential at the end-plate zone of the frog and then depolarized the presynaptic nerve terminal by applying current pulses from one barrel of an external double-barreled microelectrode. This barrel was filled with NaCl while the second barrel contained $CaCl_2$. The results showed that Ca^{2+} released during the synaptic delay had little effect but a maximal effect was found when the ion was released 10 msec before depolarization. Evidence that Ca^{2+} goes into the nerve terminal during depolarization was also obtained in the squid giant synapse by injecting aequorin into the presynaptic terminal and observing the fluorescence of the nerve ending during electrical stimulation (Llinas *et al.*, 1972). The results clearly indicated an increase in fluorescence during the stimulation of the presynaptic terminal.

Although the precise mechanism by which Ca^{2+} ions trigger the secretion process is not known, it is possible that the ion serves to bridge negatively charged groups at the surface of the vesicle and the inner surface of the nerve terminal.

For a complete review of the electrophysiology of chemical synapses, the reader must consult Eccles (1964) or Katz (1966).

2. ON THE RELEASE OF NEUROTRANSMITTERS

The release of catecholamines from synaptic vesicles involves the activation of a Mg^{2+}-activated ATPase located in the vesicle membrane (see Poisner and Trifaro, 1967).

In the hypothalamo-neurohypophyseal tract, evidence is available that vasopressin and neurophysin are synthesized in the perikarya as well as in the hypothalamus and are transported axoplasmically to the nerve endings in the posterior pituitary (see Alvarez-Buylla *et al.*, 1970).

In mammals the release of catecholamines represents an important homeostatic control of many organs. It is well known that the interaction of epinephrine or norepinephrine with adrenergic receptors located at the postsynaptic membrane regulates the heart rate, the strength of heart contractions, the hepatic metabolism, the arterial blood pressure, and the contractility of gastrointestinal smooth muscle.

In cardiac tissues the activation of adenylate cyclase by drugs that stimulate β-adrenergic receptors (e.g., epinephrine) causes a rise in the intracellular levels of cAMP which initiates a chain of events leading to increased rate and strength of heart contractions, and breakdown of glycogen to glucose-1-phosphate through the activation of phosphorylase (Robison *et al.*, 1965). An increase in junctional conductance produced by cAMP is probably involved in the enhancement of conduction velocity caused by epinephrine in cardiac muscle (see Chapter 2). The cAMP hypothesis (De Mello, 1983) proposes that hormones able to increase the intracellular concentration of cAMP control the communication between cells through a modulation of junctional conductance.

The classical view that neurons release just one neurotransmitter seems to be the exception. As shown in Fig. 6, cotransmitter neurons have been identified in vertebrates and invertebrates (O'Donohue *et al.*, 1985). Although we do not

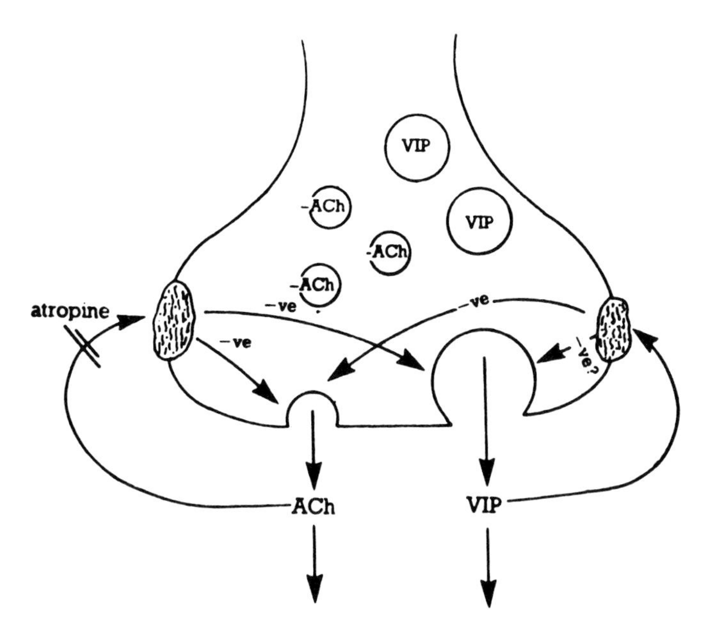

FIGURE 6. Acetylcholine–vasoactive intestinal polypeptide (VIP) cross-regulation of release in the rat submandibular gland and cerebral cortex. Courtesy of Dr. T. Bartfai.

know the factors that ultimately determine the type of neurotransmitter released by a certain neuron, it has been suggested that chemical factors from the cell environment are probably involved (Patterson, 1978).

The appearance of cotransmitters is not necessarily simultaneous. For example, in the locus coeruleus, norepinephrine and neuropeptide Y (NPY) are present but norepinephrine appears earlier during embryonic development than the peptide (Foster and Schultzberg, 1984). In vas deferens, NPY inhibits the release of norepinephrine through a negative feedback mechanism (O'Donohue *et al.*, 1985).

In some structures like the submaxillary gland, ACh and VIP (vasoactive intestinal polypeptide) are present at the nerve endings of the chorda-lingual nerve. The ratio of ACh/VIP released varies with the frequency and duration of stimulation of the nerve (Lundberg, 1981). In some cases, two neurotransmitters (Met-enkephalin and opioid peptides) bind to the same receptor which in other situations the transmitter binds to separate receptors (see O'Donohue *et al.*, 1985). It is interesting to add that cotransmitters can interact at the level of the receptor or ionic channels. VIP, for instance, seems to increase the affinity of the ACh receptor for ACh in the salivary gland (Lundberg *et al.*, 1982).

3. CELL-TO-CELL COMMUNICATION THROUGH THE EXTRACELLULAR SPACE

3.1. Electrical Interaction between Cells

The question whether the simple apposition of two excitable cells makes possible the propagation of the electrical impulse from cell to cell was a central issue in the discussion of the mechanisms of synaptic transmission.

It is well known that to generate an action potential, just ahead of the impulse enough current must flow into the membrane to reduce the resting potential of 15–20 mV. With the generation of the action potential, the change in membrane potential is markedly amplified, providing current to excite the area ahead of the impulse.

Is the electrical impulse able to jump a small gap (150–200 Å) between two apposing nerve fibers of small diameter and excite the neighboring fiber? Assuming that the membrane resistance at the area of cell contact is within the normal range, the possibility that the impulse propagates across the gap is meager because the attenuation factor, which is proportional to the inverse square of the fiber size, is about 1/10,000 for fibers of 5-μm diameter (see Katz, 1966). For giant nerve fibers, however, the attenuation factor is appreciably smaller and the possibility of impulse propagation exists.

Mann *et al.* (1981) presented a model for electrical transmission of excitation between heart cells without the requirement of low-resistance junctions. A major assumption of this model is that the pre- and postjunctional membranes at the area of cell contact are excitable. The transmission of the electrical impulse seen at electrotonic synapses implies that the two appositional membranes are inexcitable (see Bennett, 1973).

If the membrane resistance is low in the area of cell contact, then an electrical coupling could occur between the cells even when separated by a gap of 200 Å (see Bennett, 1973). Certainly, the evidence that gap junctions are required for the electrical coupling in many systems is supported by electron microscopic studies as well as by the finding that fluorescent dyes diffuse from cell to cell through intercellular channels (see Bennett, 1973; De Mello *et al.*, 1983).

Evidence is available that large external electrical fields can influence the excitability of nerve cells. A discharging axon, for instance, can alter the excitability of another axon located nearby (Katz and Schmitt, 1940). This type of cell interaction is particularly important if the membrane potential of the excitable cell is near threshold. In the tonically active crayfish stretch receptor, for instance, the rate of firing is greatly altered by an external field of a few tenths of a millivolt (Terzuolo and Bullock, 1956). Electrical interaction through the extracellular space is more likely if the cleft between cells is closed at the edge of the apposition as shown by Kusano and La Vail (1971) in the nerve fibers of the shrimp. This happens because the coupling resistance is then determined by the resistance of the membranes around the junction. It might be of interest to investigate if a pathological process, closing points of the apposition between cells, would facilitate electrical interaction between mammalian cells.

The action currents of a nerve fiber can reduce the excitability of a neighboring fiber (Katz and Schmitt, 1940) if the electrical resistance around the fibers is raised. This phenomenon is due to the fact that the large potential changes elicited outside the fiber can enter the adjacent resting cell. Furshpan and Furukawa (1962) showed that in the Mauthner cell there is a zone called axon cap which presents a high extracellular resistance, so that when the electrical impulses reach this area the potential outside the Mauthner cells is appreciably increased (about 18 mV) with the consequent entrance of current in the postsynaptic membrane and the hyperpolarization of the area normally excitable causes inhibition.

3.2. Nerve Growth Factor—An Example of Chemical Communication between Cells

The interest in the possible existence of a nerve growth factor started with some observations of Bueker in 1948. This author implanted tumors into chick embryos in order to study how rapid-growing neoplastic tissue influences the implantation of supernumerary limbs. His results indicated that as the tumor grew it was invaded by nerve fibers and that the dorsal root ganglia at the level of the tumor were greatly enlarged.

In the 1951 Levi-Montalcini and Hamburger published several studies in which the previous findings of Bueker were confirmed. These authors not only found that sympathetic ganglia in the vicinity of the tumor were enlarged, but concluded that a substance secreted by the tumor—called "nerve growth factor" (NGF)—was responsible. Ultimately, NGF purified from mouse submaxillary gland proved to be a monomer consisting of 118 amino acids with a molecular weight of 13,259 (Server and Shooter, 1977).

NGF does not affect all types of nerve cells; on the contrary, its effect is clearly seen in the sympathetic chain ganglia and sensory ganglia (Levi-Montalcini and Angeletti, 1968; Mobley *et al.*, 1977) but nonneural cells are also influenced. Treatment of perinatal rats with NGF, for instance, changes the chromaffin cells of the adrenal medulla into ordinary sympathetic neurons (see Aloe and Levi-Montalcini, 1979).

NGF is found in the convoluted tubules of the submaxillary gland of adult male mice (Levi-Montalcini and Cohen, 1960; Wallace and Partlow, 1976) but there is no explanation why the substance is present in large amounts in this tissue. NGF receptors have been found in the membrane of cells sensitive to the substance (Greene and Shooter, 1980). Evidence is available that NGF after interaction with specific receptors is internalized and transported back to the cell body (Bradshaw, 1978; Greene and Shooter, 1980; Thoenen *et al.*, 1979). Although it is known that NGF stimulates RNA synthesis, increases protein synthesis, and alters ionic fluxes (Greene and Shooter, 1980; Yanker and Shooter, 1982), it is not clear whether these effects are necessary for the trophic actions of the substance.

Evidence is available, however, that NGF is not the only substance involved in the trophic effects in nervous tissue. Local maintenance of neurites and cell death are some of several processes that are determined by trophic factors different from NGF.

3.3. A Trophic Factor Controls Skeletal Muscle Properties

There is substantial evidence that a trophic factor has an important role in determining the properties of muscle fibers. Experiments by Ginetsinskii and Shamarina (1942) and by Axelsson and Thesleff (1959) showed that mammalian muscle fibers are more responsive to ACh following denervation.

It is well known that in innervated muscle the sensitivity of the muscle fibers to the transmitter is limited to the subsynaptic area (see Fig. 7) (Fambrough, 1981). Usually, 3 days after denervation the whole muscle fiber becomes sensitive to ACh (Brookes and Hall, 1979; Fambrough, 1970; Sakmann, 1975). The longer the length of nerve left connected to the muscle, the more time it takes for denervation changes to occur (Luco and Eyzaguirre, 1955; Emmelin and Malm, 1965). These findings have been interpreted as follows: the longer stumps contain a greater amount of a possible trophic substance which maintains the normal properties of the muscle fibers for a longer period of time.

Classical studies of Miledi (1960) showed that in partially denervated skeletal muscle the area around the denervated end plate becomes supersensitive to ACh. Since these are presumably normal because they are still innervated by the other branch that was kept intact, the results indicate that nerve fibers supply a chemical agent to the muscle fibers that maintains their normal properties. An elegant way to demonstrate the presence of a trophic factor was to disrupt the microtubules and block the axoplasmic transport in motor nerves by applying colchicine to a segment of the nerve. This experimental procedure elicits denervation changes in muscle (see Hofmann and Thesleff, 1972; Albulquerque *et al.*, 1972). These observations support the view that the sensitivity of the skeletal muscle fibers to ACh is regulated by a chemical signal released by the nerve endings.

3.4. Skeletal Muscle Properties Are Dependent on the Motor Neuron Type

It is known that during the initial months of postnatal life the skeletal muscle of mammalians differentiates into "fast-twitch" and "slow-twitch" muscles

FIGURE 7. Autoradiographs of pairs of muscle fibers dissected from normal (A) and denervated (B) rat diaphragms after ACh receptors were labeled with $[^{125}I]$-α-bungarotoxin. Courtesy of Dr. Fambrough.

(see Denny-Brown, 1929; Buller *et al.*, 1960; Yellin, 1967; Jolesz and Sréter, 1981).

Strong evidence that motor nerve fibers determine the properties of muscle fibers is the observation that when slow-twitch and fast-twitch muscles are cross-innervated the fast-twitch muscles are slowed when reinnervated by nerve fibers that usually innervate slow-twitch muscles and vice versa (Fig. 8; Buller *et al.*,

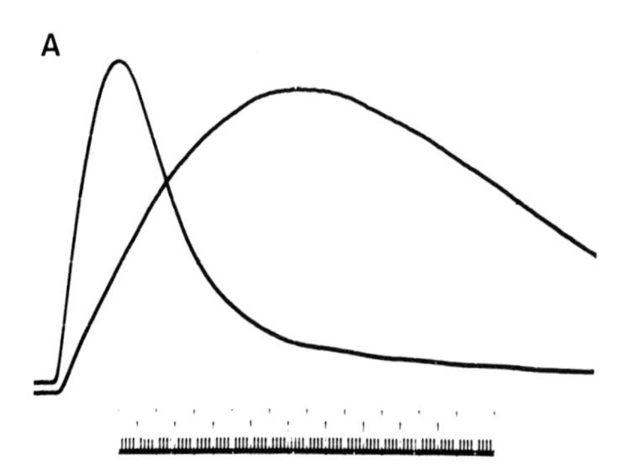

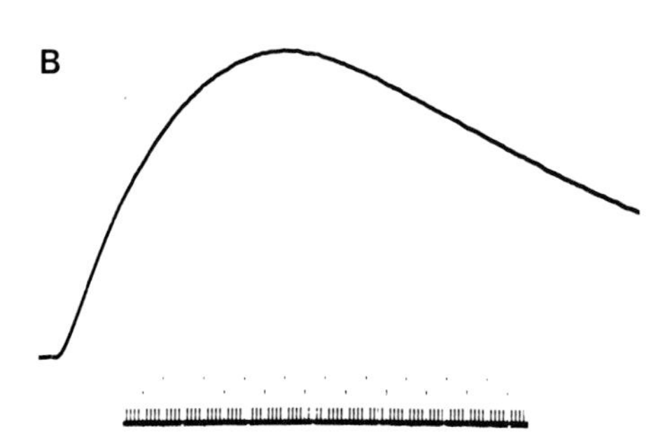

FIGURE 8. Isometric twitch contractions of left and right extensor digitorum longus (EDL) muscles (A) and right soleus muscle (B) recorded on identical time scales. The left EDL muscle (slower time course) has been stimulated at 10 Hz continuously for 20 weeks. Courtesy of Dr. S. Salmons.

1960; Close, 1965). Biochemical studies also indicated that the composition of muscle proteins is altered by changing the innervation (Guth and Watson, 1967; Close, 1972; Jolesz and Sréter, 1981).

3.5. Do Postsynaptic Cells Induce Changes in the Presynaptic Neuron?

An important type of cell communication is the ability of motor axons to recognize their former sites of termination on vertebrate skeletal muscle during the process of reinnervation. It is known that reinnervation of former end plate occurs with a high degree of precision (Gutmann and Young, 1944; Saito and Zachs, 1969; Bennett and Pettigrew, 1976). The phenomenon seems related to the presence in subsynaptic basal laminae of molecules able to induce the nerve endings to form synaptic contacts (Marshall *et al.*, 1977; Sanes *et al.*, 1978; Glickman and Sanes, 1983).

Evidence is available that when skeletal muscles are cut transversely, the fibers degenerate but the basal lamina remains intact (Sanes *et al.*, 1980). The regenerating nerves maintain their preference for old sites of innervation and establish contact with these old sites on the basal lamina as shown in Fig. 9.

Antibodies made against muscle bind specifically to antigens in the synpatic basal lamina (see Sanes and Hall, 1979).

These observations suggest that the basal lamina contains molecules that induce the nerve endings to establish synaptic contacts (see Burden *et al.*, 1979).

3.6. Neural Death during Development

During development some classes of neurons depend for survival on a protein NGF which is produced by target cells. Not only do the innervating nerve endings have specific receptors for the growth factor but they are able to transport the substance to the cell body.

A fascinating aspect of this type of intercellular communication is that in the absence of NGF the innervating neuron dies (Lamb, 1981). In some structures like the muscle of chick iris a wave of neuronal death is seen in the ciliary ganglion simultaneously with the establishment of a large number of synapses (Landmesser and Pilar, 1974; Narayanan and Narayanan, 1981). These findings suggest that synapse formation *per se* determines the death of innervating neurons.

Several observations support the view that the survival of innervating neurons is greatly dependent upon the function of the established synapse. It is known, for instance, that neuromuscular blockade avoids the death of motor neurons in chick spinal cord (Lamb, 1976; Pittman and Oppenheim, 1979; Hamburger, 1981).

The innervation seems then to control the release of a trophic factor by target cells.

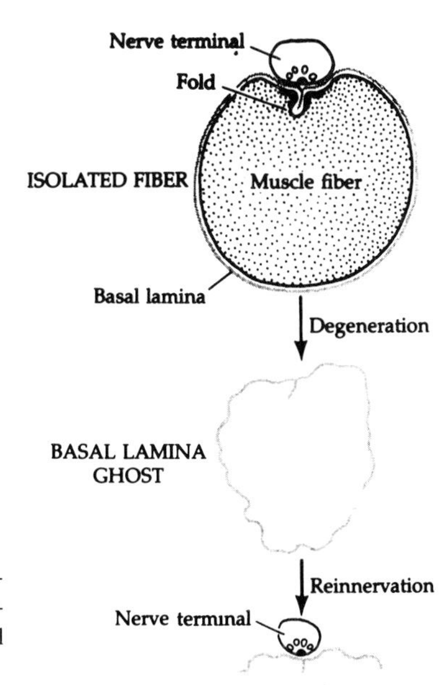

FIGURE 9. Denervation causes myofibrils to degenerate leaving a basal lamina "ghost." During reinnervation the regenerating axons "reinnervate" original synaptic sites. From Sanes (1983) with permission.

3.7. Cellular Slime Molds: Starvation Induces Communication

One of the strongest stimuli for cell communication is the need to share metabolites and nutrients.

An example of this situation is the aggregation of cells in a slime mold such as *Dictyostelium discoideum* upon exposure to chemotactic stimuli.

Isolated amoebae can be seen growing and dividing as long as bacteria,

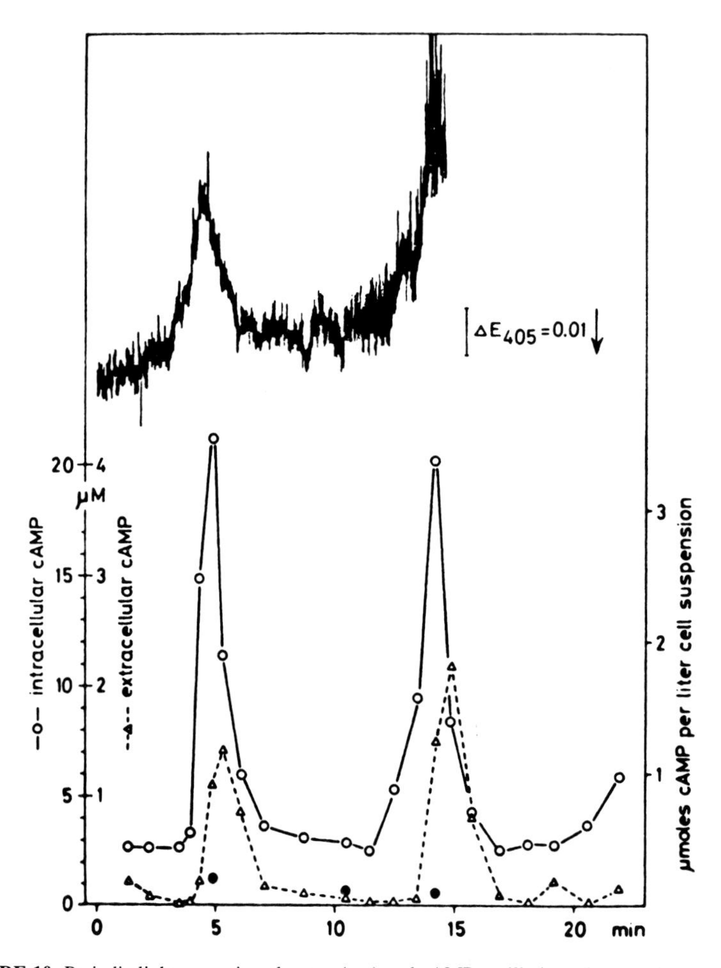

FIGURE 10. Periodic light scattering changes (top) and cAMP oscillations (bottom) in a suspension of *D. discoideum* cells. Left ordinate: intra- and extracellular concentrations of cAMP. Right ordinate: quantities of cAMP in intra- or extracellular compartment. From Gerisch and Wick (1975) with permission.

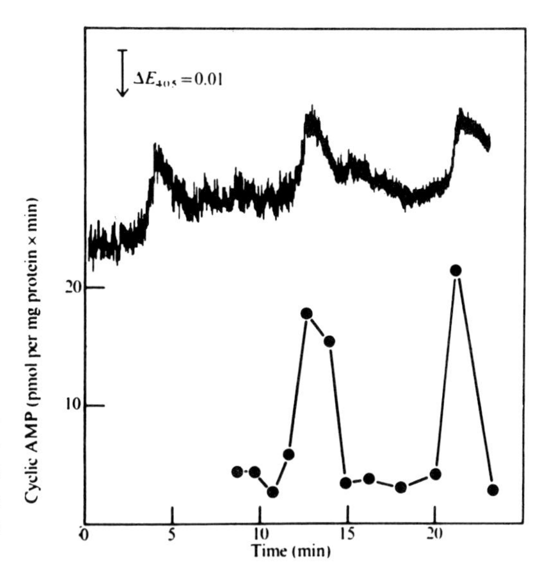

FIGURE 11. Oscillations of adenylate cyclase in *D. discoideum* probably explaining the periodic increase in concentration of cAMP. From Gerisch and Wick (1975) with permission.

present in the extracellular medium, can be ingested as food. At this stage, the cell is an independent unit of life with no apparent interest in interacting with other cells.

A decline in number of bacteria, however, leads to a process of starvation which induces the generation, in some amoebae, of rhythmic pulses of cAMP (Bonner, 1967). Although at the beginning the rate of cAMP pulses is quite low (1/10 min), gradually the rate increases to one every 2–3 min (see Gerisch and Hess, 1974).

Measurement of the variation of cAMP shows an increase in concentration of the nucleotide in the cytosol and 30–45 min later the concentration of cAMP in the extracellular fluid rises (see Fig. 10; Gerisch and Wick, 1975).

Concurrently with the generation of cAMP pulses, cAMP receptor proteins are synthesized on the cell surface of the amoeba population, making it possible for other cells to sense the chemical signal (Malchow and Gerisch, 1974; Green and Newell, 1975; Henderson, 1975). The cells receiving the cAMP pulses start moving toward the source of the nucleotide and during this process the moving cells become cAMP generators, thus attracting other cells. Aggregation territories are found controlled by centers of cAMP release (Gerisch and Hess, 1974). Oscillations of adenylate cyclase activity might explain the periodic rise in concentration of cAMP (see Figs. 11 and 12).

The addition of pulses of cAMP to the medium elicits propagated waves (Konijn *et al.*, 1967; Bonner *et al.*, 1969) which consist of a slow component resembling the spikes formed during spontaneous oscillations and a fast compo-

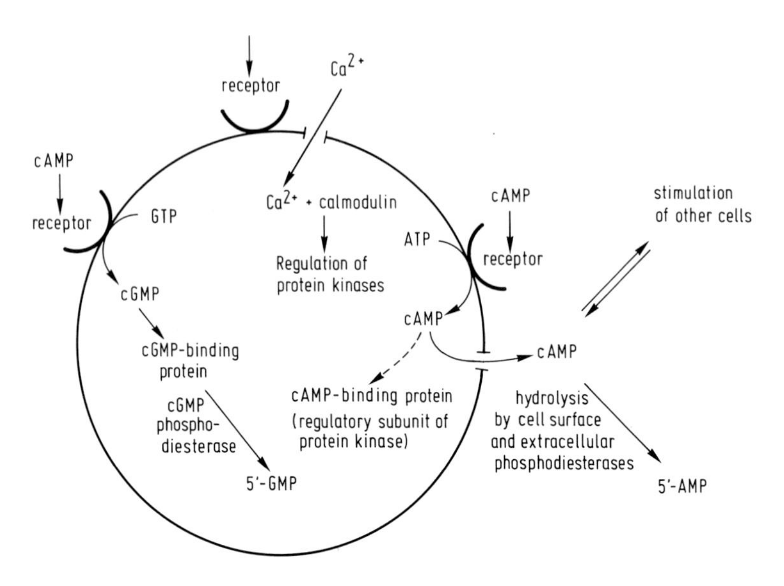

FIGURE 12. Transduction of cAMP signals in *D. discoideum.* Shown are the activation of guanylate cyclase, a net influx of Ca^{2+}, and the activation of adenylate cyclase. From Gerisch (1986) with permission.

nent with characteristics of the normal response to cAMP (Gerisch and Hess, 1974). It is important to add that the oscillation frequency of the system is insensitive to continuous infusion of cAMP, indicating that *D. discoideum* responds to rapid changes in concentration in time, but not to concentration *per se* (see Gerisch and Hess, 1974).

The spread of cAMP, is, however, limited by the presence of phosphodiesterase that is located at the cell membrane surface (Malchow and Gerisch, 1974) and able to hydrolyze 10^8 extracellular cAMP molecules per cell per min.

It is not known how the amoebae recognize the cAMP signal. A possible explanation is that the cells are able to sense a concentration gradient between the site of attachment of cAMP and the other extreme of the cell (Bonner, 1967) or that the amoebae can sense the change in concentration of cAMP as a function of time (Gerisch *et al.,* 1975c).

The process culminates with the aggregation of the amoebae and the formation of a multicellular organism. The aggregation process is achieved through glycoproteins and their receptors located at the cell membrane surface—a process that is triggered by pulses of cAMP (Gerisch *et al.,* 1975b).

ACKNOWLEDGMENTS. Supported by Grants HL-34353, HL-34148, and RR-08102.

4. REFERENCES

Ackerly, J., Blumberg, A., and Peach, M., 1976, Angiotensin interactions with myocardial sympathetic neurons: Enhanced release of dopamine-β-hydroxylase during nerve stimulation, *Proc. Soc. Exp. Biol. Med.* **151**:650–653.

Albulquerque, E. X., Warnick, J. E., Tasse, J. R., and Sansone, F. M., 1972, Effects of vinblastine and colchicine on neural regulation of the fast and slow skeletal muscle, *Exp. Neurol.* **37**:607–634.

Aloe, L., and Levi-Montalcini, R., 1979, Nerve growth factor-induced transformation of immature chromaffin cells in vivo into sympathetic neurons: Effect of antiserum to nerve growth factor, *Proc. Natl. Acad. Sci. USA* **76**:1246–1250.

Alvarez-Buylla, R., Livett, B. G., Uttenthal, O. O., Milton, S. H., and Hope, D. B., 1970, Immunohistochemical evidence for the transport of neurophysin in neurosecretory neurons of the dog, *Acta Physiol. Scand.* **55**43–552.

Axelsson, J., and Thesleff, S., 1959, A study of supersensitivity in denervated mammalian skeletal muscle, *J. Physiol. (London)* **147**:178–193.

Bennett, M. R., and Pettigrew, A. G., 1976, The formation of neuromuscular synapses, *Cold Spring Harbor Symp. Quant. Biol.* **40**:409–424.

Bennett, M. V. L., 1973, Function of electrotonic junctions in embryonic and adult tissues, *Fed. Proc.* **32**:65–75.

Boistel, J.. and Fatt, P., 1958, Membrane permeability change during inhibitory transmitter action in crustacean muscle, *J. Physiol. (London)* **144**:176–191.

Bonner, J. T., 1967, *The Cellular Slime Moulds*, 2nd ed., Princeton University Press, Princeton, N.J.

Bonner, J. T., Barkley, D. S., Hall, E. M., Konijn, T. M., Mason, J. W., O'Keefe, G., and Wolfe, P. B., 1969, Acrasin, acrasinase and the sensitivity to acrasin in *Dictyostelium discoideum*, *Dev. Biol.* **20**:72–87.

Bradshaw, R. A., 1978, Nerve growth factor, *Annu. Rev. Biochem.* **47**:191–216.

Brookes, J. P., and Hall, Z. W., 1979, Acetylcholine receptors in normal and denervated rat diaphragm muscle. II. Comparison of junctional and extrajunctional receptors, *Biochemistry* **20**:2100–2106.

Bueker, E. D., 1948, Implantation of tumors in the hind limb of the embryonic chick and the developmental response of the lumbosacral nervous system, *Anat. Rec.* **102**:369–390.

Buller, A. J., Eccles, J. C., and Eccles, R. M., 1960, Interactions between motoneurons and muscles in respect of the characteristic speeds of their responses, *J. Physiol. (London)* **150**:417–439.

Burden, S. J., Sargent, P. B., and McMahan, U. J., 1979, Acetylcholine receptors in regenerating muscle accumulate at original synaptic sites in the absence of nerve, *J. Cell Biol.* **82**:412–425.

Burgen, A. S., and Terroux, K. G., 1953, On the negative inotropic effect in the cat's auricle, *J. Physiol. (London)* **120**:449–463.

Close, R., 1965, Effects of cross-union of motor nerves to fast and slow skeletal muscles, *Nature* **206**:831–832.

Close, R., 1972, Dynamic properties of mammalian skeletal muscle, *Physiol. Rev.* **52**:129–197.

del Castillo, J., De Mello, W. C., and Morales, T., 1964, Inhibitory action of γ-aminobutyric acid (GABA) on *Ascaris* muscle, *Experientia* **20**:141–145.

De Mello, W. C., 1980, Intercellular communication and junctional permeability, in: *Membrane Structure and Function*, Vol. 3 (E. E. Bittar, ed.), pp. 128–164, Wiley, New York.

De Mello, W. C., 1983, The role of cAMP and Ca on the modulation of junctional conductance: An integrated hypothesis, *Cell Biol. Int. Rep.* **7**:1033–1040.

De Mello, W. C., and Maldonado, H., 1985, Synaptic inhibition and cell communication: Impairment of cell-to-cell coupling pronounced by γ-aminobutyric acid (GABA) in the synaptic musculature of *Ascaris lumbricoides*, *Cell Biol. Int. Rep.* **9**:813.

De Mello, W. C., Gonzalez Castillo, M., and van Loon, P., 1983, Intercellular diffusion of Lucifer Yellow CH in mammalian cardiac fibers, *J. Mol. Cell. Cardiol.* **15:**637–643.

Denny-Brown, D. E., 1929, The histological features of stripped muscle in relation to its functional activity, *Proc. R. Soc. London Ser. B* **104:**371–411.

Douglas, W. W., and Poisner, A. M., 1964, Calcium movements in the neurohypophysis of the rat and its relation to the release of vasopressin, *J. Physiol. (London)* **172:**19–30.

Du Bois-Reymond, E., 1877, Muskel-und Nervenphysisik, *Gesammelte Abhandl. d. Allegem.* **2:**700–785.

Dudel, J., and Kuffler, S. W., 1961, Presynaptic inhibition at the crayfish neuromuscular junction, *J. Physiol. (London)* **155:**543–562.

Eccles, J. C., 1964, *The Physiology of Synapses,* Springer-Verlag, Berlin.

Elliot, T. R., 1904, On the action of adrenalin, *J. Physiol. (London)* **31:**XXP.

Emmelin, N., and Malm, L., 1965, Development of supersensitivity as dependent on the length of degenerating nerve fibres, *Am J. Exp. Physiol.* **50:**142–145.

Fambrough, D. M., 1970, Acetylcholine sensitivity of muscle fiber membranes: Mechanism of regulation by motoneurons, *Science* **168:**372–373.

Fambrough, D. M., 1981, Denervation: Cholinergic receptors of skeletal muscle, in: *Receptors and Recognition Series,* Vol. 13 (R. J. Lefkowitz, ed.), pp. 125–142, Chapman & Hall, London.

Fatt, P., and Katz, B., 1950, Membrane potential changes at the motor end-plate, *J. Physiol. (London)* **111:**46–47.

Fatt, P., and Katz, B., 1952, Spontaneous subthreshold activity at motor nerve endings, *J. Physiol. (London)* **117:**109–128.

Fatt, P., and Katz, B., 1953, The effect of inhibitory nerve impulses on a crustacean muscle fibre, *J. Physiol. (London)* **121:**374–389.

Forel, A., 1887, Einige hirnanatomische Betrachtungen und Ergebnissae, *Arch. Psychiatr. Nervenheilk.* **18:**15–38.

Foster, G. A., and Schultzberg, M., 1984, Immunohistochemical analysis of the ontogeny of neuropeptide; immunoreactive neurons in foetal rat brain, *Int. J. Dev. Neurosci.* **2:**387–407.

Furshpan, E. J., and Furukawa, T., 1962, Intracellular and extracellular responses of several regions of the Mauthner cell of the goldfish, *J. Neurophysiol.* **25:**732–771.

Furshpan, E. J., and Potter, D. D., 1959, Transmission at the giant motor synapses of the crayfish, *J. Physiol. (London)* **145:**289–325.

Furukawa, T., and Furshpan, E. J., 1963, Two inhibitory mechanisms in the Mauthner neurons of goldfish, *J. Neurophysiol.* **26:**140–176.

Gerisch, G., 1986, *Dictyostelium discoideum:* A eukaryotic microorganism that develops by cell aggregation from a unicellular to a multicellular stage, in: *Cellular and Molecular Aspects of Developmental Biology* (M. Fougereau and R. Stosa, eds.), Elsevier Science Publishers, Amsterdam.

Gerisch, G., and Hess, B., 1974, Cyclic-AMP-controlled oscillations in suspended *Dictyostelium* cells: Their relation to morphogenetic cell interactions, *Proc. Natl. Acad. Sci. USA* **71:**2118–2122.

Gerisch, G., and Wick, U., 1975, Intracellular oscillations and release of cyclic AMP from *Dictyostelium* cells, *Biochem. Biophys. Res. Commun.* **65:**364–370.

Gerisch, G., Fromm, H., Huesgen, A., and Wick, U., 1975a, Control of cell contact sites by cyclic AMP pulses in differentiating *Dictyostelium* cells, *Nature* **255:**547.

Gerisch, G., Hulser, D., Malchow, D., and Wick, U., 1975b, Cell communication by periodic cyclic-AMP pulses, *Philos. Trans. R. Soc. London Ser. B* **272:**181–192.

Gerlach, J., 1871, Von dem Ruckenmarke, in: *Handbuch der Lehre von den Geweben,* Vol. 2.

Ginetsinskii, A. G., and Shamarina, N. M., 1942, The tonomotor phenomenon in denervated muscle [Department of Scientific and Industrial Research, Translation, RTS 1710], *Osp. Sourem. Biol.* **15:**283–294.

Glicksman, M. A., and Sanes, J. R., 1983, Differentiation of motor nerve terminals formed in the absence of muscle fibres, *J. Neurophysiol.* **12:**661–671.

Green, A. A., and Newell, P. C., 1975, Evidence for the existence of two types of cAMP binding sites in aggregating cells of *Dictyostelium discoideum, Cell* **6:**129–136.

Greene, L. A., and Shooter, E. M., 1980, The nerve growth factor: Biochemistry, synthesis and mechanism of action, *Annu. Rev. Neurosci.* **3:**353–402.

Guth, L., and Watson, P. K., 1967, The influence of innervation on the soluble proteins of slow and fast muscles of the rat, *Exp. Neurol.* **17:**107–117.

Gutmann, E., and Young, J. Z., 1944, Reinnervation of muscle after various periods of atrophy, *J. Anat.* **78:**15–43.

Hamburger, V., 1981, Historical landmarks in neurogenesis, *Trends Neurosci.* **4:**151–155.

Harris, E. J., and Hutter, D. F., 1956, The action of acetylcholine on the movements of potassium ions in the sinus venosus of the heart, *J. Physiol. (London)* **133:**58–59P.

Held, H., 1891, Die centralen Bahnen des Nervous acusticus bei der Katze, *Arch. Anat. Physiol.* **5:**270–291.

Held, H., 1905, Zur Kenntniss einer neurofibrillaren. Continuitat im Centralnervensystem der Wirbelthiere, *Arch. Anat. Physiol.* **43:**55–78.

Henderson, E. J., 1975, The cyclic adenosine 3′-5′-monophosphate receptor of *Dictyostelium discoideum:* Binding characteristics of aggregation-competent cells and variation of binding levels during the life cycle, *J. Biol. Chem.* **250:**4730–4736.

Hinsey, J. G., 1934, The innervation of skeletal muscle, *Physiol. Rev.* **24:**514–585.

His, W., 1886, Zur Geschichte des menslichen Rückenmarks und der Nervenwurzeln, *Abh. Saechs. Ges. (Akad.) Wiss.* **13:**477–513.

Hofmann, W. W., and Thesleff, S., 1972, Studies on the trophic influence of nerve on skeletal muscle, *Eur. J. Pharmacol.* **20:**256–260.

Jolesz, F., and Sréter, F. A., 1981, Development, innervation and activity-pattern induced changes in skeletal muscle, *Annu. Rev. Physiol.* **43:**531–552.

Katz, B., 1966, *Nerve, Muscle and Synapse,* McGraw–Hill, New York.

Katz, B., 1969, *The Release of Neural Transmitter Substances,* Liverpool University Press, Liverpool.

Katz, B., and Miledi, R., 1967, The timing of calcium action during neuromuscular transmission, *J. Physiol. (London)* **189:**535–544.

Katz, B., and Schmitt, O. H., 1940, Electrical interaction between two adjacent nerve fibres, *J. Physiol. (London)* **97:**471–488.

Koelliker, A., 1890, Zur feineren Anatomie des centralen Nervensystems. I. Das Kleinhirn, *Z. Wiss. Zool.* **49:**663–689.

Konijn, T. M., van de Meene, J. G. S., Bonner, J. T., and Barkley, D. S., 1967, The acrasin activity of adenosine-3′-5′-cyclic phosphate, *Proc. Natl. Acad. Sci. USA* **58:**1152–1154.

Kusano, K., and La Vail, M. M., 1971, Impulse conduction in the shrimp modulated giant fiber with special reference to the structure of the functionally excitable areas, *J. Comp. Neurol.* **142:**481–494.

Lamb, A. H., 1976, The projection patterns of the ventral horn to the hind limb during development, *Dev. Biol.* **54:**82–99.

Lamb, A. H., 1981, Target dependency of developing motoneurons in *Xenopus laevis, J. Comp. Neurol.* **203:**157–171.

Landmesser, L., and Pilar, G., 1974, Synaptic transmission and cell death during normal ganglionic development, *J. Physiol. (London)* **241:**737–749.

Langer, S. Z., 1976, The role of alpha and beta-presynaptic receptors in the regulation of norepinephrine release elicited by nerve stimulation, *Clin. Sci. Mol. Med.* **51:**423–426.

Levi-Montalcini, R., and Angeletti, P. U., 1968, Biological aspects of the nerve growth factor, in: *Growth of the Nervous System* (G. E. W. Wolstenholme and M. O'Conner, eds.), pp. 126–147, Churchill, London.

Levi-Montalcini, R., and Cohen, S., 1960, Effects of the extract of the mouse submaxillary salivary glands on the sympathetic system of mammals, *Ann. N. Y. Acad. Sci.* **85:**324–341.

Levi-Montalcini, R., and Hamburger, V., 1951, Selective growth-stimulation effects of mouse sarcoma on the sensory and sympathetic nervous system of the chick embryo, *J. Exp. Zool.* **116:**321–361.

Llinas, R., Blinks, J. R., and Nicholson, C., 1972, Calcium transient in presynaptic terminals in squid giant synapse: Detection with aequorin, *Science* **176:**1127–1129.

Loewi, O. L., 1933, The Ferrier Lecture on problems connected with the principle of humoral transmission of nervous impulse, *Proc. Soc. London Ser. B.* **118:**299–316.

Loewi, O. L., 1945, Edward Gamaliel Janeway Lecture: Aspects of transmission of nervous impulse; theoretical and clinical implications, *J. Mt. Sinai Hosp. N.Y.* **12:**851–865.

Löffelholz, K., and Muscholl, E., 1970, Inhibition by parasympathetic nerve stimulation of the release of the adrenergic transmitter, *Naunyn Schmiedebergs Arch. Pharmacol.* **267:**181–184.

Luco, J. V., and Eyzaguirre, C., 1955, Fibrillation and hypersensitivity to ACh in denervated muscles: Effects of length of degenerating nerve fibers, *J. Neurophysiol.* **18:**65–73.

Lundberg, J. M., 1981, Evidence for coexistence of vasoactive intestinal polypeptide (VIP) and acetylcholine in neurons of cat exocrine glands: Morphological, biochemical and functional studies, *Acta Physiol. Scand.* **496:**1–57.

Lundberg, J. M., Hedlung, B., and Bartfai, T., 1982, Vasoactive intestinal polypeptide enhances muscarinic ligand binding in cat submandibular salivary gland, *Nature* **295:**147–149.

Malchow, D., and Gerisch, G., 1974, Short-time binding and hydrolysis of cyclic 3′-5′-adenosine monophosphate by aggregating *Dictyostelium* cells, *Proc. Natl. Acad. Sci. USA* **71:**2423–2427.

Mann, J. E., Sperelakis, N., and Ruffner, J. A., 1981, Alteration in sodium channel gate kinetics of the Hodgkin–Huxley equations applied to an electric field model for interaction between excitable cells, *IEEE Trans. Biomed. Eng.* **28:**655–661.

Marshall, L. M., Sanes, J. R., and McMahan, U. J., 1977, Reinnervation of original synaptic sites on muscle fiber basement membrane after disruption of the muscle cells, *Proc. Natl. Acad. Sci. USA* **74:**3073–3077.

Miledi, R., 1960, The acetylcholine sensitivity of frog muscle fibres after complete or partial denervation, *J. Physiol. (London)* **151:**1–23.

Mobley, W. C., Server, A. C., Ishii, D. N., Riopelle, R. J., and Shooter, E. H., 1977, Nerve growth factor (Parts I, II and III), *N. Engl. J. Med.* **297:**1096–1104; 1149–1158; 1211–1218.

Narayanan, Y., and Narayanan, C. H., 1981, Ultrastructural and histochemical observations in the developing iris musculature in the chick, *J. Embryol. Exp. Morphol.* **62:**117–127.

Nonidez, J. F., 1944, The present status of the neurone theory, *Biol. Rev.* **19:**30–40.

Ochi, R., 1969, Ionic mechanism of the inhibitory postsynaptic potential of crayfish giant motor fiber, *Pfluegers Arch.* **311:**131–143.

O'Donohue, T. L., Millington, W., Handelmann, G. E., Contreras, P. C., and Chronwall, B. M., 1985, On the 50th anniversary of Dale's law: Multiple neurotransmitter neurons, *Trends Pharmacol. Sci.* **6:**305–308.

Otsuka, M., 1972, γ-Aminobutyric acid in the nervous system, in: *Structure and Function of Nervous Tissue,* Vol. 4 (G. Bourne, ed.), pp. 249–289, Academic Press, New York.

Ozawa, S., and Tsuda, K., 1973, Membrane permeability change during inhibitory transmitter action in crayfish stretch receptor cell, *J. Neurophysiol.* **36:**805–816.

Patterson, P. H., 1978, Environmental determination of autonomic neurotransmitter functions, *Annu. Rev. Neurosci.* **1:**1–17.

Pittman, R., and Oppenheim, R. W., 1979, Cell death of motoneurons in the chick embryo spinal cord. IV. Evidence that a functional neuromuscular interaction is involved in the regulation of naturally occurring cell death and the stabilization of synapses, *J. Comp. Neurol.* **187:**425–446.

Poisner, A. M., and Trifaro, J. M., 1967, The role of ATP and ATPase in the release of catecholamines from adrenal medulla. I. ATP-evoked release of catecholamines, ATP and protein from isolated chromaffin granules, *Mol. Pharmacol.* **3:**561–571.

Potter, L. T., 1970, Synthesis, storage and release of ^{14}C-acetylcholine in isolated rat diaphragm muscles, *J. Physiol. (London)* **206:**145–166.

Ramón y Cajal, S., 1909, *Histologie du system nervaux de l'homme et des vertebrés,* Vol. 1, Maloine, Paris.

Ramón y Cajal, S., 1929, *Studies on Vertebrate Neurogenesis* (translation of 1909 edition by L. Guth), Thomas, Springfield, Ill.

Robison, G. A., Butcher, R. W., Oye, I., Morgan, H. E., and Suttherland, E. W., 1965, The effect of epinephrine on adenosine 3'-5'-phosphate levels in the isolated perfused rat heart, *Mol. Pharmacol.* **1:**168–177.

Sakmann, B., 1975, Reappearance of extrajunctional acetylcholine sensitivity in denervated rat muscle after blockade with α-bungarotoxin, *Nature* **255:**415–416.

Saito, A., and Zachs, S. I., 1969, Fine structure of neuromuscular junctions after nerve section and implantation of nerve in denervated muscle, *Exp. Mol. Pathol.* **10:**256–273.

Sanes, J. R., 1983, Roles of extracellular matrix in neural development, *Annu. Rev. Physiol.* **45:**581–600.

Sanes, J. R., and Hall, Z. W., 1979, Antibodies that bind specifically to synaptic sites on muscle fiber basal lamina, *J. Cell Biol.* **83:**357–370.

Sanes, J. R., Marshall, L. M., and McMahan, U. J., 1978, Reinnervation of muscle fiber basal lamina after removal of myofibers, *J. Cell Biol.* **78:**176–198.

Sanes, J. R., Marshall, L. M., and McMahan, U. J., 1980, Reinnervation of skeletal muscle: Restoration of the normal synaptic pattern, in: *Nerve Repair and Regeneration: Its Clinical and Experimental Basis* (D. L. Gewett and H. R. McCarroll, eds.), pp. 130–138, Mosby, St. Louis.

Server, A. C., and Shooter, E. M., 1977, Nerve growth factor, *Adv. Protein Chem.* **31:**339–409.

Sherrington, C. S., 1897, The central nervous system, in: *Textbook of Physiology,* (A. Foster, ed.), 7th ed., Macmillan, London.

Sherrington, C. S., 1906, *The Integrative Action of the Nervous Systems,* Yale University Press, New Haven, Conn.

Starke, K., 1977, Regulation of nor-epinephrine release by presynaptic receptor systems, *Rev. Physiol. Biochem. Pharmacol.* **77:**1–124.

Terzulo, C. A., and Bullock, T. H., 1956, Measurements of imposed voltage gradient adequate to modulate neuronal firing, *Proc. Natl. Acad. Sci. USA* **42:**687–694.

Thoenen, H., Otten, U., and Schwab, M., 1979, Orthograde and retrograde signals for the regulation of neuronal gene expression: The peripheral sympathetic nervous system as a model, in: *The Neurosciences, 4th Study Program* (F. O. Schmitt and F. G. Worden, eds.), pp. 911–928, MIT Press, Cambridge, Mass.

Trautwein, W., and Dudel, J., 1958, Zum Mechanisms der Membranwirkung des Acetylcholin an der Herzmuskelfaser, *Pfluegers Arch.* **266:**324–334.

van Geruchten, A., 1891, La structure des centres nerveux la moelle epiniére et le cervelet, *Cellule* **7:**79–122.

Waldeyer, H. W. G., 1891, Ueber einige neuere Forschungen im Gebiete der Anatomie des Zentralnervensystems, *Dtsch. Med. Wochenschr.* **17:**1213–1218.

Wallace, L. J., and Partlow, L. M., 1976, α-Adrenergic regulation of secretion of mouse saline rich in nerve growth factor, *Proc. Natl. Acad. Sci. USA* **73:**4210–4214.

Whittaker, V. P., 1965, The application of subcellular fractionation techniques to the study of brain function, *Prog. Biophys.* **15:**39–96.

Yankner, B. A., and Shooter, E. M., 1982, The biology and mechanism of action of nerve growth factor, *Annu. Rev. Biochem.* **51:**845–868.

Yellin, H., 1967, Muscle fiber plasticity and the creation of localized motor units, *Anat. Rec.* **157:**345.

Chapter 9

Cell-to-Cell Communication in Salivary Glands

Robert Weingart
Department of Physiology
University of Berne
Berne, Switzerland

1. INTRODUCTION

Twenty-five years have passed since Burgen and Emmelin's (1961) monograph, *Physiology of the Salivary Glands,* was published. At that time, it was not yet recognized that this tissue has a syncytial structure. Nineteen years later, Petersen's (1980) new monograph, *The Electrophysiology of Gland Cells,* devoted an entire chapter to "Intercellular Communication." Obviously, exciting discoveries were made in the interim concerning the nature and importance of connections between glandular cells. In the beginning, a report by Loewenstein and Kanno (1964) demonstrated that the cells of insect salivary glands are electrically coupled. Since then, salivary glands from *Drosophila* or *Chironomus* larvae have been an experimental model for exploring problems related to various aspects of intercellular communication. The advantages of studying insect salivary glands include the large cell size (diameter ~ 80–150 μm), and the arrangement of cells in a linear array. Salivary glands from animal species other than insects also have been shown to exhibit cell-to-cell coupling. However, such preparations seldom have been used presumably because of experimental difficulties arising from the small cell size (diameter ~ 15 μm), and the complex three-dimensional structure. The resulting work on salivary glands, mainly carried out by Loewenstein's group, not only has contributed to the current understanding of cell-to-cell coupling, but also has initiated similar investigations performed with other multicellular tissues.

The aim of this chapter is to summarize the experimental work dealing with intercellular coupling in salivary glands, taking into account both structural and various functional aspects. The findings will be discussed in the context of currently available knowledge in the field. In presenting the physiological data, emphasis has been placed on the experimental approaches that were adopted. The putative mechanism(s) regulating cell-to-cell coupling also will be examined. Other reviews on intercellular communication, which include the earlier literature on insect salivary glands, have appeared (Loewenstein, 1975, 1981; Rose *et al.*, 1980).

2. ANATOMY

2.1. Appearance of Salivary Glands

Salivary glands are exocrine organs found in various animals (see, e.g., Petersen, 1980). In mammals, three pairs of glands empty into the oral cavity of the digestive tract. These glands are the parotid, submaxillary (in man, submandibular), and sublingual glands. In lower orders of animals, only one pair of glands usually is found. In most cases, the saliva excreted contains different proteins (digestive enzymes) and electrolytes. However, some animals have evolved specialized salivary glands which serve completely different purposes, e.g., silk-producing insects (silk), bloodsucking insects (anticoagulants), octopus (toxins), and snakes (venoms).

2.2. Morphology

There is considerable variation of the degree of structural organization in salivary glands, ranging from rudimentary sacs in insects to elaborate branching organs in mammals. Histologically, glands are composed of an acinus or several acini each of which consists of epithelial cell layers, a tubule system, and an excretory duct. Studies on intercellular communication have been performed exclusively on acinar cells. The gland most frequently used has been from insect larvae of *Drosophila* and *Chironomus*. This preparation is easily obtained, possesses large transparent cells (diameter $\sim$ 80–150 μm), and has a favorable geometry (flattened sac). Glands from *Drosophila* are composed of nearly identical cells (see van Venrooij *et al.*, 1974), while those from *Chironomus* have two different cell types: (1) a linear array of about 30 spherelike G-cells, and (2) two layers of flat F-cells consisting of three or four cells each (see Fig. 1A; see also Kloetzel and Laufer, 1969; Rose, 1971).

Ultrastructurally, a number of studies have demonstrated that specialized regions of apposed junctional membranes or gap junctions exist between adjacent acinar cells in vertebrate salivary glands (Kater and Galvin, 1978; Shimono *et al.*, 1980; Dunn and Revel, 1984) as well as invertebrate glands (Wiener *et al.*, 1964; Kloetzel and Laufer, 1969; Rose, 1971; Berger and Uhrik, 1972; Maxwell, 1981). Comparable to other tissues (see, e.g., Peracchia, 1980), thin

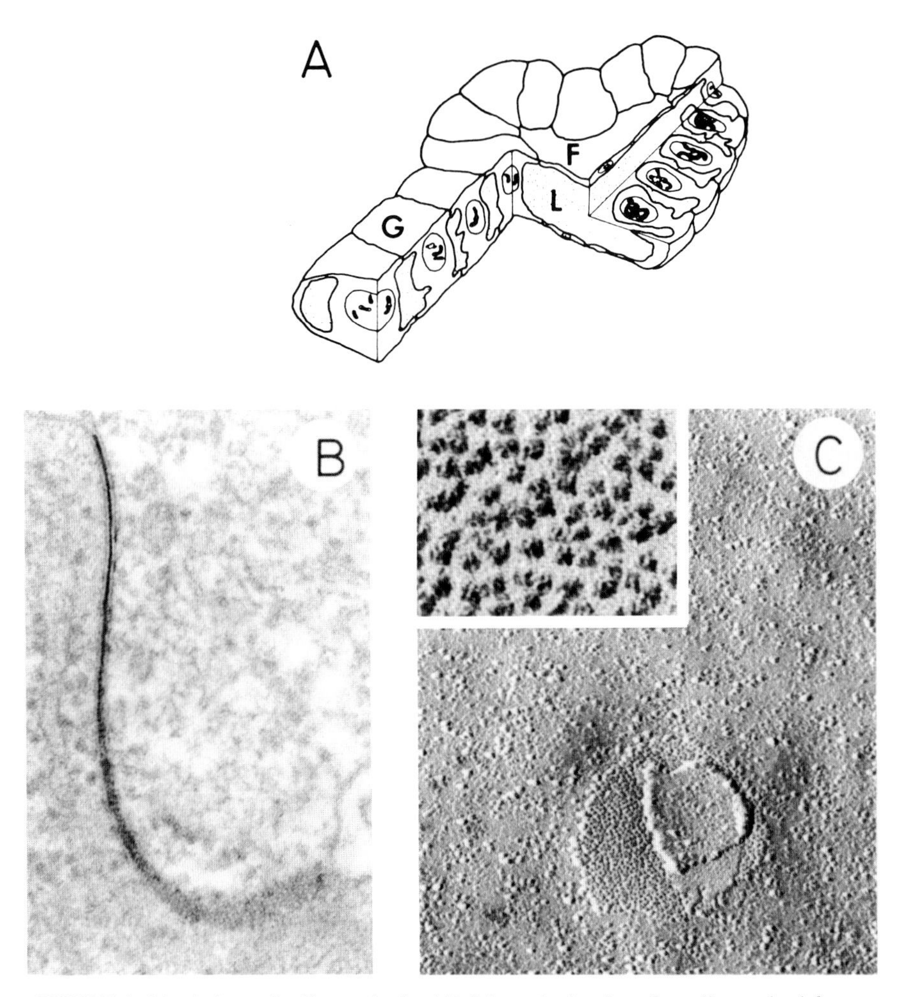

FIGURE 1. Morphology of salivary glands. (A) Schematic drawing of a salivary gland from a *Chironomus* larva. A chain of G-cells (G) and two layers of F-cells (F) form the lumen (L) of the organ. (B) gap junction between acinar cells of a rat sublingual gland. Electron microscopic thin section in the tangential plane, after contrasting with lanthanum, ×130,000. (C) Freeze-fracture replica of a gap junction from rat sublingual gland, ×74,000. The inset shows the arrangement of the gap junctional particles, ×400,000. (A) from Rose (1971); (B, C) from Shimono *et al.* (1980) with permission.

sections of gap junctions have shown the typical multilayered structure (see Fig. 1B; Wiener *et al.*, 1964; Kloetzel and Laufer, 1969; Rose, 1971; Berger and Uhrik, 1972; Shimono *et al.*, 1980; Maxwell, 1981; Dunn and Revel, 1984). Freeze-fracture replicas have revealed a quasi-hexagonal arrangement of subunits forming connexons, the membrane channels linking neighboring cells (see

Fig. 1C; Kater and Galvin, 1978; Shimono *et al.*, 1980; Dunn and Revel, 1984). Morphometric measurements performed on rat salivary glands have determined a gap junctional membrane area of 1.6% of the lateral cell surface (Dunn and Revel, 1984), and a hexagon center-to-center spacing of 9–11 nm (Shimono *et al.*, 1980). Surprisingly, no quantitative data are available regarding gap junctions in insect salivary glands.

3. ELECTRICAL EVIDENCE

3.1. Current-Clamp Studies on Intact Glands

3.1.1. Input Resistance

Input resistance measurements have provided circumstantial evidence for the existence of cell-to-cell coupling. This cell parameter may be determined using one of the following experimental approaches: (1) a double-barreled microelectrode, one barrel to pass constant-current pulses and the other to detect the resultant voltage deflection; (2) two separate microelectrodes, one to inject current and the other to measure the change in membrane potential of a nearby cell; and (3) one microelectrode used for the dual function of current injection and potential recording. With all of these methods, rather low input resistances, R_{in}, have been obtained in salivary glands: 2–18 MΩ in mammals (sublingual gland: Lundberg, 1957; Imai, 1965; submaxillary gland: Nishiyama and Kagayama, 1973; Kagayama and Nishiyama, 1974; Nishiyama and Petersen, 1974; Kater and Galvin, 1978; parotid gland: Gallacher and Petersen, 1980; Iwatsuki and Petersen, 1981) and 8–39 MΩ in mollusks (Klevets and Shuba, 1974; Kater *et al.*, 1978). However, in the larger cells from insect glands (cell diameters: 80–150 μm versus 20 μm), even lower values were found, 0.3–0.7 MΩ (Loewenstein and Kanno, 1964; Loewenstein *et al.*, 1967; Kislov and Veprintsev, 1971; Rose, 1971; Palmer and Civan, 1977; Metzger and Weingart, 1984). Such low values of R_{in} may reflect either or both of two situations: an unusually low specific resistance of the cell membrane, or electrical coupling between individual cells. On the basis of input resistance alone, it is not possible to distinguish between these cases. As a matter of fact, the early studies attributed changes in input resistance solely to the first possibility (see, e.g., Lundberg, 1957; Imai, 1965).

3.1.2. Coupling Coefficient

A significant development occurred when Loewenstein and Kanno (1964) provided the first functional evidence of electrical coupling in insect epithelia. Utilizing salivary glands from *Drosophila* larvae, they injected current pulses intracellularly via a microelectrode. The associated changes in membrane potential were monitored with two additional microelectrodes, one located in the injected cell and another in a contiguous cell. The measurements revealed voltage deflections nearly identical in amplitude in the two adjacent cells, suggesting that current flowed intercellularly via a low-resistance pathway.

Subsequently, a similar approach has been employed successfully in a number of studies, especially those dealing with salivary glands having large cellular dimensions. Such methods have enabled the degree of intercellular communication to be expressed in terms of the coupling coefficient, defined as the ratio of the voltage deflection induced in the adjacent "follower" cell to the voltage deflection elicited in the injected cell (V_2'/V_1). The coupling coefficient represents a convenient though qualitative index of cell-to-cell coupling. Coupling coefficients of salivary glands from different insect species have ranged from 0.79 to 0.96 (Loewenstein and Kanno, 1964; Loewenstein *et al.*, 1965, 1967; Politoff *et al.*, 1969; Vozhkova *et al.*, 1970; see also Ginsborg *et al.*, 1974). In molluscan (Kater *et al.*, 1978) and mammalian glands (Hammer and Sheridan, 1978; Kater and Galvin, 1978), V_2'/V_1 was 0.9 and 0.7 to ~ 1.0, respectively. To summarize, all salivary glands investigated thus far have demonstrated coupling coefficients compatible with the concept of significant electrical coupling of neighboring inexcitable cells.

3.1.3. Cable Analysis

More quantitative information about intercellular coupling has been obtained from cable analysis (see, e.g., Weidmann, 1952; Jack *et al.*, 1975). This experimental approach involves current injection at one point and measurement of the resulting voltage deflections at various distances. The decay of voltage with distance is defined by the space constant λ. In case of a one-dimensional cable, λ represents the distance over which the size of the voltage deflection decreases by a factor of e, i.e., from 100% to 37%. The first application of cable analysis to salivary glands was by Loewenstein and Kanno (1964). They observed an exponential relationship between the steady-state amplitude of the evoked voltage deflection and the distance from the site of current injection. This was the predicted relation from linear cable analysis which subsequently was confirmed by the same group and others. While most investigators have used a continuous model to interpret their data (classical cable analysis), van Venrooij *et al.* (1974) employed a more complicated discrete model. The values of λ obtained for insect glands by different investigators are summarized in Table I, column 3. Space constants ranged from 0.48 mm to 1.37 mm compared to the length of an individual cell estimated at 80 to 150 μm. In other words, cell length was up to one order of magnitude smaller than λ. Therefore, intracellular current spreads over many cell lengths, implying that low-resistance barriers exist between individual cells.

Table I
Electrical Properties of Insect Salivary Glands as Revealed by Linear Cable Analysis[a]

Coupling coefficient V'_2/V_1	Input resistance R_{in} (MΩ)	Space constant λ (mm)	Intracellular axial resistance r_i (MΩ/cm)	Gap junction resistance r_j (kΩ)	Animal species	Reference
0.96	0.4–0.5	1.2	7.5	60	*Drosophila*	Loewenstein and Kanno (1964)
0.93	0.42	1.1	7.75	62	*Drosophila*	Loewenstein *et al.* (1965)
0.80	0.19	0.75	5	75	*Chironomus*	Loewenstein *et al.* (1965)
0.82	0.54			89–190	*Chironomus*	Loewenstein *et al.* (1967)
	0.4–0.7	1.0	11.0		*Drosophila*	Kislov and Veprintsev (1971)
	0.6	1.37	8.75	54	*Chironomus*	Rose (1971)
		0.6		230–370	*Drosophila*	van Venrooiij *et al.* (1974)
	0.27	0.48	11.4	131	*Chironomus*	Metzger and Weingart (1984)

[a]The parameters listed are relevant to cell-to-cell coupling.

The basic cable parameters may be calculated from input resistance, R_{in}, and λ (see Table I, columns 2 and 3), i.e., membrane resistance r_m and intracellular axial resistance r_i. The latter parameter represents a series combination of two resistive elements, cytoplasmic resistance r_c and gap junctional membrane resistance r_j ($r_i = r_c + nr_j$; n = number of cells per cm). For convenience, r_c will be neglected as a separate entity and lumped together with r_j. Considering this combination, Table I, column 5, summarizes the available values of r_j for insect glands. Individual r_j values range from 54 kΩ to 370 kΩ, and thus cluster within one order of magnitude. An estimate of the specific resistance of the gap junctional membrane, R_j, could be obtained if both r_j and the appropriate preparation geometry were known. However, no accurate morphometric data are available so the specific resistance was not included in Table 1.

Unlike *Drosophila* larvae, the salivary glands of *Chironomus* are composed of different types of cells, a linear array of large G-cells and two layers of small F-cells (see Section 2.2). Rose (1971) demonstrated electrical coupling not only between G-cells (V_2'/V_1 = 1.0, approx.), but also between F-cells and between G- and F-cells ($V_2'/V_1 < 0.7$). Coupling between the two different cell populations would seem to render *Chironomus* glands unsuitable for one-dimensional cable analysis. Yet, the original data by Loewenstein's group were compatible with the predictions from cable theory (see above). According to Rose (1971), this discrepancy may be explained as follows. In earlier times, the glands usually were immobilized in the tissue bath by a wisp of glass pressing against the F-cells. This procedure must have destroyed the F-cells. As a consequence, the damaged cells were electrically insulated from the intact G-cells (see Section 5.1.1), thus yielding an ideal preparation for linear cable analysis, i.e., a chain of G-cells. As a matter of fact, this is exactly what has been achieved in later studies, in which the F-cells were eliminated on purpose by puncture with a blunt microelectrode (Rose, 1971; Metzger and Weingart, 1984).

So far no cable analysis has been carried out on mammalian salivary glands. Presumably, this is because of geometrical complexities. The cells are of small size and are arranged in a three-dimensional structure within the tissue. The associated difficulties in collecting and analyzing experimental data hinder the interpretation of any electrical studies that might be performed. Fortunately, the development of investigative techniques on pairs of cells may enable these problems to be surmounted.

3.2. Current-Clamp Studies on Cell Pairs

To further characterize the gap junctional membrane of salivary glands, Metzger and Weingart (1984) performed current clamp studies on two-cell preparations. Intact glands from *Chironomus* larvae were reduced to two-cell preparations (G-cells) by destroying neighboring cells mechanically. Such cell pairs represent the smallest cellular system for studying cell-to-cell coupling. Each cell

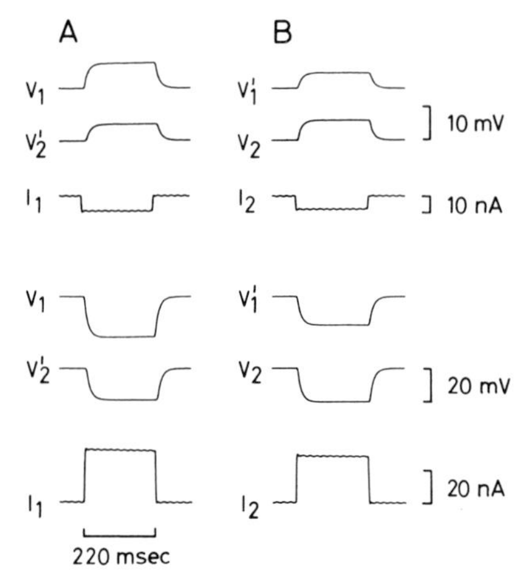

FIGURE 2. Current spread in a two-cell preparation of a *Chironomus* salivary gland. (A) A small rectangular current pulse (220-msec duration) was injected into cell 1 (I_1) in either depolarizing (upper panel) or hyperpolarizing (lower panel) direction. Traces V_1 and V_2' represent the resulting voltage deflections in cell 1 and cell 2, respectively. (B) Records of the symmetrical case of current injection into cell 2 (I_2). From Metzger and Weingart (1984) with permission.

of a cell pair was impaled with two microelectrodes, one to pass current pulses, and the other to monitor resulting voltage deflections. Figure 2 shows an example of analog records from such an experiment.

Metzger and Weingart's cell pair experiments revealed a V_2'/V_1 of 0.7. This was somewhat lower than the coupling coefficient reported previously for intact glands (see Section 3.1.2 and Table I, column 1). However, more interestingly, the shape of the current–voltage relationship for the gap junctional membrane could be elucidated. It was found to be ohmic over the voltage range tested, i.e., ± 10 mV. Unfortunately with this method, no larger voltage deflections could be established across the r_j because of the low ratio of junctional membrane resistance to nonjunctional membrane resistance (0.55 approx.). The analysis did yield values of r_j ranging from 20 kΩ to 3800 kΩ, with an overall mean of 1100 kΩ. According to the authors, the lower values (average of four experiments: r_j = 100 kΩ; coupling coefficient = 0.88) may represent the more physiological state of intercellular coupling, whereas the higher ones could have been affected by partial uncoupling from local damage and subsequent Ca^{2+} entry during impalement (see Section 5.1.1). Significantly, the cell pair experiments revealed no sign of rectification of the gap junctional membrane. This finding agrees with previous qualitative observations (Loewenstein and Kanno, 1964; Kater and Galvin, 1978).

3.3. Voltage-Clamp Studies on Cell Pairs

A major limitation of the current clamp method described above is that it does not allow *direct* study of the electrical properties of the gap junctional membrane. In 1981, Spray *et al.* proposed a double voltage clamp approach for isolated cell pairs which helps to overcome this restriction. Each cell of a cell pair was impaled with two microelectrodes, one to inject current and the other to monitor voltage. Each electrode was connected to a voltage clamp circuit so as to individually control the membrane potential of both cells and, hence, the voltage gradient across the gap junction. From the latter and the associated junctional current flow, it was possible to *directly* characterize the gap junctional membrane.

Recently, Obaid *et al.* (1983) have applied this method to cell pairs enzymatically isolated from salivary glands of *Chironomus* larvae. They succeeded in determining the current–voltage relationship of the gap junctional membrane. This was accomplished using the following experimental protocol: Starting from a common holding potential, V_H, cell 1 was pulsed to various voltage levels while the membrane potential of cell 2 remained constant. The results demonstrated a family of I_j versus V_j curves (junctional current to junctional voltage), all of which were nonlinear (see Fig. 3). For a given holding potential, a maximal conductance was observed when cell 1 was pulsed to negative potentials and a minimal conductance when stepping cell 1 to positive levels. In addition, the contour of the I_j/V_j relationships was potential-dependent with the steepest I_j/V_j curves being at negative holding potentials.

In another kind of experiment, small test pulses were applied to one of the cells while the holding potential of both cells was varied. A sigmoidal relationship was found between junctional membrane conductance g_j and nonjunctional membrane potential V_m (see Fig. 4). At negative potentials (i.e., around -80 mV), g_j reached a maximum. Upon depolarization toward positive potentials, g_j fell by two to three orders of magnitude, asymptotically diminishing to zero. Obaid *et al.* (1983) concluded from these experiments that g_j is dependent on the membrane potential of each cell rather than on the imposed transjunctional voltage. Similar inferences, although based on less reliable observations, have been made in the past (see, e.g., Rose, 1970; Socolar and Politoff, 1971).

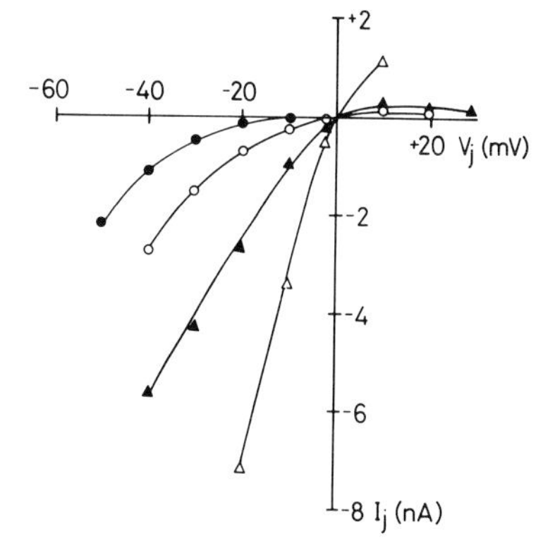

FIGURE 3. Current–voltage relationship of the gap junctional membrane, determined in a cell pair isolated from *Chironomus* salivary gland. One of the cells was clamped at a particular potential (V_2), while the other cell's potential (V_1) was stepped to various levels, negative and positive to V_2. The curves depict the steady-state junctional current I_j versus transjunctional voltage gradient V_j at various levels of V_2: -40 mV (△), -30 mV (▲), -20 mV (○), and -10 mV (●) Redrawn from Obaid *et al.* (1983) with permission.

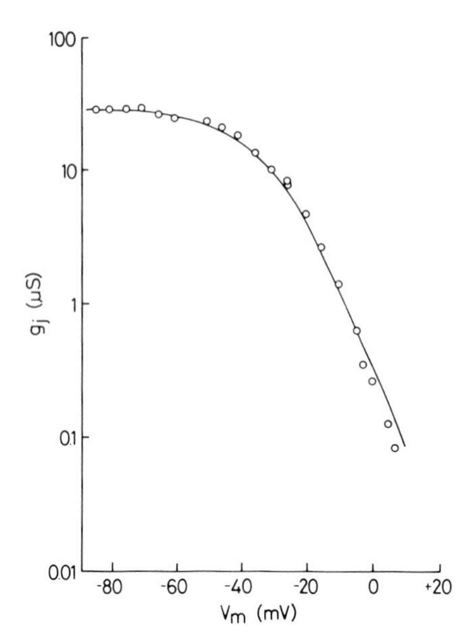

FIGURE 4. Relationship between junctional conductance g_j and nonjunctional membrane potential V_m, from voltage-clamp studies performed on a salivary gland cell pair of *Chironomus*. Both cells were clamped to various but equal potentials (V_m), while g_j was determined through application of small voltage steps to one of the cells. Redrawn from Obaid *et al.* (1983) with permission.

For a more quantitative approach to the problem, Obaid *et al.* (1983) adopted a physicochemical model involving shifting of the dipole moment of connexons. It was based on the following assumptions: A cell-to-cell channel consists of two hemichannels, one associated with each cell face of the junction. Each hemichannel has a gate which is controlled by the intrinsic membrane potential. Therefore, g_j must be proportional to the product of the opening probability of the two gates arranged in series. The model enabled the following parameters to be calculated: g_j(max) = 3–30 μS (equivalent to r_j = 30–300 kΩ; see Table I, column 5 to compare with values obtained by different methods); gating charge z = 2.0–2.2; voltage sensitivity of the gates A = 0.077–0.085 mV^{-1} ($A = zF/RT$); potential (V_0) at which half the gates on each junction face were open, $V_0 = -10$ to -35 mV. Changes in g_j developed with time constants on the order of tenths of a second.

Interestingly, dependence of r_j on the nonjunctional membrane potential appears to be a unique property of insect salivary glands. At present, there is no evidence for such a mechanism in other tissues (for review, see Spray and Bennett, 1985). From a mechanistic point of view, it seems difficult to imagine how an electric field across the nonjunctional membrane is sensed by the voltage-sensitive gates of a connexon. In other words, this mechanism is not easy to reconcile with current knowledge of gap junctional architecture.

3.4. Insect Development and Intercellular Coupling

Insect larvae undergo a sequence of developmental stages which eventually lead to formation of a pupa (see, e.g., Kloetzel and Laufer, 1969). There have been reports about correlation between developmental stages of insect larvae and cell-to-cell coupling in their salivary glands. In all studies to date, larvae were used from early prepupal stages (third and fourth instar stage). Coupling coefficients on the order of 0.79 to almost 1.0 have been observed (see Section 3.1.2) as compared to much smaller coupling coefficients of 0.04 or less (Loewenstein *et al.*, 1965; Vozhkova *et al.*, 1970; Kislov and Veprintsev, 1971) at later stages prior to pupation. Such findings suggest progressive electrical uncoupling during development. In fact, the smaller discrepancies in coupling data obtained by Loewenstein *et al.* (1965) and van Venrooij *et al.* (1974) have been attributed to this phenomenon (see Table I; see also van Venrooij *et al.*, 1974). Perhaps changes in cell-to-cell coupling are mediated by the hormones responsible for insect metamorphosis. Testing this hypothesis may well be a fruitful area for future investigation.

4. DIFFUSIONAL EVIDENCE

Syncytia may be examined from the perspective of being diffusional as well as electrical systems. As a matter of fact, for nonexcitable tissues, the diffusional function may reflect more precisely the physiological role of intercellular coupling (see Section 6). This section deals with experiments prompted by questions regarding intercellular diffusion. Methodological problems underlying the various experimental approaches have been reviewed elsewhere (see, e.g., Stewart, 1978; Brink and Dewey, 1981; Meech, 1981).

4.1. Static Approach

The experimental evidence presented thus far clearly demonstrates that cells are electrically coupled in salivary glands (see Section 3). Obvious candidates for charge carriers of intercellular current are the small cytosolic electrolytes, such as K^+ and Cl^-. However, unlike other tissues (see, e.g., Weidmann, 1966), no diffusional data for salivary glands exist which would verify directly this point. Instead, the diffusional behavior of a variety of larger substances has been studied quite extensively.

The first attempt to explore cell-to-cell coupling via diffusion of large molecules was by Loewenstein and Kanno (1964). Fluorescein (332 daltons) was pressure-injected via a microelectrode into a single cell of an intact salivary gland from *Drosophila* larvae. The subsequent intercellular redistribution of the tracer was followed under the dark field of a microscope. Within 10 to 20 min, almost all cells of the organ (100 to 200 cells) were fluorescent. Thus, this study demonstrated that junctional membranes are permeable not only to small ions but to larger substances as well.

Shortly thereafter, Kanno and Loewenstein (1966) performed a more systematic study of cell-to-cell diffusion employing a similar experimental procedure and preparation. They chose a set of colored and fluorescent compounds (azure B, orange G, solantine turquoise, trypan blue, Evans blue, labeled serum albumin, and polylysine), covering a wide spectrum of molecular sizes [range: 305 daltons (azure B) to 127,000 daltons (polylysine)]. Semiquantitative analysis visually or on film revealed that the smaller compounds (up to 450 daltons) spread over many cell lengths while larger ones (700 to 1000 daltons) diffused merely over a few cells. The largest molecule found to permeate gap junctions was serum albumin (69,000 daltons; equivalent hydrodynamic diameter: 7.2 nm). However, the possibility of intracellular degradation of the protein could not be excluded with intercellular diffusion of a labeled fragment.

Additional studies, similar in approach and focus to those described above, were performed by Rose (1971) and Rose *et al.* (1980). The former described the intercellular passage of a new fluorescent compound, Procion Yellow M-4RS (697 daltons), the latter demonstrated the transfer of fluorescent angiotensin II (1444 daltons). Examining histological sections, Maxwell (1981) also demonstrated that Procion Yellow M-4RS diffuses from cell to cell in salivary glands from *Nauphoeta cinerea,* a cockroach.

The diffusional studies discussed so far do not permit a precise cutoff limit to be established for intercellular permeation. This is because the repertoire of probes does not contain enough compounds within the critical range, i.e., 1000 to 70,000 daltons. Furthermore, the probes are not members of a homologous series. Instead, they belong to a wide spectrum of chemical classes and possess quite different molecular structures and net electrical charge. Hence, the effective pore size could not be deduced for a gap junctional channel. Therefore, previous diffusion studies needed to be extended to address these deficiencies.

Simpson *et al.* (1977) set out to rectify the problems and explore the topic of pore size. Their aim was to determine more accurately the size limit of molecules permeating the gap junctional membrane of insect salivary glands. A series of linear hydrophilic molecules was selected to perform the investigation including amino acids and synthetic or naturally occurring peptides. Easy optical detection visually or using a videotape system was achieved by covalently labeling with fluorophores, such as fluorescein isothiocyanate, dansyl chloride, and lissamine rhodamine B. In this way, a set of conjugates was obtained with defined chemical structures. Pressure-injection of the molecules led to the following observations: peptides with a molecular size less than or equal to 1158 daltons readily passed through the gap junctions; substances larger than 1926 daltons did not exhibit detectable permeation. The spread of fluorescence from the site of injection into neighboring cells occurred at rates inversely proportional to the size of the diffusing probes. The synthesized probes showed different emission spectra depending on the conjugated fluorophore. This property allowed simultaneous study of the diffusion of two injected probes of differing molecular sizes. Such an approach had the additional virtue of demonstrating that impermeable molecules do not impair diffusion of permeable compounds. This suggested that the impermeable molecules *per se* do not block the channels.

The studies by Simpson *et al.* (1977) predicted a critical limit of 1200 to 1900 daltons for permeation of the junctional membrane channels in insects. Subsequently, a putative estimate of pore size can be obtained from the structure of the permeable molecules themselves. Thus, channel diameter may be interposed between the sizes of the two extreme geometries for the largest permeant particle. Transformation of molecular mass into either a sphere, or a prolate spheroid with a major diameter of 3 nm, reveals that the effective channel diameter lies between 1.0 and 1.4 nm.

To this point, gap junctions have been probed by means of hydrophilic linear peptide molecules carrying several net electrical charges. Unfortunately, such studies do not allow one to distinguish between steric and polar constraints in the channel. In other words, the actual channel diameter may well be larger than previously calculated above. Therefore, Schwarzmann *et al.* (1981) performed studies to overcome these limitations to the precise assessment of junctional channel diameter in insect salivary glands. Their approach involved probing the junction with a series of *neutral* sugar moieties. A set of linear oligosaccharides and branched glycopeptides was chosen for labeling with the fluorophores fluorescein isothiocyanate, rhodamine B, or lissamine rhodamine B. The various compounds were hydraulically microinjected into isolated salivary glands. Optical examination led to the following observations: all linear oligosaccharides ranging from 732 to 3002 daltons penetrated the gap junctions. An effective pore size of 1.6 to 2.0 nm was deduced on the basis of space-filling structures (Corey–Pauling–Koltun models) for the largest compounds of the series. The studies employing branched compounds yielded interesting additional findings. Branched glycopeptides with four galactose-terminated ends (3097 daltons) did not pass from cell to cell. However, the same molecule did transfer after clipping-off the galactose residues (2449 daltons). The former molecule had a permeation-limiting width of 3.0 nm, the latter 2.1 nm (see Fig. 5). This suggests that the permeation-limiting diameter of the channels must lie between 2 and 3 nm.

Relatively little work has been done on diffusion in mammalian salivary glands. Hammer and Sheridan (1978), experimenting with rat parotid and submandibular glands, microinjected fluorescein (332 daltons) and Procion Yellow M-4RS (697 daltons). Examinations of both living and fixed tissue revealed that the compounds permeated the intercellular junctions. Kater and Galvin (1978), carrying out similar studies in living and fixed mouse submaxillary glands,

3 nm
2 nm

confirmed the diffusion of an intermediate-sized probe, a fluorescent probe with high quantum yield, Lucifer Yellow CH (443 daltons). The only diffusional study in molluscan glands (Kater, 1977) reported intercellular passage of Fast Green and Lucifer Yellow.

Because of the paucity of data from salivary glands of mammals and molluscs, it is impossible to decide whether or not genuine differences in connexon size exist in the animal kingdom. However, experiments on mammalian tissue other than salivary glands strongly implicate such a difference between invertebrates and vertebrates. Vertebrates in general seem to possess smaller channels, i.e., 1.6 to 2.0 nm in diameter (for references, see Weingart, 1974; Schwarzmann *et al.*, 1981).

4.2. Dynamic Approach

All studies described so far have been based on the single criterion of whether a given compound penetrates the gap junction or not. This static approach may be subject to certain criticisms; e.g., conceivably, cell-to-cell diffusion may not have been manifest because of limits to detection of a particular molecule. If this were true, previous values for channel diameter would be inaccurate. An alternative means of analysis is to measure the *time course* of intercellular diffusion. This dynamic approach enables determination of the effective channel size experienced by the moving particle itself. In analyzing the motion of molecules, intercellular diffusion in a multicellular preparation essentially involves diffusion through two elements in series: the cytoplasm and the gap junctional membrane. Therefore, a quantitative analysis has to take into account the contribution of each element in order to ultimately derive a value for permeability of the gap junctional membrane.

In insect salivary glands, this approach was initially used by van Venrooij *et al.* (1975). Fluorescein or dansylated amino acids (range: 379 to 482 daltons) were pressure-injected into cells of an intact gland from *Drosophila* larvae. Intracellular redistribution of the fluorescence was assayed by means of high-sensitivity film. Densitometry was used to subsequently evaluate the diffusion profiles. The formalism adopted for data analysis assumed two diffusional obstacles, the junctional and nonjunctional membranes. van Venrooij *et al.* (1975) found that these two types of membranes were about equally permeable to the different compounds tested. This conclusion contrasted sharply with the view

FIGURE 5. Space-filling model of branched glycopeptide tracer molecules. The complete molecule has four terminal galactoses (3097 daltons). The planes (arrows) indicate where galactose molecules are clipped off in order to obtain the thinner agalactopeptide (2449 daltons). The widths are indicated for the primary permeation-limiting dimension of the two molecules; other dimensions are to scale. From Schwartzmann *et al.* (1981) with permission.

previously deduced from electrical studies (see Section 3). The discrepancy may be explained by either the developmental stage of the glands (late third-instar stage), the limited resolution of the detection system, or the simplified assumptions used for the analysis. The relative importance of each of these explanations to the observed differences is still unknown.

Subsequently, Zimmerman and Rose (1985) improved the dynamic approach adopted by van Venrooij *et al.* (1975). A set of tracers of different size (range: 376 to 2327 daltons) and fluorescence wavelength were chosen. Based on previous studies, all probes were expected to pass through the gap junctions. Two tracers, distinguishable in molecular mass and fluorescence spectrum, were coinjected into a cell. Fluorescence was monitored in the injected cell and an adjacent one by means of a multiwavelength microfluorimeter. Rate constants for tracer diffusion from cell to cell were derived after accounting for loss from the cells, binding to cytoplasmic components, and mobility in the cytosol. Control experiments in single cells revealed that intracellular diffusion was too rapid to be rate-limiting for cell-to-cell diffusion. The rate constants were inversely related to the size of the tracers but not to their net charge. The latter may not be so surprising since the permeant molecules had low surface charge densities. In each case, the permeability ratio for paired, variably-sized substances was constant. This suggested that only one type of cell-to-cell channel exists since all channels exhibited the same selectivity.

In conjunction with investigating permeability of capillary walls, models have been developed to evaluate the restricted diffusion of large molecules through pores of molecular dimension (see, e.g., Pappenheimer, 1953; House, 1974). These theories permit estimates to be made of the effective pore size of the permeating molecules. A dynamic pore diameter of 2.9 nm is obtained (see Fig. 6) if the formalism of Faxén (1922) is applied to the kinetic diffusion results of Zimmerman and Rose (1985). Such a treatment sharpens previous estimates which were based simply on static approaches.

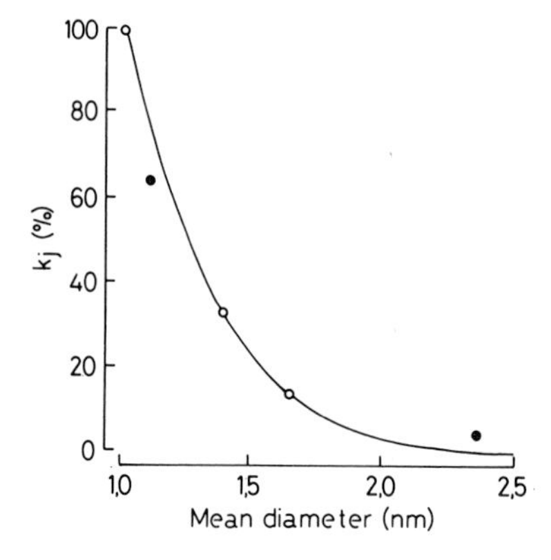

FIGURE 6. Diffusion of fluorescent tracers in an intact salivary gland from *Chironomus* larvae. Dependence of the junctional diffusion rate constant (k_j) on the mean diameter of the tracer molecule. Values of k_j were normalized with respect to that of the smallest molecule tested. Unfilled circles: the smallest molecule was coinjected as reference; the curve for a 2.9-nm-diameter particle most closely describes the relative k_j's for these cases (curve drawn according to the formalism by Faxén). Filled circles: another molecule was coinjected as reference. Redrawn from Zimmerman and Rose (1985) with permission.

Zimmerman and Rose (1985) also have investigated the influence of nonjunctional membrane potential on intercellular tracer diffusion. Depolarization of the membrane by exposure to high-K^+ solutions slowed down the cell-to-cell diffusion of both small and large tracers. Significantly, the ratio of rate constants remained the same for coinjected tracers of different size. Such equal impairment of both tracers suggested that the relative permeability of the junction was unaffected by depolarization. Therefore, it appears that depolarization affects junctional permeability by reducing the number of open channels rather than by changing the effective pore size or pore selectivity.

5. PHARMACOLOGY

In a variety of tissues, the transfer properties of gap junctions have been shown to be variable (see, e.g., Spray and Bennett, 1985). However, salivary glands are similar to other tissues in that the physiological state of gap junctional conductance is determined primarily by the ionic composition of the intracellular milieu. Electrical and diffusional investigations have been undertaken in order to elucidate the mechanism(s) underlying functional modulation of gap junctions. The experiments discussed in this section were carried out on insect salivary glands exclusively, usually from *Chironomus* or *Drosophila* larvae.

5.1. Electrical Measurements on Intact Glands

For the studies described below, current-clamp techniques have been adopted almost exclusively (see Section 3.1). Thus, alterations in intercellular coupling were assessed by monitoring the coupling coefficient. Exceptional cases will be stated as such.

5.1.1. Calcium Ions

Circumstantial evidence for involvement of intracellular Ca^{2+} originated from the observation that damaged cells isolate themselves from intact cells and do not short-circuit the entire gland to the extracellular space (see, e.g., Nakas *et al.*, 1966). In cardiac tissue, this phenomenon long has been known as ''healing-over'' (Engelmann, 1877).

5.1.1a. Membrane Perforation. Direct evidence of calcium involvement was provided by experiments aimed at elevating free $[Ca^{2+}]_i$ in intact cells. This may be achieved by equilibrating the low cytosolic $[Ca^{2+}]$ (100 nM approx.; Berridge, 1980) with the high $[Ca^{2+}]$ of the medium via a hole made in the nonjunctional membrane. This approach led to intercellular uncoupling in the presence of 1 mM $[Ca^{2+}]_o$, but not in 0.1 mM (Loewenstein *et al.*, 1967). A more accurate assessment was obtained when attention was paid to the size of the membrane hole fabricated (diameter: 10 μm; Oliveira-Castro and Loewenstein, 1971). In this case, electrical coupling was reduced drastically by a $[Ca^{2+}]_o >$ 40–80 nM.

5.1.1b. Ca^{2+} Injection. Alternatively, $[Ca^{2+}]_i$ may be increased via microelectrode injection of Ca^{2+}. This method produced marked electrical uncoupling (Loewenstein *et al.*, 1967). In an attempt to quantify the resultant changes in $[Ca^{2+}]_i$, Rose and Loewenstein (1975b, 1976) employed an optical recording system utilizing aequorin as a Ca^{2+} indicator. The photoprotein was microinjected into a cell of an intact gland and Ca^{2+}-dependent light emission detected by a photomultiplier. An image intensifier–TV system enabled the investigators to scan the spatial distribution of the emission spectrum. Preliminary measurements demonstrated that Ca^{2+} injection leads to a nonuniform $[Ca^{2+}]_i$ profile (Rose and Loewenstein, 1975a). Obviously, this complicates quantification of the relationship of pCa_i versus junctional conductance. Nevertheless, subsequent experiments revealed a decrease in coupling coefficient whenever $[Ca^{2+}]_i$ rose in the vicinity of the junction but not otherwise (see Fig. 7; Rose and Loewenstein, 1975b, 1976; Délèze and Loewenstein, 1976). The changes in coupling occurred at a $[Ca^{2+}]_i$ of 500 nM, i.e., just above the detection limit for aequorin. Therefore, this figure may be regarded as an upper limit for the threshold concentration of Ca^{2+} resulting in electrical uncoupling.

5.1.1c. Ca^{2+} Ionophores. Exposure to Ca^{2+} ionophores, such as X537A and A23187, may be expected to elevate the $[Ca^{2+}]_i$ through enhanced Ca^{2+} influx. In salivary glands, these agents increased $[Ca^{2+}]_i$ as visualized by aequorin. Similar to before, changes in $[Ca^{2+}]_i$ were accompanied by a decrease in coupling coefficient (Rose and Loewenstein, 1975b, 1976).

5.1.1d. Metabolic Poisons. Poisoning cellular metabolism also might be anticipated to modify intercellular communication, the rationale being that the cell maintains a low $[Ca^{2+}]_i$ via energy expenditure (see, e.g., Rose and Loewenstein, 1975a). Indeed, exposure of salivary glands to cyanide, 2,4-dinitrophenol, *N*-ethylmaleimide (a sulfhydryl reagent), or oligomycin (an antibiotic blocking ATP synthesis) has been found to impair intercellular coupling (Politoff *et al.*, 1967, 1969). Furthermore, injection of ATP delayed uncoupling or even restored uncoupled cells in some cases. These observations were confirmed by Rose and Loewenstein (1975b, 1976), performing combined measurements of coupling coefficient and aequorin luminescence. Once again, uncoupling correlated with a rise in $[Ca^{2+}]_i$ at the junctional locale. Inhibition of metabolism by lowering the temperature to 6–8°C also uncoupled the cells eventually (Politoff *et al.*, 1969).

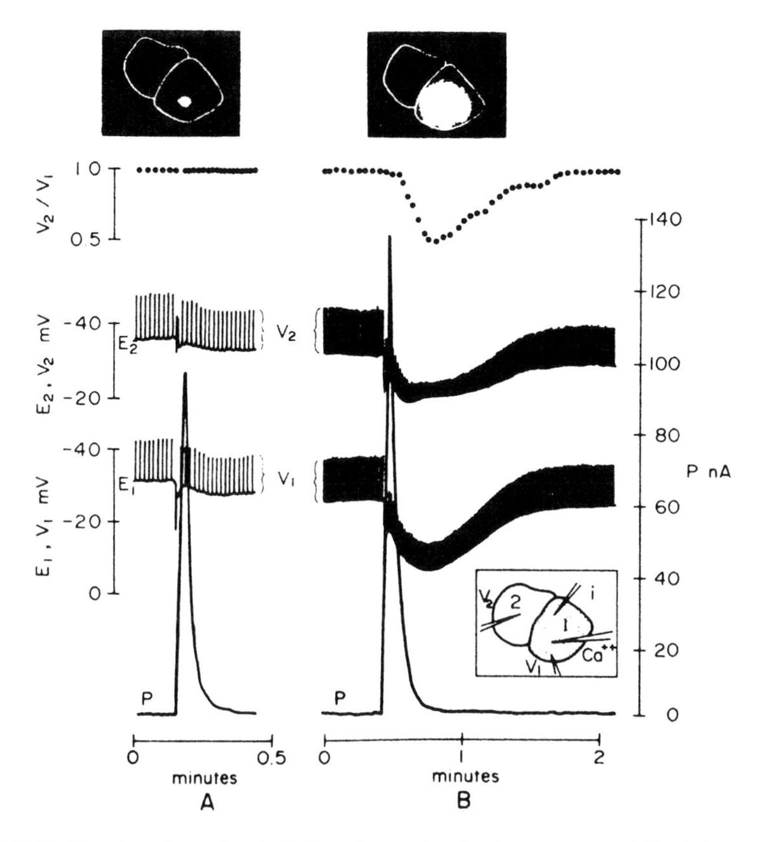

FIGURE 7. Elevation of cytoplasmic Ca^{2+} at the gap junction interrupts current flow between cells in an intact *Chironomus* salivary gland. (Top) TV pictures of aequorin luminescence at the time of maximal spatial spread, corresponding to peaks in *P* curve. (Bottom) A pulse of Ca^{2+} was pressure-injected into cell 1 (see inset in the lower right corner) while the coupling coefficient (V_2/V_1) and aequorin light signal (*P*) were determined. (A) and (B) compare the effects of a small and large Ca^{2+} pulse, respectively. From Rose and Loewenstein (1976) with permission.

5.1.1e. Na^+-Free Solution. Exposure to Na^+-free solution represents an indirect method whereby $[Ca^{2+}]_i$ may be elevated. The underlying mechanism presumably involves the impairment of a Na^+/Ca^{2+} countertransport system (see, e.g., Sulakhe and St. Louis, 1980). Concordant with this view, substitution of Li^+ for extracellular Na^+ has been reported to decrease the coupling coefficient (Rose and Loewenstein, 1971).

5.1.1f. Ca^{2+},Mg^{2+}-Free Solution. Prolonged exposure of salivary glands to solutions deficient in divalent cations also produces a decrease in coupling coefficient (Rose and Loewenstein, 1971). Combined studies involving Ca^{2+} measurements with aequorin have demonstrated that this intervention was not always associated with a rise in $[Ca^{2+}]_i$ (Rose and Loewenstein, 1976, 1975b). The reasons for this discrepancy remain unclear.

5.1.1g. Ca^{2+} Chelators. It is well known that cell membranes become leaky and may depolarize when exposed to Ca^{2+}-free solutions (for literature on the Ca^{2+} paradox, see Ruano-Arroyo *et al.*, 1984). For example, exposure of salivary glands to solutions containing EGTA (binding Ca^{2+}) has been shown to cause electrical uncoupling (Loewenstein *et al.*, 1967). However, injection of EDTA (binding Ca^{2+} and Mg^{2+}) or even EGTA, did not delay significantly the uncoupling caused by the Ca^{2+},Mg^{2+}-free medium, or 2,4-dinitrophenol (Rose and Loewenstein, 1971).

5.1.2. Protons

Early experiments by Loewenstein *et al.* (1967) revealed that increasing pH_o produces cellular uncoupling. Later, Rose and Rick (1978) explored more systematically the possibility of H^+ involvement in the control of cell-to-cell coupling. Using aequorin and ion-selective microelectrodes, pCa_i and pH_i were measured under experimental conditions known to primarily affect $[Ca^{2+}]_i$ or $[H^+]_i$. The fall in coupling coefficient caused by Ca^{2+} injection, treatment with CN^-, 2,4-dinitrophenol, or Ca^{2+} ionophore A23187 was correlated with changes in pCa_i rather than pH_i. From this evidence it was inferred that the action of Ca^{2+} on gap junctions may *not* be mediated by H^+. Intracellular acidification, brought about by H^+ injection, exposure to CO_2, or the H^+ ionophore nigericin, produced uncoupling associated with a concomitant rise in $[Ca^{2+}]_i$ in some cases. The lack of rise in $[Ca^{2+}]_i$ during some cases of acidification was explained by a pH-dependent depression of the Ca^{2+} sensitivity of aequorin. The authors concluded that their experiments do not support a role for H^+ in regulating the cell-to-cell channels of *Chironomus* salivary glands, at least not within the range of pH_i tested (6.5 to 8.2).

5.1.3. Multivalent Cations

Inspired by the role of Ca^{2+}, it is natural to wonder whether other divalent cations might exert similar effects on cell-to-cell coupling. Oliveira-Castro and Loewenstein (1971) have explored this hypothesis in *Chironomus* salivary glands. Gap junctions were exposed to divalent cations of the alkaline earth series via a hole punctured in the nonjunctional membrane. Pronounced uncoupling developed from divalent cation exposure, the order of potency being $Ca^{2+} > Mg^{2+} > Sr^{2+} > Ba^{2+} > Mn^{2+}$.

5.1.4. Other Interventions

A number of other interventions have been reported which also uncouple cells, e.g., exposure to trypsin, or solutions of twice normal tonicity. Presumably, such impairment developed through a different mechanism, namely loss of mechanical stability between the cells (Loewenstein *et al.*, 1967).

In two studies, linear cable analysis was performed (see Section 3.1.3). In one, the $[cAMP]_i$ was elevated by exposure to cAMP, dibutyryl cAMP, theophylline, or ecdysterone. These interventions evoked moderate *decreases* in axial resistance, r_i, thus *improving* intercellular communication (Hax *et al.*, 1974). The rationale offered for such responses was that cAMP acts as an intermediate in the action of many hormones, and thus may control cell-to-cell coupling via regulation of cell metabolism. In the other study, treatment with lysolecithin, a membrane lipid derivative, produced a moderate increase in r_i (van Venrooij *et al.*, 1974).

5.2. Diffusional Studies on Intact Glands

Knowledge about the mechanism(s) regulating intercellular uncoupling also may be obtained from diffusional studies. This approach has been applied during interventions previously shown to uncouple cells. Some examples which blocked intercellular diffusion of dye-tracers are Ca^{2+} injection, membrane perforation, and exposure to Mg^{2+},Ca^{2+}-free medium (Kanno and Loewenstein, 1966; Oliveira-Castro and Loewenstein, 1971; Rose and Loewenstein, 1971; Délèze and Loewenstein, 1976; Simpson *et al.*, 1977). Délèze and Loewenstein (1976) further extended these studies. They injected divalent cations (either Ca^{2+}, Mg^{2+}, or Sr^{2+}) while cell-to-cell coupling was assessed by simultaneous measurement of the coupling coefficient and fluorescein diffusion. On the one hand, cation injections that blocked fluorescein passage also were associated with electrical uncoupling. However, injections which merely hindered fluorescein redistribution (albeit quite markedly) were accompanied by little if any change in electrical coupling. This observation was interpreted to reflect either total closure of a fraction of channels (former case) or partial closure of all channels (latter case). Rose *et al.* (1977) explored the latter hypothesis of graded junctional permeability during a perturbation using a different approach. Intercellular diffusion was studied with pairs of fluorescent molecules of differing size (from 380 to 1158 daltons) after coinjection. Following moderate elevation of $[Ca^{2+}]_i$, retardation of diffusion of the larger molecule through the gap junction exceeded that of the smaller-sized probe. This suggested that the size limit for junctional permeation may depend on $[Ca^{2+}]_i$ in a graded fashion. The authors speculated that such discrimination may be attributed to stepwise changes in pore diameter or permeability. However, in view of more recent data by Zimmerman and Rose (1985), this interpretation may need to be reconsidered (see Section 4.2).

5.3. Complications Due to Changes in Membrane Potential

As mentioned above (see Section 3.3), the resistance of the junctional membrane in insect salivary glands is a sensitive function of the potential across the nonjunctional membrane, V_m (see also Obaid *et al.*, 1983). This concept has some relevance to the interpretation of the uncoupling experiments just described. On the one hand, interventions which *impair* intercellular coupling, very often have been associated with *depolarization* of V_m (see, e.g., Loewenstein *et al.*, 1967; Politoff *et al.*, 1967, 1969; Oliveira-Castro and Loewenstein, 1971; Rose and Loewenstein, 1971, 1975b, 1976; van Venrooij *et al.*, 1974). Conversely, interventions which *improve* cell-to-cell communication have been accompanied by *hyperpolarization* (Hax *et al.*, 1974). Hence, the question: What is the primary cause of uncoupling under these experimental conditions, a change in $[\text{cation}]_i$, or a decrease in V_m?

Techniques of controlling V_m have proven to be helpful in clarifying this situation. In conjunction with several interventions (e.g., exposure to DNP, Na^+-free or Ca^{2+},Mg^{2+}-free medium), electrical coupling has been restored simply by injection of repolarizing current (Rose, 1970; Rose and Loewenstein, 1971). Alternatively, uncoupling by ions of the alkaline earth series has been partially reversed by inward current (Oliveira-Castro and Loewenstein, 1971). In such cases, uncoupling must have been linked causally to the depolarization of V_m. This view corroborates the observation that depolarization of V_m through injection of outward current is accompanied by uncoupling (Socolar and Politoff, 1971). However, not all uncoupling is strictly related to changes in V_m. Rose and Loewenstein (1975b, 1976) reinvestigated uncoupling associated with Ca^{2+} injection. In these experiments, uncoupling occurred whenever $[Ca^{2+}]_i$ increased in the locale of the junction, whether V_m was controlled or not. Conversely, depolarization of V_m secondary to high $[K^+]_o$ was not accompanied by uncoupling. Under these conditions, depolarization of V_m seems not to be a prerequisite for uncoupling to develop.

5.4. Electrical Measurements on Cell Pairs

In the studies described in Section 5.1, intercellular coupling has been expressed in terms of coupling coefficients, i.e., the ratio V_2'/V_1 (see Section 3.1.2). However, use of this index may be criticized on several grounds. The coupling coefficient contains information not only concerning junctional membrane resistance, r_j, but also about nonjunctional membrane resistance, r_m ($V_2'/V_1 = r_{m,2}/(r_j + r_{m,2})$ (see, e.g., Socolar and Loewenstein, 1979). This implies that alterations in V_2'/V_1 are ambiguous in that they may reflect changes of either r_j, r_m, or both. Furthermore, at low values of r_j, V_2'/V_1 is insensitive to alterations in intercellular coupling. This may be the reason why small changes in coupling could be missed with milder interventions, such as exposure to high $[K^+]_o$ (Rose and Loewenstein, 1971, 1976).

5.4.1. Current-Clamp Experiments

In order to handle the complexities described above, one answer is to use isolated cell pairs instead of intact glands. Weingart and Metzger utilized this preparation to explore quantitatively the effects of metabolic poisons on electrical coupling (unpublished experiments; for methods, see Metzger and Weingart, 1984). Their current-clamp technique enabled electrical changes to be expressed in terms of changes in junctional and nonjunctional membrane resistance, as well as coupling coefficients. Figure 8 illustrates the result of an experiment in which a tenfold increase in r_j occurred after 0.4 mM 2,4-dinitrophenol was administered, a change which was readily reversible upon washout. This increase in r_j was accompanied by a decrease in coupling coefficient from 0.9 to 0.4. No significant changes in either r_m or V_m developed at this drug concentration.

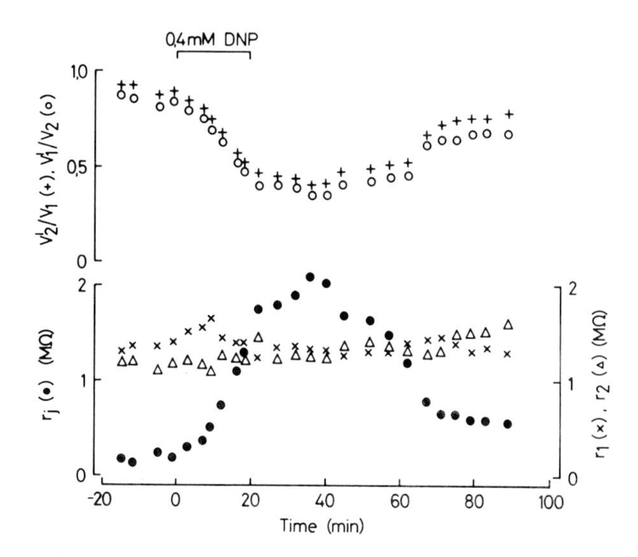

FIGURE 8. Effects of a metabolic inhibitor on intercellular current flow within a two-cell preparation of *Chironomus* salivary gland. Exposure to 0.4 mM 2,4-dinitrophenol produced a tenfold increase in junctional resistance r_j (●), which was readily reversible. The resistances of the nonjunctional membranes $r_{m,1}$ (X) and $r_{m,2}$ (△) did not change significantly. The change in r_j was accompanied by a decrease in coupling coefficient (pulsing cell 1, +; pulsing cell 2, ○). Weingart and Metzger (unpublished data).

5.4.2. Voltage-Clamp Experiments

The voltage-clamp technique provides a convenient means of separating the putative effects on g_j of V_m and [cation]$_i$ on g_j (see Section 5.3). With this method, V_m can be maintained while g_j is determined during interventions inducing a change in $[Ca^{2+}]_i$ or $[H^+]_i$. Obaid *et al.* (1983) utilized this approach in isolated cell pairs from *Chironomus* larvae. An increase in $[Ca^{2+}]_i$ evoked by Ca^{2+} injection, or treatment with CN^-, caused a decrease in g_j. The reduction in g_j seemed to correspond to a shift of the g_j versus V_m relationship (see Section 3.3 and Fig. 4) to more negative values of V_m. Conversely, a decrease in $[Ca^{2+}]_i$ via EGTA injection shifted the curve to more positive values of V_m. Neither g_j (max) nor the slope of the relationship was affected. Analogous experiments with H^+ produced a similar picture. Intracellular acidosis induced by exposure to CO_2 or propionate, was associated with a decrease in g_j. Once again, the changes in g_j could be accounted for entirely by a shift of the g_j versus V_m curve along the voltage axis (i.e., to more negative values with acidosis). Ca^{2+} and H^+ thus seem to determine the *position* of the g_j versus V_m curve along the voltage axis rather than altering g_j (max) *per se*.

6. PHYSIOLOGICAL ROLE OF INTERCELLULAR COUPLING

There have been a number of suggestions as to the biological role of cell-to-cell coupling in multicellular tissues. However, in salivary glands as in other tissues (see, e.g., Sheridan and Atkinson, 1985), the main problem with identifying physiological functions has been the lack of *direct* experimental confirmation. The fact that a hypothesis seems to make sense within a given context does not render it true.

Salivary glands of some mollusks have been reported to show Ca^{2+}-dependent action potentials (Kater *et al.*, 1978; Hadley *et al.*, 1980). Therefore, cell-to-cell coupling in these tissues may serve to propagate impulses within a secretory unit, or from one secretory unit to another. However, salivary secretory cells of other animal species are composed of inexcitable membranes. Thus, there would be no need for intercellular current flow in the absence of electrical signal transmission.

Cell-to-cell coupling also could provide a means for electrotonic spread of secretory potentials evoked in single cell (House, 1975). Depolarizing secretory potentials elicited by nerve stimulation have been observed in a variety of salivary glands (see, e.g., House, 1975; Gallacher and Petersen, 1980). In addition, adrenergic and cholinergic innervation of salivary glands has been reported to occur in a nonuniform fashion (for references, see Hammer and Sheridan, 1978). Therefore, it is conceivable that cell-to-cell coupling may act to coordinate nervous control of salivation. However, the weakness of this argument may be that membrane depolarization tends to uncouple cells, at least in insect salivary glands (see Section 3.3).

Gap junctions are permeable not only to small ions, but also to molecules that are below a critical size (cutoff limit in invertebrates: around 2000 daltons; in vertebrates: presumably around 1000 daltons; see Section 4). Therefore, cell-to-cell coupling may provide the basis for transfer of functionally relevant molecules normally present in the cytoplasm. For example, gap junctions must be permeable to most molecules involved in cellular metabolism and thus may allow metabolic cooperation among cells. Alternatively, it has been proposed that cell-to-cell passage of regulatory intermediates may be involved in controlling tissue growth and differentiation (Caveney, 1985). Salivary glands of insects are known to undergo developmental changes in cell-to-cell coupling (see Section 3.4). This makes it credible that such processes are triggered by a molecular signal which is transmitted from cell to cell through intercellular channels.

7. CONCLUSIONS

Salivary glands have been used widely in studies focusing on intercellular communication. In fact, the development of current concepts of cell-to-cell coupling has been closely linked to the experimental work performed on arthropod salivary glands. Among these, glands from insect larvae have served a crucial role. They have been employed mainly in functional studies covering both electrical and diffusional topics. Consequently, a large number of investigations have been performed involving insect glands, but only a few concerning glands from other animal species.

Considerable progress has been made over the last few years since new experimental techniques and methods have become available. As a consequence, the gap junction of insect salivary glands now may be one of the best understood intercellular junctions. Earlier studies, more qualitative in nature, were performed on intact glands, while recent quantitative investigations have been carried out in isolated cell pairs. The emerging picture of the properties of a functional unit are as follows: gap junctions of insects consist of intercellular channels, named connexons, which are arranged in parallel. The conductance of a connexon is determined by the state of two independent gates in series. These gates are sensitive to nonjunctional membrane potential (V_m), $[Ca^{2+}]_i$, and $[H^+]_i$. Neither Ca^{2+} nor H^+ modifies the voltage sensor responsible for gating. So far, dependency of g_j on nonjunctional membrane potential has only been observed in insect sailvary glands. In contrast, the involvement of Ca^{2+} and H^+ in regulating intercellular channels has been established for most multicellular tissues studied to date (see Spray and Bennett, 1985).

Nevertheless, detailed information about the properties of gap junctions of vertebrate salivary glands is still unavailable. However, considering the recent progress in developing enzymatic techniques of cell isolation (see, e.g., Mangos, 1979; Quissell and Redman, 1979), and electrical recording techniques suitable for small cells (see Hamill *et al.*, 1981), certainly these gaps in our knowledge will soon be filled in.

ACKNOWLEDGMENTS. I am grateful to Dr. M. Pressler for reading the manuscript and making valuable suggestions.

8. REFERENCES

Berger, W. K., and Uhrik, B., 1972, Membrane junctions between salivary gland cells of *Chironomus thummi, Z. Zellforsch.* **127:**116–126.

Berridge, M. J., 1980, Preliminary measurements of intracellular calcium in an insect salivary gland using a calcium-sensitive microelectrode, *Cell Calcium* **1:**217–227.

Brink, P. R., and Dewey, M. M., 1981, Diffusion and mobility of substances inside cells, *Techniques in Cellular Physiology* **P104:**1–17.

Burgen, A. S. V., and Emmelin, N. G., 1961, *Physiology of the Salivary Glands,* Physiol. Soc. Monogr. Vol. 8, Arnold, London.

Caveney, S., 1985, The role of gap junctions in development, *Annu. Rev. Physiol.* **47:**319–335.

Délèze, J., and Loewenstein, W. R., 1976, Permeability of a cell junction during intracellular injection of divalent cations, *J. Membr. Biol.* **28:**71–86.

Dunn, J., and Revel, J.-P., 1984, Association of gap junctions with endoplasmic reticulum in rat parotid glands, *Cell Tissue Res.* **238:**589–594.

Engelmann, T. W., 1877, Vergleichende Untersuchungen zur Lehre von der Muskel- und Nervenelektricität, *Pfluegers Arch.* **11:**465–480.

Faxén, H., 1922, Die Bewegung einer starren Kugel längs der Achse eines mit zäher Flüssigkeit gefüllten Rohres, *Ark. Mat. Astron. Fys.* **17:**1–28.

Gallacher, D. V., and Petersen, O. H., 1980, Electrophysiology of mouse parotid acini: Effects of electrical field stimulation and iontophoresis of neurotransmitters, *J. Physiol. (London)* **305:**43–57.

Ginsborg, B. L., House, C. R., and Silinsky, E. M., 1974, Conductance changes associated with the secretory potential in the cockroach salivary gland, *J. Physiol. (London)* **236:**723–731.

Hadley, R. D., Murphy, A. D., and Kater, S. B., 1980, Ionic basis of resting and action potentials in salivary gland acinar cells of the snail *Helisoma, J. Exp. Biol.* **84:**213–225.

Hamill, O. P., Marty, A., Neher, E., Sakmann, B., and Sigworth, F. J., 1981, Improved patch-clamp techniques for high-resolution current recording from cells and cell-free membrane patches, *Pfluegers Arch.* **391:**85–100.

Hammer, M. G., and Sheridan, J. D., 1978, Electrical coupling and dye transfer between acinar cells in rat salivary glands, *J. Physiol. (London)* **275:**495–505.

Hax, W. M. A., van Venrooij, G. E. P. M., and Vossenberg, J. B. J., 1974, Cell communication: A cyclic-AMP mediated phenomenon, *J. Membr. Biol.* **19:**253–266.

House, C. R., 1974, *Water Transport in Cells and Tissues,* Physiol. Soc. Monogr. Vol. 24, Arnold, London.

House, C. R., 1975, Intracellular recording of secretory potentials in a 'mixed' salivary gland, *Experientia* **31:**904–906.

Imai, Y., 1965, Studies on the secretory mechanism of the submaxillary gland of dog. Part 1. Electrophysiological studies with microelectrodes, *J. Physiol. Soc. Jpn.* **27:**304–312.

Iwatsuki, N., and Petersen, O. H., 1981, Dissociation between stimulant-evoked acinar membrane resistance change and amylase secretion in the mouse parotid gland, *J. Physiol. (London)* **314**:79–84.

Jack, J. J. B., Noble, D. and Tsien, R. W., 1975, *Electric Current Flow in Excitable Cells*, Clarendon Press, Oxford.

Kagayama, M., and Nishiyama, A., 1974, Membrane potential and input resistance in acinar cells from cat and rabbit submaxillary glands in vivo: Effects of autonomic nerve stimulation, *J. Physiol. (London)* **242**:157–172.

Kanno, Y., and Loewenstein, W. R., 1966, Cell-to-cell passage of large molecules, *Nature* **212**:629–630.

Kater, S. B., and Galvin, N. J., 1978, Physiological and morphological evidence for coupling in mouse salivary gland acinar cells, *J. Cell Biol.* **79**:20–26.

Kater, S. B., Rued, J. R., and Murphy, A. D., 1978, Propagation of action potentials through electrotonic junctions in the salivary glands of the pulmonate molluscs, *Helisoma trivolvis*, *J. Exp. Biol.* **72**:77–90.

Kislov, A. N., and Veprintsev, B. N., 1971, Electric characteristics of the cellular and nuclear membranes of the salivary gland cells of *Drosophila funebris* larvae, *Comp. Biochem. Physiol.* **39A**:521–529.

Klevets, M. J., and Shuba, M. F., 1974, Electrical characteristics of plasma membranes from salivary gland cells of *Helix pomatia* [in Ukrainian], *Fiziol. Zh. SSSR* **20**:540–542.

Kloetzel, J. A., and Laufer, H., 1969, A fine-structure analysis of larval salivary gland function in *Chironomus thummi*, *J. Ultrastruct. Res.* **29**:15–36.

Loewenstein, W. R., 1975, Permeable junctions, *Cold Spring Harbor Symp. Quant. Biol.* **40**:49–63.

Loewenstein, W. R., 1981, Junctional intercellular communication: The cell-to-cell membrane channel, *Physiol. Rev.* **61**:829–913.

Loewenstein, W. R., and Kanno, Y., 1964, Studies on an epithelial (gland) cell junction. I. Modifications of surface membrane permeability, *J. Cell Biol.* **22**:565–586.

Loewenstein, W. R., Socolar, S. J., Higashino, S., Kanno, Y., and Davidson, N., 1965, Intercellular communication: Renal, urinary bladder, sensory, and salivary gland cells, *Science* **149**:295–298.

Loewenstein, W. R., Nakas, M., and Socolar, S. J., 1967, Junctional membrane uncoupling: Permeability transformations at a cell membrane junction, *J. Gen. Physiol.* **50**:1865–1891.

Lundberg, A., 1957, The mechanism of establishment of secretory potentials in sublingual gland cells, *Acta Physiol. Scand.* **40**:35–58.

Mangos, J. A., 1979, Morphological and functional characterization of isolated human parotid acinar cells, *J. Dent. Res.* **58**:2028–2035.

Maxwell, D. J., 1981, The presence of gap junctions in the septate desmosomes of the salivary apparatus of the cockroach *Nauphoeta cinerea*, *J. Exp. Biol.* **94**:341–344.

Meech, R., 1981, Microinjection, *Techniques in Cellular Physiology* **P109**:1–16.

Metzger, P., and Weingart, R., 1984, Electric current flow in a two-cell preparation from *Chironomus* salivary glands, *J. Physiol. (London)* **346**:599–619.

Nakas, M. S., Higashino, S., and Loewenstein, W. R., 1966, Uncoupling of an epithelial cell membrane junction by calcium-ion removal, *Science* **151**:89–91.

Nishiyama, A., and Kagayama, M., 1973, Biphasic secretory potentials in cat and rabbit submaxillary glands, *Experientia* **29**:161–163.

Nishiyama, A., and Petersen, O. H., 1974, Membrane potential and resistance measurement in acinar cells from salivary glands in vitro: Effect of acetylcholine, *J. Physiol. (London)* **242**:173–188.

Obaid, A. L., Socolar, S. J., and Rose, B., 1983, Cell-to-cell channels with two independently regulated gates in series: Analysis of junctional conductance modulation by membrane potential, calcium, and pH, *J. Membr. Biol.* **73**:69–89.

Oliveira-Castro, G. M., and Loewenstein, W. R., 1971, Junctional membrane permeability: Effects of divalent cations, *J. Membr. Biol.* **5:**51–77.

Palmer, L. G., and Civan, M. M., 1977, Distribution of Na^+, K^+ and Cl^- between nucleus and cytoplasm in *Chironomus* salivary gland cells, *J. Membr. Biol.* **33:**41–61.

Pappenheimer, J. R., 1953, Passage of molecules through capillary walls, *Physiol. Rev.* **33:**387–423.

Peracchia, C., 1980, Structural correlates of gap junction permeation, *Int. Rev. Cytol.* **66:**81–146.

Petersen, O. H., 1976, Electrophysiology of mammalian gland cells, *Physiol. Rev.* **56:**535–577.

Petersen, O. H., 1980, *The Electrophysiology of Gland Cells,* Physiol. Soc. Monogr. Vol. 36, Academic Press, New York.

Politoff, A. L., Socolar, S. J., and Loewenstein, W. R., 1967, Metabolism and the permeability of cell membrane junctions, *Biochim. Biophys. Acta* **135:**791–793.

Politoff, A. L., Socolar, S. J., and Loewenstein, W. R., 1969, Permeability of a cell membrane junction: Dependence on energy metabolism, *J. Gen. Physiol.* **53:**489–515.

Quissell, D. O., and Redman, R. S., 1979, Functional characteristics of dispersed rat submandibular cells, *Proc. Natl. Acad. Sci. USA* **76:**2789–2793.

Rose, B., 1970, Junctional membrane permeability: Restoration by repolarizing current, *Science* **169:**607–609.

Rose, B., 1971, Intercellular communication and some structural aspects of membrane junctions in a simple cell system, *J. Membr. Biol.* **5:**1–19.

Rose, B., and Loewenstein, W. R., 1971, Junctional membrane permeability: Depression by substitution of Li for extracellular Na, and by long-term lack of Ca and Mg; restoration by cell repolarization, *J. Membr. Biol.* **5:**20–50.

Rose, B., and Loewenstein, W. R., 1975a, Calcium ion distribution in cytoplasm visualized by aequorin: Diffusion in cytosol restricted by energized sequestering, *Science* **190:**1204–1206.

Rose, B., and Loewenstein, W. R., 1975b, Permeability of cell junction depends on local cytoplasmic calcium activity, *Nature* **254:**250–252.

Rose, B., and Loewenstein, W. R., 1976, Permeability of a cell junction and the local cytoplasmic free ionized calcium concentration, *J. Membr. Biol.* **28:**87–119.

Rose, B., and Rick, R., 1978, Intracellular pH, intracellular free Ca, and junctional cell–cell coupling, *J. Membr. Biol.* **44:**377–415.

Rose, B., Simpson, I., and Loewenstein, W. R., 1977, Calcium ion produces graded changes in permeability of membrane channels in cell junction, *Nature* **267:**625–627.

Rose, B., Simpson, I., and Loewenstein, W. R., 1980, The cell-to-cell membrane channel: Permeability and its regulation, in: *Advances in Chemistry. Bioelectrochemistry: Ions, Surface Membranes,* Vol. 188 (M. Blank, ed.), pp. 391–408, American Chemical Society, Washington, D.C.

Ruano-Arroyo, G., Gerstenblith, G., and Lakatta, E. G., 1984, Calcium paradox in the heart is modulated by cell sodium during the calcium-free period, *J. Mol. Cell. Cardiol.* **16:**783–793.

Schwarzmann, G., Wiegandt, H., Rose, B., Zimmerman, A. L., Ben-Haim, D., and Loewenstein, W. R., 1981, Diameter of the cell-to-cell junctional channels as probed with neutral molecules, *Science* **213:**551–553.

Sheridan, J. D., and Atkinson, M. M., 1985, Physiological roles of permeable junctions: Some possibilities, *Annu. Rev. Physiol.* **47:**337–353.

Shimono, M., Yamamura, T., and Fumagalli, G., 1980, Intercellular junctions in salivary glands: Freeze fracture and tracer studies of normal rat sublingual gland, *J. Ultrastruct. Res.* **72:**286–299.

Simpson, I., Rose, B., and Loewenstein, W. R., 1977, Size limit of molecules permeating the junctional membrane channels, *Science* **195:**294–296.

Socolar, S. J., and Loewenstein, W. R., 1979, Methods for studying transmission through permeable cell-to-cell junctions, in: *Methods in Membrane Biology*, Vol. 10 (E. D. Korn, ed.), pp. 123–179, Plenum Press, New York.

Socolar, S. J., and Politoff, A. L., 1971, Uncoupling cell junctions of a glandular epithelium by depolarizing current, *Science* **172:**492–494.

Spray, D. C., and Bennett, M. V. L., 1985, Physiology and pharmacology of gap junctions, *Annu. Rev. Physiol.* **47:**281–303.

Spray, D. C., Harris, A. L., and Bennett, M. V. L., 1981, Equilibrium properties of a voltage-dependent junctional conductance, *J. Gen. Physiol.* **77:**77–93.

Stewart, W. W., 1978, Functional connections between cells as revealed by dye-coupling with a highly fluorescent naphthalimide tracer, *Cell* **14:**741–759.

Sulakhe, P. V., and St. Louis, P. J., 1980, Passive and active calcium fluxes across plasma membranes, *Prog. Biophys. Mol. Biol.* **35:**135–195.

van Venrooij, G. E. P. M., Hax, W. M. A., van Dantzig, G. F., Prijs, V., and Denier van der Gon, J. J., 1974, Model approaches for the evaluation of electrical cell coupling in the salivary gland of the larva of *Drosophila hydei:* The influence of lysolecithin on the electrical coupling, *J. Membr. Biol.* **19:**229–252.

van Venrooij, G. E. P. M., Hax, W. M. A., Schouten, V. J. A., Denier van der Gon, J. J., and van der Vorst, H. A., 1975, Absence of cell communication for fluorescein and dansylated amino acids in an electrotonic coupled cell system, *Biochim. Biophys. Acta* **394:**620–632.

Vozhkova, V. P., Kovalev, S. A., Mittelman, L. A., and Shilianskaia, E. N., 1970, Changes in the conductivity of intercellular contacts in the process of cell differentiation in the salivary gland of the *Drosophila virilis* larvae at the 3rd stage of development [in Russian], *Tsitologiya* **12:**1108–1115.

Weidmann, S., 1952, The electrical constants of Purkinje fibres, *J. Physiol. (London)* **118:**348–360.

Weidmann, S., 1966, The diffusion of radiopotassium across intercalated disks of mammalian cardiac muscle, *J. Physiol (London)* **187:**323–342.

Wiener, J., Spiro, D., and Loewenstein, W. R., 1964, Studies on an epithelial (gland) cell junction. II. Surface structure, *J. Cell Biol.* **22:**587–598.

Weingart, R., 1974, The permeability to tetraethylammonium ions of the surface membrane and the intercalcated disks of sheep and calf myocardium, *J. Physiol. (London)* **240:**741–762.

Zimmerman, A. L., and Rose, B., 1985, Permeability properties of cell-to-cell channels: Kinetics of fluorescent tracer diffusion through a cell junction, *J. Membr. Biol.* **84:**269–283.

Chapter 10

Intercellular Communication in Arthropods

Biophysical, Ultrastructural, and Biochemical Approaches

Robert C. Berdan

Department of Medical Physiology
The University of Calgary
Health Sciences Centre
Calgary, Alberta T2N 4N1, Canada

Scientific progress depends on the prudent selection of a basic problem which is susceptible to being solved at the current stage of scientific advancement and the selection of a proper system to study this problem.

—R. P. Rubin (1974)

1. INTRODUCTION

The gap junction is one form of membrane specialization that appears in the electron microscope as a closely apposed region of plasma membrane separating adjacent cells in contact by a 2- to 5-nm-wide "gap," from which the name *gap junction* is derived (Revel and Karnovsky, 1967). In freeze-fracture replicas viewed by electron microscopy, gap junctions are composed of tightly packed clusters of particles (Kreutziger, 1968). Each particle in a gap junction presumably represents the site of a single intramembrane channel that connects with the adjacent cell allowing a direct exchange or "communication" of ions and other small molecules. The functions of these "communicating channels" and the spectrum of activities they regulate are now beginning to be realized. In excitable tissues, gap junctions are synonymous with electrical synapses and are sometimes referred to as a nexus (Dewey and Barr, 1962). Gap junctions allow rapid signaling and the coordination of smooth and cardiac muscle contraction. It is apparent, however, that the role of gap junctions in the nervous system can be more complex and that they may also modify or regulate the behavior of discrete populations of neurons (Getting, 1974; Getting and Willows, 1974; Rayport and Kandel, 1980; Dudek *et al.*, 1983; Marder, 1984; Bennett *et al.*, 1985). The functions of gap junctions in nonexcitable tissues include, but are not limited to, tissue homeostasis (metabolic cooperation, i.e., the sharing of metabolites), coordination of a response to hormones or neurotransmitters, and, in lower invertebrates, adhesion. Since many tissues appear capable of regulating the permeability of their channels, the specificity of gap junction formation, and the timing at which gap junctions appear or disappear during development, cells may regulate the passage of signal molecules (morphogens) via gap junctions, which in turn could influence cell growth, differentiation, and pattern formation. For reviews on gap junctions in development, see Griepp and Revel (1977), Powers and Tupper (1977), Loewenstein (1978), Bennett *et al.* (1981), Schultz (1985), Caveney (1985), and Green and Gilula (1986).

The evidence that gap junctions provide a pathway for cell-to-cell communication has until recently been based primarily on correlations of the presence of gap junctions and the ability of cells to exchange small molecules (reviewed by Hertzberg *et al.*, 1981). Direct evidence for gap junctions providing a cell-to-cell pathway has now been demonstrated by specifically blocking this pathway by microinjection of antisera directed against the rat liver 27k gap junction polypeptide (Warner *et al.*, 1984) or against isolated gap junctions (Hertzberg *et al.*, 1985). These immunological probes are now being used to determine the potential roles this communication pathway may play in development (Warner *et al.*, 1984). The factors that may regulate the permeability of the individual channels, their biosynthesis, posttranslational modifications and assembly are, however, still largely unknown. Furthermore, if morphogens do pass through these channels, their identity remains a mystery. For more detailed information concerning the history, ultrastructure, structural diversity, and proposed functions of gap junctions, see reviews by Bennett and Goodenough (1978), Loewenstein (1979, 1981), Peracchia (1980), Hertzberg *et al.* (1981), and Larsen (1983).

Since the discovery of the electrical synapse between crayfish axons (Furshpan and Potter, 1957, 1959), and the detection of similar low-resistance pathways (channels) between insect salivary gland cells (Loewenstein and Kanno, 1963, 1964), studies carried out using arthropod tissues have contributed significantly to our understanding of gap junction-mediated communication. Arthropod tissues are particularly suitable for studying this form of cell communication since their cells are often large, readily accessible, and exhibit a simple geometrical organization. These properties are particularly relevant since they facilitate direct visualization of microelectrode impalements used to detect and quantify changes in intercellular communication. The rapid generation time of arthropods also facilitates monitoring changes in cell communication during development (Lopresti *et al.*, 1974; Caveney, 1976; Goodman and Spitzer, 1979; Weir and Lo, 1984) and introduces the possibility of studying intercellular communication from a genetic and molecular perspective (Palka, 1982; Wyman and Thomas, 1983; Ryerse and Nagel, 1984; Thomas and Wyman, 1984; Weir and Lo, 1985).

Arthropod gap junctions can be distinguished from those in vertebrates on the basis of their channel diameter, freeze-fracture morphology, solubility in detergents, and immunological reactivity. The purpose of this chapter is to present an overview of the properties of arthropod gap junctions and the range of techniques used to determine their biophysical, ultrastructural, and biochemical properties. Previous reviews on arthropod gap junctions have only dealt with their morphology, distribution, and physiology; however, they may be utilized as resources to supplement this chapter (Satir and Gilula, 1973; Flower, 1977; Gilula, 1978; Lane, 1978, 1981a, 1984; Rose, 1980; Peracchia, 1980; Telfer *et al.*, 1982; Caveney and Berdan, 1982; Lo, 1985; Caveney, 1986).

2. DETECTION AND ANALYSIS OF LOW-RESISTANCE PATHWAYS

The presence of channels connecting adjacent cells was first inferred on the basis of electrophysiological measurements. Furshpan and Potter (1959) observed that action potentials recorded with glass microelectrodes spread from one crayfish axon to the next with a 0.1-msec delay, or about 50 times faster than a chemical synapse. They also observed that even subthreshold depolarizations of the prefiber were accompanied by appreciable but smaller depolarizations of the postfiber. Similar low-resistance pathways were first observed in a nonexcitable tissue, the insect salivary gland, by Loewenstein and Kanno (1964). The presence of such low-resistance pathways has subsequently been demonstrated in almost all organized tissues from sponges to man.

2.1. Coupling Ratio

The most sensitive method for the detection of low-resistance pathways between cells due to the presence of interconnecting channels or cytoplasmic bridges involves multiple intracellular microelectrode recordings. The injection of a square pulse of current into a cell via a fine-tipped (0.2- to 0.5-μm diameter) glass microelectrode filled with an electrolyte (0.1–3 M KCl), results in a quantifiable change in membrane potential (Fig. 1).* If one considers the simplest

*Steady-state membrane potential changes due to current injection with one microelectrode may be measured with a second independent microelectrode inserted into the same cell, a double-barreled electrode, or even the same electrode used to inject current (Fig. 1). If the single-microelectrode

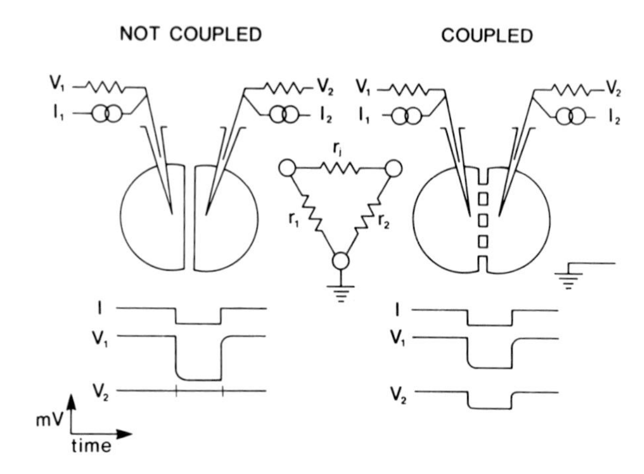

FIGURE 1. Diagram illustrating the detection of low-resistance pathways (''coupling'') between a pair of cells in contact. Microelectrodes are inserted into the cells; each microelectrode is capable of both injecting current and recording the change in membrane potential due to the drop in current as it passes from the cell through the membrane to the grounded saline bath. If the cells are not coupled by low-resistance pathways, a potential difference is only detected in the cell injected with current. If the cells are coupled by low-resistance pathways, some of the current injected into cell 1 travels into the adjacent cell, resulting in a smaller voltage drop in that cell. The ratio of the voltage drop in cell 2 to that in cell 1, into which the current was injected, V_2/V_1, is termed the coupling ratio. The lower the junctional resistance (r_j) between the cells and/or the higher the nonjunctional membrane resistances, the greater is the coupling ratio. Between the cell pairs is a schematic diagram of the equivalent circuit for a coupled cell pair. r_j represents the junctional resistance and r_1 and r_2 the nonjunctional membrane resistances of the respective cells. The membrane capacitance has been ignored in this simplified schematic and there is no evidence for a significant junctional capacitance.

arrangement, a cell pair, where the cells are connected by channels, part of the current injected into cell 1 enters cell 2. Passage of current into the adjacent ''coupled cell'' results in a change in its membrane potential, detected with a second microelectrode inserted in cell 2 (Fig. 1). If the cells are not coupled,

---

method is used, the amplifier must be equipped with a Wheatstone bridge, a circuit for measuring an unknown resistance by comparing it with known resistances. Using this circuit it is possible to nullify the voltage drop due to the resistance of the microelectrode which is in series with the membrane resistance. This procedure, referred to as ''balancing the bridge,'' is limited to cells with a high input resistance and long time constant, i.e., 10–100× greater than the microelectrode time constant. When these conditions are satisfied, recordings from a single microelectrode are comparable to those using a second independent microelectrode or a double-barreled microelectrode (Merickel *et al.*, 1977; Purves, 1981; Blatt and Slayman, 1983). It is also important, particularly with high-resistance microelectrodes, to determine the range over which they exhibit a linear current–voltage relationship.

Of further note, coupling measurements made between nerve cells are often made in low-calcium high-magnesium saline (Bulloch *et al.*, 1984). This saline prevents chemical synaptic activity on the neuron from acting as a shunt for the injected current.

current injection only results in a voltage drop in the current-injected cell (Fig. 1). Coupling therefore indicates that ions can move freely from cell to cell.

When cells are electrically coupled, the ratio of the voltage change in cell 2 relative to that in cell 1, when current is injected into cell 1, is referred to as the coupling coefficient or coupling ratio (Watanabe and Grundfest, 1961; Bennett, 1966). The magnitude of the coupling ratio is strongly influenced by the surface area of the coupled cells, the specific membrane resistance of each cell, the topology of the connections, and the number of channels connecting the cells. The usefulness of the coupling coefficient as a quantitative index of junctional resistance (r_j, also called r_c, coupling resistance) or its inverse junctional conductance ($1/r_j$) is limited to systems in which the topology of the cell connections is well defined and the coupling relatively weak (Socolar, 1977). From the equivalent circuit diagram of a coupled cell pair (Fig. 1), it is evident that the coupling ratio is a function of both the junctional resistance and the nonjunctional membrane resistances (r_1, r_2). Plotting the coupling coefficient as a function of the junctional and nonjunctional membrane conductances, Socolar (1977) shows that the coupling ratio is only sensitive to changes in the junctional conductance below values of about 0.8. Therefore, in well-coupled systems the coupling ratio is a rather insensitive index of junctional resistance. In many studies, however, primarily for reasons of convenience, or where the topology of the connections is unknown, only the coupling coefficient is determined.

Where coupling is restricted to a cell pair, it is possible to determine the junctional resistance (r_j). This involves three measurements (Watanabe and Grundfest, 1961).

1. Input resistance of cell 1:

$$R_1 = \frac{V_1}{I_1} = \frac{r_1\,(r_j + r_2)}{r_1 + r_2 + r_j}$$

2. Input resistance of cell 2:

$$R_2 = \frac{V_2}{I_2} = \frac{r_2\,(r_j + r_1)}{r_1 + r_2 + r_j}$$

3. Transfer resistance:

$$R_t = \frac{V_1}{I_2} = \frac{V_2}{I_1} = \frac{r_1 r_2}{r_1 + r_2 + r_j}$$

Note that the transfer resistances in the two directions are equal. From the above mesurements one can determine the following:

4. Membrane resistance of cell 1:

$$r_1 = \frac{R_1 R_2 - (R_t)^2}{R_2 - R_t}$$

5. Membrane resistance of cell 2:

$$r_2 = \frac{R_1 R_2 - (R_t)^2}{R_1 - R_t}$$

6. Junctional resistance:

$$r_j = \frac{R_1 R_2 - (R_t)^2}{R_t}$$

The coupling ratio has been shown to be related to the junctional resistance (Bennett, 1966) by the following relationship:

$$\frac{V_2}{V_1} = \frac{R_t}{R_1} = \frac{r_2}{r_j + r_2}$$

What is the minimum number of channels necessary to detect electrical coupling using standard electrophysiological methods? The major limiting factors are the nonjunctional membrane resistances and the sensitivity of the electrometers used. For computational purposes, if the sensitivity of the electrometers is considered to be 1 mV and the nonjunctional membrane resistances of the two cells are equal ($r_1 = r_2$), the magnitude of the nonjunctional resistances will be the major limiting factor. If r_j for a single channel is taken as $10^{10}\ \Omega$, a value consistent with experimentally determined single-channel resistances in the insect epidermis (Berdan and Caveney, 1985) and in vertebrate cells (Sheridan *et al.*, 1978; Neyton and Trautmann, 1985), then for $r = 10^7\ \Omega$, a single channel would be sufficient ($V_2/V_1 = 0.001$). If $r = 10^6\ \Omega$, at least 100 channels would be necessary to detect a 1-mV potential difference in cell 2. The sensitivity of this method of detecting electrical coupling is illustrated by the observation that coupling can be detected often within minutes of two cells coming into contact (Loewenstein, 1967).

Can electrical coupling occur in the absence of gap junctions, or interconnecting channels? The answer is yes. Electrical coupling can be detected if cytoplasmic bridges are present, for example usually only temporarily following mitosis. If cytoplasmic bridges are present, they may be revealed by electron microscopy or by observing the intercellular exchange of a microinjected tracer that is impermeant to gap junctions, such as horseradish peroxidase (Lo and Gilula, 1979a). Even in the absence of cytoplasmic bridges, electrical coupling has been detected in tissues in which gap junctions have not been found by freeze-fracture or in thin sections (Daniel *et al.*, 1976; Hardie, 1978; Dahl *et al.*, 1980; Williams and DeHaan, 1981). This is not entirely surprising since it is often difficult to detect small gap junctions by electron microscopy. Furthermore, there is no *a priori* reason for the channels to be clustered in order to be functional. Presumably, clustering functions to stabilize the site of contact, although the mechanisms by which clustering occurs are unknown (Weinbaum, 1980). Pseudocoupling (capacitive coupling) can also occur between cells or tissues that exhibit marked asymmetries in their nonjunctional membrane resistances. This occurs most notably in some embryos (Potter *et al.*, 1966; Bennett and Trinkaus, 1970; Regen and Steinhardt, 1986). In addition to the asymmetric membrane resistances, the cells may enclose a blastocoel cavity which may exhibit a substantial transepithelial resistance with respect to the bathing solution. Under these conditions, all or a major component of electrical coupling may occur via a low-resistance pathway in the blastocoel cavity. The magnitude of this component can be determined by short-circuiting the transepithelial resistance by poking a hole in the embryo. Alternative explanations for electrical coupling in some systems also exist (Politoff, 1977; Ramon and Moore, 1978; Somlyo, 1979).

2.2. Cable Analysis

Cells in arthropod tissues do not usually occur naturally in pairs, but rather are arranged into tubes or sheets. When tissues are arranged into tubes or linear arrays such as for example in the Malpighian tubules or tracheal oenocytes of insects (Fig. 2), the average intercellular resistance between the cells can be estimated by examining the intercellular electrotonic exponential decay of current. This analysis assumes the coupled cells behave as a core conductor and the electrotonic decay of injected current is given by the expression:

$$V_x = V_0 e^{-\lambda/x}$$

where V_0 is the electronic potential at $x = 0$ the point of current injection, V_x is the potential at distance x from this point, e is the natural logarithm 2.7182, and λ is the one-dimensional space constant, namely distance over which the electrotonic potential falls to a value of $1/e$ or 0.37 of its source value (Cole and Curtis, 1938; Hodgkin and Rushton, 1946; Weidmann, 1952; Loewenstein and Kanno, 1963).

To carry out a cable analysis on a linear array of cells, one introduces a microelectrode for current injection into one cell in the chain. Pulses of current are passed between the electrode and the grounded bath electrode, while a second "roving" microelectrode is used to measure the transient changes in membrane potential induced by the current injection. The attenuation of voltage drops

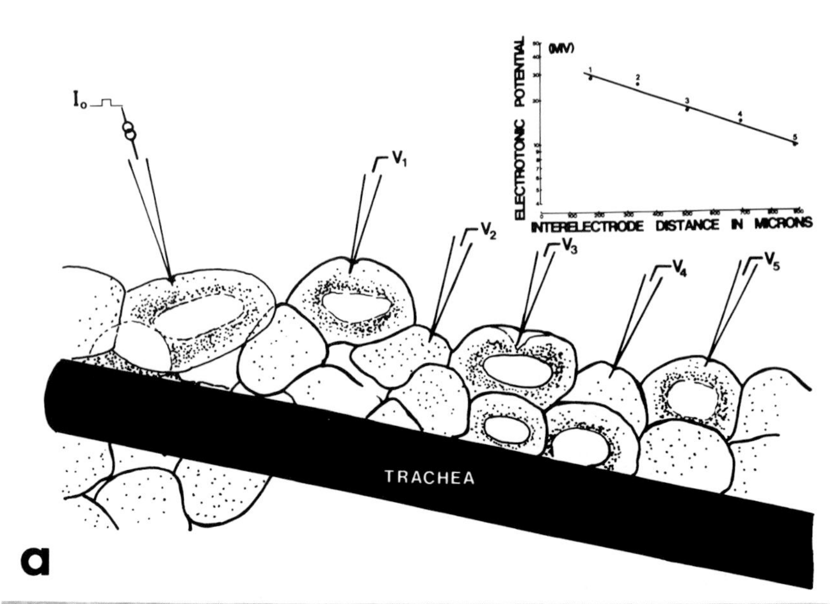
ELECTROTONIC POTENTIAL
(MV)
INTERELECTRODE DISTANCE IN MICRONS
I_o
V_1
V_2
V_3
V_4
V_5
TRACHEA
a

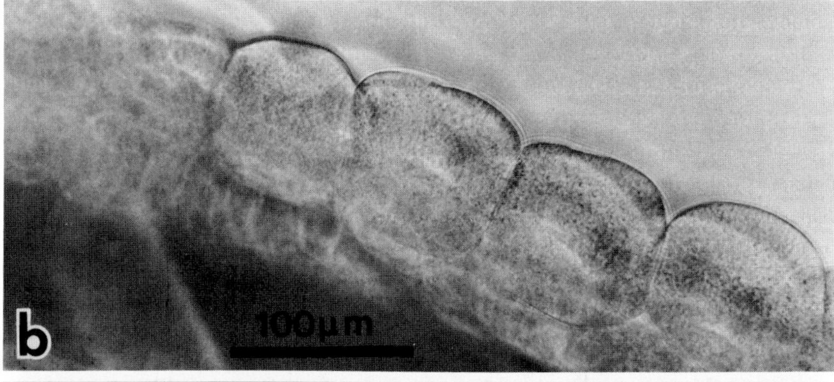
b
100 µm

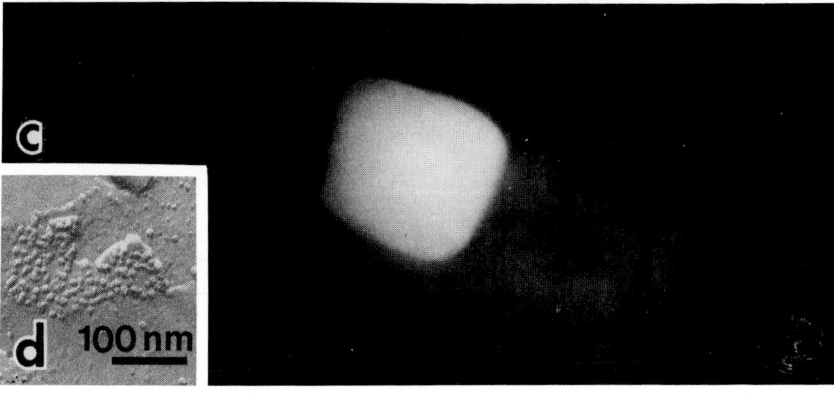
c
d
100 nm

recorded should be a function of distance from the current electrode (Fig. 2a). From the electrotonic potential decay curve, the input resistance and the space constant can be determined. A low input resistance (typically less than 1 MΩ) indicates strong coupling. The input resistance for the chain of oenocytes shown in Fig. 2 was 0.7 MΩ and the space constant was 700 μm. Space constants which are many times the length of single cells are common in insect tissues, indicating the ease with which inorganic ions pass from cell to cell (Caveney, 1986). For a detailed discussion of the cable properties of arthropod tissues, see Loewenstein (1966) and Socolar and Loewenstein (1979).

2.3. Ionic Coupling in Cell Sheets

Arthropods possess a hard cuticular exoskeleton secreted by an underlying monolayer sheet of cells, the epidermis. The arrangement of cells into sheets facilitates quantitative analysis of electrical coupling (Fig. 3). It has therefore been of interest to examine coupling in the insect epidermis during development (Caveney, 1976), in response to hormones (Caveney, 1978), drugs (Caveney and Blennerhassett, 1980; Lees-Miller and Caveney, 1982), and at compartmental boundaries (Warner and Lawrence, 1982; Caveney, 1974). The theory for the electrotonic spread in cell monolayers has been developed by Eisenberg and Johnson (1970), Shiba (1971), and Siegenbeek van Heukelom *et al.* (1972) and can be described by the Bessel function:

$$V_l = \frac{I_0 R_i}{2\pi t} K_0 \left(\frac{l}{\lambda'} \right)$$

where V_l is the electrotonic potential recorded at a distance l from the point of current injection, I_0 is the injected current at $l = 0$, R_i is the resistivity of the core material, and t is the sheet thickness. K_0 is a tabulated zero-order Bessel function of the second kind with an imaginary argument (Jahnke and Emde, 1960). The above equation holds true provided $l > t$. The two-dimensional space constant is defined as

FIGURE 2. (a) Analysis of ionic coupling in the tracheal oenocytes of *Tenebrio molitor*. An electrical current injected into the cell on the left ($I_0 = 60$ nA) causes a change in membrane potential in the other cells along the row. The electrotonic potential decay is an exponential function of the distance from the injected cell (inset upper right). (b) Phase-contrast picture of oenocytes lying on tracheae. (c) Epifluorescence image of tracheal oenocytes after the central cell had been iontophoretically injected with carboxyfluorescein. The bright (green) fluorescence of the tracer fills the cell into which it was injected; the adjacent cells have only a low (red) autofluorescence. Although ionic coupling between the cells is strong (coupling ratio $V_2/V_1 > 0.7$), the diffusion of carboxyfluorescein into the adjacent cells occurs at a rate too low to be visually detectable. The sensitivity of electronic methods for detecting coupling is normally much greater than fluorescent tracers. (d) Freeze-fracture replicas demonstrated only small gap junctions connecting the oenocytes.

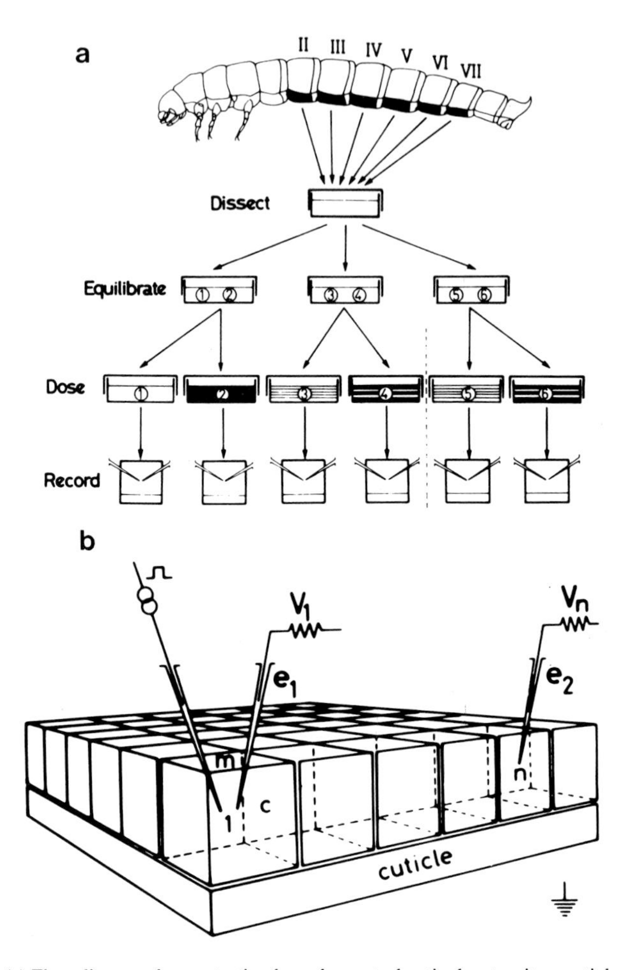

FIGURE 3. (a) Flow diagram demonstrating how the ventral cuticular sternites, cuticle, and attached epidermis from the larva of the beetle *Tenebrio molitor* are dissected and placed *in vitro*. The epidermal preparations can be maintained *in vitro* for several weeks. Such peparations can be exposed to a variety of drugs or hormones and the electrophysiological effects on membrane potential and coupling quantified. Subsequently, the same preparations can be prepared for histological or ultrastructural examination. From Caveney and Blennerhassett (1980) with permission.

(b) Diagram of a portion of the epidermal sheet, with microelectrodes positioned for electrotonic measurements. The current-injecting electrode (left) is inserted into cell 1. The recording electrode e is then positioned at fixed distances e_1 to e_2 and the electrotonic potential decay of current measured. Therefore, the average intercellular resistance between the cells is measured. The preparation in saline is grounded. From Caveney (1976) with permission.

$$\lambda' = \sqrt{R_m t/2R_i}$$

where R_m is the specific resistance of the membrane bounding the cytoplasm. In a hexagonally packed monolayer of cells, the specific resistance of the junctional membranes (R_j), in series with the cytoplasmic resistivity (R_c), determine the overall resistivity of the intercellular route:

$$R_i = R_c + R_j/d\sqrt{3}$$

(Siegenbeek van Heukelom *et al.*, 1972). R_c can be determined experimentally with a single microelectrode (Schanne and DeCeretti, 1971; Caveney and Blennerhassett, 1980; Caveney *et al.*, 1986) by calibrating the microelectrode tip resistance in salt solutions of different resistivities and then within the cell cytoplasm. In the beetle epidermis R_c = 60 Ω-cm (Caveney and Blennerhassett, 1980), the denominator $d\sqrt{3}$ takes into account the cell density where d is the length of one side of a hexagonal cell in the sheet. The effective intercellular resistance $r_i = R_i/t$ is proportional to the slope of the electrotonic potential decay (Fig. 4). Changes in the nonjunctional membrane resistance are detected as parallel shifts in the linear regions of the electrotonic potential decay curves. Solving for the junctional resistivity gives:

$$R_j = [(r_i t) - R_c]d\sqrt{3}$$

This equation assumes that the entire area of the membrane separating the cells in the sheet contains the low-resistance pathways. Assuming the channels are restricted to the gap junctions, the specific resistance of the gap junctional membrane may be obtained. The junctional resistivity (R_j), multiplied by the mean length of gap junction per cell interface (determined by morphometric analysis of electron micrographs) and divided by the sheet thickness, gives the specific gap junction resistance (Ω-cm^2). The inverse of this value is the specific gap junctional conductance (S/cm^2, S = siemens = mho). It is often more convenient to deal with conductance values since they are additive when in parallel. The mean conductance of a single gap junction channel can be calculated by dividing the specific gap junctional conductance by the density of the gap junction channels determined from freeze-fracture replicas (Berdan and Caveney, 1985). The electrophysiological and morphometric data necessary to determine the single gap junction channel conductance in the beetle epidermis are summarized in Table I. The single-channel conductance after lowering the intercellular resistance by exposure to the insect molting hormone (Caveney and Blennerhassett, 1980) is also shown in Table I. The calculation of the single-channel conductance rests on several assumptions. It assumes for example that each particle in a gap junction plaque is an open channel and only the particles within gap junction plaques are considered to be functional channels.

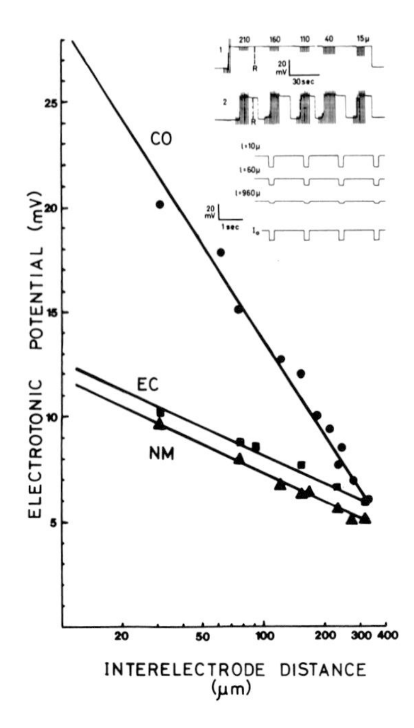

FIGURE 4. Electrotonic spreads from *Tenebrio* epidermis. Representative chart recordings for electrotonic potentials are shown at the upper right (with permission from Caveney, 1976). The values when plotted against interelectrode distance, fit a Bessel function, which for practical purposes between experimentally determined values is approximated by a straight line in a logarithmic plot. 20-Hydroxyecdysone treatment (EC, 2 μg/ml for 18 hr *in vitro*, $r_i = 2.08 \times 10^5\ \Omega$) exhibits shallow electrotonic potential decays, a characteristic of a highly conductive epidermal cell sheet. Control epidermis (CO) untreated *in vitro* for 48 hr from 6-day-old animals (from time of last molt) exhibits a more rapid electrotonic potential decay ($r_i = 6.85 \times 10^5\ \Omega$). Note that the intercellular resistance in newly molted animals (NM) is similar to that in ecdysone-treated epidermis. In newly molted epidermis, however, the increased conductivity is associated with a threefold increase in amount of gap junction, while in hormone-treated epidermis it appears to be due to preexisting closed channels opening.

Table I
Conductance of a Single Gap Junction Channel[a]

	Control epidermis	Hormone-treated epidermis
Intercellular resistance, r_i (Ω)[b]	5.03×10^5	2.75×10^5
Sheet thickness, t (μm)	8.4	9.3
Intercellular resistivity, R_i (Ωcm)[b]	423	256
Cytoplasmic resistivity, R_c (Ωcm)[c]	64	65
Cell density (per mm^2)[c]	10,948	11,878
Cell density factor ($d\sqrt{3}$)	1.08×10^{-3}	1.03×10^{-3}
Cell girth (μm)[d]	40	40
Length of gap junction per interface (μm)[b]	0.92	0.87
Junctional resistivity (Ωcm^2)	0.39	0.20
Gap junctional specific resistivity (Ωcm^2)	0.043	0.019
Gap junctional specific conductance (S/cm^2)	23	53
Channel packing (per cm^2)	2.456×10^{11}	2.490×10^{11}
Number of channels per cell	9.0×10^4	8.7×10^4
Conductance of a single channel (pS)	94	213

[a]From Berdan and Caveney (1985).
[b]Data from Caveney *et al.* (1980, Table 5).
[c]Data from Caveney and Blennerhassett (1980, Tables 5 and 6).
[d]Data from Caveney and Podgorski (1975, Table 1).

The conductance value for a single channel is useful (if all the assumptions are fulfilled) since it can be used to estimate the channel diameter (Socolar and Loewenstein, 1979). From the conductance values in Table I and a consideration of the theoretical channel conductance (Berdan and Caveney, 1985), we have estimated that the gap junction channel diameter in the beetle, *Tenebrio molitor*, epidermis is between 1.6 and 2.3 nm, which is consistent with the channel diameters estimated in other insect tissues on the basis of microinjected fluorescent tracer molecules of varying size (Rose, 1980; Schwarzmann *et al.*, 1981; Berdan, 1982).

Estimates of the single gap junction channel conductance are rare since it involves obtaining quantitative electrophysiological and ultrastructural data from the same cells. However, the theoretical conductance of a gap junction channel has been estimated to be around 100 pS (Loewenstein, 1979). In vertebrate tissues the single gap junction channel conductance between Novikoff hepatoma cells was found to be 113 pS (Sheridan *et al.*, 1978) and by use of a double patch clamp on isolated cell pairs from rat lacrimal glands, 120 pS (Neyton and Trautmann, 1985).

2.4. Specificity of Cell–Cell Coupling

Electrical coupling, under most circumstances, suggests that cells are connected by intercellular channels. In most tissues this form of communication appears to be mediated by gap junctions. Previous suggestions that the septate junction may also mediate electrical coupling in arthropods (Bullivant and Loewenstein, 1968; Gilula *et al.*, 1970) seem unlikely since coupling is absent in some insect midguts that have extensive septate junctions, but lack gap junctions (Blankemeyer and Harvey, 1978; Berdan *et al.*, 1985). Arthropod tissues lacking septate junctions but possessing gap junctions, however, are electrically coupled. Examples include the insect fat body (Caveney and Berdan, 1982), trachael oenocytes (Fig. 2), and electrically coupled neurons such as the crayfish septate axon (Payton *et al.*, 1969). Not all cells within a tissue are necessarily coupled, but rather some degree of selectivity occurs. This topic has been reviewed (Caveney and Berdan, 1982) and only a few representative examples are given here. In the beetle epidermis, other specialized cells such as dermal glands and oenocytes may be embedded (Wigglesworth, 1948; Quennedey *et al.*, 1983). Coupling within the epidermis is uniform but is reduced between epidermal cells and gland cells, and absent between oenocytes and epidermal cells (Caveney and Berdan, 1982). The absence of coupling between the oenocytes (which are thought to be involved in steroid biosynthesis; Locke, 1969) and the epidermis, allows the oenocytes to operate metabolically independent of the epidermis. Another example of selective coupling occurs in the rectum of the cockroach, *Periplaneta americana* (Caveney and Berdan, 1982). The insect rectum forms a complex ion-transporting organ which can be divided into physiologically distinct compartments. Only cells within each compartment are electrically and dye coupled. Compartmentalization in the rectum presumably assists in the formation of extracellular ion gradients involved in the reabsorption of water (Berridge and Oschman, 1972). The greatest specificity of coupling, however, occurs within the nervous system. Here, coupling may be restricted to a single pair of neurons or alternatively to a small group of neurons. The identification of coupled neurons in arthropod nervous tissue is usually accomplished by iontophoresis of fluorescent tracers (Glantz and Kirk, 1980; Arechiga *et al.*, 1985). In the molluscan and leech nervous systems, however, the specificity of electrical coupling has been elegantly demonstrated between readily identifiable neurons (Kuffler and Potter, 1964; Bulloch and Kater, 1982; Bulloch *et al.*, 1984). It is interesting to note that whereas neurons couple with other neurons, they do not appear to couple with the often closely apposed glial cells (Kuffler and Potter, 1964). The glial cells are, however, coupled and connected by numerous gap junctions (Kuffler and Potter, 1964; Lane, 1981a; Pentreath, 1982; Berdan and Bulloch, unpublished observations). It is also notable that gap junctions between glia can be distinguished morphologically from those between neurons (in vertebrates, Brightman and Reese, 1969; in crayfish, Peracchia and Dulhunty, 1976). Although dye coupling has been observed between crayfish axons and the surrounding glia (Viancour *et al.*, 1981), it is not clear whether this occurs via cytoplasmic bridges (Peracchia, 1981), possibly as a result of a fixation artifact (Bennett, 1973), or is due to actual gap junctions (Caudras *et al.*, 1985, 1986) or some other transport mechanism (Lasek and Tytell, 1981). The mechanism of intercellular exchange and communication between glia and neurons appears to be unique and is not restricted to small molecules and certainly deserves closer examination. Coupling measurements have also been used to demonstrate that whereas a variety of vertebrate cells in culture will couple nonspecifically with heterologous vertebrate cells, they do not couple with insect cells (Epstein and Gilula, 1977). Furthermore, these investigators showed that among different insect cell lines, coupling would not occur between cells from different insect orders (Homoptera, Lepidoptera, and Diptera). The only other instance where species specificity of coupling has been shown is between sponge cells from the phylum Porifera (Loewenstein, 1967).

2.5. Modulation of Coupling

Changes in coupling or intercellular resistance have been documented during the development of the insect epidermis (Caveney, 1976), insect salivary glands (Loewenstein, 1966), and during the differentiation of insect neuroblasts (Goodman and Spitzer, 1979). Changes in electrical coupling that are independent of the changes in the nonjunctional membrane resistance reflect changes in gap junction permeability. Several mechanisms shown in Fig. 5 may account for these changes. An increase in coupling (drop in r_i or r_j) can occur by an increase in the number of channels connecting adjacent cells. If the channels are clustered into gap junctions, this would be detected as increase in the amount of gap

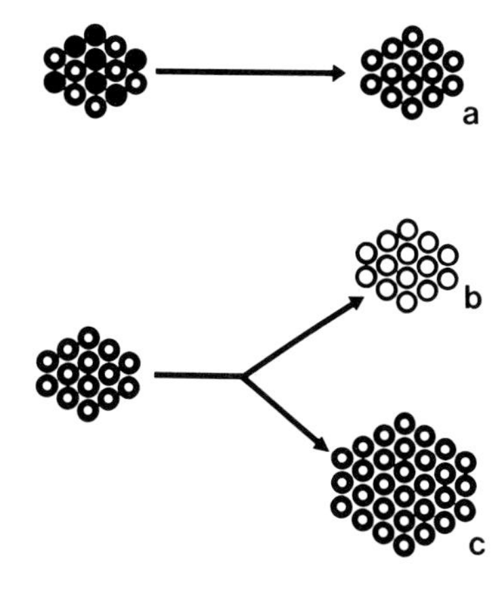

FIGURE 5. Three possible mechanisms which could account for an increase in intercellular conductivity between cells in the absence of changes in nonjunctional membrane resistance. (a) Preexisting closed gap junction channels could open. This type of change is not detectable on the basis of morphological criteria alone. (b) The diameter of the gap junction channels could increase. This mechanism would allow a population of larger molecules to pass intercellularly. (c) Additional gap junction channels could be incorporated between the cells. Even this mechanism would only be detected morphologically if newly formed channels were clustered into discrete plaques.

junction per cell interface by an ultrastructural morphometric analysis. Such a mechanism appears responsible for the drop in intercellular resistance ($r_i = 2.14 \times 10^5\ \Omega$, Fig. 4) in the newly molted beetle epidermis (Caveney and Podgorski, 1975). In "control epidermis" taken from animals 6 days into the instar (stage between molts), $r_i = 6.85 \times 10^5\ \Omega$ (Fig. 4) and there is 0.9–1.0 μm total length of gap junction per cell interface (Caveney and Podgorski, 1975; Caveney *et al.*, 1980). In the newly molted epidermis, this value increases to 3.0 μm. A drop in intercellular resistance can also be induced in the epidermis by exposing 6-day epidermis *in vitro* to the insect molting hormone (2 μg/ml 20-hydroxyecdysone, Caveney, 1978). This increase in cell communication occurs within 18 hr and in the absence of protein synthesis or an increase in the amount of gap junction per cell interface (Caveney *et al.*, 1980). Since the drop in intercellular resistance ($r_i = 2.08 \times 10^5\ \Omega$) cannot be explained by changes in the nonjunctional membrane resistance (Caveney and Blennerhassett, 1980), two plausible explanations are that the gap junction channel diameters get larger or that preexisting closed channels open in response to hormonal stimulation (Fig. 5). Kinetic studies using microinjected fluorescent tracers (discussed in the following section) suggest that the latter mechanism best explains the available data.

A decrease in cell communication or electrical coupling may come about by the removal of gap junction channels, either by dispersal (Lane and Swales, 1980), by internalization of the gap junctions (Szollosi and Marcaillou, 1980), or by closure of the channels (Loewenstein, 1966). Cell injury is often the easiest manner in which to induce uncoupling (Loewenstein, 1966). Closing of the gap junction channels following injury prevents metabolite leakage from adjacent cells and allows the injured cell to die in isolation. The presence of divalent cations in the extracellular medium, in particular Ca^{2+}, appears necessary for injury-induced uncoupling (Loewenstein, 1966). Ca^{2+} therefore has been hypothesized to act directly on the gap junction membrane (Loewenstein, 1966). Incubating insect cells in the presence of the Ca^{2+} ionophore A23187, or micro-

Table II
Agents That Abolish Ionic Coupling between Arthropod Cells[a]

Agent	Concentration	Time to uncoupling	Species, tissue	Reference
1. *Ionophores*				
A23187	2×10^{-6} M	20 min	*Chironomus* salivary gland	Rose and Rick (1978)
	2×10^{-6} M	10–15 min	*Trichoplusia* cell line	Gilula and Epstein (1976)
	2×10^{-6} M	60 min	*Tenebrio* epidermis	Caveney (1978)
X537A	1×10^{-5} M	6 min	*Chironomus* salivary gland	Rose and Loewenstein (1976)
Nigericin[b]	2×10^{-6} M	2 min	*Chironomus* salivary gland	Rose and Rick (1978)
2. *Ion substitution in the external medium*				
Li_e^+ for Na_e^+	8.4×10^{-2} M	180 min	*Chironomus* salivary gland	Rose and Loewenstein (1971)
	8.0×10^{-2} M	60 min	*Tenebrio* epidermis	Popowich and Caveney (1976)
3. *Direct introduction of cations into cells*				
H^+ injection	pH 6.73	4 min	*Chironomus* salivary gland	Rose and Rick (1978)
Divalent series	$^iCa > Mg > Sr > Ba > Mn > 5 \times 10^{-5}$ M		*Chironomus* salivary gland	Oliveira-Castro and Loewenstein (1971)
Ca^{2+} injection			*Chironomus* salivary gland	Loewenstein *et al.* (1967)
	$\sim 10^{-4}$ M	<3 min	*Tenebrio* epidermis	Popowich and Caveney (1976)
4. *Anticalmodulin drugs*				
Trifluoperazine	1×10^{-4} M	60 min	*Tenebrio* epidermis	Lees-Miller and Caveney (1982)
Chlorpromazine	7×10^{-4} M	20–30 min	*Chironomus* salivary gland	Suzuki *et al.* (1978)
	2.2×10^{-4} M	60 min	*Tenebrio* epidermis	Lees-Miller and Caveney (1982)
Chlorprothixine	2.3×10^{-4} M	60 min	*Tenebrio* epidermis	Lees-Miller and Caveney (1982)
Thioridazine	1.3×10^{-4} M	60 min	*Tenebrio* epidermis	Lees-Miller and Caveney (1982)
d-Butaclamol	1.5×10^{-4} M	60 min	*Tenebrio* epidermis	Lees-Miller and Caveney (1982)
l-Butaclamol	3.2×10^{-4} M	60 min	*Tenebrio* epidermis	Lees-Miller and Caveney (1982)
5. *Anesthetics*				
Procaine	2×10^{-3} M	15–30 min	*Chironomus* salivary gland	Suzuki *et al.* (1978)

	2×10^{-2} M[c]	80–180 min	*Tenebrio* epidermis	Lees-Miller (unpublished)
	5×10^{-3} M	no effect	*Procambarus* septate axon	Johnston *et al.* (1980)
Dibucaine	9×10^{-4} M	60 min	*Tenebrio* epidermis	Lees-Miller and Caveney (1982)
Tetracaine	3.7×10^{-3} M	60 min	*Tenebrio* epidermis	Lees-Miller and Caveney (1982)
6. *Metabolic inhibitors*				
Cyanide	5×10^{-3} M	14 min	*Chironomus* salivary gland	Rose and Loewenstein (1976)
Dinitrophenol	$1–2 \times 10^{-4}$ M	5–40 min	*Chironomus* salivary gland	Politoff *et al.* (1969)
Oligomycin	$1–2 \times 10^{-4}$ M	1–2 hr	*Chironomus* salivary gland	Politoff *et al.* (1969)
Iodoacetate	2×10^{-3} M	23–140 min	*Tenebrio* epidermis	Lees-Miller and Caveney (1982)
7. *Steroids*				
Corticosterone	2.9×10^{-5} M	10 min	*Chironomus* salivary gland	Suzuki and Higashino (1977)
Aldosterone	5.6×10^{-7} M	10–15 min	*Chironomus* salivary gland	Suzuki and Higashino (1977)
8. *Temperature*				
Dropped from 19° to 6–8°C		60 min	*Chironomus* salivary gland	Politoff *et al.* (1969)
Dropped from 12° to 2°C		70 min	*Procambarus* septate axon	Ramon and Zampighi (1980)
9. *Agents that perturb membrane structure*				
n-Alkanol series				
Ethanol	20%	5–20 min	*Chironomus* salivary gland	Suzuki *et al.* (1978)
Heptanol	3×10^{-3} M	50 min	*Procambarus* septate axon	Johnston *et al.* (1980)
	7×10^{-3} M	10 min	*Tenebrio* epidermis	Lees-Miller (unpublished)
Octanol	1×10^{-3} M	(1 hr?)	*Procambarus* septate axon	Johnston *et al.* (1980)
Nananol	3×10^{-4} M	slight effect	*Procambarus* septate axon	Johnston *et al.* (1980)
	1×10^{-3} M	7–22 min	*Tenebrio* epidermis	Lees-Miller (unpublished)
Dodecanol	1×10^{-3} M	3 hr	*Tenebrio* epidermis	Lees-Miller (unpublished)
Lipids				
Tea[d]	2×10^{-3} M	10–20 min	*Chironomus* salivary gland	Suzuki *et al.* (1978)
cis-Oleic acid	5×10^{-4} M	18–27 min	*Tenebrio* epidermis	Lees-Miller (unpublished)

[a]From Caveney (1986).
[b]pH of external medium was lowered from 7.4 to 6.5.
[c]Uncoupling was only seen when pH of medium raised from 6.7 to 7.8.
[d]TEA, tetraethylammonium.

injecting Ca^{2+} into the cells, also rapidly uncouples them (Rose and Loewenstein, 1975; Popowich and Caveney, 1976; Caveney, 1978). Measurements conducted on the salivary gland cells of the midge, *Chironomus*, with the fluorescent probe aequorin and Ca^{2+}-sensitive microelectrodes (Rose and Rick, 1978), have correlated a rise in intracellular Ca^{2+} with uncoupling. In other tissues, including the crayfish electrical synapse, the internal pH (i.e., H^+ ion concentration) has been implicated in regulating gap junction channel permeability (Turin and Warner, 1980; Spray *et al.*, 1981; Campos de Carvalho *et al.*, 1984). The difficulty in distinguishing which (if either) ion, Ca^{2+} or $H^+{}_i$, exerts a primary effect on the gap junction channels is that changes in $[Ca^{2+}]_i$ affect the internal pH and vice versa. By way of further complication, cell acidification induced by exposure to elevated CO_2 may induce either a rise or a drop in free Ca^{2+} (Baker and Honerjager, 1978; Lea and Ashley, 1978; Alvarez-Leefmans *et al.*, 1981). In addition, different cells and tissues exhibit marked differences in their tolerance to uncoupling treatments. In one study in which the junctional membrane between coupled pairs of *Fundulus* blastomeres was perfused with known H^+ and Ca^{2+} concentrations while electrical coupling was monitored, the sensitivity to H^+ was 10,000 times greater than that to Ca^{2+} (Spray *et al.*, 1982). However, in crayfish axons, Johnston and Ramon (1981) were unable to uncouple adjacent axons by perfusing the junctional membranes with either elevated H^+ or Ca^{2+}. These investigators hypothesize that an additional diffusible factor that was removed by perfusion is required for gap junction uncoupling. A potential candidate for this diffusible factor is calmodulin (Lees-Miller and Caveney, 1982; Peracchia *et al.*, 1983; Peracchia, 1984; Peracchia and Bernardini, 1984). Although calmodulin inhibitors have been shown to block electrical and dye coupling in the beetle epidermis in a reversible manner (Lees-Miller and Caveney, 1982; Table II), they inhibit uncoupling between vertebrate cells (Peracchia *et al.*, 1983) and fail to specifically block coupling between electrically coupled snail neurons (Bulloch and Berdan, unpublished observation). Other agents reported to block electrical coupling between arthropod cells are summarized in Table II. Two other factors implicated to regulate coupling that may be of physiological significance are transjunctional voltage (reviewed by Harris *et al.*, 1983) and membrane potential (Obaid *et al.*, 1983). Taken together, the available data suggest that the regulation of gap junctional conductance in different organisms and different tissues may be regulated by different mechanisms and/or that junctional conductance may be regulated by several mechanisms.

3. JUNCTIONAL PERMEABILITY TO MOLECULAR TRACERS

The permeability properties of intercellular channels have been probed with a variety of tracers including colorant dyes, fluorescent tracers, radioactive metabolites, and organic catalysts. Tracers may be microinjected into cells using

Table III
Junctional Permeability of *Tenebrio* Epidermis Probed with Fluorescent Tracers[a]

	Molecular weight	Charge	Newly molted epidermis	Intermolt epidermis	20-Hydroxyecdysone-treated epidermis
5-Aminofluorescein	347	–	+	+	+
Sodium fluorescein	376	–	+	+	+
6-Carboxyfluorescein	376	–	+	+	+
Dans(Glu)OH	380	–	+	+	+
FITC	389	–	+	+	+
Dichlorofluorescein	401	–	+	+	+
Lucifer Yellow CH	457	–	+	+	+
Rhodamine-6G	474	+	+	+	+
Dibromofluorescein	490	–	+	IF	IF
FITC(Glu)OH	536	–	+	+	+
LRB-SO_2Cl	578	–	+	+	–
LRB-200	581	–	+	+	+
Diiodofluorescein	584	–	IF	IF	IF
Fluorexon	623	–	+	+	+
Procion Yellow[b]	625	–	+	+	NT
FITC(Glu_2)OH	655	–	+	+	+
FITC(Gly_6)OH	749	–	+	IF	IF
FITC(Val-Gly-Asp-Glu)OH	808	–	IF	IF	IF
Tridansyl spermidine	843	0	IF	IF	IF
Rose bengal	1017	–	IF	IF	IF
FITC-proctolin	1038	–	IF	IF	IF
Actinomycin D	1255	–	+	+	+

[a]Abbreviations: Asp, aspartic acid; Dans, dansyl chloride; FITC, fluorescein isothiocyanate; (Glu)OH, glutamic acid; (Gly)OH, glycine; IF, insufficient fluorescence; LRB, Lissamine rhodamine B; NT, not tested; protcolin, (Arg-Tyr-Leu-Pro-Thr)OH; Val, valine.
[b]Caveney and Podgorski (1975).

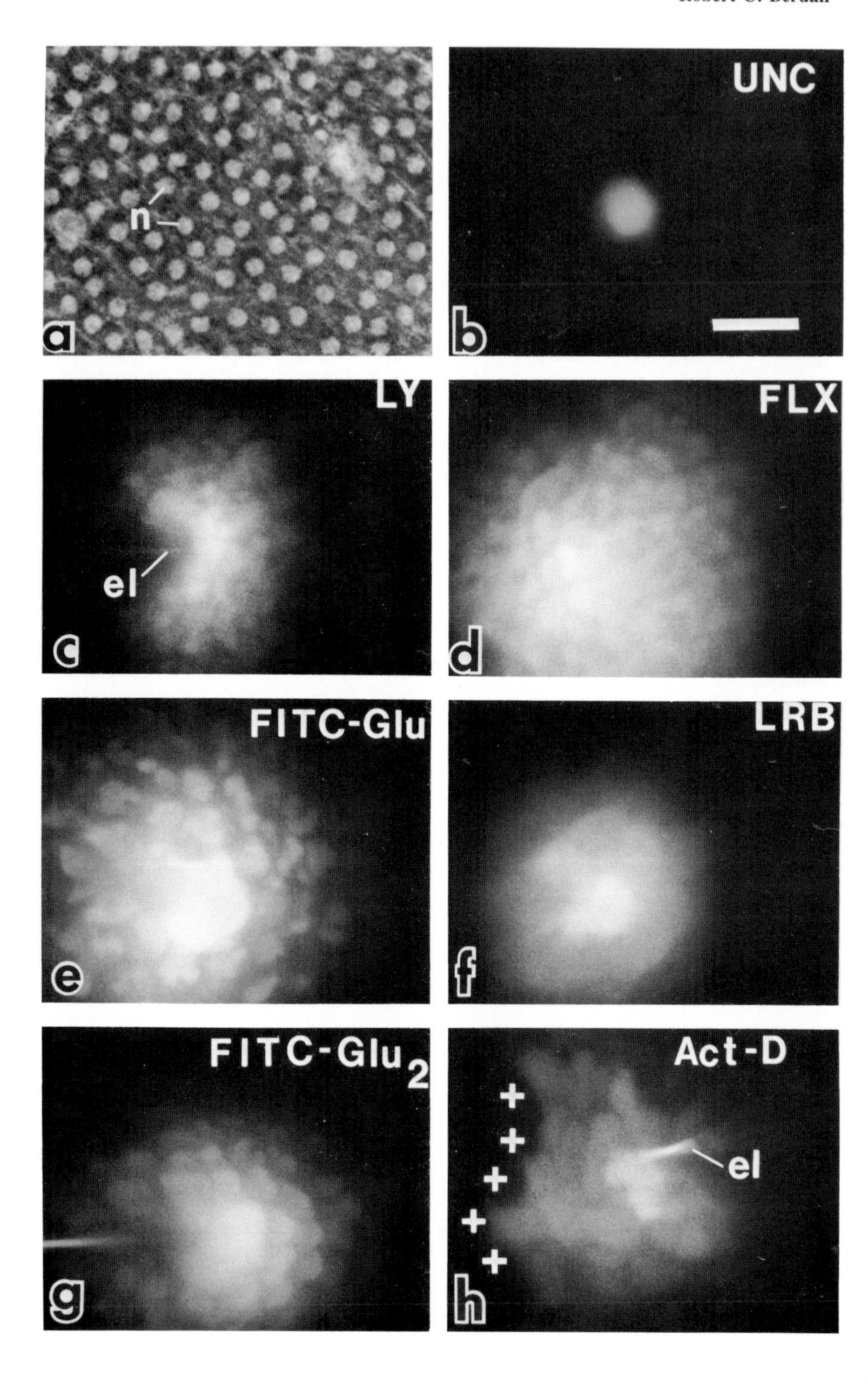
n
a
UNC
b
LY
el
c
FLX
d
FITC-Glu
e
LRB
f
FITC-Glu
2
g
Act-D
el
h

capillary filled glass microelectrodes similar to those normally used for recording membrane potentials. Delivery of the tracers may be via pulses of air (pressure injection) or by current injection (iontophoresis). Pressure injection may be utilized for large molecules which have no net charge, although its use is generally limited to large cells. Iontophoresis involves injection of current pulses or ramps of opposite polarity to the charged molecule one wishes to eject from the microelectrode. Iontophoresis is often aided by backfilling the microelectrodes with a conducting electrolyte such as 0.1 M KCl on 0.1 M $NaHCO_3$. Large tracers such as horseradish peroxidase (40,000 daltons) may be injected iontophoretically; however, larger tracers or viscous solutions often require filling the microelectrode tips with buffer first (2–5 μliters) and then with the tracer. Diffusion of the tracer solution into the microelectrode tips then occurs by storing them at 4°C, usually overnight. For a comprehensive review of iontophoretic injection methodology, see Hicks (1984).

Fluorescent tracers have been particularly valuable since they are detectable at concentrations several orders of magnitude lower than colorants, are relatively nontoxic, and their movement can be seen in living cells. Another major advantage of the use of fluorescent tracers for the detection of intercellular channels is that only a single microelectrode is required. In complex tissues such as in the nervous system where communication may be restricted to only a few cells, often by tortuous pathways, microinjected fluorescent tracers are particularly useful.

3.1. Fluorescent Tracers

The identification of coupled cells has been facilitated by the introduction of several highly fluorescent tracers (Table III). The two most commonly used fluorescent tracers are carboxyfluorescein and Lucifer Yellow CH (Figs. 6c, 7a, and 8). Both tracers have two negatively charged side groups and have very low nonjunctional membrane permeabilities (Weinstein *et al.*, 1977; Stewart, 1978). The major advantage of Lucifer Yellow is that it has an -NH_2 group which can be cross-linked to proteins by formaldehyde so that it can be observed in tissues after fixation, dehydration, and clearing. Carboxyfluorescein is significantly

←

FIGURE 6. Light micrographs of *Tenebrio* epidermis following *in vitro* culture and iontophoretic injection with fluorescent tracers. Photographs are at approximately the same magnification. Bar = 20 μm; el, microelectrode. (a) Negative phase contrast, intermolt epidermis; n, cell nuclei. (b–h) Epifluorescence optics. (b) Chlorpromazine (10^{-4} M)-uncoupled (UNC) epidermis. Carboxyfluorescein stays within the injected cell. Tracer coupling can be restored by placing cells in fresh medium. (c) Intermolt epidermis, Lucifer Yellow (LY). (d) Intermolt epidermis, fluorexon (FLX). (e) Newly molted epidermis, FITC-glutamic acid (FITC-Glu). (f) Newly molted epidermis, Lissamine rhodamine B-200 (LRB). (g) Intermolt epidermis, FITC-diglutamic acid (FITC-Glu_2). (h) Newly molted epidermis, actinomycin D (Act-D). +, cells intentionally uncoupled by electrode-induced injury. The tracer does not pass into the uncoupled cells, suggesting it is not passing through nonjunctional membranes at detectable rates.

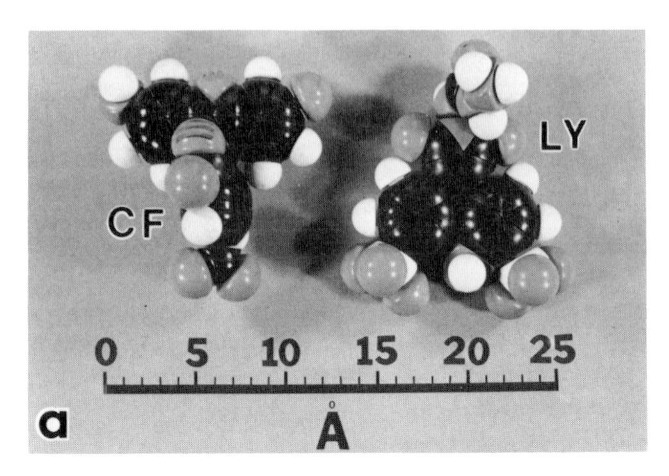
LY
CF
0 5 10 15 20 25
a
Å

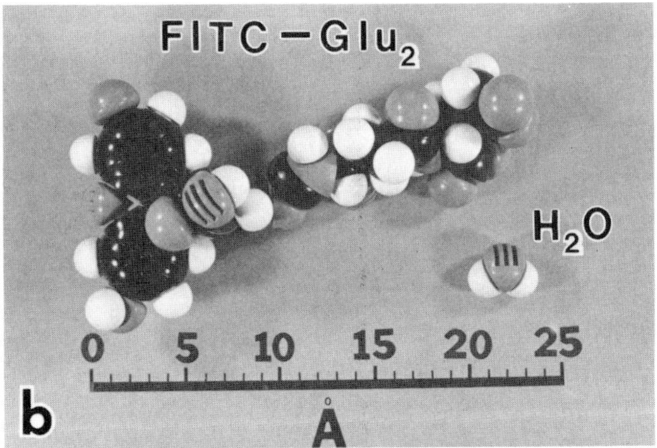
FITC-Glu2
H2O
0 5 10 15 20 25
b
Å

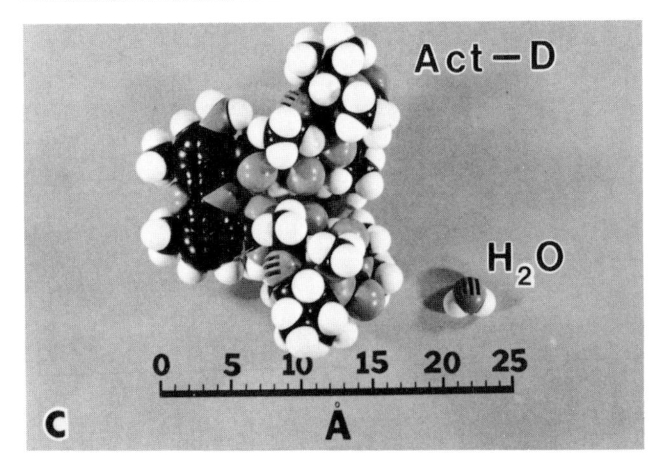
Act-D
H2O
0 5 10 15 20 25
c
Å

more fluorescent than Lucifer Yellow, more economical to use, but cannot be observed after fixation. Unfortunately, the most highly fluorescent tracer available, sodium fluorescein, has a relatively high nonjunctional membrane permeability, passing through membranes with a half-time of only 5 min (Weinstein *et al.*, 1977; Socolar and Loewenstein, 1979). In any event, the nonjunctional permeability of a fluorescent tracer can be readily tested by bathing the tissue in a 1–50 mM concentration of the probe in saline, or medium, for various periods of time and subsequently examining the tissue for uptake. It is not surprising to find considerable variability in the nonjunctional membrane permeability in different tissues of the same organism. Even small alterations in the net charge of the tracer, for instance as induced by pH, may significantly alter the permeability of the tracer through nonjunctional membranes. Rhodamine-6G, for example, which is similar in structure to carboxyfluorescein and rhodamine but has a net positive charge, crosses insect epidermal cell membranes and is sequestered within mitochondria within seconds (Berdan, 1982). Fluorescein and dansylated amino acids also cross insect cell membranes rapidly and can be detected within tissues bathed in 1 mM solutions for several minutes (van Venrooij *et al.*, 1975; Berdan, 1982).

The detection of the diffusion of a tracer into adjacent cells (dye coupling) is considerably less sensitive than electrical coupling measurements. Therefore, the absence of dye coupling cannot be regarded as proof for the absence of electrical coupling or the structural correlate the gap junction. The trachael oenocytes of *Tenebrio molitor* serve as an example. These giant cells are electrically coupled and possess small gap junctions (Fig. 2d); however, even after prolonged observation, neither Lucifer Yellow nor carboxyfluorescein can be detected in the adjacent cells. If the injected cell is examined an hour later, most of the tracer fluorescence has disappeared. Although it is possible that some leakage occurred through the nonjunctional membrane, no significant fluorescence is detected within the cells bathed in saline with the tracers for the same period of time. Also, if the injected cell is uncoupled by damaging it during microelectrode withdrawal, the tracer fluorescence fails to fade significantly over an hour (Berdan, unpublished observations). Therefore, the observations suggest that the fluorescent tracer is passing into the adjacent cells, but at a rate too low to be detectable.

There are several other tissues in which electrical coupling but not dye coupling has been detected. For instance, during the differentiation of grasshop-

←

FIGURE 7. Molecular space-filling models of four fluorescent tracers. (a) Carboxyfluorescein (CF), Lucifer Yellow (LY). (b) Fluorescein isothiocyanate conjugated to the dipeptide, diglutamic acid, FITC–Glu_2. Note that the addition of longer peptides does not alter the channel limiting dimension. (c) Actinomycin D (Act-D) has a channel limiting dimension of 17 Å. Note the models appear approximately 0.5–1.0 Å larger in the photographs with respect to the scale than actual measurements carried out on the models indicate due to parallax errors during photography.

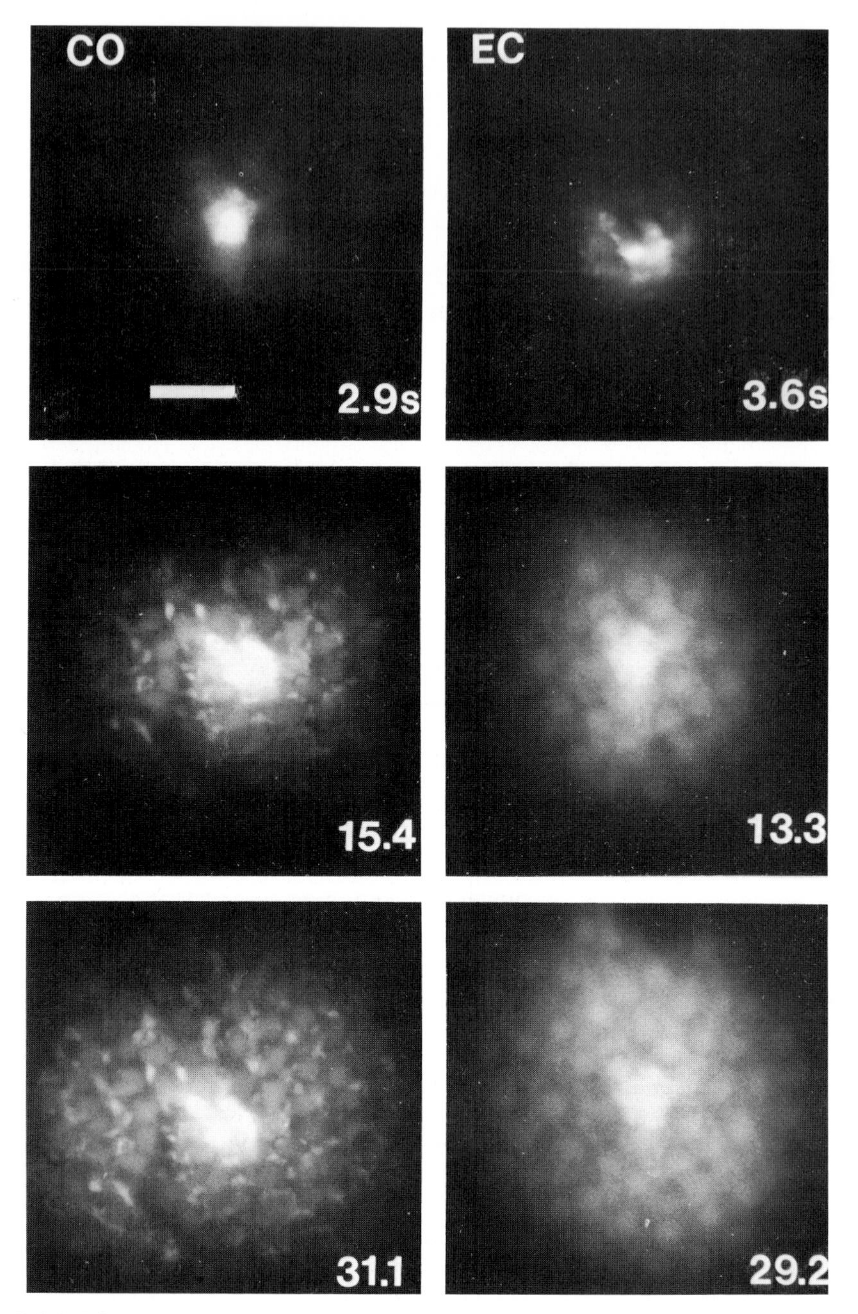

FIGURE 8. Time sequences of intercellular diffusion of carboxyfluorescein in control (CO) and 20-hydroxyecdysone-treated (EC; 2 μg/ml, 18 hr *in vitro*) *Tenebrio* epidermis. The sequences are arranged vertically. The time (seconds) when each exposure was taken is indicated in the lower right

per neuroblasts, dye uncoupling precedes ionic uncoupling (Goodman and Spitzer, 1979). Dye coupling also becomes restricted during development of mammalian embryos, but electrical coupling persists (Lo and Gilula, 1979b). Dye coupling has also been shown to be reduced at some insect intersegmental borders (Berdan, 1982; Warner and Lawrence, 1982; Blennerhassett and Caveney, 1984) at which there is no interruption in the electrotonic decay of current (Warner and Lawrence, 1982; Caveney, 1974). Furthermore, dye coupling between specific neurons (LR, R4) in the buccal ganglia of the snail, *Helisoma*, is only detectable when coupling coefficients are greater than 0.7 (Murphy *et al.*, 1983). It is also important to realize that dye coupling, which is an index of molecular coupling, may be significant even when electrical coupling is absent since the permeability of nonjunctional membrane to current-carrying ions and small molecules may be significantly different.

3.2. Channel Diameter

The synthesis or tagging of variously sized molecules with fluorescent labels and their subsequent microinjection into cells has allowed the functional diameter of the intercellular channels to be estimated. The injection of fluorochrome-tagged peptides and oligosaccharides of known dimension and molecular weight into fly salivary gland cells has allowed the steric limit of the insect intercellular channel to be estimated at greater than 20.5 nm but less than 30 nm (Rose, 1980; Schwarzmann *et al.*, 1981). Molecules with maximum abaxial dimensions in this range (or lower) may diffuse from cell to cell, and transfer of molecules of up to 3000 daltons has been reported in arthropods. The mammalian channel appears to have a smaller pore diameter, since weakly charged peptides of less than 900 daltons pass, but larger or more strongly charged peptides do not pass through gap junctions (Flagg-Newton *et al.*, 1979; Flagg-Newton, 1980). The channel limiting dimension in mammalian cells appears to be about 1.6 nm.

Estimates of the maximum gap junction channel diameter based on microinjection experiments with fluorescent tracers of various size are limited. This is owing to technical considerations related to the labeling and purification of discrete size classes of molecules which retain sufficient solubility and fluorescent intensity to be detected after microinjection into cells. Earlier investigations attempted to probe the gap junction channel diameter by synthesizing a series of fluorescent peptides of increasing size. This was accomplished by tagging pep-

---

of each panel. Total exposure time for each photograph was 1 sec with 1600 ISO speed film. Densitometry of the negatives or analysis of images on videotape allows estimates for the diffusion rates to be determined. Note the rate of tracer exchange is extremely rapid in this tissue traveling more than six cell diameters within 30 sec. Qualitative differences of the tracer spread are also apparent. Scale bar = 20 μm.

tides of various length with fluorescent molecules such as fluorescein isothiocyanate (FITC), rhodamine isothiocyanate (RITC), or Lissamine rhodamine sulphoryl chloride (LRB–SO_3Cl; Simpson *et al.*, 1977; Simpson, 1978; Berdan, 1982). After building molecular models of these tracers, it became apparent (Rose, 1980; Berdan, 1982; Fig. 7) that the addition of these fluorescent probes to linear peptides did not alter the channel limiting dimension. Schwarzmann *et al.* (1981), however, were able to fluorescently tag a set of branched glycopeptides. These consisted of a fluorescent labeled molecule with four terminal galactoses and a second molecule, the same glycopeptide, but with the galactoses enzymatically cleaved. The permeation-limiting widths of these molecules are 3.0 and 2.05 nm, respectively; both are neutral. The complete molecule will not pass through the junctions of insect or vertebrate cells. The enzymatically cleaved molecule, however, will pass through insect but not vertebrate cells. The effective pore diameter of insect cells was therefore determined to be between 2.05 and 3.0 nm. On the other hand, mammalian cells were not permeated even by the smaller 2.05-nm tracer but were by the ~ 1.6-nm-diameter LRB(Glu)OH. In conclusion, as demonstrated with fluorescent tracers, the effective pore diameter of channels between vertebrate cells appears smaller than that between insect cells.

In preliminary experiments, I have also probed vertebrate cells with the fluorescent probe actinomycin D, which is structurally similar to fluorescein but contains two covalently attached rings consisting of five amino acids each (Fig. 7c). This tracer does not appear to permeate vertebrate gap junctions. This observation is based on studies with eight-cell-stage mouse embryos which were ionically and dye coupled (Lucifer Yellow; McLachlin *et al.*, 1983) but the intercellular transfer of actinomycin D could not be detected (Berdan, unpublished observations). Since the channel limiting dimension of this tracer is 1.7 nm (Fig. 7c), it suggests that these mouse embryo gap junctions have functional channel diameters close to this size. The possibility that this tracer, which has a relatively low fluorescent yield, permeates vertebrate gap junctions but at a rate that is too low to be detected, cannot be ruled out. This tracer does, however, pass between beetle epidermal cells (Fig. 6h, Table III), but does not cross the nonjunctional membranes at detectable rates if these cells are uncoupled by microelectrode-induced injury. The synthesis and availability of probes similar to actinomycin D with higher fluorescent yields and slightly larger channel limiting dimensions would be useful for probing arthropod gap junctions. It would also be useful to have a series of varying-size electron-dense probes in order to determine the permeability of gap junctions at the ultrastructural level. In this respect, only "cobalt coupling" has been convincingly demonstrated, after precipitation with ammonium sulfide, between crayfish axons (Politoff *et al.*, 1974) and between neurons of the giant fiber system in Diptera (Strausfeld and Bassemir, 1983a,b). While it is generally accepted that the catalyst horseradish peroxidase does not permeate gap junction channels, one report indicated microperoxidase (M_r 1800; Feder, 1971) did transfer between crayfish axons (Reese *et al.*, 1971). The significance of this observation was later questioned by these investigators since they were able to detect the appearance of new low-resistance pathways after initial glutaraldehyde-induced uncoupling and fixation (Bennett, 1973). The subsequent formation of low-resistance pathways was attributed to microscopic tears or punctures in the membrane. Therefore, the possibility that larger tracers may move from cell to cell after fixation must be realized. In studies on the beetle epidermis, however, we could not detect low-resistance or dye-coupled pathways even several hours after fixation in 2.5% glutaraldehyde (Berdan and Caveney, 1985). It is possible, however, that artifactual pathways may occur as a result of poor fixation, rough handling of fixed tissue, or osmotic stress induced during fixation. Nevertheless, it is the opinion of this author that suitable electron-dense probes for channel permeation studies (diameters between 1.0 and 3.0 nm) are needed and perhaps could be derived from peroxidase peptide fragments (Plattner *et al.*, 1977).

3.3. Kinetic Analysis of Tracer Diffusion

Another approach that has been used to estimate the intercellular gap junction channel diameter, and its possible modulation, has been to carry out a kinetic analysis of the diffusion rates of two differently sized fluorescent probes. The rationale for this approach is that if the channel diameter changes in a graded fashion when the junctional conductance is experimentally lowered or elevated, the diffusion rate of the larger molecule would be affected to a greater extent than that of the smaller molecule. The ratio of the diffusion rates would therefore change if the diameter of the channels was altered. If the channels open and close in an all-or-none fashion, the ratio of the diffusion rates should remain constant. Using carboxyfluorescein as the smaller probe and rhodamine as the larger probe, kinetic studies in *Chironomus* salivary glands (Zimmerman and Rose, 1985) and *Tenebrio* epidermis (Safranyos and Caveney, 1985; Caveney and Safranyos, 1985; Caveney *et al.*, 1986, Table IV) indicate that gap junction channels open and close in an all-or-none fashion. Analysis of the diffusion kinetics in the beetle epidermis is summarized in Figs. 8 and 9 and Table IV. Further details can be found in the review by Caveney *et al.* (1986). In one other kinetic study using fluorescein derivatives, Brink and Dewey (1980) suggested that the gap junction channels between the median giant axons of the earthworm, *Lumbricus terrestris,* are at least 1.2 nm in diameter and contain a fixed anionic charge.

In summary, the use of fluorescent tracers has been particularly valuable in identifying communicating cells in complex tissues with tortuous connections such as occurs in the nervous system (Stewart, 1978), and in revealing "metabolic compartments" in insect tissues (Berdan, 1982; Weir and Lo, 1982; Caveney and Berdan, 1982; Blennerhassett and Caveney, 1984; Fraser and Bryant,

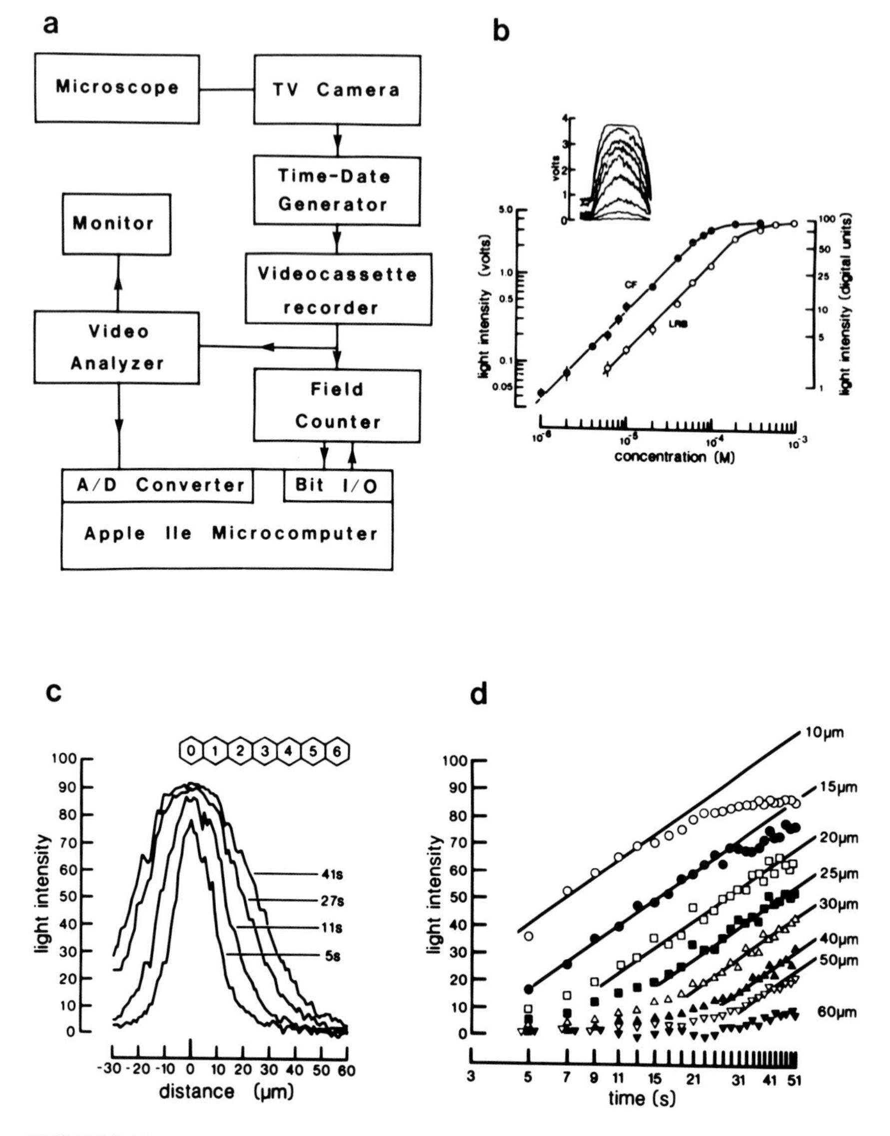

FIGURE 9. Kinetic analysis of fluorescent tracer diffusion. (a) Arrangement of equipment for video recording and analyzing the diffusion of fluorescent tracers in *Tenebrio* epidermis. (b) Response of a silicon-intensified target (SIT) television camera to increasing concentrations of fluorescent dye. Curve CF is for carboxyfluorescein (in medium at pH 6.5). Curve LRB is for Lissamine rhodamine B. (c) Topographic plot of carboxyfluorescein spread through the epidermis. The spatial distribution of CF fluorescence, plotted as digital light intensities, is sampled along a line through the center of the dye spread. The data are plotted at 1-μm intervals for 5, 11, 27, and 41 sec of injection into newly molted tissue. A row of idealized cells drawn to scale above these curves shows that CF had entered the sixth-order cells by 41 sec. (d) The digital light intensities plotted against log (time) show the

Table IV
Increased Rates of Tracer Movement after *Tenebrio molitor* Epidermis Is Exposed to the Insect Molting Hormone 20-Hydroxyecdysone[a]

Molecule	Effective diffusion coefficient (D_e, $\times 10^{-7}$ cm^2/sec)	
	Control ($n = 15$)	Hormone-treated ($n = 14$)
Carboxyfluorescein	3.72 ± 0.95[b]	4.85 ± 1.00
Lissamine rhodamine B	1.50 ± 0.37	2.00 ± 1.00
CF/LRB ratio	2.54 ± 0.68	2.52 ± 0.61

[a]From Caveney and Safranyos (1985).
[b]Mean ± S.D.

1985). The use of different-sized fluorescent tracers has shown that only small molecules (less than 3.0 nm in one dimension) may permeate intercellular channels of arthropods, whereas vertebrate gap junction channels appear restricted to molecules less than 1.7 nm in one dimension. Why the channels are larger in arthropods is unknown, but it may relate to the lower packing densities of the channels within arthropod gap junctions and it generally correlates with larger gap junction intramembrane particle diameters seen by freeze-fracture (Berdan and Caveney, 1985; next section). While the use of radioactive tracers and autoradiography in metabolic cooperation studies with vertebrate cells *in vitro* (reviewed by Hooper and Subak-Sharpe, 1981) are consistent with fluorescent tracer studies, no similar studies appear to have been carried out in arthropod tissues. In conclusion, a number of variables can influence the detection of intercellular communicating channels with fluorescent tracers, including: the number of interconnecting channels, nonjunctional membrane permeability, cytoplasmic binding of the tracer, the solubility, charge, size, and fluorescent yield of the tracer, cell size, tissue autofluorescence (Aubin, 1978; Benson *et al.*, 1979), and the sensitivity of the optics and equipment used to detect fluorescence (Nairn, 1976; Reynolds and Taylor, 1980).

Thus, the application of fluorescent tracers should continue to play an important role in both mapping and defining discrete cell communication pathways in both arthropod and vertebrate tissues.

---

following linear relationship at selected distances from the source cell injected with CF. The mean diffusion coefficient (D_e) calculated from the lines drawn through the sets of points is 3.4×10^{-7} cm²/sec. Near the source cell (such as 10 and 15 μm), where high light intensities are seen, the points fall away from the predicted relationship because of the response characteristics of the camera (b). From Safranyos and Caveney (1985) with permission.

4. ULTRASTRUCTURE OF ARTHROPOD GAP JUNCTIONS

4.1. Thin Sections

In thin sections of plastic-embedded tissue examined by electron microscopy, gap junctions appear as closely apposed regions of plasma membrane separated by a narrow (2–5 nm wide) gap (Fig. 10, Table V). Gap junctions may

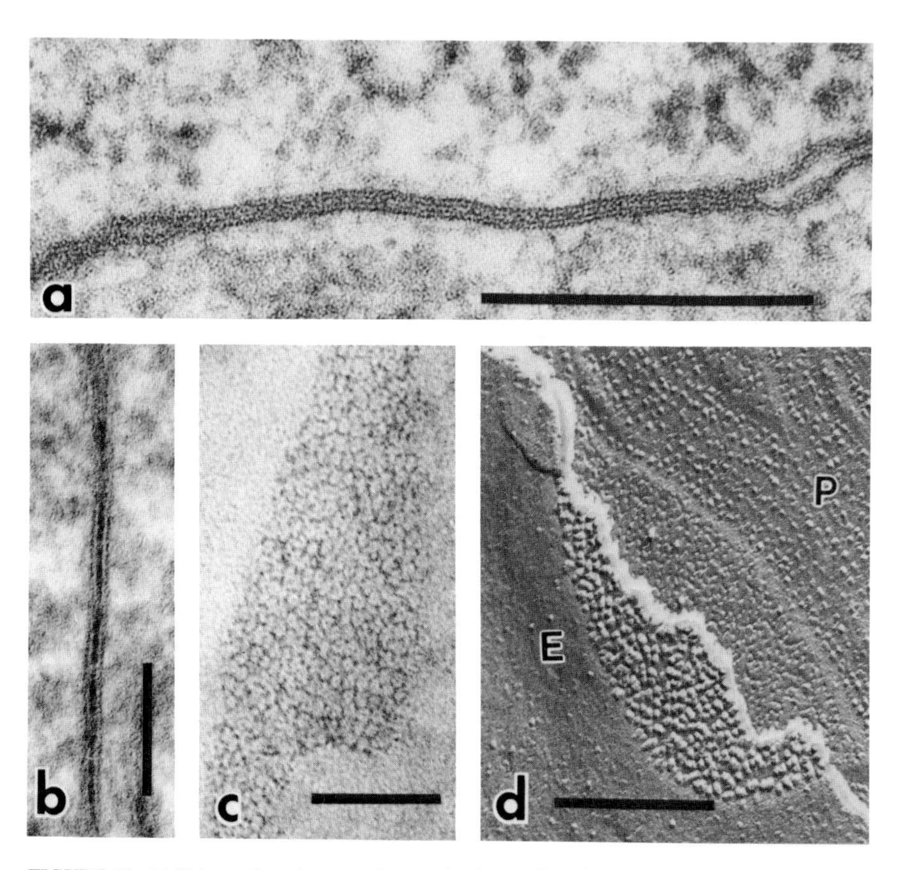

FIGURE 10. (a) Thin-section electron micrograph of a gap junction from the midgut of the beetle *Sitophilus granarius;* conventional glutaraldehyde and osmium fixation. Note the gap between the membranes is narrowed to 2–4 nm. Scale bar = 250 nm. (b) Gap junction from *Tenebrio* epidermis in which the gap is filled with lanthanum, an extracellular tracer. Scale bar = 100 nm. (c) Tangential view of a gap junction from *Tenebrio* epidermis fixed in the presence of lanthanum. Note the subunits with a central electron-dense core. Scale bar = 100 nm. (d) Freeze-fracture replica of a gap junction from *Tenebrio* epidermis (chlorpromazine uncoupled) showing the E (exoplasmic) fracture face particles and P (protoplasmic) fracture face pits. The rows of particles on the P face are components of the septate junction. Scale bar = 250 nm.

Table V
Arthropod and Vertebrate Gap Junction Dimensions in Thin Plastic Sections

Tissue	Overall gap junction width (nm)	Gap width (nm)	Reference
Arthropods			
Crayfish; neuron	18	3–4	Pappas *et al.* (1971)
	20	3–4	Peracchia (1973a)
	18–22	4–5	Zampighi *et al.* (1978)
Crayfish; hepatopancreas	18–22	2–3	Berdan (1985)
Beetle; epidermis	15–17	2–4	Berdan (1982)
Fly; salivary gland	13–14	4	Rose (1971
Moth; germ cells	15–18	4.5–5.5	Szollosi and Marcaillou (1980)
Fly; photoreceptors	—	3.8–5.8	Shaw and Stowe (1982)
Vertebrates			
Mouse, guinea pig, cat; myocardium	17–19	2–3	McNutt and Weinstein (1970)
Rat; various tissues	15–21	2–4	Friend and Gilula (1972)
Mouse; liver	14–16	2–4	Goodenough and Revel (1970)
Mouse, goldfish, chicken; astrocytes, ependymal cells	13.5–15.0	2.3	Brightman and Reese (1969)
——; neurons	13–14	—	Brightman and Reese (1969)
——; neuron–glia pentalaminar contacts	12.0	—	Brightman and Reese (1969)

vary in length from small punctate contacts to stretches greater than 5 μm in overall length. In most instances it is difficult, if not impossible, to distinguish between arthropod and vertebrate gap junctions based on their appearance in thin sections. Although it has been reported that the gap width is slightly larger in arthropod tissues, this does not appear to be a reliable distinguishing feature (Table V). The width of the gap is close to the resolution of biological material in the electron microscope and its dimensions may be significantly altered by fixation, dehydration, compression during sectioning, and shrinkage under the electron beam. The appearance of gap junctions is also markedly affected by fixation and staining (Brightman and Reese, 1969). The addition of *en bloc* uranyl acetate appears to provide the best preservation of the gap and adjacent trilaminar membrane profiles. The addition of tannic acid, $KMnO_4$, K^+-pyroantimonate, or lanthanum during processing often accentuates the gap region (Fig. 10b).

The interneuronal gap junctions between crayfish septate axons are exceptional in that they can be readily distinguished morphologically in thin sections from other gap junctions. These gap junctions are bordered by 50- to 80-nm-diameter vesicles that appear to be involved with the turnover of the junctional membrane (Peracchia, 1973a; Hanna *et al.*, 1978). Also, if the tissue is fixed in the presence of H_2O_2 (Peracchia and Mittler, 1972; Peracchia, 1973a), discrete intramembranous particles can be seen bridging the gap between junctional membranes. The clarity with which these particles can be seen between crayfish neurons has not been observed elsewhere.

In tangential views of gap junctions fixed in the presence of an extracellular tracer such as lanthanum, both vertebrate and arthropod gap junctions can be seen to contain small subunits (Revel and Karnovsky, 1967; McNutt and Weinstein, 1970; Hudspeth and Revel, 1971; Peracchia, 1973b; Lane, 1978; Berdan and Caveney, 1985; Fig. 10c). The subunits sometimes contain a central electron-dense spot. This electron-dense spot presumably represents the site of the intercellular channel, although it is not understood how the stain penetrates the channel from the extracellular space. Similar images of the subunits have been obtained with isolated gap junctions from the crayfish hepatopancreas (Berdan and Gilula, 1986a; Figs. 18b and 19). Rotational photographic reinforcement of the subunits (Markham *et al.*, 1963) suggests that they possess an underlying sixfold symmetry in crayfish neuronal gap junctions (Peracchia, 1973a), crayfish hepatopancreas gap junctions (Berdan and Gilula, 1986a; Fig. 18c), and isolated rat liver gap junctions (Benedetti and Emmelot, 1965).

Quantitative estimates of the amount of gap junction, or the spatial distribution of gap junctions throughout a tissue are most easily performed by morphometric analysis of thin sections. Although thin-section analysis is tedious and time-consuming, it is reliable and reproducible. Quantitative changes in the amount of gap junction have been detected between intermolt and newly molted beetle epidermis (Caveney and Podgorski, 1975; Caveney *et al.*, 1980), during the development of *Drosophila* embryos and imaginal disks (Eichenberger-Glinz, 1979), between wild-type and mutant *Drosophila* imaginal disks (Ryerse and Nagel, 1984), and during the development of neuronal connections in *Daphnia* (Lopresti *et al.*, 1974). Although freeze-fracture may be a more sensitive technique for the detection of small gap junctions, it is difficult to control the orientation of the fracture plane which may follow preferred pathways. In spite of these difficulties, quantitative estimates of the amount of gap junction have been determined by freeze-fracture in the guinea pig ileum (Gabella and Blundell, 1979) and also vertebrate cells *in vitro* (Sheridan *et al.*, 1978; Radu *et al.*, 1982).

Finally, it is important to point out that not all closely apposed regions of plasma membrane are necessarily gap junctions. Punctate membrane contacts may be tight junctions, although they are rare in arthropod tissues (reviewed by Lane, 1981b). Some membrane contacts resembling gap junctions may arise artifactually during fixation. Brightman and Reese (1969) observed that with certain preparative methods, "labile appositions" approximately 12 nm in width containing a dense median lamina appeared between neuron and glia processes. Similar close membrane appositions are also often observed within cells inside myelin figures or between secretory vesicles (Neutra and Schaeffer, 1977; Specian and Neutra, 1980; Tanaka *et al.*, 1980). Therefore, structures reported to be gap junctions in thin sections should also include measurements of their overall thickness and/or electrophysiological evidence for their presence.

4.2. Freeze-Fracture

In freeze-fracture replicas, arthropod and vertebrate gap junctions are easily distinguishable. In vertebrate cells, the intramembrane particles preferentially adhere to the protoplasmic fracture face and are rather uniform in diameter (6–9 nm, Fig. 11b). The intramembrane particles are often tightly packed, but may range in density between 2600 and 12,500 particles/μm^2 (Gabella and Blundell, 1979; Hirokawa and Heuser, 1982; Meda *et al.*, 1983; Miller and Baldridge, 1985). In contrast, the particles which comprise arthropod gap junctions generally exhibit lower packing densities (2000–3000 particles/μm^2), are larger and more heterogeneous in size, and range from 8 to 22 nm in diameter (Caveney and Berdan, 1985, Figs. 10d, 11a, and 12). Many of the particles often appear to be fused (Fig. 12a). In addition, the particles preferentially adhere to the exoplasmic fracture face in arthropods (Flower, 1977) although a few exceptions have been reported (Graf, 1978; Hall *et al.*, 1985). The relative scarcity of intramembrane particles on the exoplasmic fracture face facilitates the detection of gap junction particles in arthropods. In both vertebrates and arthropods, the complementary fracture faces of gap junctions also often exhibit smaller pits (Figs. 10d and 11).

4.3. Ultrastructure in Different States of Conductance

The arrangement of gap junction particles into crystalline lattices is attractive from a structural standpoint; however, it has been difficult to relate their size and arrangement with a particular functional state. It has been hypothesized, however, that gap junction particles arranged in highly ordered crystalline arrays are characteristic of uncoupled gap junctions (Peracchia and Dulhunty, 1976; Peracchia, 1977). On the basis of a growing number of studies (summarized in Table VI), it is clear that there is no consistent correlation between gap junction particle packing order and the junctional conductance. Discrepancies exist even when the same tissue is examined by different investigators (Peracchia and Dulhunty, 1976; Dahl and Isenberg, 1980; Délèze and Herve, 1983; Hanna *et al.*, 1984).

It is also important to note that only a few studies actually monitored the effect and timing of the uncoupling reagents electrophysiologically. Furthermore, only in the crayfish (Peracchia and Dulhunty, 1976) and beetle epidermis (Berdan and Caveney, 1985) has the junctional ultrastructure been examined in the same cells in which uncoupling was detected. One can also argue that in conventional freeze-fracture utilizing glutaraldehyde-fixed tissues, differences in

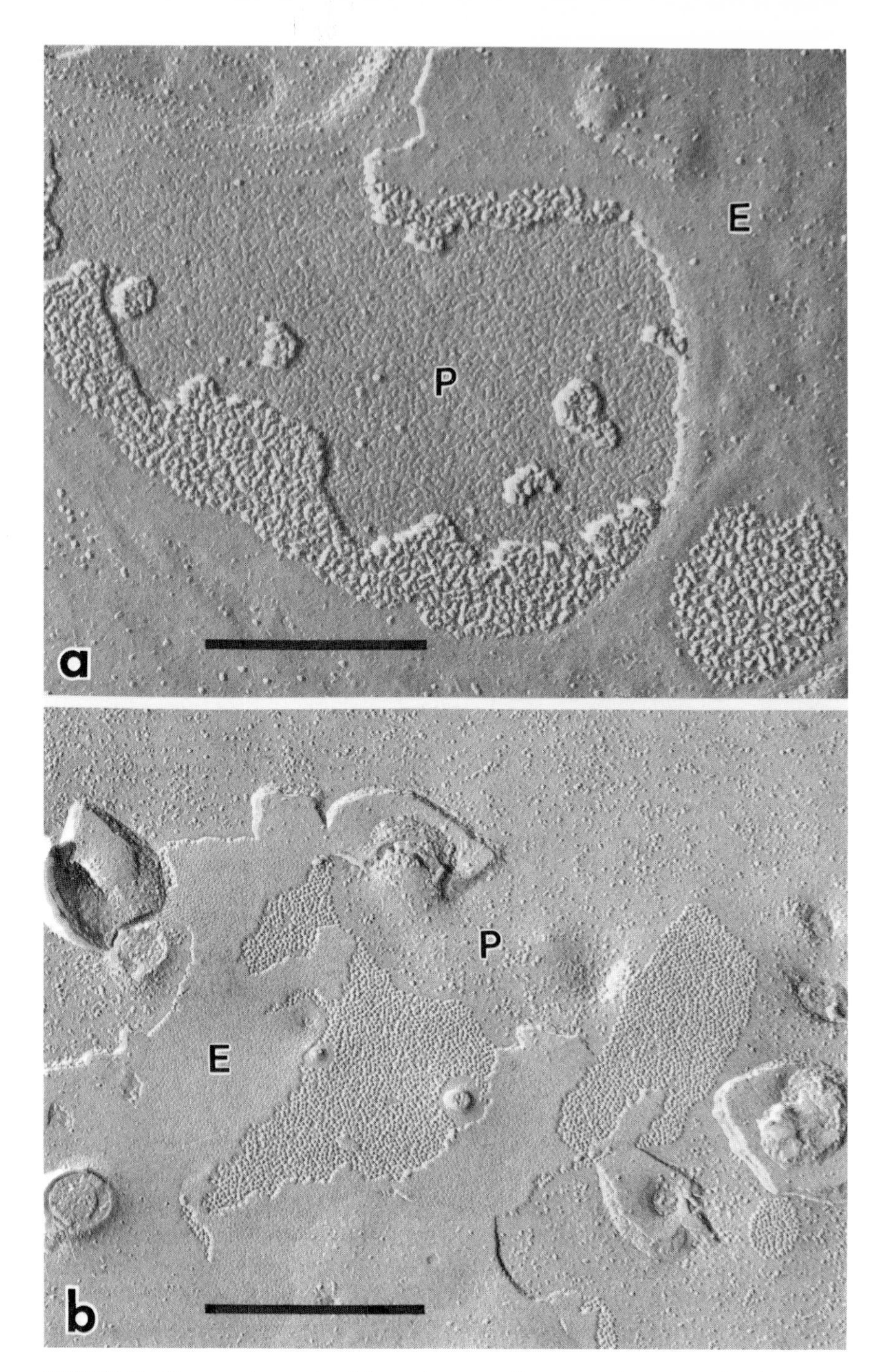

FIGURE 11. Electron micrographs of freeze-fracture replicas comparing arthropod and vertebrate gap junctions. (a) Gap junctions from epidermis of the beetle *Tenebrio molitor*. (b) Gap junctions from liver of the rabbit *Oryctolagus caniculus*. P, protoplasmic fracture face; E, exoplasmic fracture face. Scale bars = 0.5 μm.

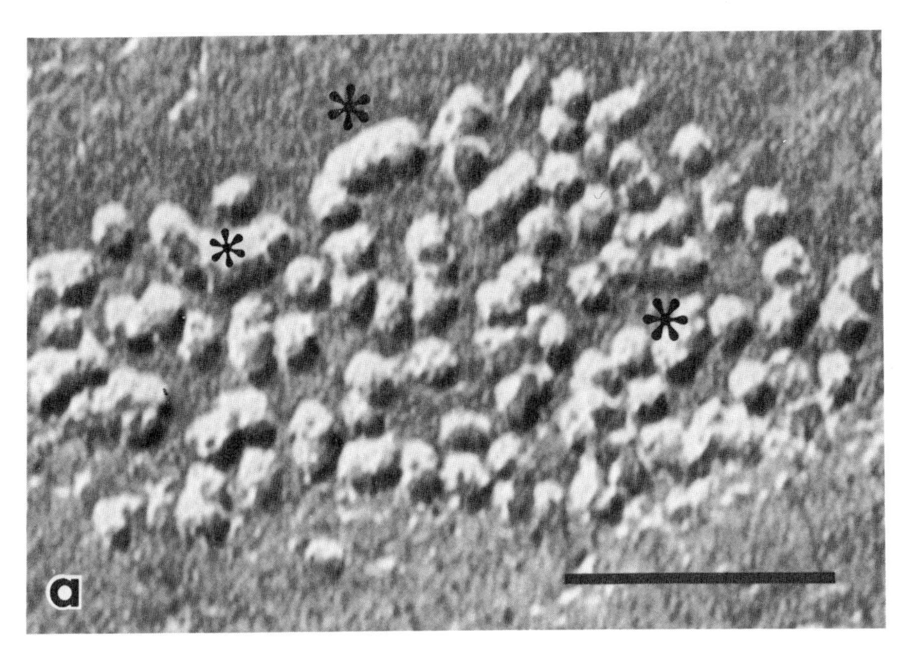

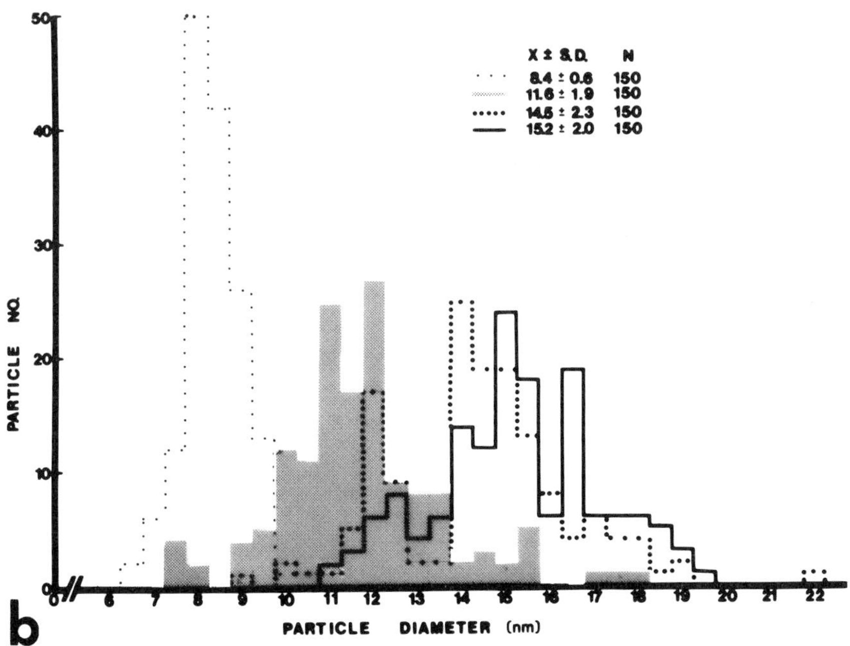

FIGURE 12. (a) High-magnification electron micrograph of a gap junction from epidermis of the beetle *Tenebrio molitor* demonstrating particle fusions (*), particle size heterogeneity, and the rather loose organizational packing of the particles. Scale bar = 100 nm. (b) Gap junction particle diameter histograms comparing particle size from rabbit liver (· · · ·), crayfish hepatopancreas (░), *Tenebrio* fat body (• • • •), and *Tenebrio* epidermis (——).

Table VI
Changes in Gap Junction Structure Induced by Uncoupling Agents as Revealed by Freeze-Fracture[a]

Animal, tissue	Uncoupling agent	Particle packing order/ density	Reference
Arthropods			
Crayfish, neuron	EDTA, DNP[b]	↑	Peracchia and Dulhunty (1976
Crayfish, glia	EDTA, DNP	NC[c]	Peracchia and Dulhunty (1976
Crayfish, neuron	Injury	NC	Hanna *et al.* (1984)
Beetle, epidermis	Chlorpromazine	NC	Berdan and Caveney (1985)
Vertebrates			
Rat, stomach, liver	Hypoxia, DNP	↑	Peracchia (1977)
Axolotl, blastomers	CO_2	↑	Hanna *et al.* (1978)
Rabbit, myocardium	Injury	↑	Baldwin (1979)
	Anoxia	↓	Green and Severs (1984)
Sheep, myocardium	DNP	↑	Dahl and Isenberg (1980)
	Injury	↓	Shibata and Page (1981)
	DNP, heptanol, hypotonic CA^{2+}-free saline	↓	Délèze and Herve (1983)
Sheep, follicle	CO_2	↓	Lee *et al.* (1982)
Mouse, liver	Glutaraldehyde	↑	Sikerwar and Malhotra (1981)
	Hypertonic sucrose, azide, deoxyglucose, cyanide, homogenization	↑	Hirokawa and Heuser (1982)
Mouse, pancreas	CO_2	NC	Meda *et al.* (1983)
Mouse, stomach, liver	Anoxia, CO_2	↑	Raviola *et al.* (1980)
	Glutaraldehyde	NC	Raviola *et al.* (1980)
Chick, lens epithelium	CO_2, A23187, heptanol, Na acetate, cyanide	NC	Miller and Goodenough (198
	Glutaraldehyde	↑	Miller and Goodenough (198
Rabbit, corneal and ciliary epithelium	Anoxia, CO_2, glutaraldehyde	↑	Raviola *et al.* (1980)
Goldfish, horizontal cell			
Soma	Dopamine, light	↓	Miller and Baldridge (1985)
Axon	Dopamine, light	NC	

[a]Effect of uncoupling by glutaraldehyde was deteremined by comparison with unfixed tissue rapidly cooled in liquid helium.
[b]EDTA, ethylenediaminetetraacetic acid (Ca^{2+} chelator); DNP, dinitrophenol (uncouples oxidative phosphorylation); A23187 (Ca^{2} ionophore).
[c]NC, no change.

particle size and distribution may be due to glutaraldehyde cross-linking and uncoupling. Delayed fixation for example (30 to 60 min) of rat muscle fibers is associated with clustering of intramembrane particles of the sarcolemma and formation of bandlike particle-free areas (Schmalbruch, 1980). Glutaraldehyde alone irreversibly uncouples cells, but may leave membrane potentials intact for several hours (Politoff and Pappas, 1972; Ramon and Zampighi, 1980; Berdan and Caveney, 1985). In order to characterize the effects of glutaraldehyde on gap junctions, several investigators (Raviola *et al.*, 1980; Sikerwar and Malhotra, 1981; Miller and Goodenough, 1985) have utilized the rapid-freezing technique of Heuser *et al.* (1979), which avoids fixation and cryoprotectants. The effect of glutaraldehyde on gap junction particle packing as assessed by rapid freezing, appears to vary with different tissues (Table VI). Nevertheless, it is clear that uncoupling may occur without any apparent change in particle packing order. It seems probable, however, that highly crystalline gap junctions are uncoupled although it is not clear whether crystallization is a secondary event that occurs after uncoupling.

In conclusion, the particle packing arrangement of gap junctions as observed by freeze-fracture, whether by conventional fixation or rapid freezing, is of little predictive value in determining their functional state. One should also consider any future gap junction morphological correlations by freeze-fracture with functional state of coupling with skepticism until the current discrepancies can be resolved.

In addition to changes in particle arrangement within gap junctions, some investigators have attempted to measure changes in particle diameter. Here again the data are ambiguous. Délèze and Herve (1983) reported that when they uncoupled sheep Purkinje fibers with dinitrophenol, they detected a reduction in the mean gap junction particle diameter from 8.18 nm to 7.46 nm. In the same tissue using the same uncoupling agent, Dahl and Isenberg (1980) found the mean particle diameter increased from 8.3 nm to 10.8 nm. In another study on mouse pancreatic acinar cells, no significant change in particle diameter was detected in the uncoupled state (Meda *et al.*, 1983). Quantitative measurements of particle diameter in arthropods in different states of coupling are restricted to crayfish neurons (Peracchia and Dulhunty, 1976) and the beetle epidermis (Berdan and Caveney, 1985). In crayfish neurons, adjacent gap junctions have been observed to possess different diameter particles (12.1 nm versus 15.2 nm) (Peracchia and Dulhanty, 1976). Peracchia has correlated the smaller particles with the uncoupled state. It should be noted that no ultrastructural changes were observed in gap junctions of surrounding glia exposed to the same uncoupling agents. In the beetle epidermis, no significant change in particle diameter could be detected in the uncoupled state or hormone-elevated conductance state (Berdan and Caveney, 1985). Therefore, potential structural and biochemical differences between gap junctions from different tissues, especially excitable and nonexcitable tissues, must be recognized. There is in fact no *a priori* reason why a change in particle diameter need occur at all when a channel closes; even if a change in particle diameter occurred, it would have to be greater than 2.0 nm before it could be reliably detected by conventional freeze-fracture. The reason for this is that the resolution of freeze-fracture is limited primarily by the thickness of the platinum–carbon coating (2.0 nm) and the platinum grain size (also 2.0 nm) (Margaritis *et al.*, 1977; Sheffield, 1979; LeMaire *et al.*, 1981). In addition, the inherent variability of the technique and a number of other factors might potentially alter particle size, including (1) the level of focus, (2) angle of shadowing, (3) plastic deformation, (4) decoration, (5) surface contamination of the replica, (6) etching, (7) fixation, (8) the interaction between closely packed particles, and (9) errors in magnification or measurements (Amberman *et al.*, 1972; Zingsheim, 1972; Margaritis *et al.*, 1977; Rash *et al.*, 1979; Sheffield, 1979; LeMaire *et al.*, 1981). It appears unlikely that changes in particle diameter much less than 2.0 nm as determined by freeze-fracture, can be considered significant.

More recently, freeze-fracture in conjunction with markers that perturb the membrane ultrastructure (reviewed by Severs and Robenek, 1983) has been used to identify microdomains in biomembranes. The polyene antibiotic filipin and the saponis digitonin and tomatin have been used to localize sterols such as cholesterol in membranes by freeze-fracture. Filipin forms complexes with 3β-hydroxysterols and can be visualized as large (~ 25-nm diameter) circular protrusions by freeze-fracture (Fig. 13). Although cholesterol has been reported to be enriched in detergent-treated isolated vertebrate gap junctions (Henderson *et al.*, 1979), attempts to localize cholesterol within vertebrate (Elias *et al.*, 1979; Severs, 1981; Robenek *et al.*, 1982; Risinger and Larsen, 1983; Severs and Robenek, 1983) and arthropod (Berdan and Shivers, 1985) gap junctions have been unsuccessful. Whereas positive results with filipin are conclusive in demonstrating a high sterol content, the absence of filipin–cholesterol complexes is not proof of the absence of cholesterol.

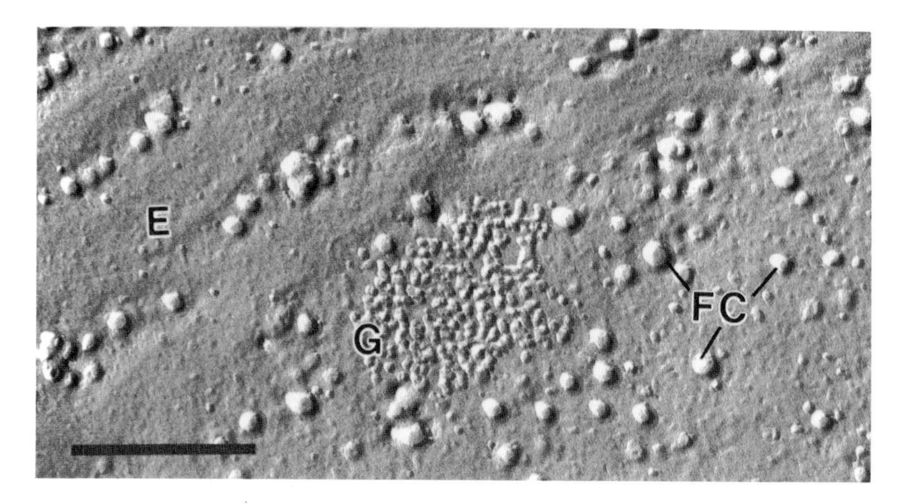

FIGURE 13. Freeze-fracture replica of a gap junction from epidermis of the beetle *Tenebrio molitor* surrounded by filipin–cholesterol complexes (FC). Complexes have not been located within gap junction plaques. Scale bar = 250 nm.

5. ISOLATION AND BIOCHEMICAL CHARACTERIZATION OF GAP JUNCTIONS

5.1. Isolation of Vertebrate Gap Junctions

In recent years, procedures to isolate vertebrate gap junctions for biochemical characterization and also for studies on the biosynthesis and regulation of the gap junction channel permeability have been developed. Thus far, gap junctions have been isolated from the following vertebrate tissues: mammalian liver (Benedetti and Emmelot, 1965; Goodenough and Revel, 1971; Goodenough and Stoeckenius, 1974; Goodenough, 1974; Culvenor and Evans, 1977; Hertzberg and Gilula, 1979; Henderson *et al.*, 1979; Finbow *et al.*, 1980; Ehrhart, 1981; Nicholson *et al.*, 1981; Sikerwar and Malhotra, 1983; Hertzberg, 1984), mammalian heart (Kensler and Goodenough, 1980; Manjunath *et al.*, 1982, 1984), and uterus (Zervos *et al.*, 1985). Most studies have used differential centrifugation and ultracentrifugation in sucrose gradients to first isolate a plasma membrane fraction containing cell junctions. Subsequently, gap junctions have been enriched by solubilizing nonjunctional membrane components with detergents, chaotropic agents, alkali, or by proteolytically degrading nonjunctional components with proteases (summarized for vertebrate liver gap junctions in Table VII). Proteases have been omitted in many recent protocols with one exception (Finbow *et al.*, 1984), since the polypeptides recovered tended to be of lower molecular weight. Despite the apparent reduction in the molecular weight of the polypeptides after protease digestion, the ultrastructure of the gap junctions remains apparently unaffected (Henderson *et al.*, 1979). Therefore, in the absence of immunoprecipitation studies on the gap junction polypeptide in an *in vitro* translation system or cloning of the gap junction gene(s), the polypeptide correlated with the isolation of gap junctions cannot be assumed to be of native molecular weight. With this in mind, the major consensus at the present time is that the isolated vertebrate liver gap junction contains a polypeptide component of 26,000–28,000 daltons that upon boiling in SDS sample buffer, dimerizes to a polypeptide of about 47,000 daltons. The presence of immunological cross-reacting polypeptides (47,000 and 27,000 daltons) has recently been demonstrated in a wide variety of vertebrate tissues with the exception of the lens (Hertzberg and Skibbens, 1984).

Whereas antiserum to the rat liver gap junction cross-reacts with polypeptides in tissues from a variety of vertebrates, mollusks, and even *Hydra* (Green *et al.*, personal communication), it does not recognize arthropod gap junctions (Hertzberg and Skibbens, 1984; Berdan, 1985; Berdan and Gilula, 1986a; Fig. 22d).

So far, no report has demonstrated glycosylation of gap junction components. Lipid analysis of isolated gap junctions (Hertzberg and Gilula, 1979; Henderson *et al.*, 1979) that have been extracted with alkali or detergent is

Table VII
Agents Used to Enrich Gap Junctions from Mouse or Rat Liver by Selective Solubilization or Degradation of Nonjunctional Components

Agents	Major polypeptides after SDS–PAGE	Reference
Sarkosyl (0.5%)	34, 20, 18, 10k	Goodenough (1974)
Collagenase, hyaluronidase, sonication		Goodenough and Stoeckenius (1974)
Sarkosyl (2.0%), 7 M urea, collagenase, trypsin	40, 38, 34, 25, 18, 17, 12k	Culvenor and Evans (1977)
Sarkosyl (0.45%), 0.1 M Na_2CO_3, (pH 11.0), sonication	47, 27k	Hertzberg and Gilula (1979)
Triton X-100 (5%), 6 M urea, 0.1 mM EDTA, 0.1% deoxycholate, sonication	50–45, 26k	Henderson *et al.* (1979)
Sarkosyl (0.55%), 50 mM Na_2CO_3, trypsin	52, 46, 26, 10k	Finbow *et al.* (1980)
Deoxycholate (2.0%), 1 mM Lubrol-WX	~55, 45, 30k	Zampighi and Unwin (1979)
Deoxycholate (1.2%), clostripain (protease)	26, 18k	Ehrhart (1981)
Triton X-100 (1.0%), Sarkosyl (0.5%), 6 M urea, trypsin	16k	Finbow *et al.* (1983)
Sodium hydroxide (20 mM, pH 12)	47, 27k	Hertzberg (1984)

unlikely to reflect their *in vivo* composition but rather their differential solubility in the extractor. Further discussion of different isolation approaches and the chemistry of gap junctions can be found in the review by Revel *et al.* (1985).

5.2. Enrichment of Arthropod Cell Junctions

A significant enrichment of gap junctions has recently been achieved using the crayfish (*Procambarus clarkii*) hepatopancreas as a tissue source (Berdan and Gilula, 1986a). The procedures utilized to isolate crayfish junctions are summarized in Fig. 15. The crayfish hepatopancreas as a source from which to isolate arthropod cell junctions has several advantages. The histology and physiology of this tissue have been investigated (McWhinnie and Kirchenberg, 1962; Davis and Burnett, 1964; Bunt, 1968; Loizzi, 1971; Berdan and Shivers, 1980). Gap junctions are extremely abundant in this tissue (Gilula, 1978; Fig. 14) and have been estimated to occupy 5% of the total plasma membrane surface area (McVicar and Shivers, 1985). The cells are ionically coupled and this form of

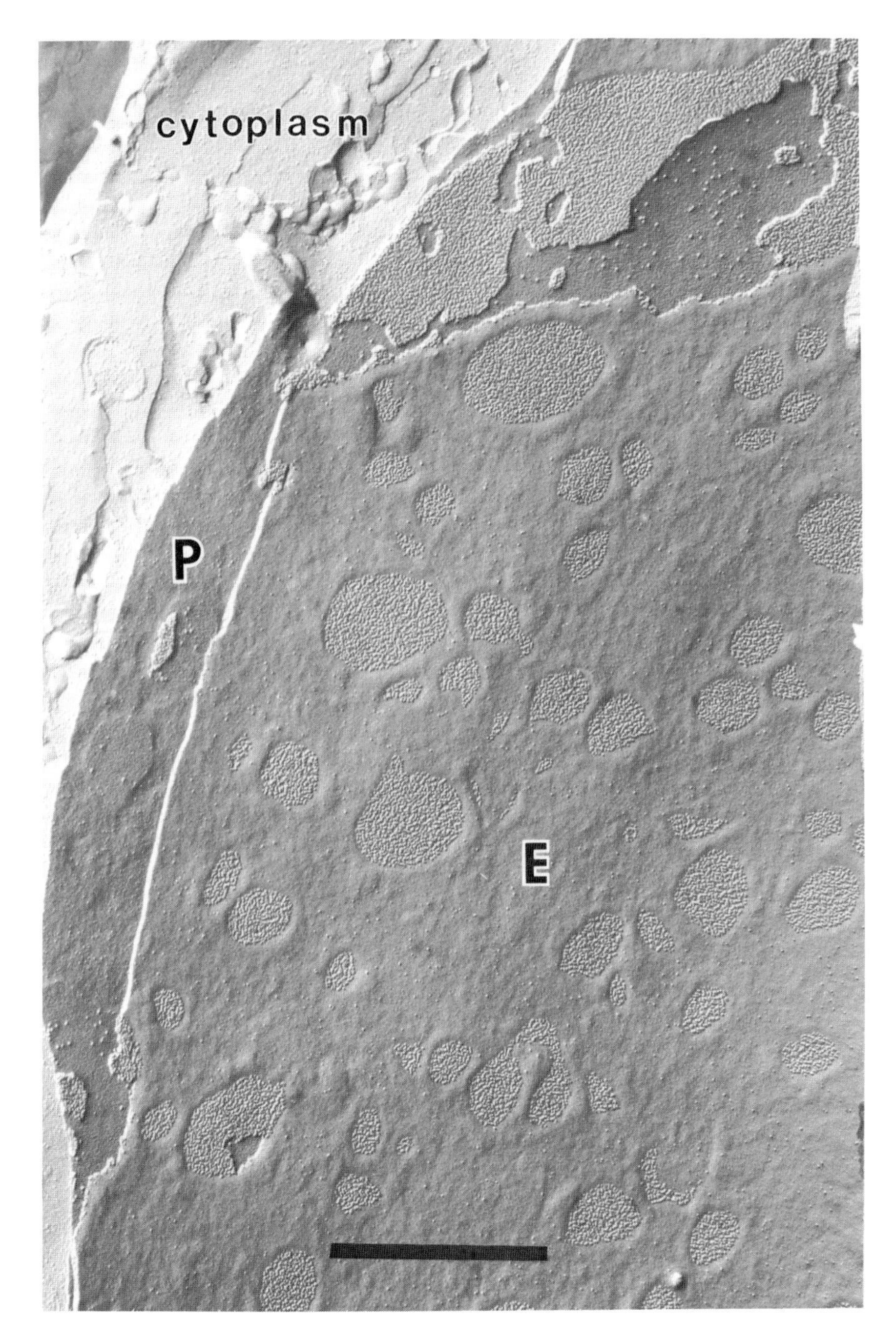

FIGURE 14. Freeze-fracture replica from hepatopancreas of the crayfish *Procambarus clarkii* showing gap junctions *in situ*. The lateral membranes below the septate junction possess an unusual abundance of gap junctions. P, protoplasmic fracture face; E, exoplasmic fracture face. Scale bar = 1 μm.

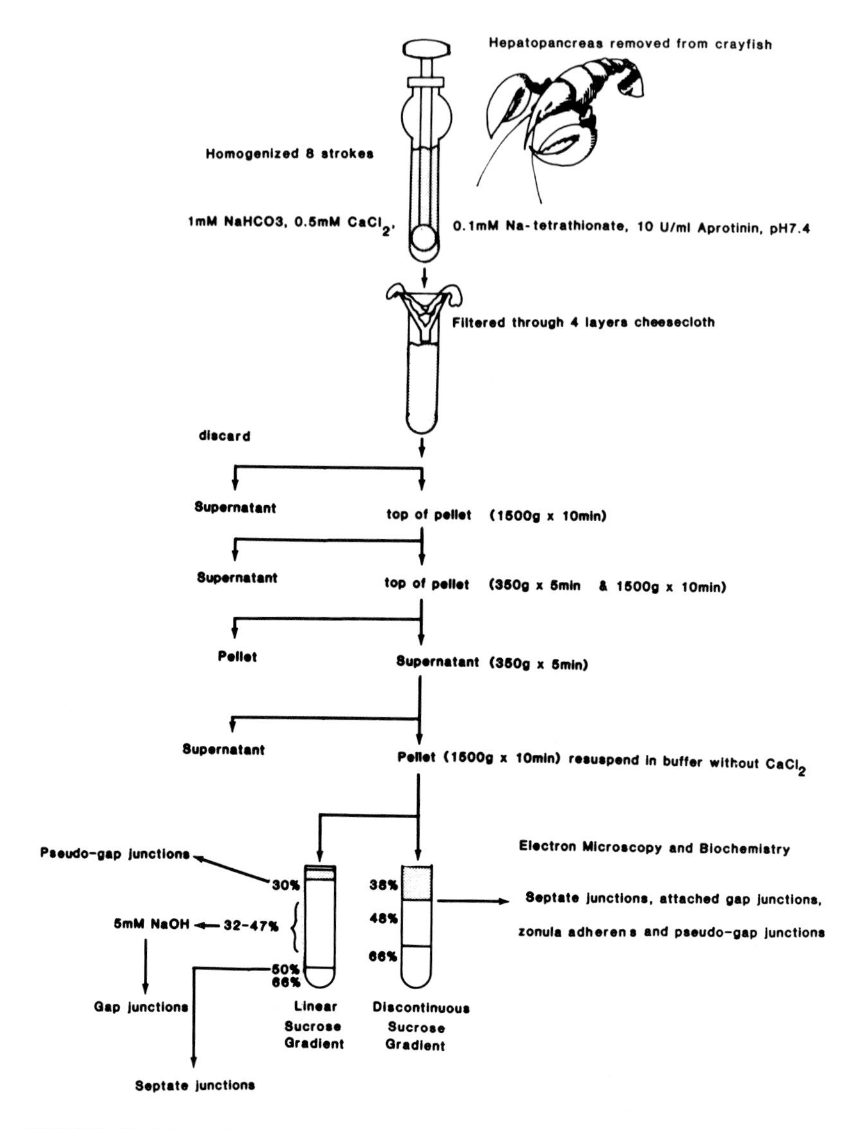

FIGURE 15. Scheme for the enrichment of septate and gap junctions from the crayfish hepatopancreas. Following differential centrifugation, the crude membrane fraction was separated on either a linear or a discontinuous sucrose gradient by ultracentrifugation. On linear gradients, pseudo-gap junctions were concentrated between 30 and 32% (w/w) sucrose. Gap junctions that were detached from septate junctions were dispersed between 32 and 47% (w/w) sucrose, and septate junctions with attached zonula adherens and gap junctions were concentrated in a narrow region (47–50% w/w sucrose) of the gradient. Gap junctions were further enriched by extraction of the material between 32 and 47% (w/w) sucrose with 5 mM NaOH after the gradient spin. For preparative purposes, a highly enriched cell junction fraction was obtained at the 38%/48% (w/w) sucrose interface of a discontinuous sucrose gradient.

cell communication is under hormonal regulation (McVicar and Shivers, 1984). Septate junctions are extremely long (5–7 μm) in this tissue and upon mild homogenization can be disrupted into large sheets along with attached gap junctions (Fig. 16). The hepatopancreas weighs 1–5 g and is therefore more suitable for biochemical investigation than, for example, most insect tissues. The major disadvantages of using the crayfish hepatopancreas include the secretion of proteases utilized in digestion, the abundance of lipid, and the presence of structures, apparently of intracellular origin, that resemble gap junctions (Figs. 17, 20c, and 21).

Since there is no known enzymatic activity associated with gap junctions, their subsequent enrichment from a plasma membrane fraction is monitored by morphological criteria (negative staining, thin sections, and freeze-fracture) and potential junctional polypeptides identified by SDS–PAGE. The identification of a plasma membrane cell junction fraction from arthropod tissues is made easier by the presence of extensive septate junctions containing an attached apical *zonula adherens* and plasma membrane-linked gap junctions (Figs. 16 and 17).

Septate junctions are a family of related junctions restricted in their distribution to invertebrates [but see Sotelo and Llinas (1972) and Connell (1978) for possible exceptions]. Septate junctions were first described by Wood (1959). The major unifying morphological characteristic of this family of cell junctions is a 13- to 15-nm uniform space between adjacent plasma membranes which is partitioned by more or less regularly spaced septa. These junctions most often occur as a belt around the apical region of the cells where one of their functions appears to be the maintenance of a transepithelial ion barrier analogous to that maintained by tight junctions. Since microfilaments are often seen attached to their cytoplasmic surfaces, a role in cell adhesion also seems likely.

Arthropods are unusual in that they may possess two types of septate junctions. One type of septate junction contains distinct ladderlike septa between the cells (pleated septate) and was first described by Locke (1965). The other type of septate junction is characterized by a dense homogeneous-staining material between the adjacent cells (smooth septate or *zonula continua*) and was first described in the insect midgut by Noirot and Noirot-Timothee (1967). The large size of the septate junctions which may extend several (5–7) micrometers deep in some tissues can be easily recognized in homogenates by light microscopy using phase contrast. If the tissue is homogenized gently in a buffer containing calcium, these junctions form large sheets after cell disruption containing attached plasma membrane and gap junctions. A significant enrichment of the plasma and attached membrane cell junctions can thus be achieved by differential centrifugation and ultracentrifugation (Fig. 15). So far only smooth septate junctions have been isolated from arthropods (Cioffi and Wolfersberger, 1983; Green *et al.*, 1983; Berdan *et al.*, 1986). Since these junction fractions still contain attached plasma membranes and other cell junctions and are correlated with a heterogeneous range of polypeptides (Fig. 15), the unambiguous identity of septate junction polypeptides awaits the availability of immunological probes.

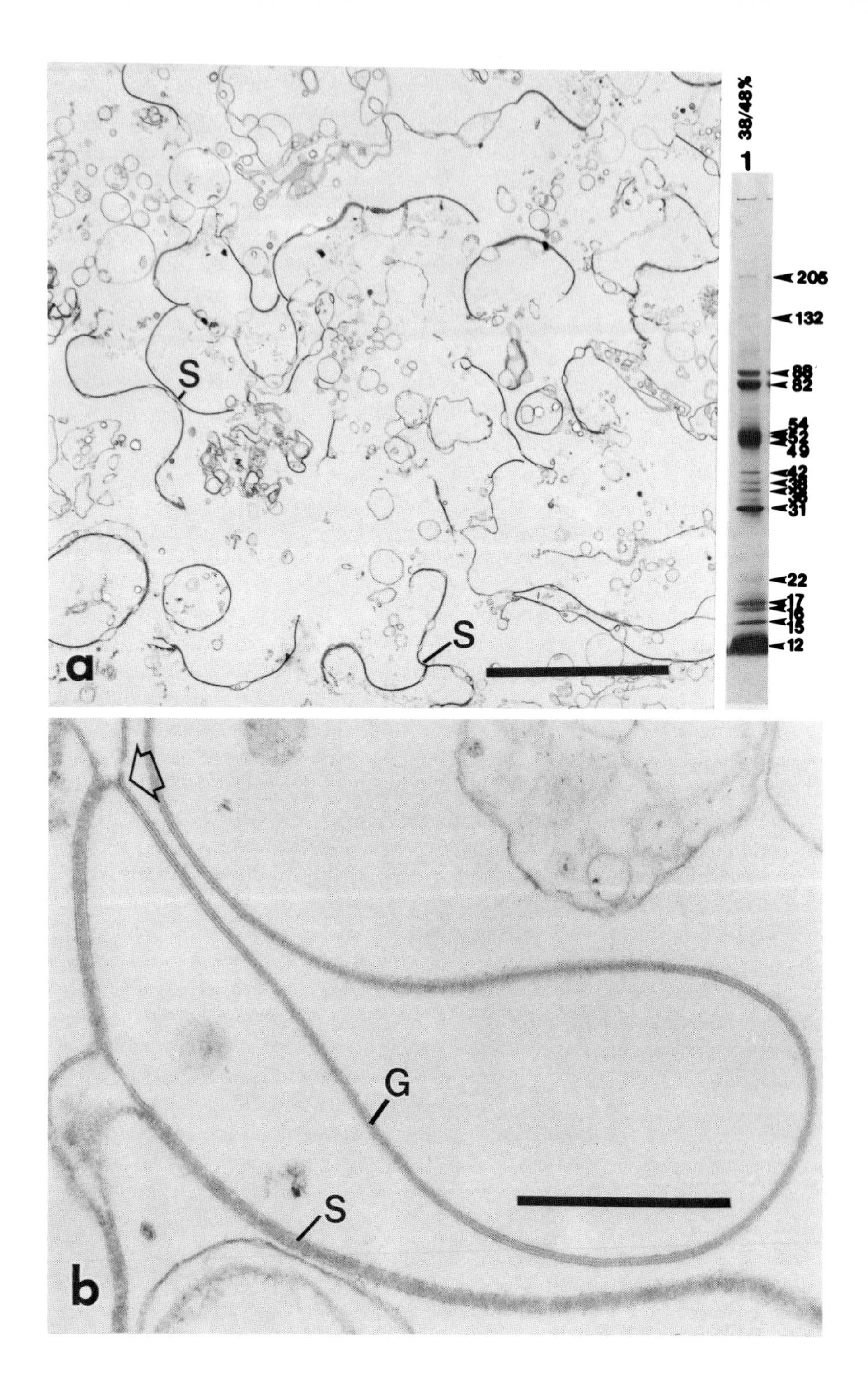

38/48%
1
205
132
88
82
54
42
31
22
17
16
15
12
S
S
a
G
S
b

Using a plasma membrane junction fraction from the crayfish hepatopancreas obtained from a 38%/48% (w/w) sucrose step gradient after ultracentrifugation, we have attempted to selectively enrich for gap junctions by extraction with detergents, alkali, and chaotropic agents which have been used successfully for vertebrate gap junctions (Table VII). The rationale of this approach is based on the observation that vertebrate gap junctions, which are presumably stabilized by protein–protein interactions, are relatively insoluble in these agents, whereas the majority of plasma membrane proteins and lipids are solubilized. We found that extraction with nonionic detergents (Triton X-100 5% w/v; Triton X-114 5% w/v; deoxycholate 2.5% w/v; CHAPS 0.6% w/v) resulted in about a similar enrichment of arthropod gap junctions; treatment with the ionic detergent Sarkosyl (0.5% w/v) or sodium hydroxide ($>$ 20 mM) disrupted the arthropod gap junctions. Therefore, at the level of protein–protein or protein–lipid interactions, arthropod and vertebrate gap junctions appear to be different. The greater susceptibility of arthropod gap junctions to disruption by these agents correlates with their generally larger intramembrane particle diameter and less tightly packed arrays seen in freeze-fracture replicas.

Following extraction of crayfish plasma membrane junction fractions with nonionic detergents, besides gap junctions, a considerable amount of amorphous material (delipidated membrane proteins?) and "pseudo-gap junctions" remains (Fig. 20b,c). Despite further extraction procedures and separations on additional sucrose gradients, it has not been possible to remove all the "contaminants." The greatest enrichment of crayfish gap junctions, however, has been obtained by extracting a gap junction plasma membrane fraction devoid of septate junctions, with 5 mM NaOH (Fig. 18). In linear sucrose gradients centrifuged to equilibrium (275,00*g* for 17 hr), gap junctions linked by plasma membranes, but not septate junctions, band between 32 and 47% (w/w) sucrose. The septate junction fraction, however, bands lower in the gradient (47–50% w/w sucrose, $\rho = 1.21–1.23$ g/cm^3). The most detergent- and alkali-resistant structures, pseudo-gap junctions, are concentrated primarily (but not exclusively) between 30 and 35% (w/w) sucrose. Although the resultant 5 mM NaOH-insoluble gap junction fraction is not morphologically pure, it is sufficiently enriched to allow high-resolution electron microscopic analysis (Fig. 18). Preliminary studies using Markham photographic rotational reinforcement for example suggest that

←

FIGURE 16. (a) Low-magnification electron micrograph of a section through pelleted material from the 38%/48% (w/w) sucrose gradient interface enriched in septate junctions (S), attached cell junctions, and plasma membrane. Scale bar = 5 μm. At the right is an SDS–PAGE profile of the major polypeptides (75 μg loaded) and stained with Coomassie blue. The apparent molecular masses of the major polypeptides are indicated in kilodaltons. The major staining band at 12 is due to the addition of Aprotinin, a low-molecular-weight peptide proteolytic inhibitor. (b) High-magnification electron micrograph taken from a region in (a) showing a gap junction (G) attached to a septate junction (S, arrow). Scale bar = 0.5 μm.

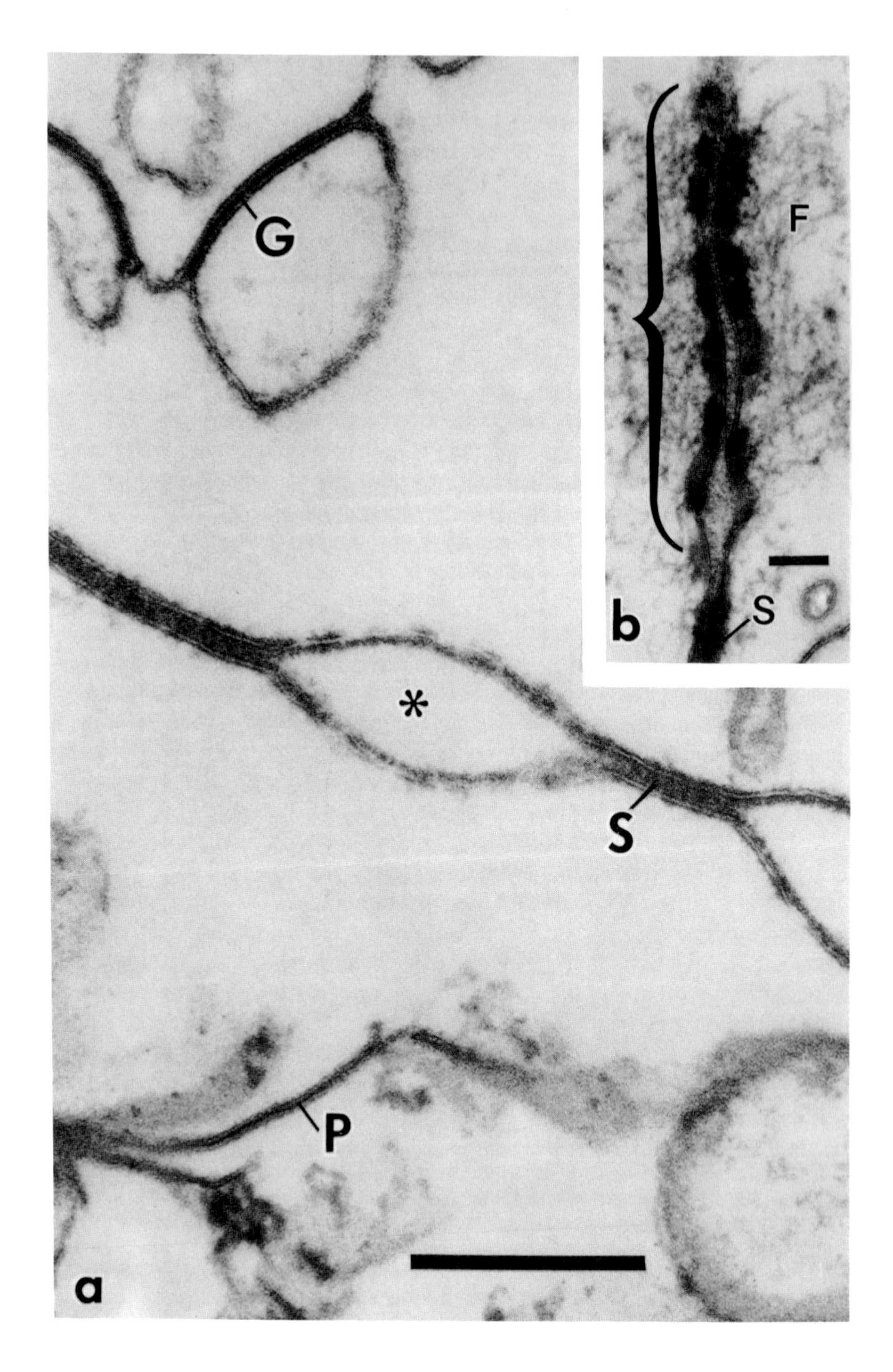
G
F
*
S
b
S
P
a

the gap junction subunits seen in tangential sections exhibit a sixfold symmetry (Fig. 18c). Similarly, negative staining of the isolated plaques reveals a double membrane bilayer (Fig. 19) and a population of relatively uniform 9.6-nm-diameter subunits. The uniformity of subunit size is in surprising contrast to the results seen by freeze-fracture. However, the uniformity of subunit size does not appear to be an artifact induced during extraction, since a similar size range of subunits is seen by negative staining of gap junctions in membrane fractions (Fig. 19b). Furthermore, subunits seen after extracellular staining of arthropod tissues fixed *in situ,* also display a less heterogeneous size range (Fig. 10b). The possibility that the particle heterogeneity of arthropod gap junctions seen by freeze-fracture is an artifact must therefore be considered. Alternatively, the heterogeneity of particle size in freeze-fracture replicas may be a result of randomly but tightly bound lipid. Certainly, differences in protein–lipid interactions might explain why arthropod and vertebrate gap junction intramembrane particles preferentially adhere to different fracture faces. Differences in lipid content might also account for the greater density of hepatopancreas gap junctions, $\rho = 1.20\ g/cm^3$, as compared to those isolated from mouse and rat liver, $\rho = 1.17\ g/cm^3$ (Henderson *et al.*, 1979; Hertzberg and Gilula, 1979).

5.3. Biochemical Analysis of Arthropod Gap Junctions

Gap junctions have recently been enriched from the crayfish hepatopancreas (Berdan and Gilula, 1987a). In the regions of the sucrose gradients, where crayfish hepatopancreas gap junctions are most enriched, five to ten polypeptides are detected by SDS–PAGE even after 5 mM NaOH or Triton X-100 extraction (Fig. 22c). Since it has not yet been possible to purify arthropod gap junctions, the unambiguous identification of the arthropod gap junction polypeptide(s) is not possible. Nevertheless, a 31,000-dalton polypeptide appears to be a component of the crayfish gap junction for the following reasons. First, a 31,000-dalton polypeptide from Triton X-100-enriched crayfish hepatopancreas junction fractions has been observed to bind calmodulin in gel overlays in a calcium-dependent fashion (van Eldik *et al.* 1985; Fig. 23). An interaction of calmodulin with arthropod gap junctions has been postulated on the basis of electrophysiological experiments with calmodulin inhibitors (Lees-Miller and Caveney, 1982). Calmodulin has also been demonstrated to bind the vertebrate liver gap junction

←

FIGURE 17. (a) High-magnification electron micrograph of a plasma membrane junction fraction showing three different membrane specializations. Note the similarity between gap junctions (G) and pseudo-gap junctions (P). The latter structure is never found attached to septate junctions and is ~ 2 nm thinner than gap junctions. Septate junctions (S) lose their filamentous attachments when isolated in the absence of calcium and sometimes split (*) retaining a fuzzy coat on their external surfaces. Scale bar = 250 nm. (b) Zonula adherens (bracket) attached to the septate junction (S). Filaments (F) appear to attach via dense-staining thickenings on the membranes. Scale bar = 100 nm.

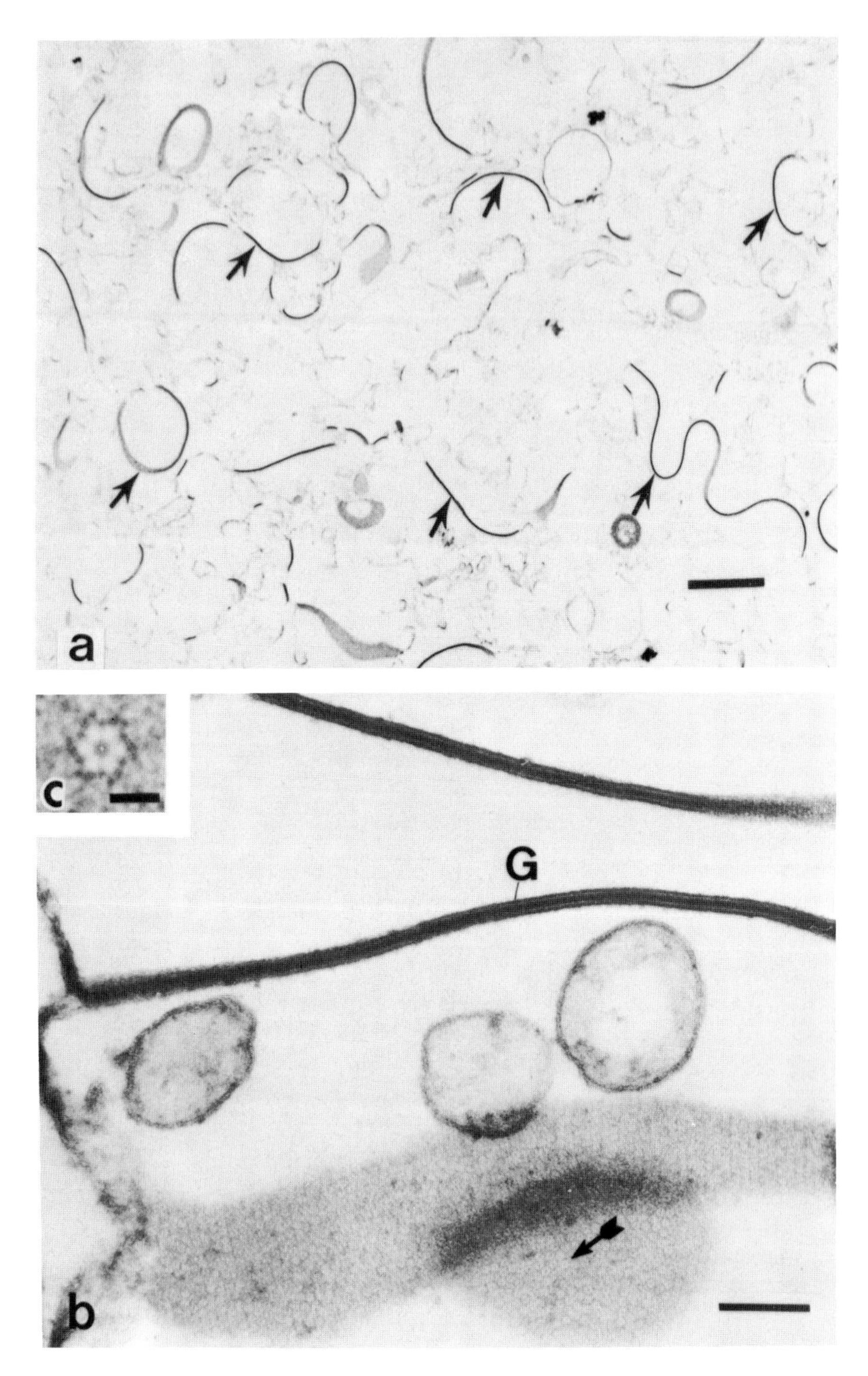
a
c
G
b

polypeptide and the binding is reduced in the absence of calcium (Hertzberg and Gilula, 1982; van Eldik *et al.*, 1985). Very few other polypeptides in the hepatopancreas junction fractions, with the exception of the 16 and 17k polypeptides, exhibit any significant calmodulin binding (Fig. 23). The calmodulin binding experiments, however, need to be repeated with 5 mM NaOH-enriched gap junctions. Second, a 31k polypeptide enriched by 5 mM NaOH extraction gradually diminishes in staining intensity when examined by SDS–PAGE after extraction with 10 mM NaOH and was absent after 20 mM NaOH (Fig. 22b). Electron microscopic analysis of duplicate fractions showed the disappearance of recognizable gap junctions was correlated with the loss of this polypeptide. Third, comparison of the polypeptides by SDS–PAGE after extraction of gap junctions enriched on linear sucrose gradients with Triton X-100 or 5 mM NaOH, revealed that five polypeptides (205, 42, 31, 17, and 16k) were common (Fig. 22c). Since the 17 and 16k polypeptides could be correlated with the further enrichment of the pseudo-gap junctions and the 205 and 42k polypeptides are strong candidates for myosin and actin, the data suggest that a 31k polypeptide may be a component of the arthropod gap junction.

One other study that has dealt with the isolation of septate junctions from the midgut of the insect *Tenebrio molitor* reported one fraction ($\rho = 1.14–1.16$ g/cm^3) to be more enriched in gap junctions than others (Green *et al.*, 1983). These investigators correlated the enrichment of gap junctions with a 36k polypeptide by SDS–PAGE. We have also observed a 36k polypeptide after NaOH enrichment in the crayfish, but not after Triton X-100 enrichment. Furthermore, the studies of Green *et al.* (1983) also show the presence of a 31 and 32k polypeptide in all of their junctional fractions.

Whereas the identity of arthropod gap junction polypeptide(s) awaits immunological clarification, it is important to point out that if the samples have not been separated by two-dimensional SDS–PAGE one cannot assume that a single band on a gel represents a single polypeptide. Unfortunately, the relative insolubility of gap junctions in nonionic detergents makes analysis by two-dimensional SDS–PAGE difficult, but not impossible (Johnson *et al.*, 1986). Therefore, antiserum generated to a single polypeptide band cut from a one-dimensional polyacrylamide gel may be heterogeneous.

←

FIGURE 18. (a) Low-magnification electron micrograph of a pellet of enriched gap junctions (arrows) from a linear sucrose gradient (32–47% w/w sucrose) after 5 mM NaOH extraction. Note the length of some gap junctions. Scale bar = 1 μm. (b) High-magnification section of several gap junctions (G) revealing both cross sections and tangential views. Individual subunits with electron-dense cores can be visualized (arrow). *En bloc* uranyl acetate and tannic acid were used to enhance membrane contrast. Scale bar = 100 nm. (c) A single subunit from a tangential section of a gap junction that has undergone photographic image enhancement exhibits an apparent sixfold symmetry. Scale bar = 10 nm.

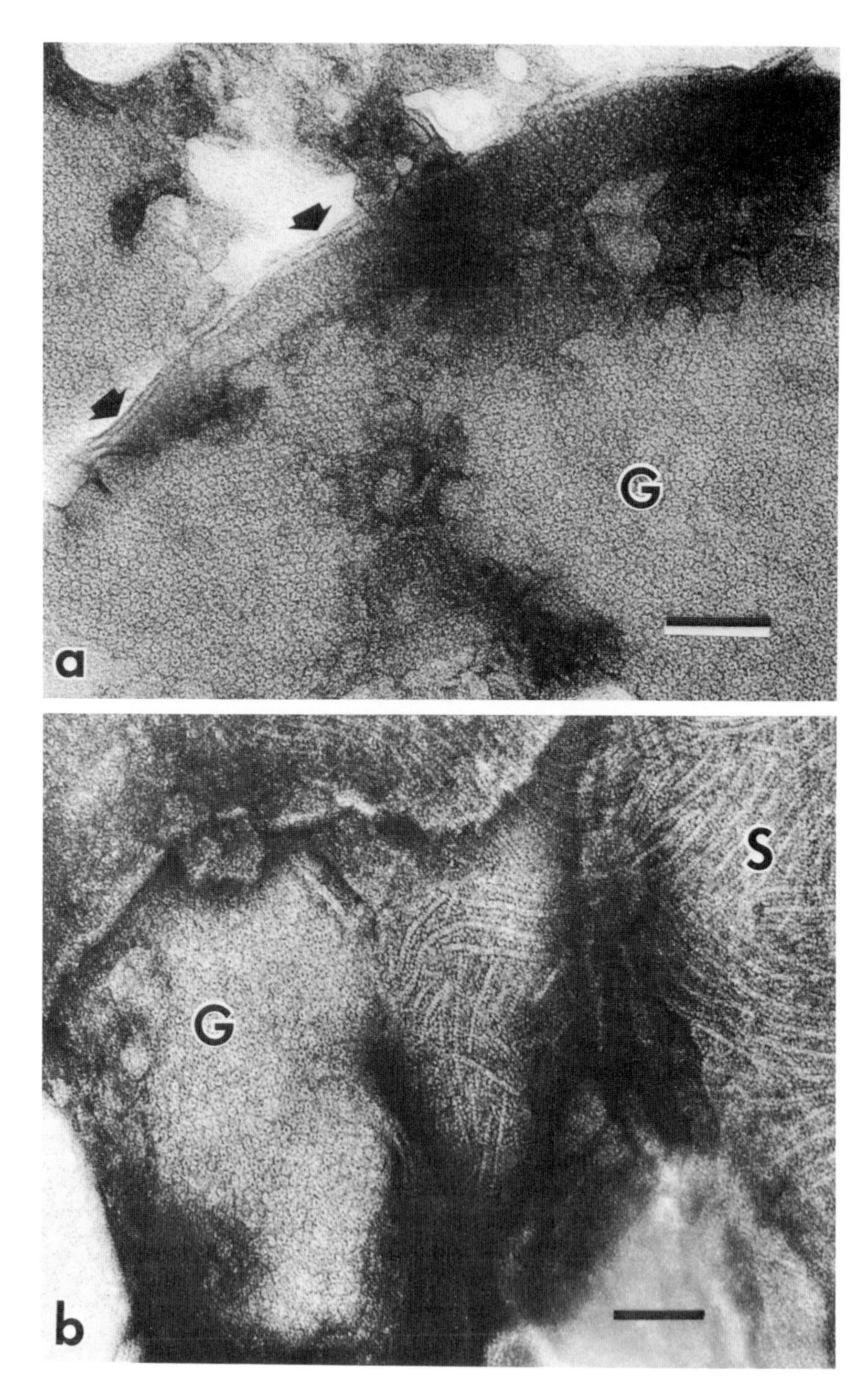
G
a
S
G
b

5.4. Pseudo-Gap Junctions

During preliminary experiments aimed at isolating arthropod gap junctions, we first isolated from the crayfish hepatopancreas a structure that closely resembled gap junctions. The resemblance of this structure to isolated gap junctions was so striking that we mistakenly identified it as a gap junction in a preliminary report (Berdan and Gilula, 1982). Subsequent to further studies we have tentatively named them "pseudo-gap junctions" (Berdan and Gilula, 1987b). The pseudo-gap junctions isolated from the crayfish hepatopancreas appear to be analogous in morphology and molecular weight of its constituent polypeptides to another structure recently isolated from the lobster hepatopancreas and identified as an arthropod gap junction (Finbow *et al.*, 1984).

Several features of pseudo-gap junctions allow them to be distinguished from "authentic" gap junctions. Pseudo-gap junctions in thin sections are about 2 nm thinner in cross-sectional width than crayfish gap junctions both *in situ* and *in vitro*. We have also observed numerous intracellular myelinlike figures in the intact hepatopancreas and after homogenization which are composed of structures having the same thickness and appearance of pseudo-gap junctions. Pseudo-gap junctions, unlike authentic gap junctions, have never been observed attached to septate junctions (i.e., associated with the cell surface membrane) *in vitro*. Furthermore, following extraction of a membrane junction fraction containing both gap junctions and pseudo-gap junctions with 0.5% Sarkosyl or greater than 20 mM NaOH, only the latter structures are identifiable. No intermediate structures containing both thick and thin regions have been observed. Finally, pseudo-gap junctions tend to band in less dense regions of the sucrose gradients as compared to gap junctions. If the thinner pseudo-gap junctions had arisen as a result of lipid extraction, we would expect the resultant protein-enriched structures to band at a greater density than gap junctions, assuming an average density for lipid (0.91 g/cm^3) and that for protein (1.35 g/cm^3); however, the reverse is observed.

Pseudo-gap junctions can be readily isolated and purified on sucrose gradients after membrane preparations are extracted with greater than 20 mM NaOH or 0.5% (w/v) Sarkosyl (Fig. 21). The isolated structures correlate with two polypeptides having apparent molecular weights of 16 and 17k. Tryptic peptide mapping demonstrates that both polypeptides are similar, the 16k most likely representing a proteolytic degradation product (Berdan, 1985). Both of these

←

FIGURE 19. Electron micrographs of crayfish hepatopancreas gap junctions on Formvar–carbon-coated grids negatively stained with 1% uranyl formate. (a) Part of a gap junction (G) from a 5 mM NaOH enriched fraction. Note the apparent uniformity of subunit diameter and the double membrane bilayer (arrows) at the edge of the junction. Scale bar = 100 nm. (b) Gap junction (G) attached to septate junction (S) from a membrane fraction before alkali extraction. Note similarity of the subunits to those in (a). Scale bar = 100 nm.

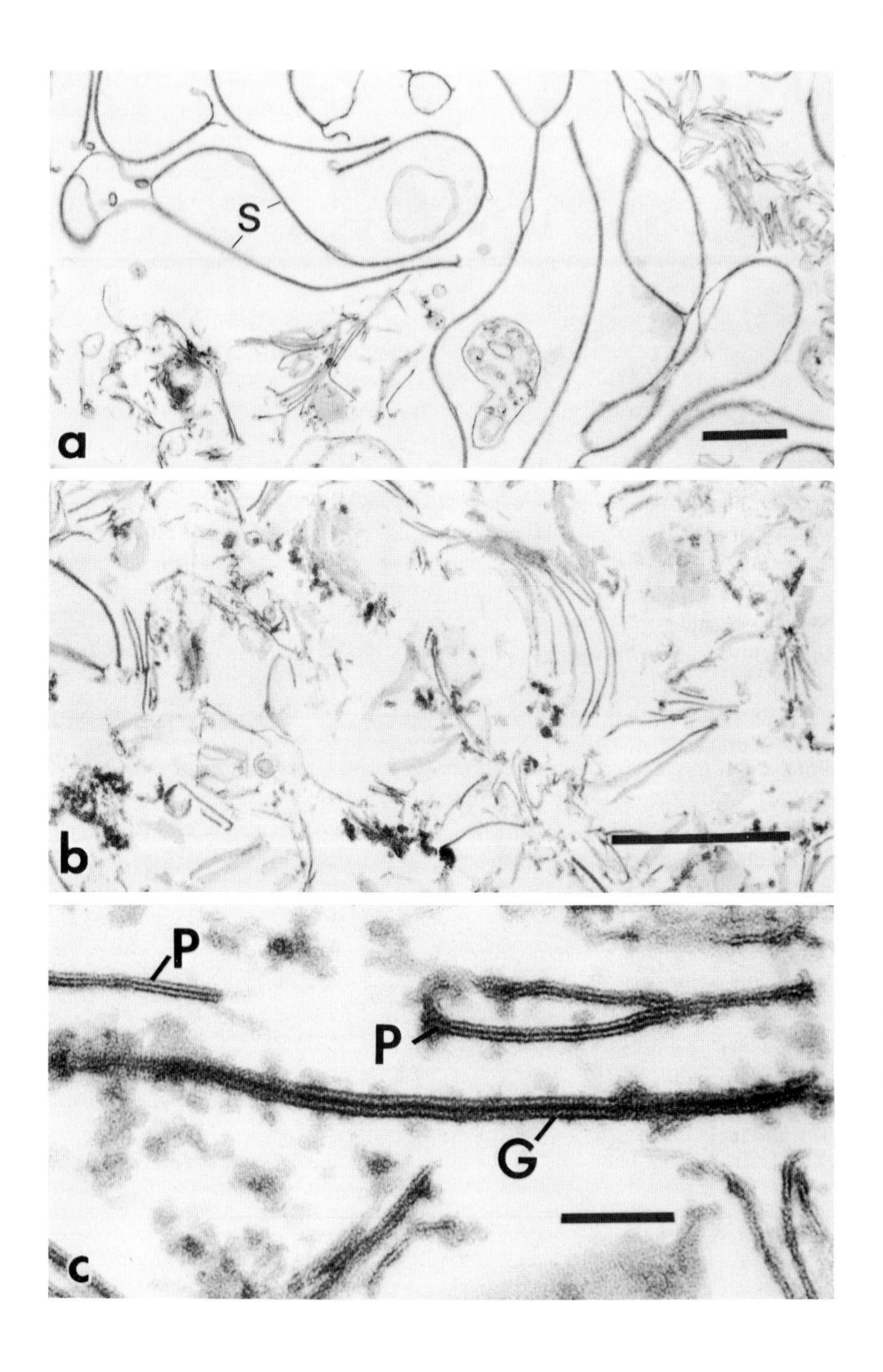
S
a
b
P
P
G
c

polypeptides bind calmodulin in gel overlays in a calcium-independent fashion (van Eldik *et al.*, 1985; Fig. 23). After negative staining, the isolated structures appear plaquelike containing smaller subunits (8–10 nm in diameter) often arranged in a crystalline array (Fig. 21a,b). The crystalline arrays, however, could not be identified in membrane fractions prior to detergent or alkali extraction.

The reasons we believe the structures isolated from the lobster hepatopancreas (Finbow *et al.*, 1984) may be analogous to pseudo-gap junctions, include their similar thickness (14–16 nm), and similarity in polypeptides correlated with their isolation, 18k versus 16 and 17k. That these investigators see only one band (18k) by SDS–PAGE can easily be explained if their gel was overloaded (Fig. 22a, lane 7 versus 8). Furthermore, these investigators having examined their isolated structures by freeze-fracture were unable to demonstrate intramembrane particles characteristic of gap junctions. The isolation procedure utilized by Finbow *et al.* (1984) also skips isolation of a plasma membrane fraction and utilizes the ionic detergent Sarkosyl which we have found disrupts crayfish hepatopancreas gap junctions.

What are pseudo-gap junctions and what is their significance? At present, it is not possible to rule out this structure as an artifact induced during isolation, fixation, or even whether it is somehow related metabolically to gap junctions. Similar structures have, however, been observed in a wide variety of tissues often between membrane regions that appear to be undergoing fusion (Lawson *et al.*, 1977; Neutra and Schaeffer, 1977; Specian and Neutra, 1980; Tanaka *et al.*, 1980) or in myelinlike figures (Lawson *et al.*, 1977; McGookey and Anderson, 1983; Barondes *et al.*, 1985). Their abundance in secretory tissues suggests that these regions of membrane may be involved in membrane fusion, recycling, or biosynthesis.

Thin structures resembling gap junctions have also been observed in detergent-, urea-, and protease-treated membranes obtained from the vertebrate lens; along with a population of thicker gap-junction-like structures (Kistler and Bullivant, 1980; Zampighi *et al.*, 1982; Nicholson *et al.*, 1983). The thicker structures are 15–17 nm in width while the thinner structures are 13–14 nm (Zampighi *et al.*, 1982). Analysis of the isolated thin structures by negative staining and freeze-fracture revealed highly crystalline lattices and the absence of intramembrane particles on their fracture faces (Kistler and Bullivant, 1980;

←

FIGURE 20. (a) Electron micrograph of a thin section through an enriched plasma membrane junction fraction from a sucrose gradient prior to detergent extraction. S, septate junctions. Scale bar = 1 μm. (b) Thin section through a Triton X-100 (5% w/v)-insoluble pellet derived from the plasma membrane junction fraction. Detergent extraction solubilizes the septate junctions and most nonjunctional membranes. Scale bar = 1 μm. (c) At higher magnification, the detergent-insoluble fraction contains gap junctions (G), pseudo-gap junctions (P), and amorphous material which probably represents insoluble delipidated membrane-associated proteins. Scale bar = 100 nm.

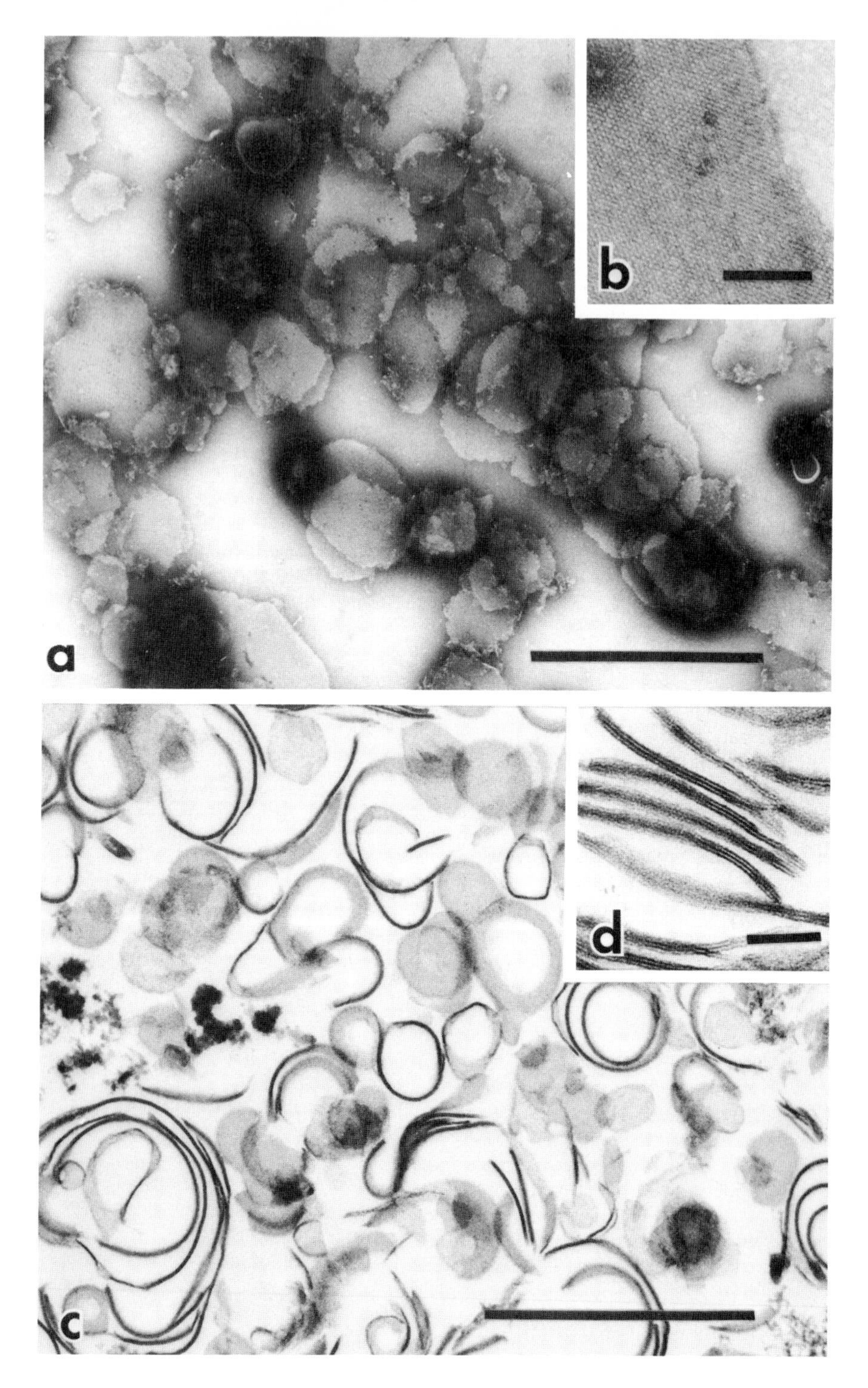

a
b
c
d

Zampighi *et al.*, 1982). Both of these features are similar to the "pseudo-gap junctions" isolated from the crayfish (Berdan, 1985) and the lobster (Finbow *et al.*, 1984). Kistler and Bullivant (1980) concluded that the thin structures from the lens are formed as a consequence of protease digestion of the lens plasma membranes, and these structures correlate with the appearance of a 16k polypeptide. It is noteworthy that polypeptides (16–20k) obtained by proteolytic digestion of lens junction preparations, bind calmodulin in gel overlay experiments in a calcium-independent fashion like the 16 and 17k polypeptides from the crayfish (van Eldik *et al.*, 1985). The possibility that the thin structures (pseudo-gap junctions) may arise from the proteolytic digestion of membranes either intracellularly or in response to exogenously added proteases is intriguing in light of several studies implicating proteases in membrane fusion and exocytosis (Ahkong *et al.*, 1978; Couch and Strittmatter, 1983, 1984; Mundy and Strittmatter, 1985). Structures resembling pseudo-gap junctions have also been observed adjacent to isolated rat liver gap junctions. While isolated rat liver gap junctions could be labeled heavily with antiserum to the 27k rat liver gap junction polypeptide, no binding was observed on adjacent structures resembling pseudo-gap junctions (Gilula, unpublished observations).

In conclusion, the biochemical characterization of arthropod gap junctions has only just begun. While the identity of arthropod gap junction polypeptide awaits immunological verification, it appears that immunological probes to vertebrate gap junctions do not cross-react. Also, it is important that in future gap junction isolation procedures, particularly those carried out in arthropods or secretory tissues, careful attention be paid to the thickness of the isolated structures and to whether the structures originated from the plasma membrane. The significance of pseudo-gap junctions will depend on whether the morphologically similar structures seen in other tissues are also biochemically similar and ultimately on the functions of this structure. I would also point out that despite an enormous variety of immunization protocols with milligrams of 16 and 17k polypeptides, it has not been possible to elicit an immune response that recognizes the pseudo-gap junctions (Berdan, 1985), suggesting that these polypeptides may be highly conserved.

←

FIGURE 21. Electron micrographs of enriched pseudo-gap junctions. (a) Low-power micrograph of pseudo-gap junctions negatively stained with 2% phosphotungstic acid on Formvar–carbon-coated grid. Pseudo-gap junctions were enriched by detergent extraction (Triton X-100) and subsequently purified on a linear sucrose gradient. Scale bar = 1 μm. (b) Inset shows crystalline arrangement of subunits observed frequently on pseudo-gap junctions. Scale bar = 100 nm. (c) Thin section through an enriched fraction of pseudo-gap junctions prepared by detergent (0.5% w/v Sarkosyl) extraction of a plasma membrane junction fraction. Thicker gap junctions are not found after extraction with this detergent. Scale bar = 1 μm. (d) Inset shows higher-magnification view of pseudo-gap junctions and demonstrates their apparent similarity to authentic gap junctions. Scale bar = 100 nm.

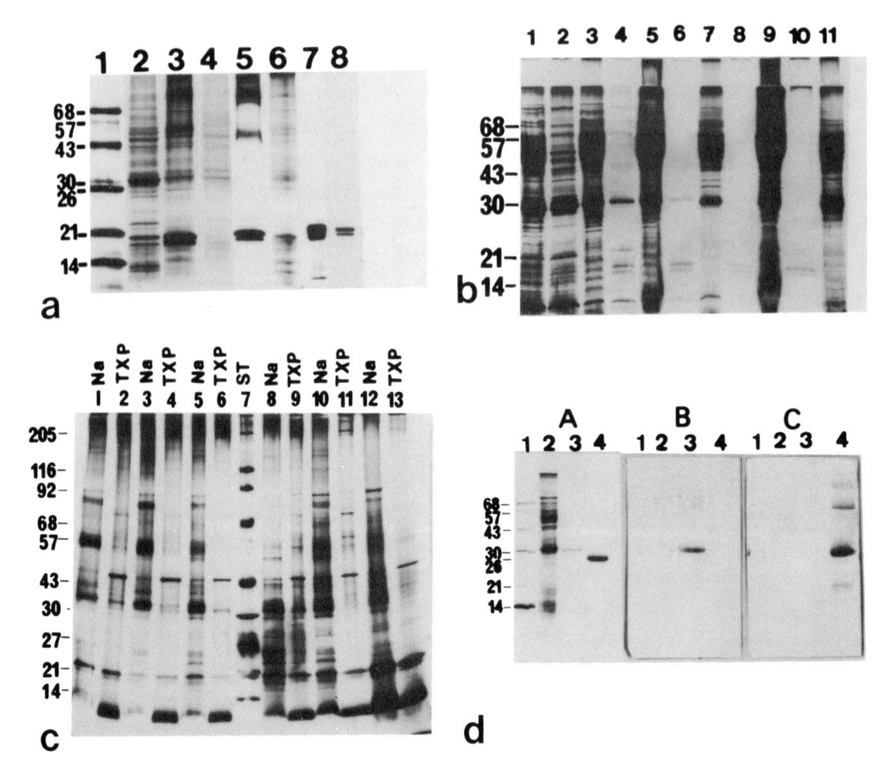

FIGURE 22. (a) SDS–PAGE analysis of fractions obtained during the enrichment of pseudo-gap junctions. Approximately 10 μg of protein was loaded into each lane and the polypeptides visualized after electrophoresis by silver staining. Lane 1—molecular mass standards in kilodaltons; lane 2—plasma membrane cell junction fraction; lane 3—Triton X-100-insoluble plasma membrane pellet; lane 4—trichloroacetic acid (TCA)-precipitated Triton X-100-soluble membrane fraction; lane 5—NaOH (50 mM)-insoluble pellet derived from the Triton X-100-insoluble pellet (bulk isolation); lane 6—TCA-precipitated NaOH-soluble polypeptides; lane 7—similar to lane 5 except prepared on a smaller scale and showing the absence of higher-molecular-mass bands; lane 8—same as lane 7 but with lower protein load showing two polypeptide bands with apparent molecular masses of 16,000 and 17,000 daltons. These polypeptides correlate with the purification of pseudo-gap junctions.

(b) SDS–PAGE of membrane junction fractions after extraction with 1–50 mM NaOH. Seventy-five micrograms of membrane protein was extracted with 10 volumes of NaOH for 30 min at room temperature and pelleted (16,000*g* for 10 min). Material remaining in the supernatant (sup) was precipitated with 10% TCA. After several washes in 1 mM $NaHCO_3$, samples were solubilized in SDS, electrophoresed, and the polypeptides visualized by silver staining. Duplicate samples were analyzed by electron microscopy. Lane 1—control unextracted membrane junction fraction; lane 2—1 mM NaOH pellet; lane 3—1 mM NaOH sup; lane 4—5 mM NaOH pellet (most enriched fraction of gap junctions after analysis by electron microscopy); lane 5—5 mM NaOH sup; lane 6—10 mM NaOH pellet; lane 7—10 mM NaOH sup; lane 8—20 mM NaOH pellet; lane 9—20 mM NaOH sup; lane 10—50 mM NaOH pellet; lane 11—50 mM NaOH sup.

(c) One-milliliter fractions from six linear sucrose gradients having approximately the same sucrose concentration were pooled and extracted with either 5 mM NaOH (Na) or 5% (w/v) Triton X-100 (TXP). The resultant insoluble material was analyzed by SDS–PAGE on 7.5–15% poly-

FIGURE 23. ^{125}I-labeled calmodulin gel overlays. Panel a is a Coomassie blue-stained polyacrylamide gel in which (1) 75 μg of a crayfish Triton X-100-insoluble plasma membrane fraction was loaded; and (2) 0.5 μg of eluted rat liver 27K gap junction polypeptide was loaded. Panel b is an autoradiograph of polyacrylamide gel as in "a" except incubated in [^{125}I]calmodulin in the presence of 1 mM calcium. Panel c is an autoradiograph as "b" except in the presence of 5 mM EDTA. Calmodulin bound to three polypeptides in the crayfish—16, 17, and 31K. Binding of calmodulin to the rat liver gap junction polypeptide and a 31K crayfish polypeptide was reduced in the absence of calcium.

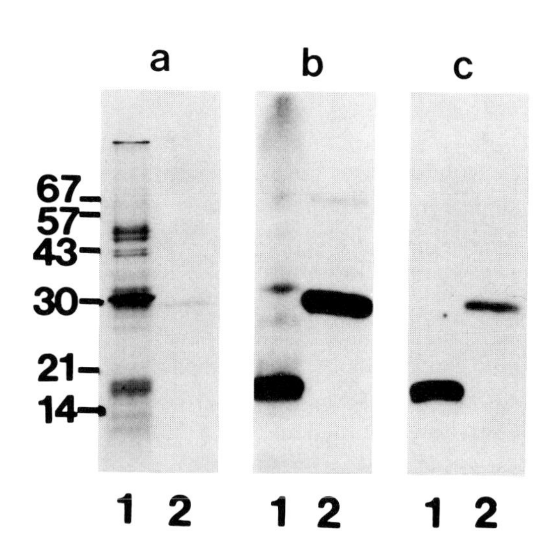

acrylamide gel gradients after silver staining. The sucrose concentrations (w/w) from which the samples were taken were: lane 1—23%; lane 2—29%; lane 3—34%; lane 4—35%; lane 5—37%; lane 6—39%; lane 7—ST (molecular weight standards); lane 8—40%; lane 9—42%; lane 10—42%; lane 11—46%; lane 12—48%; lane 13—50%. The major remaining insoluble polypeptides common to both samples after extraction were: > 205, 42, 31, 16, 17K in the 32–42% region of the gradient where gap junctions were found to be most abundant after electron microscopic analysis. Samples for this analysis were solubilized in SDS sample buffer for 30 min at room temperature.

(d) Triton X-100-enriched junction fractions were separated by SDS–PAGE and electrophoretically transferred to nitrocellulose (Burnette, 1981). Nonspecific binding was blocked with 5% w/v powdered milk in saline (Johnson *et al.*, 1984). The blots were incubated in antiserum overnight, washed, then incubated with second anti-rabbit antibody conjugated with horseradish peroxidase (2 hr), rinsed, and antibody binding visualized by development in chloronaphthol/H_2O_2 (Hawkes *et al.*, 1982). (A) Polyacrylamide gel stained with Coomassie brilliant blue. (B) Immunoblot incubated with anti-27K rat liver gap junction antiserum. (C) Immunoblot incubated with anti-26K bovine lens junction antiserum. Lane 1—molecular weight standards (negative control); lane 2—Triton X-100-insoluble fraction, 75 μg; lane 3—27K eluted rat liver gap junction polypeptide, 0.5 μg; lane 4—26K eluted bovine lens polypeptide, 2.0 μg. All samples were boiled for 90 sec prior to loading. Neither the 27 nor 26K affinity-purified antisera cross-reacted with any polypeptides in the crayfish fraction. Similar observations were obtained with 5 mM NaOH-enriched crayfish gap junction fractions and the use of iodinated protein A to detect primary antibody binding.

6. SUMMARY AND PROSPECTIVES

The use of intracellular microelectrodes first permitted the detection of low-resistance pathways in both excitable and nonexcitable tissues in arthropods. The large size, ready accessibility, and simple geometric organizations of arthropod tissues subsequently allowed quantitative estimates of junctional conductance. In some instances, changes in intercellular junctional conductance have been shown to occur both in a temporal and spatial manner and in response to drugs and developmentally important hormones. The iontophoresis and microinjection of dyes and fluorescent tracers of various size first established in arthropods and subsequently in other tissues, that in most instances low-resistance pathways could be attributed to discrete channels of finite diameter. Although the channel diameter appears to be slightly larger in arthropod than in vertebrate tissues, the significance of this difference is unknown. The size range of channel diameters present in other organisms has yet to be examined. It is reasonable to suggest, however, that a larger channel diameter may compensate for the lower overall packing density of gap junction particles in arthropods versus that in vertebrate tissues. Alternatively, arthropods may utilize larger molecules for intercellular communication. The signal molecules or morphogens in arthropods that could pass through gap junctions have not yet been identified. In *Hydra,* however, morphogens with a molecular weight range of 500–1100 that could pass through gap junctions have been found (Schaller and Bodenmuller, 1981).

Analysis of the diffusion patterns of fluorescent tracers has revealed how some metabolites and potential morphogens might distribute themselves in compact tissues. In this regard, tracer studies have revealed distinct functional compartments in some insect tissues. Compartments are of obvious functional importance in ion-transporting epithelia, but they may also play a role in pattern formation during development. Kinetic studies of different sized fluorescent tracers in insect epidermis and salivary glands also suggest that gap junction channels may flicker between open and closed states in an all-or-none fashion. In this respect, gap junction channels appear to resemble other ion channels. Furthermore, the available evidence indicates that the gating of gap junction channels may be regulated by several mechanisms ($Ca^{2+}{}_i$, $H^+{}_i$, membrane potential, transjunctional voltage), although it is not certain whether these effectors are acting directly or indirectly on the channels. Reconstitution studies with the vertebrate lens junction protein into liposomes, for example, have shown that the calcium ion has no effect on the channel permeability in the absence of calmodulin (Girsch and Peracchia, 1985). Future reconstitution studies with gap junction proteins from other sources are necessary to establish the generality of the mechanisms regulating channel gating.

At the ultrastructural level, arthropod gap junctions can be distinguished by freeze-fracture from those in vertebrates and many other invertebrates. It is unknown why arthropods have evolved gap junctions that are morphologically and physiologically distinct from other organisms. It is possible, as suggested by the appearance of uniform sized subunits in isolated arthropod gap junctions, that the apparent differences seen by freeze-fracture may be due to differences in membrane lipid composition. Septate junctions, for example, presumably have similar functions in different organisms yet they exhibit an even greater ultrastructural diversity. One should also be aware that arthropods exhibit a tremendous adaptive diversity and over three-quarters of a million species have been described, i.e., more than three times the number of all other animal species combined! (Barnes, 1974).

The isolation and biochemical characterization of vertebrate gap junctions has preceded that in arthropod tissues. This in part is a reflection of the larger size of some vertebrate tissues and of the potential relevance of gap junctions in medicine. It is relevant to point out that the purification and identification of the vertebrate gap junction polypeptide (26–28k) has required more than a decade and is still today not universally accepted (Table VII). Cloning of the gap junction gene(s) in the near future should clarify any ambiguities.

Procedures successful in purifying vertebrate gap junctions have so far only resulted in the enrichment of arthropod gap junctions. The greater lability of arthropod gap junctions to agents designed to extract primarily nonjunctional membranes (detergents, alkali, chaotropic agents) suggests that new or novel approaches may be necessary for their isolation. Alternatively, the enriched gap junction fractions from the crayfish hepatopancreas could be used to immunize mice for the production of hybridomas secreting monoclonal antibodies (Kohler and Milstein, 1975). The monoclonal antibodies could then be used to identify the junctional polypeptides and their distribution in different tissues and/or organisms. Another approach to identify the arthropod gap junction polypeptide that would not require purification would be to utilize a genetic and molecular approach. This approach would involve identifying behavioral mutants in *Drosophila* which could then be screened by electrophysiological techniques. This approach has been utilized to identify mutations altering synaptic connectivity between identified neurons in *Drosophila* (Thomas and Wyman, 1983; Wyman and Thomas, 1983). The mutation could then be localized to specific regions of the polytene chromosomes by meiotic recombination and deletion mapping. The DNA within this mapped region could subsequently be microdissected and cloned (Scalenghe *et al.*, 1981). It may also be possible to utilize molecular probes to vertebrate gap junctions when they become available to screen copy DNA libraries constructed from *Drosophila* (Maniatis *et al.*, 1978). The observation that vertebrate gap junction antiserum does not cross-react with arthropod gap junctions does not preclude the existence of some conserved sequence homology that may be detected by a molecular probe. Ultimately the more widespread availability of molecular and immunological probes should allow us to gain a greater insight into the structure and function of gap junctions as well as the potential significance of gap junction-mediated cell communication.

In conclusion, studies on intercellular communication in arthropods from a biophysical, ultrastructural, biochemical, and in the future a molecular and genetics perspective, will continue to contribute significantly to our understanding of gap junction-mediated intercellular communication.

ACKNOWLEDGMENTS. I wish to thank Stan Caveney, Richard Shivers, Norton Gilula, and Andrew Bulloch for their guidance and opportunity to conduct research on gap junctions in their laboratories. I am also indebted to Colin Green, Eric Eastman, Peter Jones, Bob Pardue, and Katrina Waymire for many helpful discussions and suggestions regarding technical considerations. Support for this research was provided by the Natural Sciences and Engineering Research Council of Canada and the Alberta Heritage Foundation for Medical Research. The author sincerely appreciates the excellent secretarial assistance of Ms. Caroline Collins.

7. REFERENCES

Ahkong, Q. F., Blow, A. M., Botham, G. M., Launder, J. M., Quirk, S. J., and Lucy, J. A., 1978, Proteinase and cell fusion, *FEBS Lett.* **9S:**147–159.

Alvarez-Leefmans, F. J., Rink, T. J., and Tsien, R. Y., 1981, Free calcium ions in neurons of *Helix aspersa*, measured with ion-sensitive microelectrodes, *J. Physiol. (London)* **315:**531–548.

Amberman, R., Salpeter, M. M., and Bachmann, L., 1972, High resolution shadowing, in: *Principles and Techniques of Electron Microscopy* (M. A. Hyat, ed.), pp. 197–217, Van Nostrand, Princeton, N.J.

Arechiga, H., Chavez, B., and Glantz, R. M., 1985, Dye coupling and gap junctions between crustacean neurosecretory cells, *Brain Res.* **326:**183–187.

Aubin, J. E., 1979, Autofluorescence of viable cultured mammalian cells, *J. Histochem. Cytochem.* **27:**36–43.

Baker, P. F., and Honerjager, P., 1978, Influence of carbon dioxide on level of ionized calcium in squid axons, *Nature* **273:**160–161.

Baldwin, K. M., 1979, Cardiac gap junction configuration after an uncoupling treatment as a function of time, *J. Cell Biol.* **82:**66–75.

Barnes, R. D., 1974, *Invertebrate Zoology*, 3rd ed., p. 434, Saunders, Philadelphia.

Barondes, S. H., Haywood-Reid, P. L., and Cooper, D. N. W., 1985, Discoidin 1, an endogenous lectin, is externalized from *Dictyostelium discoideum* in multilamellar bodies, *J. Cell Biol.* **100:**1825–1833.

Benedetti, E. L., and Emmelot, P., 1965, Electron microscopic observations on negatively stained plasma membranes isolated from rat liver, *J. Cell Biol.* **26:**299–305.

Bennett, M. V. L., 1966, Physiology of electrotonic junctions, *Ann. N.Y. Acad. Sci.* **137:**509–539.

Bennett, M. V. L., 1973, Permeability and structure of electrotonic junctions and intercellular movements of tracers, in: *Intracellular Staining in Neurobiology* (S. B. Kater and C. Nicholson, eds.), pp. 115–134, Springer-Verlag, Berlin.

Bennett, M. V. L., and Goodenough, D. A., 1978, Gap junctions, electrotonic coupling and intercellular communication, *Neurosci. Res. Program Bull.* **16:**377–488.

Bennett, M. V. L., and Trinkaus, J. P., 1970, Electrical coupling between embryonic cells by way of extracellular space and specialized junctions, *J. Cell Biol.* **44:**592–610.

Bennett, M. V. L., Spray, D. C., and Harris, A. L., 1981, Electrical coupling in development, *Am. Zool.* **21:**413–427.

Bennett, M. V. L., Zimering, M. B., Spira, M. E., and Spray, D. C., 1985, Interaction of electrical and chemical synapses, in: *Gap Junctions* (M. V. L. Bennett and D. C. Spray, eds.), pp. 355–366, Cold Spring Harbor Laboratory, Cold Spring Harbor, N.Y.
Benson, R. C., Meyer, R. A., Zaruba, M. E., and McKhann, G. J., 1979, Cellular autofluorescence—Is it due to flavins? *J. Histochem. Cytochem.* **27**:44–48.
Berdan, R. C., 1982, Gap junctions: An ultrastructural, electrophysiological and fluorescent tracer study, MSc. thesis, Department of Zoology, University of Western Ontario, London, Ontario.
Berdan, R. C., 1985, Crayfish hepatopancreas cell junctions: Subcellular fractionation, ultrastructural and biochemical analysis, Ph.D. thesis. Baylor College of Medicine, Houston, Texas.
Berdan, R. C., and Caveney, S., 1985, Gap junction ultrastructure in three states of conductance, *Cell Tissue Res.* **239**:111–122.
Berdan, R. C., and Gilula, N. B., 1982, Isolation and preliminary biochemical characterization of an arthropod gap junction, *J. Cell Biol.* **95**:94a.
Berdan, R. C., and Gilula, N. B., 1987a, Arthropod gap junction: Enrichment, ultrastructure and preliminary biochemical analysis, submitted for publication.
Berdan, R. C., and Gilula, N. B., 1987b, II. Pentalaminar membranes "pseudo gap junctions" isolation, ultrastructure and biochemical analysis, submitted for publication.
Berdan, R. C., and Shivers, R. R., 1980, Localization of glycogen in the heart and hepatopancreas of the crayfish *Procambarus clarkii*, *J. Cell Biol.* **87**:259a.
Berdan, R. C., and Shivers, R. R., 1985, Filipin–cholesterol complexes in plasma membranes and cell junctions of *Tenebrio molitor* epidermis, *Tissue Cell* **17**:177–187.
Berdan, R. C., Lees-Miller, J. P., and Caveney, S., 1985, Lack of cell communication in an epithelium: Ultrastructure and electrophysiology of the midgut epithelium of the larval mealworm *Tenebrio molitor*, *J. Ultrastruct. Res.* **90**:55–70.
Berdan, R. C., Shivers, R. R., and Gilula, N. B., 1987, I. Smooth septate junction: Ultrastructure, enrichment and preliminary biochemical analysis, submitted for publication.
Berridge, M. J., and Oschman, J. L., 1972, *Transporting Epithelia*, pp. 20–21, Academic Press, New York.
Blankemeyer, J. T., and Harvey, W. R., 1978, Identification of active cell in potassium transporting epithelium, *J. Exp. Biol.* **77**:1–13.
Blatt, M. R., and Slayman, C. L., 1983, KCl leakage from microelectrodes and its impact on the membrane parameters of a nonexcitable cell, *J. Membr. Biol.* **72**:223–234.
Blennerhassett, M. G., and Caveney, S., 1984, Separation of developmental compartments by a cell type with reduced junctional permeability, *Nature* **309**:361–364.
Brightman, M. W., and Reese, T. S., 1969, Junctions between intimately apposed cell membranes in the vertebrate brain, *J. Cell Biol.* **40**:648–677.
Brink, P. R., and Dewey, M. M., 1980, Evidence for fixed charge in the nexus, *Nature* **285**:101–102.
Bullivant, S., and Loewenstein, W. R., 1968, Structure of coupled and uncoupled cell junctions, *J. Cell Biol.* **37**:621–632.
Bulloch, A. G. M., and Kater, S. B., 1982, Neurite outgrowth and selection of new electrical connections by adult *Helisoma* neurons, *J. Neurophysiol.* **48**:569–583.
Bulloch, A. G. M., Kater, S. B., and Miller, H. R., 1984, Stability of new electrical connections between adult *Helisoma* neurons is influenced by preexisting neuronal interactions, *J. Neurophysiol.* **52**:1094–1105.
Bunt, A. H., 1968, An ultrastructural study of the hepatopancreas of *Procambarus clarkii* (Girard) (Decapoda, Astacidea), *Crustaceana* **15**:282–288.
Burnette, W. N., 1981, "Western blotting": Electrophoretic transfer of proteins from sodium dodecyl sulphate–polyacrylamide gels to unmodified nitrocellulose and radiographic detection with antibody and radioiodinated protein A, *Anal. Biochem.* **112**:195–203.
Campos de Carvalho, A., Spray, D. C., and Bennett, M. V. L., 1984, pH dependence of transmission at electronic synapses of the crayfish septate axon, *Brain Res.* **321**:279–286.

Caudras, J., 1986, Neuron–glia communicatory structures in crustaceans, *Comp. Biochem. Physiol.* **83A:**9–12.

Caudras, J., Martin, G., Czternasty, G., and Bruner, J., 1985, Gap-like junctions between neuron cell bodies and glial cells of the crayfish, *Brain Res.* **326:**149–151.

Caveney, S., 1974, Intercellular communication in a positional field: Movement of small ions between insect epidermal cells, *Dev. Biol.* **40:**311–322.

Caveney, S., 1976, The insect epidermis: A functional syncytium, in: *The Insect Integument* (H. R. Hepburn, ed.), pp. 259–274, Elsevier, Amsterdam.

Caveney, S., 1978, Intercellular communication in insect development is hormonally controlled, *Science* **199:**192–195.

Caveney, S., 1985, The role of gap junctions in development, *Annu. Rev. Physiol.* **47:**319–335.

Caveney, S., 1986, Intercellular communication, in: *Comprehensive Insect Physiology, Biochemistry and Pharmacology,* Vol. 2 (G. A. Kerkut and L. I. Gilbert, eds.), pp. 319–370, Pergamon Press, Elmsford, N.Y.

Caveney, S., and Berdan, R. C., 1982, Selectivity in junctional coupling between cells of insect tissues, in: *Insect Ultrastructure,* Vol 1 (R. C. King and H. Akai, eds.), pp. 434–465, Plenum Press, New York.

Caveney, S., and Blennerhassett, M. G., 1980, Elevation of ionic conductance between insect epidermal cells by β-ecdysone in vitro, *J. Insect Physiol.* **26:**13–25.

Caveney, S., and Podgorski, C., 1975, Intercellular communication in a positional field, *Tissue Cell* **7:**559–574.

Caveney, S., and Safranyos, R., 1985, Control of molecular movement within a developmental compartment, in: *Gap Junctions* (M. V. L. Bennett and D. C. Spray, eds.), pp. 265–273, Cold Spring Harbor Laboratory, Cold Spring Harbor, N.Y.

Caveney, S., Berdan, R. C., and McLean, S., 1980, Cell-to-cell ionic communication stimulated by 20-hydroxyecdysone occurs in the absence of protein synthesis and gap junction growth. *J. Insect Physiol.* **26:**557–567.

Caveney, S., Berdan, R. C., Blennerhassett, M. G., and Safranyos, R. G. A., 1986, Cell-to-cell coupling via membrane junctions: Methods that show its regulation by a developmental hormone in an insect epidermis, in: *Techniques in Cell Biology,* Vol. 2 (E. Kurstak, ed.), pp. 1–23, Elsevier, Amsterdam.

Cioffi, M., and Wolfersberger, M. G., 1983, Isolation of separate apical, lateral and basal membranes from cells of an insect epithelium: A procedure based on tissue organization and ultrastructure, *Tissue Cell* **15:**781–803.

Cole, K. S., and Curtis, H. J., 1938, Electric impedance of *Nitella* during activity, *J. Gen. Physiol.* **22:**37–64.

Connell, C. J., 1978, A freeze fracture and lanthanum tracer study of the complex junction between Sertoli cells of the canine testis, *J. Cell Biol.* **76:**57–75.

Couch, C. B., and Strittmatter, W. J., 1983, Rat myoblast fusion requires metalloendoprotease activity, *Cell* **32:**257–265.

Couch, C. B., and Strittmatter, W. J., 1984, Specific blockers of myoblast fusion inhibit a soluble and not the membrane-associated metalloendoprotease in myoblasts, *J. Biol. Chem.* **259:**5396–5399.

Culvenor, J. G., and Evans, H. W., 1977, Preparation of hepatic gap (communicating) junctions, *Biochem. J.* **168:**475–481.

Dahl, G., and Isenberg, G., 1980, Decoupling of heart muscle cells: Correlation with increased cytoplasmic calcium activity with changes of nexus ultrastructure, *J. Membr. Biol.* **53:**63–75.

Dahl, G., Azarnia, R., and Werner, R., 1980, De novo construction of cell-to-cell channels, *In Vitro* **16:**1068–1075.

Daniel, E. E., Daniel, V. P., Duchon, R. E., Garfield, R. E., Nichols, M., Malhotra, S. K., and Oki, M., 1976, Is the nexus necessary for cell-to-cell coupling of smooth muscle? *J. Membr. Biol.* **28:**207–239.

Davis, L. E., and Burnett, A. L., 1964, A study of growth and cellular differentiation in the hepatopancreas of the crayfish, *Dev. Biol.* **10:**122–153.

Délèze, J., and Herve, J. C., 1983, Effect of several uncouplers of cell-to-cell communication on gap junction morphology in mammalian heart, *J. Membr. Biol.* **74:**203–215.

Dewey, M. M., and Barr, L., 1962, Intercellular connection between smooth muscle cells: The nexus, *Science* **137:**670–672.

Dudek, F. E., Andrew, R. D., MacVicar, B. A., Snow, R. W., and Taylor, C. P., 1983, Recent evidence for and possible significance of gap junctions and electrotonic synapses in the mammalian brain, in: *Basic Mechanisms of Neuronal Hyperexcitability* (H. H. Jasper and N. M. van Gelder, eds.), pp. 31–73, Liss, New York.

Ehrhart, J. C., 1981, Further purification of mouse liver gap junctions with deoxycholate and protein composition, *Cell Biol. Int. Rep.* **5:**1055–1061.

Eichenberger-Glinz, S., 1979, Intercellular junctions during development and in tissue cultures of *Drosophila melanogaster:* An electron microscopic study, *Wilhelm Roux Arch.* **186:**333–349.

Eisenberg, R. S., and Johnson, E. A., 1970, Three dimensional electrical field problems in physiology, *Prog. Biophys. Mol. Biol.* **20:**1–65.

Elias, P. M., Friend, D. S., and Goerke, J., 1979, Membrane sterol heterogeneity, *J. Histochem. Cytochem.* **27:**1247–1260.

Epstein, M. L., and Gilula, N. B., 1977, A study of communication specificity between cells in culture, *J. Cell Biol.* **75:**769–787.

Feder, N., 1971, Microperoxidase, an ultrastructural tracer of low molecular weight, *J. Cell Biol.* **51:**339–343.

Finbow, M., Yancey, B. S., Johnson, R., and Revel, J.-P., 1980, Independent lines of evidence suggesting a major gap junctional protein with a molecular weight of 26,000, *Proc. Natl. Acad. Sci. USA* **77:**970–974.

Finbow, M. E., Shuttleworth, J., Hamilton, A. E., and Pitts, J. D., 1983, Analysis of vertebrate gap junction protein, *EMBO J.* **2:**1479–1486.

Finbow, M. E., Eldridge, T., Bumltjens, J., Lane, N. J., Shuttleworth, J., and Pitts, J. D., 1984, Isolation and characterization of arthropod gap junctions, *EMBO J.* **3:**2271–2278.

Flagg-Newton, J. L., 1980, The permeability of the cell-to-cell membrane channel and its regulation in mammalian cell junctions, *In Vitro* **16:**1043–1049.

Flagg-Newton, J., Simpson, I., and Loewenstein, W. R., 1979, Permeability of the cell-to-cell channels in mammalian cell junctions, *Science* **205:**404–407.

Flower, N. E., 1977, Invertebrate gap junctions, *J. Cell Sci.* **25:**163–171.

Fraser, S. E., and Bryant, P. J., 1985, Patterns of dye coupling in the imaginal wing disk of *Drosophila melanogaster, Nature* **317:**533–536.

Friend, D. S., and Gilula, N. B., 1972, Variations in tight and gap junctions in mammalian tissues, *J. Cell Biol.* **53:**758–776.

Furshpan, E. J., and Potter, D. D., 1957, Mechanisms of nerve-impulse transmission at a crayfish synapse, *Nature* **180:**342–343.

Furshpan, E. J., and Potter, D. D., 1959, Transmission at the giant motor synapses of the crayfish, *J. Physiol. (London)* **145:**289–325.

Gabella, G., and Blundell, D., 1979, Nexuses between the smooth muscle cells of the guinea pig ileum, *J. Cell Biol.* **82:**239–247.

Getting, P. A., 1974, Modification of neuron properties by electrotonic synapses. I. Input resistance, time constant, and integration, *J. Neurophysiol.* **37:**847–857.

Getting, P. A., and Willows, A. O. D., 1974, Modification of neuron properties by electrotonic synapses. II. Burst formation by electrotonic synapses, *J. Neurophysiol.* **37:**858–868.

Gilula, N. B., 1978, Structure of intercellular junctions, in: *Intercellular Junctions and Synapses* (J. Feldman, N. B. Gilula, and J. D. Pitts, eds.), pp. 1–22, Chapman & Hall, London.

Gilula, N. B., and Epstein, M. L., 1976, Cell-to-cell communication, gap junctions and calcium, *Symp. Soc. Exp. Biol.* **30:**257–272.

Gilula, N. B., Branton, D., and Satir, P., 1970, The septate junction: A structural basis for intercellular coupling, *Proc. Natl. Acad. Sci. USA* **67:**213–220.

Girsch, S. J., and Peracchia, C., 1985, Lens cell-to-cell channel protein. I. Self-assembly into liposomes and permeability regulation by calmodulin, *J. Membr. Biol.* **83:**217–225.

Glantz, R. M., and Kirk, M. D., 1980, Intercellular dye migration and electrotonic coupling within neuronal networks of the crayfish brain, *J. Comp. Physiol.* **140:**121–133.

Goodenough, D. A., 1974, Bulk isolation of mouse hepatocyte gap junctions, *J. Cell Biol.* **61:**557–563.

Goodenough, D. A., and Revel, J.-P., 1970, A fine structural analysis of intercellular junctions in the mouse liver, *J. Cell Biol.* **45:**272–290.

Goodenough, D. A., and Revel, J.-P., 1971, The permeability of isolated and *in situ* mouse hepatic gap junctions, *J. Cell Biol.* **50:**81–91.

Goodenough, D. A., and Stoeckenius, W., 1974, The isolation of mouse hepatocyte gap junctions, *J. Cell Biol.* **54:**646–656.

Goodman, C. S., and Spitzer, N. C., 1979, Embryonic development of identified neurones: Differentiation from neuroblast to neurone, *Nature* **280:**208–214.

Graf, F., 1978, Diversite structurale des jonctions intercellulaires communicantes (gap junctions) de l'epithelium des caecums posterieurs du Crustace *Orchestia, C.R. Acad. Sci.* **287:**41–44.

Green, C. R., and Gilula, N. B., 1986, Gap junctional communication between cells during development, in: *NATO Advanced Study Institute* (ASI Series), Life Sciences, Plenum Press, New York.

Green, C. R., and Severs, N. J., 1984, Gap junction connexon configuration in rapidly frozen myocardium and isolated intercalated disks, *J. Cell Biol.* **99:**453–463.

Green, C. R., Noirot-Timothee, C., and Noirot, C., 1983, Isolation and characterization of invertebrate smooth septate junctions, *J. Cell Sci.* **62:**351–370.

Griepp, E. B., and Revel, J.-P., 1977, Gap junctions in development, in: *Intercellular Communication* (W. C. De Mello, ed.), pp. 1–32, Plenum Press, New York.

Hall, D. H., Marder, E., and Bennett, M. V. L., 1985, Interneuronal and interglial gap junctions in the stomatogastric ganglion of the rock crab, *Cancer borealis, J. Neurosci.* **11:**506.

Hanna, R. B., Keeter, J. S., and Pappas, G. D., 1978, The fine structure of a rectifying electrotonic synapse, *J. Cell Biol.* **79:**764–773.

Hanna, R. B., Pappas, G. D., and Bennett, M. V. L., 1984, The fine structure of identified electrotonic synapses following increased coupling resistance, *Cell Tissue Res.* **235:**243–249.

Hardie, J., 1978, Electrotonic coupling between supercontracting body-wall muscle fibres and their attachment morphology in the blowfly larva, *J. Insect Physiol.* **24:**647–655.

Harris, A. L., Spray, D. C., and Bennett, M. V. L., 1983, Control of intercellular communication by voltage dependence of gap junctional conductance, *J. Neurosci.* **3:**79–100.

Hawkes, R., Niday, E., and Gordon, J., 1982, A dot-immunobinding assay for monoclonal and other antibodies, *Anal. Biochem.* **119:**142–147.

Henderson, D., Eibl, H., and Weber, K., 1979, Structure and biochemistry of mouse hepatic gap junctions, *J. Mol. Biol.* **132:**193–218.

Hertzberg, E. L., 1984, A detergent-independent procedure for the isolation of gap junctions from rat liver, *J. Biol. Chem.* **259:**9936–9943.

Hertzberg, E. L., and Gilula, N. B., 1979, Isolation and characterization of gap junctions from rat liver, *J. Biol. Chem.* **254:**2138–2147.

Hertzberg, E. L., and Gilula, N. B., 1982, Liver gap junctions and lens fiber junctions: Comparative analysis and calmodulin interaction, *Cold Spring Harbor Symp. Quant. Biol.* **46:**639–645.

Hertzberg, E. L., and Skibbens, R. V., 1984, A protein homologous to the 27,000 dalton liver gap junction protein is present in a wide variety of species and tissues, *Cell* **39:**61–69.

Hertzberg, E. L., Lawrence, T., and Gilula, N. B., 1981, Gap junctional communication, *Annu. Rev. Physiol.* **43:**479–491.

Hertzberg, E. L., Spray, D. C., and Bennett, M. V. L., 1985, Reduction of gap junctional conductance by microinjection of antibodies against the 27-KD liver gap junction polypeptide, *Proc. Natl. Acad. Sci. USA* **82:**2412–2416.

Heuser, J. E., Reese, T. S., Dennis, M. J., Jan, Y., Jan, L., and Evans, L., 1979, Synaptic vesicle exocytosis captured by quick-freezing and correlated with quantal transmitter release, *J. Cell Biol.* **81:**275–300.

Hicks, T. P., 1984, The history and development of microiontophoresis in experimental neurobiology, *Prog. Neurobiol.* **22:**185–240.

Hirokawa, N., and Heuser, J., 1982, The inside and outside of gap-junction membranes visualized by deep etching, *Cell* **30:**395–406.

Hodgkin, A. L., and Rushton, W. A. H., 1946, The electrical constants of a crustacean nerve fibre, *Proc. R. Soc. London Ser. B.* **133:**444–479.

Hooper, M. L., and Subak-Sharpe, J. H., 1981, Metabolic cooperation between cells, *Int. Rev. Cytol.* **69:**45–104.

Hudspeth, A. J., and Revel, J.-P., 1971, Coexistence of gap and septate junctions in an invertebrate epithelium, *J. Cell Biol.* **50:**92–101.

Jahnke, E., and Emde, F., 1960, *Tables of Higher Functions,* 6th ed., McGraw–Hill, New York.

Johnson, D. A., Gautsch, J. W., Sportsman, J. R., and Elder, J. H., 1984, Improved technique utilizing non-fat dry milk for analysis of proteins and nucleic acid transferred to nitrocellulose, *Gene Anal. Tech.* **1:**3–8.

Johnson, K. R., Lampe, P. D., Jur, K. C., Louis, C. F., and Johnson, R. G., 1986, A lens intercellular junction protein, MP26, is a phosphoprotein, *J. Cell Biol.* **102:**1334–1343.

Johnston, M. F., and Ramon, F., 1981, Electrotonic coupling in internally perfused crayfish segmented axons, *J. Physiol. (London)* **317:**509–518.

Johnston, M. F., Simon, S. A., and Ramon, F., 1980, Interaction of anaesthetics with electrical synapses, *Nature* **286:**498–499.

Kensler, R. W., and Goodenough, D. A., 1980, Isolation of mouse myocardial gap junctions, *J. Cell Biol.* **86:**755–764.

Kistler, J., and Bullivant, S., 1980, Lens gap junctions and orthogonal arrays are unrelated, *FEBS Lett.* **111:**73–78.

Kohler, G., and Milstein, C., 1975, Continuous cultures of fused cells secreting antibody of predefined specificity, *Nature* **256:**495–497.

Kreutziger, G. O., 1968, Freeze-etching of intracellular junctions of mouse liver, *Proc. Electron Microsc. Soc. Am.* **26:**234.

Kuffler, S. W., and Potter, D. D., 1964, Glia in the leech central nervous system: Physiological properties and neuron–glia relationship, *J. Neurophysiol.* **27:**290–320.

Lane, N. J., 1978, Intercellular junctions and cell contacts in invertebrates, in: *Electron Microscopy,* Vol. 3 (J. M. Sturgess, ed.), pp. 673–691, Imperial Press, Toronto.

Lane, N. J., 1981a, Invertebrate neuroglia: Junctional structure and development, *J. Exp. Biol.* **95:**7–33.

Lane, N. J., 1981b, Tight junctions in arthropod tissues, *Int. Rev. Cytol.* **73:**243–318.

Lane, N. J., 1982, Insect intercellular junctions: Their structure and development, in: *Insect Ultrastructure,* Vol. 1 (R. C. King and H. Akai, eds.), pp. 402–433, Plenum Press, New York.

Lane, N. J., 1984, A comparison of the construction of intercellular junctions in the CNS of vertebrates and invertebrates, *Trends Neurosci.* **7:**95–99.

Lane, N. J., and Swales, L. S., 1980, Dispersal of junctional particles, not internalization, during the *in vivo* disappearance of gap junctions, *Cell* **19:**579–586.

Larsen, W. J., 1983, Biological implications of gap junction structure, distribution and composition: A review, *Tissue Cell* **15:**645–671.

Lasek, R. J., and Tytell, M. A., 1981, Macromolecular transfer from glia to the axon, *J. Exp. Biol.* **95:**153–165.

Lawson, D., Raff, M. C., Gomperts, B., Fewtrell, C., and Gilula, N. B., 1977, Molecular events during membrane fusion, *J. Cell Biol.* **72:**242–259.

Lea, T. J., and Ashley, C. C., 1978, Increase in free Ca^{+2} in muscle after exposure to CO_2, *Nature* **275:**236–238.

Lee, W. M., Cran, D. G., and Lane, N. J., 1982, Carbon dioxide induced disassembly of gap junctional plaques, *J. Cell Sci.* **57:**215–228.

Lees-Miller, J., and Caveney, S., 1982, Drugs that block calmodulin activity inhibit cell-to-cell coupling in the epidermis of *Tenebrio molitor, J. Membr. Biol.* **69:**223–245.

LeMaire, M., Moller, J. V., and Gulik-Krzywicki, T., 1981, Freeze-fracture study of water soluble, standard proteins and of detergent-solubilized forms of sarcoplasmic reticulum Ca^{++}-ATPase, *Biochim. Biophys. Acta* **643:**115–125.

Lo, C. W., 1985, Communication compartmentation and pattern formation in development, in: *Gap Junctions* (M. V. L. Bennett and D. C. Spray, eds.), pp. 251–263, Cold Spring Harbor Laboratory, Cold Spring Harbor, N.Y.

Lo, C. W., and Gilula, N. B., 1979a, Gap junctional communication in the preimplantation mouse embryo, *Cell* **18:**399–409.

Lo, C. W., and Gilula, N. B., 1979b, Gap junctional communication in the postimplantation stage mouse embryo, *Cell* **18:**411–422.

Locke, M., 1965, The structure of septate desmosomes, *J. Cell Biol.* **25:**166–169.

Locke, M., 1969, The ultrastructure of the oenocytes in the molt/intermolt cycle of an insect, *Tissue Cell* **1:**103–154.

Loewenstein, W. R., 1966, Permeability of membrane junctions, *Ann. N.Y. Acad. Sci.* **137:**441–472.

Loewenstein, W. R., 1967, On the genesis of cellular communication, *Dev. Biol.* **15:**503–520.

Loewenstein, W. R., 1978, The cell-to-cell membrane channel in development and growth, in: *Differentiation and Development* (F. Ahmad, R. Russell, R. M. Schultz, and W. R. Werner, eds.), Vol. 15, pp. 339–409, Academic Press, New York.

Loewenstein, W. R., 1979, Junctional intercellular communication and the control of growth, *Biochim. Biophys. Acta* **560:**1–65.

Loewenstein, W. R., 1981, Junctional intercellular communication: The cell-to-cell membrane channel, *Physiol. Rev.* **61:**829–913.

Loewenstein, W. R., and Kanno, Y., 1963, Some electrical properties of a nuclear membrane examined with a microelectrode, *J. Gen. Physiol.* **46:**1123–1140.

Loewenstein, W. R., and Kanno, Y., 1964, Studies on an epithelial (gland) cell junction. I. Modifications of surface membrane permeability, *J. Cell Biol.* **22:**565–568.

Loewenstein, W. R., Nakas, M., and Socolar, S. J., 1967, Permeability transformations at a cell membrane junction, *J. Gen. Physiol.* **50:**1865–1891.

Loizzi, R. F., 1971, Interpretation of a crayfish hepatopancreatic function based on fine structural analysis of epithelial cell lines and muscle network, *Z. Zellforsch.* **113:**420–440.

Lopresti, V., Macagno, E. R., and Levinthal, C., 1974, Structure and development of neuronal connections in isogenic organisms: Transient gap junctions between growing optic axons and lamina neuroblasts, *Proc. Natl. Acad. Sci. USA* **71:**1098–1102.

McGookey, D. J., and Anderson, R. G. W., 1983, Morphological characterization of the cholesteryl ester cycle in cultured mouse macrophage foam cells, *J. Cell Biol.* **97:**1156–1168.

McLachlin, J. R., Caveney, S., and Kidder, G. M., 1983, Control of gap junction formation in early mouse embryos, *Dev. Biol.* **98:**155–164.

McNutt, N. S., and Weinstein, R. S., 1970, The ultrastructure of the nexus: A correlated thin-section and freeze-cleave study, *J. Cell Biol.* **47:**666–688.

McVicar, L. K., and Shivers, R. R., 1984, Crustecdysone-induced modulation of electrical coupling and gap junction structure in crayfish hepatopancreas, *Tissue Cell* **16:**917–928.

McVicar, L. K., and Shivers, R. R., 1985, Gap junctions and intercellular communication in the

hepatopancreas of the crayfish (*Orconectes propinquus*) during molt, *Cell Tissue Res.* **240:**261–269.
McWhinnie, M. A., and Kirchenberg, R. J., 1962, Crayfish hepatopancreas metabolism and the intermoult cycle, *Comp. Biochem. Physiol.* **6:**117–128.
Maniatis, T., Harison, R. C., Lacy, E., Lauer, J., O'Connell, C., and Quon, D., 1978, The isolation of structural genes from libraries of eucaryotic DNA, *Cell* **15:**687–701.
Manjunath, C. K., Goings, G. E., and Page, E., 1982, Isolation and protein composition of gap junctions from rabbit hearts, *Biochem. J.* **205:**189–194.
Manjunath, C. K., Goings, G. E., and Page, E., 1984, Detergent sensitivity and splitting of isolated liver gap junctions, *J. Membr. Biol.* **78:**147–155.
Marder, E., 1984, Roles for electrical coupling in neural circuits as revealed by selective neuronal deletions, *J. Exp. Biol.* **112:**147–167.
Margaritis, L. H., Elgsaeter, A., and Branton, D., 1977, Rotary replication for freeze-etching, *J. Cell Biol.* **72:**47–56.
Markham, R., Frey, S., and Hills, G. J., 1963, Methods for the enhancement of image detail and accentuation of structure in electron microscopy, *Virology* **20:**88–102.
Meda, P., Findlay, I., Kolod, E., Orci, L., and Peterson, O. H., 1983, Short and reversible uncoupling evokes little change in the gap junctions of pancreatic acinar cells, *J. Ultrastruct. Res.* **83:**69–84.
Merickel, M. B., Eyman, E. D., and Kater, S. B., 1977, Analysis of a network of electrically coupled neurons producing rhythmic activity in the snail *Helisoma trivolis, Biomed. Eng.* **24:**277–287.
Miller, R. G., and Baldridge, W. H., 1985, Morphological effects of dopamine on goldfish horizontal cell gap junctions, *J. Neurosci.* **11:**240.
Miller, T. M., and Goodenough, D. A., 1985, Gap junction structures after experimental alteration of junctional channel conductance, *J. Cell Biol.* **101:**1741–1748.
Mundy, D. I., and Strittmatter, W. J., 1985, Requirement for metalloendoprotease in exocytosis: Evidence in mast cells and adrenal chromatin cells, *Cell* **40:**645–656.
Murphy, D. A., Hadley, R. D., and Kater, S. B., 1983, Axotomy-induced parallel increases in electrical and dye coupling between identified neurons of *Helisoma, J. Neurosci.* **3:**1422–1429.
Nairn, R. C., 1976, *Fluorescent Protein Tracing,* 4th ed., Churchill Livingston, Edinburgh.
Neutra, M., and Schaeffer, S. F., 1977, Membrane interactions between adjacent mucous secretion granules, *J. Cell Biol.* **74:**983–991.
Neyton, J., and Trautmann, A., 1985, Single-channel currents of an intercellular junction, *Nature* **317:**331–335.
Nicholson, B. J., Hunkapiller, M. W., Grim, L. B., Hood, L. E., and Revel, J.-P., 1981, Rat liver gap junction protein: Properties and partial sequence, *Proc. Natl. Acad. Sci. USA* **78:**7594–7598.
Nicholson, B. J., Takemoto, L. J., Hunkapiller, M. W., Hood, L. E., and Revel, J.-P., 1983, Difference between liver gap junctions and lens MIP26 from rat: Implications for tissue specificity of gap junctions, *Cell* **22:**967–978.
Noirot, C., and Noirot-Timothee, C., 1967, Un nouveau type de jonction intercellulaire (*zonula continua*) dans l'intestin moyen des insectes, *C.R. Acad. Sci.* **264:**2796–2798.
Noirot-Timothee, C., and Noirot, C., 1967, Septate and scalariform junctions in arthropods, *Int. Rev. Cytol.* **63:**97–139.
Obaid, A. L., Socolar, S. J., and Rose, B., 1983, Cell-to-cell channels with two independently regulated gates in series: Analysis of junctional conductance modulation by membrane potential, calcium and pH, *J. Membr. Biol.* **73:**69–89.
Oliveira-Castro, G. M., and Loewenstein, W. R., 1971, Junctional membrane permeability: Effects of divalent cations, *J. Membr. Biol.* **5:**51–77.

Palka, J., 1982, Genetic manipulation of sensory pathways in *Drosophila,* in: *Neuronal Development* (N. C. Spitzer, ed.,), pp. 121–170, Plenum Press, New York.

Pappas, G. D., Asada, Y., and Bennett, M. V. L., 1971, Morphological correlates of increased coupling resistance at an electrotonic synapse, *J. Cell Biol.* **49:**173–188.

Payton, B. W., Bennett, M. V. L., and Pappas, G. D., 1969, Permeability and structure of junctional membranes at an electrotonic synapse, *Science* **166:**1641–1643.

Pentreath, V. W., 1982, Potassium signalling of metabolic interactions between neurons and glial cells, *Trends Neurosci.* **5:**341–345.

Peracchia, C., 1973a, Low resistance junctions in crayfish. I. Two arrays of globules in junctional membranes, *J. Cell Biol.* **57:**54–65.

Peracchia, C., 1973b, Low resistance junctions in crayfish. II. Structural details and further evidence for intercellular channels by freeze-fracture and negative staining, *J. Cell Biol.* **56:**66–76.

Peracchia, C., 1977, Gap junctions: Structural changes after uncoupling procedures, *J. Cell Biol.* **72:**628–641.

Peracchia, C., 1980, Structural correlates of gap junction permeation, *Int. Rev. Cytol.* **66:**81–146.

Peracchia, C., 1981, Direct communication between axons and sheath glial cells in crayfish, *Nature* **290:**597–598.

Peracchia, C., 1984, Communicating junctions and calmodulin: Inhibition of electrical uncoupling in *Xenopus* embryo by calmidazolium, *J. Membr. Biol.* **81:**49–58.

Peracchia, C., and Bernardini, G., 1984, Gap junction structure and cell-to-cell coupling regulation: Is there a calmodulin involvement? *Fed. Proc.* **43:**2681–2691.

Peracchia, C., and Dulhunty, A. F., 1976, Low resistance junctions in crayfish, *J. Cell Biol.* **70:**419–439.

Peracchia, C., and Mittler, R. S., 1972, Fixation by means of glutaraldehyde–hydrogen peroxide reaction products, *J. Cell Biol.* **53:**234–238.

Peracchia, C., Bernardini, G., and Peracchia, L., 1983, Is calmodulin involved in the regulation of gap junctional permeability? *Pfluegers Arch.* **399:**152–154.

Plattner, H., Wachter, E., and Grobner, P., 1977, A heme-nonapeptide tracer for electron microscopy, *Histochemistry* **53:**223–252.

Politoff, A. L., 1977, Protein semiconduction: An alternative explanation of electrical coupling, in: *Intercellular Communication* (W. C. De Mello, ed.), pp. 127–143, Plenum Press, New York.

Politoff, A., and Pappas, G. D., 1972, Mechanisms of increase in coupling resistance at electronic synapses of the crayfish septate axon, *Anat. Rec.* **172:**384.

Politoff, A. L., Socolar, S. J., and Loewenstein, W. R., 1969, Permeability of cell membrane junction: Dependence on energy metabolism, *J. Gen. Physiol.* **53:**498–515.

Politoff, A., Pappas, G. D., and Bennett, M. V. L., 1974, Cobalt ions cross an electrotonic synapse if cytoplasmic concentration is low, *Brain Res.* **76:**343–346.

Popowich, J., and Caveney, S., 1976, An electrophysiological study of epidermal cell membranes in larval *Tenebrio molitor* L., *J. Insect Physiol.* **22:**1617–1622.

Potter, D. D., Furshpan, E. J., and Lennox, E. S., 1966, Connections between cells of the developing squid as revealed by electrophysiological methods, *Proc. Natl. Acad. Sci. USA* **55:**328–335.

Powers, R. D., and Tupper, J. T., 1977, Intercellular communication in the early embryo, in: *Intercellular Communication* (W. C. De Mello, ed.), pp. 231–251, Plenum Press, New York.

Purves, R. D., 1981, *Microelectrode Methods for Intracellular Recording and Iontophoresis,* pp. 86–91, Academic Press, New York.

Quennedey, A., Quennedey, B., Delbecque, J.-P., and Delachambre, J., 1983, The in vitro development of the pupal integument and the effects of ecdysteroids in *Tenebrio molitor* (Insecta, Coleoptera), *Cell Tissue Res.* **232:**493–511.

Radu, A., Dahl, G., and Loewenstein, W. R., 1982, Hormonal regulation of cell junction permeability: Up regulation by catecholamine and prostaglandin E_1, *J. Membr. Biol.* **70:**239–251.

Ramon, F., and Moore, J. W., 1978, Ephaptic transmission in squid giant axons, *Am. Physiol. Soc.* **78:**C162–C169.
Ramon, F., and Zampighi, G., 1980, On the electrotonic coupling mechanism of crayfish segmented axons: Temperature dependence of junctional conductance, *J. Membr. Biol.* **54:**165–171.
Rash, J. E., Graham, W. F., and Hudson, S. C., 1979, Sources and rates of contamination in a conventional Balzers freeze-etch device, in: *Freeze-fracture Methods, Artifacts and Interpretations* (J. E. Rash and C. S. Hudson, eds.), pp. 111–122, Raven Press, New York.
Raviola, E., Goodenough, D. A., and Raviola, G., 1980, Structure of rapidly frozen gap junctions, *J. Cell Biol.* **87:**273–279.
Rayport, S. G., and Kandel, E. R., 1980, Developmental modulation of an identified electrical synapse: Functional uncoupling, *J. Neurophysiol.* **44:**559–568.
Reese, T. S., Bennett, M. V. L., and Feder, N., 1971, Cell-to-cell movement of peroxidases injected into the septate axon of crayfish, *Anat. Rec.* **169:**409.
Regen, C. M., and Steinhardt, R. A., 1986, Global properties of the *Xenopus* blastula are mediated by a high-resistance epithelial seal, *Dev. Biol.* **113:**147–154.
Revel, J.-P., and Karnovsky, M. J., 1967, Hexagonal array of subunits in intercellular junctions of the mouse heart and liver, *J. Cell Biol.* **33:**C7–C12.
Revel, J.-P., Nicholson, B. J., and Yancey, S. B., 1985, Chemistry of gap junctions, *Annu. Rev. Physiol.* **47:**263–279.
Reynolds, G. T., and Taylor, D. L., 1980, Image intensification applied to light microscopy, *BioScience* **30:**586–592.
Risinger, M. A., and Larsen, W. J., 1983, Interaction of filipin with junctional membrane at different stages of the junction's life history, *Tissue Cell* **15:**1–15.
Robenek, H., Jung, W., and Gebhardt, R., 1982, The topography of filipin–cholesterol complexes in the plasma membrane of cultured hepatocytes and their relation to cell junction formation, *J. Ultrastruct. Res.* **78:**95–106.
Rose, B., 1971, Intercellular communication and some structural aspects of membrane junctions in a simple cell system, *J. Membr. Biol.* **5:**1–19.
Rose, B., 1980, Permeability of the cell-to-cell membrane channel and its regulation in an insect cell junction, *In Vitro* **16:**1029–1042.
Rose, B., and Loewenstein, 1971, Junctional membrane permeability. Depression by substitution of Li for extracellular Na, and by long-term lack of Ca and Mg; restoration by cell polarization, *J. Membr. Biol.* **5:**20–50.
Rose, B., and Loewenstein, W. R., 1975, Permeability of cell junction depends on local cytoplasmic calcium activity, *Nature* **254:**250–252.
Rose, B., and Loewenstein, W. R., 1976, Permeability of a cell junction and the local cytoplasmic free ionized calcium concentration: A study with aequorin, *J. Membr. Biol.* **28:**87–119.
Rose, B., and Rick, R., 1978, Intracellular pH, intracellular free Ca, and junctional cell–cell coupling, *J. Membr. Biol.* **44:**337–415.
Rubin, R. P., 1974, *Calcium and the Secretory Process*, p. 28, Plenum Press, New York.
Ryerse, J., and Nagel, B. A., 1984, Gap junction distribution in the *Drosophila* wing disc mutants vg, lgd, 1c43, and lgl, *Dev. Biol.* **105:**396–403.
Safranyos, R. G. A., and Caveney, S., 1985, Rates of diffusion of fluorescent molecules via cell-to-cell membrane channels in a developing tissue, *J. Cell Biol.* **100:**736–747.
Satir, P., and Gilula, N. B., 1973, The fine structure of membranes and intercellular communication in insects, *Annu. Rev. Entomol.* **18:**143–166.
Scalenghe, F., Turco, E., Edstrom, J. E., Pirrotta, V., and Meilli, M., 1981, Microdissection and cloning of DNA from a specific region of *Drosophila melanogaster* polytene chromosomes, *Chromosoma* **82:**205–216.
Schaller, H. C., and Bodenmuller, H., 1981, Isolation and amino acid sequence of a morphogenetic peptide from *Hydra*, *Proc. Natl. Acad. Sci. USA* **78:**7000–7004.

Schanne, O. F., and DeCeretti, E. R. P., 1971, Measurement of input impedance and cytoplasmic resistivity with a single microelectrode, *Can. J. Physiol. Pharmacol.* **49:**713–716.

Schmalbruch, H., 1980, Delayed fixation alters the pattern of intramembrane particles in mammalian muscle fibers, *J. Ultrastruct. Res.* **70:**15–20.

Schultz, R. M., 1985, Roles of cell-to-cell communication in development, *Biol. Reprod.* **32:**27–42.

Schwarzmann, G., Weigandt, R., Rose, B., Zimmerman, A., Ben-Haim, D., and Loewenstein, W. R., 1981, Diameter of the cell-to-cell junctional channels as probed with neutral molecules, *Science* **213:**551–553.

Severs, N. J., 1981, Plasma membrane cholesterol in myocardial muscle and capillary endothelial cells: Distribution of filipin-induced deformations in freeze-fracture, *Eur. J. Cell Biol.* **25:**289–299.

Severs, N. J., and Robenek, H., 1983, Detection of microdomains in biomembranes: An appraisal of recent developments in freeze-fracture cytochemistry, *Biochim. Biophys. Acta* **737:**373–408.

Shaw, S. R., and Stowe, S., 1982, Freeze-fracture evidence for gap junctions connecting the axon terminals of dipteran photoreceptors, *J. Cell Sci.* **53:**758–776.

Sheffield, J. B., 1979, Contribution of carbon to the image in freeze-fracture replication, in: *Freeze-fracture Methods, Artifacts and Interpretations* (J. E. Rash and C. S. Hudson, eds.), pp. 169–173, Raven Press, New York.

Sheridan, J. D., Hammer-Wilson, M., Preus, D., and Johnson, R. G., 1978, Quantitative analysis of low-resistance junctions between cultured cells and correlation with gap-junctional areas, *J. Cell Biol.* **76:**532–544.

Shiba, H., 1971, Heavisides' 'Besselcable' as an electric model for flat simple epithelial cells with low resistive junctional membranes, *J. Theor. Biol.* **30:**59–68.

Shibata, Y., and Page, E., 1981, Gap junctional structure in intact and cut sheep cardiac Purkinje fibers: A freeze-fracture study of Ca^{2+}-induced resealing, *J. Ultrastruct. Res.* **75:**195–204.

Siegenbeek van Heukelom, J., Denier van der Gon, J. J., and Prop, F. J. A., 1972, Model approaches for evaluation of cell coupling in monolayers, *J. Membr. Biol.* **7:**88–102.

Sikerwar, S., and Malhotra, S., 1981, Structural correlates of glutaraldehyde induced uncoupling in mouse liver gap junctions, *Eur. J. Cell Biol.* **25:**319–323.

Sikerwar, S. S., and Malhotra, S. K., 1983, A structural characterization of gap junctions isolated from mouse liver, *Cell Biol. Int. Rep.* **7:**897–903.

Simpson, I., 1978, Labelling of small molecules with fluorescein, *Anal. Biochem.* **89:**304–305.

Simpson, I., Rose, B., and Loewenstein, W. R., 1977, Size limit of molecules permeating the junctional membrane channels, *Science* **195:**294–296.

Socolar, S. J., 1977, The coupling coefficient as an index of junctional conductance, *J. Membr. Biol.* **34:**29.

Socolar, S. J., and Loewenstein, W. R., 1979, Methods for studying transmission through permeable cell-to-cell junctions, in: *Methods in Membrane Biology,* Vol. 10 (E. D. Korn, ed.), pp. 123–179, Plenum Press, New York.

Somlyo, A. V., 1979, Bridging structures spanning the junctional gap at the triad of skeletal muscle, *J. Cell Biol.* **80:**743–750.

Sotelo, C., and Llinas, R., 1972, Specialized membrane junctions between neurons in the vertebrate cerebellar cortex, *J. Cell Biol.* **53:**271–289.

Specian, R. D., and Neutra, M. R., 1980, Mechanism of rapid mucous secretion in goblet cells stimulated by acetylcholine, *J. Cell Biol.* **85:**626–640.

Spray, D. C., Harris, A. L., and Bennett, M. V. L., 1981, Gap junctional conductance is a simple sensitive function of intracellular pH, *Science* **211:**712–715.

Spray, D. C., Stern, J. H., Harris, A. L., and Bennett, M. V. L., 1982, Gap junctional conductance: Comparison of sensitivities to H and Ca ions, *Proc. Natl. Acad. Sci. USA* **79:**441–445.

Stewart, W. W., 1978, Functional connections between cells as revealed by dye-coupling with a highly fluorescent naphthalimide tracer, *Cell* **14:**741–759.

Strausfeld, N. J., and Bassemir, U. K., 1983a, Cytology of cobalt-filled neurons in flies: Cobalt deposits at presynaptic and postsynaptic sites, mitochondria and cytoskeleton, *J. Neurocytol.* **12:**949–970.

Strausfeld, N. J., and Bassemir, U. K., 1983b, Cobalt-coupled neurons of a giant fibre system in Diptera, *J. Neurocytol.* **12:**971–991.

Suzuki, K., and Higashino, S., 1977, Protective effect of corticosteroids on intercellular junction, *Exp. Cell Res.* **109:**263–268.

Suzuki, K., Sangworasil, M., and Higashino, S., 1978, On correlation between gland stiffness and cell coupling in salivary gland of *Chironomus plumosus* larva, *Cell Struct. Funct.* **3:**161–172.

Szollosi, A., and Marcaillou, C., 1980, Gap junctions between germ and somatic cells in the testis of the moth, *Anagasta kuehniella* (Insecta: Lepidoptera), *Cell Tissue Res.* **213:**137–147.

Tanaka, Y., DeCamilli, P., and Melodolesi, J., 1980, Membrane interactions between secretion granules and plasmalemma in three exocrine glands, *J. Cell Biol.* **84:**438–453.

Telfer, W. H., Huebner, E., and Smith, D. S., 1982, The cell biology of vitellogenic follicles in *Hyalophora* and *Rhodnius*, in: *Insect Ultrastructure*, Vol. 1 (R. C. King and H. Akai, eds.), pp. 118–149, Plenum Press, New York.

Thomas, J. B., and Wyman, R. J., 1983, Normal and mutant connectivity between identified neurons in *Drosophila, Trends Neurosci.* **6:**214–219.

Thomas, J. B., and Wyman, R. J., 1984, Mutations altering synaptic connectivity between identified neurons in *Drosophila, J. Neurosci.* **4:**530–538.

Turin, L., and Warner, A. E., 1980, Intracellular pH in early *Xenopus* embryos: Its effect on current flow between blastomeres, *J. Physiol. (London)* **300:**489–504.

van Eldik, L. J., Hertzberg, E. L., Berdan, R. C., and Gilula, N. B., 1985, Interaction of calmodulin and other calcium-modulated proteins with mammalian and arthropod junctional membrane proteins, *Biochem. Biophys. Res. Commun.* **126:**825–832.

van Venrooij, G. E. P. M., Hax, W. M. A., Schouten, V. J. A., Denier van der Gon, J. J., and van der Vorst, H. A., 1975, Absence of cell communication for fluorescein and dansylated amino acids in an electrotonic coupled cell system, *Biochim. Biophys. Acta* **394:**620–632.

Viancour, T. A., Bittner, G. D., and Ballinger, M. L., 1981, Selective transfer of Lucifer Yellow CH from axoplasm to adaxonal glia, *Nature* **293:**65–67.

Warner, A. E., and Lawrence, P. A., 1982, Permeability of gap junctions at the segmental border in insect epidermis, *Cell* **28:**243–252.

Warner, A. E., Guthrie, S. C., and Gilula, N. B., 1984, Antibodies to gap junctional protein selectively disrupt junctional communication in early amphibian embryo, *Nature* **311:**127–131.

Watanabe, A., and Grundfest, H., 1961, Impulse propagation at the septal and commissural junctions of crayfish lateral giant axons, *J. Gen. Physiol.* **45:**267–308.

Weidmann, S., 1952, The electrical constants of Purkinje fibres, *J. Physiol. (London)* **118:**348–360.

Weinbaum, S., 1980. Theory for the formation of intercellular junctions based on intramembranous particle patterns observed in the freeze-fracture technique, *J. Theor. Biol.* **83:**63–92.

Weinstein, J. N., Yoshikami, S., Henkart, P., Blumenthal, R., and Haggins, W. A., 1977, Liposome–cell interaction: Transfer and intercellular release of a trapped fluorescent marker, *Science* **195:**489–492.

Weir, M. P., and Lo, C. W., 1982, Gap junctional communication compartments in *Drosophila* wing imaginal disk, *Proc. Natl. Acad. Sci. USA* **79:**3232–3235.

Weir, M. P., and Lo, C. W., 1984, Gap-junctional communication compartments in the *Drosophila* wing imaginal disk, *Dev. Biol.* **102:**130–146.

Weir, M. P., and Lo, C. W., 1985, An anterior/posterior communication compartment border in engrailed wing discs: Possible implications for *Drosophila* pattern formation, *Dev. Biol.* **110:**84–90.

Wigglesworth, V. B., 1948, The structure and deposition of the cuticle in the adult mealworm, *Tenebrio molitor* L. (Coleoptera), *Q. J. Microsc. Sci.* **89:**197–221.

Williams, E. H., and DeHaan, R. L., 1981, Electrical coupling among heart cells in the absence of ultrastructurally defined gap junctions, *J. Membr. Biol.* **60:**237–248.

Wood, R. L., 1959, Intercellular attachment in the epithelium of *Hydra* as revealed by electron microscopy, *J. Biophys. Biochem. Cytol.* **6:**343–352.

Wyman, R. J., and Thomas, J. B., 1983, What genes are necessary to make an identified synapse? *Cold Spring Harbor Symp. Quant. Biol.* **48:**641–652.

Zampighi, G., and Unwin, P. N. T., 1979, Two forms of isolated gap junctions, *J. Mol. Biol.* **135:**451–464.

Zampighi, G., Ramon, F., and Duran, W., 1978, Fine structure of the electrotonic synapse of the lateral giant axons in a crayfish (*Procambarus clarkii*), *Tissue Cell* **10:**413–426.

Zampighi, G., Simon, S. A., Robertson, J. D., McIntosh, T. J., and Costello, M. J., 1982, On the structural organization of isolated bovine lens fiber junctions, *J. Cell Biol.* **93:**175–189.

Zervos, A. S., Hope, J., and Evans, W. H., 1985, Preparation of a gap junction fraction from uteri of pregnant rats: The 28-Kd polypeptides of uterus, liver and heart gap junctions are homologous, *J. Cell Biol.* **101:**1363–1370.

Zimmerman, A. L., and Rose, B., 1985, Permeability properties of cell-to-cell channels: Kinetics of fluorescent tracer diffusion through a cell junction, *J. Membr. Biol.* **84:**269–283.

Zingsheim, H. P., 1972, Membrane structure and electron microscopy: The significance of physical problems and techniques (freeze-etching), *Biochim. Biophys. Acta* **265:**339–366.

Index